# ENVIRONMENTAL CONSEQUENCES OF ENERGY PRODUCTION:
Problems and Prospects

The Pennsylvania Academy of Science Publications

Books and Proceedings

Book Editor: Shyamal K. Majumdar
Professor of Biology
Lafayette College
Easton, Pennsylvania 18042

1. *Energy, Environment, and the Economy,* 1981. ISBN: 0-9606670-0-8. Editor: Shyamal K. Majumdar.

2. *Pennsylvania Coal: Resources, Technology and Utilization,* 1983. ISBN: 0-9606670-1-6. Editors: Shyamal K. Majumdar and E. Willard Miller.

3. *Hazardous and Toxic Wastes: Technology, Management and Health Effects,* 1984. ISBN: 0-9606670-2-4. Editors: Shyamal K. Majumdar and E. Willard Miller.

4. *Solid and Liquid Wastes: Management, Methods and Socioeconomic Considerations,* 1984. ISBN: 0-9606670-3-2. Editors: Shyamal K. Majumdar and E. Willard Miller.

5. *Management of Radioactive Materials and Wastes: Issues and Progress,* 1985. ISBN: 0-9606670-4-0. Editors: Shyamal K. Majumdar and E. Willard Miller.

6. *Endangered and Threatened Species Programs in Pennsylvania and Other States: Causes, Issues and Management,* 1986. ISBN: 0-9606670-5-9. Editors: Shyamal K. Majumdar, Fred J. Brenner, and Ann F. Rhoads.

7. *Environmental Consequences of Energy Production: Problems and Prospects.* ISBN: 0-9606670-6-7. Editors: Shyamal K. Majumdar, Fred J. Brenner and E. Willard Miller.

8. *Proceedings* of the Pennsylvania Academy of Science. Two issues per year; current volume (1987) is 61. ISSN: 0096-9222. Editor: Daniel Klem, Jr.

ENVIRONMENTAL CONSEQUENCES OF ENERGY PRODUCTION: Problems and Prospects

EDITED BY
SHYAMAL K. MAJUMDAR, Ph.D.
Professor of Biology
Lafayette College
Easton, Pennsylvania 18042

FRED J. BRENNER, Ph.D.
Professor of Biology
Grove City College
Grove City, Pennsylvania 16127

E. WILLARD MILLER, Ph.D.
Professor of Geography and
  Associate Dean for Resident
  Instruction (Emeritus)
The Pennsylvania State University
University Park, Pennsylvania 16802

Founded on April 18, 1924

A Publication of
The Pennsylvania Academy of Science

Library of Congress Cataloging in Publication Data

Environmental Consequences of Energy Production: Problems and Prospects

  Bibliography
  Index

Library of Congress Catalog Card No.: 87-061248

ISBN 0-9606670-6-7
  Copyright © 1987 By The Pennsylvania Academy of Science

  All rights reserved. No part of this book may be reproduced in any form without written consent from the publisher, The Pennsylvania Academy of Science. For information write to The Pennsylvania Academy of Science, Attention: Dr. S.K. Majumdar, Editor, Department of Biology, Lafayette College, Easton, Pennsylvania 18042.

  Printed in the United States of America by

Typehouse of Easton
Phillipsburg, New Jersey 08865

PREFACE

The economies of the world since the beginning of the Industrial Revolution more than two centuries ago have been built on energy. The nonrenewable energy sources of coal, petroleum, natural gas and radioactive materials have thus provided the power base for the world's people. Throughout most of the history of the exploitation of these earth resources little or no attention was given to the environmental problems created in energy utilization. With a continually rising demand, the primary objective was to fuel the expanding industrial societies. Only in recent times is there recognition that the destruction of the land in the mining process and the pollution of the environment in the utilization of the energy resources cannot be tolerated in a modern society.

In this volume, national and international experts address the environmental impacts of each stage in the production of energy from the mining of fossil fuels to the generation and transmission of electrical energy. The possible consequences of alternatives to fossil fuels including hydro, nuclear, and waste conversion are also addressed as to their regional, national and global impact.

The objective of this volume is not only to describe the environmental problems of energy production and utilization, but also to provide solutions to these problems. The technology is available to solve the environmental problems and the societal framework of laws and regulations are now being developed.

The volume is divided into five parts. The first part considers environmental problems created in the mining of coal, petroleum, natural gas and radioactive materials. Parts 2, 3, and 4 consider consequences of utilization of fossil fuels, hydro power production and nuclear power generation. The final part treats the political, economic and humanistic aspects of energy utilization.

This is a comprehensive collection of technical papers written by scientists, engineers and social scientists on the important environmental issues involved in supplying energy to a complex industrial world. With continuous scientific and technological advancements combined with a society that no longer will tolerate the destruction of our natural environment, there is an assurance that not only will our energy needs be met in the future but a livable world will be created.

This book will be of interest to a wide audience. Such individuals as scientists, engineers, environmentalists, conservationists, social scientists, lawyers will find the chapters informative. They provide a wide perspective so that specialists in one field can be informed about developments and trends in another branch of the subject. The volume will also be of interest to individuals who want to be informed about some of the most critical problems of the day.

We express our deep appreciation for the excellent cooperation and dedication of the contributors, who recognize the importance of solving the critical problems of the environment. Many other individuals in addition to the authors made contributions and we are appreciative to those persons. Gratitude is extended to Lafayette and Grove City Colleges and The Pennsylvania State University for providing facilities to the editors for this work. The editors extend heartfelt thanks to their wives for their encouragement and help in the preparation of this book.

Shyamal K. Majumdar, Ph.D.
Fred J. Brenner, Ph.D.
E. Willard Miller, Ph.D.
Editors
May 1987

Environmental Consequences of Energy Production: Problems and Prospects

Table of Contents

Preface V

Chapter 1: ENERGY RESOURCES OF THE WORLD: NON-RENEWABLE AND RENEWABLE
E. Willard Miller 1

Chapter 2: INTERNATIONAL ASPECTS OF THE MITIGATION OF THE ENVIRONMENTAL EFFECTS OF ENERGY PRODUCTION—AN INTERNATIONAL PERSPECTIVE
Heinz G. Pfeiffer 21

Chapter 3: INTERNATIONAL COOPERATION IN REDUCTION OF ENERGY RELATED ATMOSPHERIC POLLUTION
John W. Landis 25

Part One: Environmental Problems, Plannings and Control When Producing Energy Resources

Chapter 4: ABANDONED MINE AND COAL REFUSE FIRES: A BURNING PROBLEM
Raja V. Ramani 34

Chapter 5: SURFACE SUBSIDENCE
Lee W. Saperstein 49

Chapter 6: RECLAMATION OF DEEP COAL MINING WASTES WITH PARTICULAR REFERENCE TO BRITAIN AND WESTERN EUROPE
Martin Kent 61

Chapter 7: ENVIRONMENTAL PLANNING FOR SURFACE MINING OF COAL
Raja V. Ramani 78

Chapter 8: COAL REFUSE PILES: PROBLEMS AND PROSPECTS
Raja V. Ramani 102

Chapter 9: ALTERNATIVE RECLAMATION STRATEGIES FOR MINED LANDS
Fred J. Brenner and Richard P. Steiner 115

Chapter 10: ACID MINE DRAINAGE
L. Barry Phelps 131

Chapter 11: OIL AND NATURAL GAS DRILLING AND TRANSPORTATION: ENVIRONMENTAL PROBLEMS AND CONTROL
M.A. Adewumi and T. Ertekin 141

Chapter 12: THE INFLUENCE OF OIL DRILLING OPERATIONS AND CRUDE OIL ON THE BIOLOGICAL COMMUNITY
E.C. Masteller 164

Chapter 13: RECLAMATION OF URANIUM MINING AND MILLING
           DISTURBANCES
E.E. Farmer and Gerald E. Schuman 182

Part Two: Environmental Aspects of Energy Production from Fossil Fuels

Chapter 14: DEVELOPMENT OF PROJECT APPRAISAL
           METHODOLOGY IN AIR POLLUTING INDUSTRIES IN
           EASTERN EUROPE
Nenad Starc 198

Chapter 15: ENVIRONMENTAL ASPECTS OF THERMAL DISCHARGES
           TO AQUATIC HABITATS
Robert Domermuth 212

Chapter 16: ENVIRONMENTAL IMPACT OF FLY ASH DISPOSAL
James P. Miller, Jr. 223

Chapter 17: ACID RAIN
John J. Cahir 232

Chapter 18: ACID DEPOSITION ABATEMENT AND REGIONAL COAL
           PRODUCTION
D. G. Arey, J. A. Crenshaw and G. D. Parker 245

Chapter 19: IMPACTS OF ATMOSPHERIC DEPOSITION ON FOREST-
           STREAM ECOSYSTEMS IN PENNSYLVANIA
James A. Lynch, David R. DeWalle, and William E. Sharpe 261

Chapter 20: MANAGEMENT AND ENVIRONMENTAL IMPACTS OF
           ELECTRIC POWER TRANSMISSION RIGHTS-OF-WAY
W.E. Bramble, W.R. Byrnes, and R.J. Hutnik 288

Part Three: Environmental Impacts and Plannings of Energy Production from Hydropower

Chapter 21: IMPACTS OF HYDROPOWER DEVELOPMENT ON RIVER
           RESPONSE
Daryl B. Simons and Robert K. Simons 301

Chapter 22: IMPACTS OF HYDROPOWER DEVELOPMENT ON
           ANADROMOUS FISH IN THE NORTHEAST UNITED
           STATES
Gordon W. Russell and Richard A. St. Pierre 319

Chapter 23: POTENTIAL IMPACTS OF HYDRO AND TIDAL POWER
           DEVELOPMENTS ON THE ECOLOGY OF BAYS AND
           ESTUARIES
G.R. Daborn 334

Part Four: Environmental Consequences and Management of Nuclear Power Generation

Chapter 24: A "BATHTUB" PRIMER—LESSONS FROM WEST VALLEY,
           NEW YORK
Robert F. Schmalz 349

Chapter 25: ENVIRONMENTAL IMPACTS OF NUCLEAR POWER
           PRODUCTION
Warren Witzig, Kanaga Sahadewan, and James K. Shillenn 369

Chapter 26: EFFECTS OF THERMAL EFFLUENTS FROM NUCLEAR REACTORS
William D. McCort 386

Chapter 27: GENERATION, PROCESSING AND DISPOSAL OF LOW LEVEL RADIOACTIVE WASTE IN THE UNITED STATES
Keith G. Mattern 402

Chapter 28: RADIATION PROTECTION ASPECTS OF THE TMI-2 ACCIDENT AND CLEANUP
James E. Hildebrand and M.J. Slobodien 406

Part Five: Environmental Legislation, Economic, and Humanistic Considerations

Chapter 29: ENVIRONMENTAL AND LEGISLATIVE CONSEQUENCES OF STRIP MINING
L.W. Saperstein 419

Chapter 30: ENERGY PRODUCTION AND FOREST ECOSYSTEM HEALTH
William H. Smith 431

Chapter 31: EFFECTS OF INDUSTRIAL CHEMICALS AND RADIOACTIVE MATERIALS IN BIOLOGICAL SYSTEMS
A. Gangopadhyay and S. Chatterjee 445

Chapter 32: MAJOR ENVIRONMENTAL IMPACTS OF RESOURCE RECOVERY FACILITIES
Michael D. Brown 464

Chapter 33: SITING LOW-LEVEL RADIOACTIVE WASTE IN PENNSYLVANIA: SOCIO-POLITICAL PROBLEMS AND THE SEARCH FOR SOLUTIONS
Richard J. Bord 474

Chapter 34: ECONOMICS OF ENERGY SUPPLY AND DEMAND IN DEVELOPING COUNTRIES
Wen S. Chern 487

Chapter 35: ENVIRONMENTAL LEGISLATION AND THE COAL MINING INDUSTRY
E. Willard Miller 501

Subject Index 523

Academy Officers 531

Environmental Consequences of Energy Production: Problems and Prospects. Edited by S. K. Majumdar, F. J. Brenner and E. W. Miller. © 1987, The Pennsylvania Academy of Science.

Chapter One
ENERGY RESOURCES OF THE WORLD: NONRENEWABLE AND RENEWABLE

E. WILLARD MILLER
Professor of Geography and Associate Dean for Resident Instruction (Emeritus)
College of Earth and Mineral Sciences
The Pennsylvania State University
University Park, PA 16802

Energy is one of the foundations on which our civilization rests, and one for which there is no substitute (Sassin, 1980). In the nineteenth and early twentieth centuries, energy from nonrenewable resources became abundant. At that time the environmental aspects of energy utilization was not a matter of consideration. This has changed. It is now recognized that satisfying energy demands in the future by employing environmentally safe technological procedures has become one of the greatest problems facing humanity.

To solve the energy problems, a worldwide perspective must be assumed. Assuming that energy production is environmentally sound requires strenuous scientific, technical, economical, sociological and political efforts. Many aspects of energy utilization in the future will require international cooperation. The geographical distribution of primary energy sources is very uneven. A tremendous capital investment will be required to develop new energy technology. This chapter provides background perspectives on the availability of energy and the consumption patterns of nonrenewable and renewable energy resources.

CLASSIFICATION OF WORLD ENERGY RESERVES AND RESOURCES

The availability of energy can be determined if the demand for primary energy is known along with the reserves known to exist at a given time. The techniques to determine reserves have improved greatly in recent years so that the estimates of primary energy reserves/resources differ only slightly from each other (Grathwohl, 1982). An internationally uniform method for classification of energy now exists (Govett, 1974).

These guidelines distinguish between resources and reserves. According to the World Energy Conference: Survey of Energy Resources 1974, the definitions are: In the broadest sense, *resources* of nonrenewable raw materials are the total quantities available in the earth that may be successfully exploited and used by man within the foreseeable future. *Reserves*, however, are the corresponding fraction of resources that have been carefully measured and accessed as being exploitable in a particular nation or region under present local economic conditions using existing available technology. Recoverable reserves are that fraction of reserves-in-place that can be recovered under the above economic and technical limits (Bauer, et al, 1977).

NONRENEWABLE ENERGY SOURCES

The fossil and nuclear fuels that are found in the earth's crust constitute the nonrenewable energy sources of the earth. The fossil fuels thus include peat, lignite, bituminous coal, anthracite, petroleum, natural gas, oil shales and oil sands. The nuclear fuels for the present-day fission nuclear reactors are uranium and thorium and for the fusion reactors of the future the fuels will be deuterium and lithium. Each of the nonrenewable fuels has the potential to cause environmental damage.

The present-day available recoverable reserves and the additional resources vary greatly for each of the nonrenewable energy resources. The rate that coal will be utilized to supply the world's demand for energy depends upon such factors as available technology, economic conditions and political atmosphere.

Coal

Coal makes up the largest proportion of the fossil fuel reserves. Because coal has been mined for centuries, estimates of available reserves and potential resources have long been made (Fettweis, 1979). In 1913 the world's coal reserves were estimated to be $6400 \cdot 10^9$ tce and to be $7000 \cdot 10^9$ tce in 1936. The World Energy Conference (WED) in 1980 estimated the proved recoverable reserves

TABLE 1

Proved Recoverable Reserves of Coal

Region	Bituminous Coal and Anthracite		Subbituminous Coal		Lignite		Total	
	10^9tce	%	10^9tce	%	10^9tce	%	10^9tce	%
Africa	32.5	6.7	0.1	0.1	0.0	0.0	32.6	4.8
America	111.4	22.8	75.3	67.4	13.2	15.0	199.9	29.1
Asia	113.9	23.4	0.8	0.7	1.4	1.6	116.1	16.9
Soviet Union	104.0	21.3	32.8	29.4	28.7	32.6	165.5	24.0
Europe	100.5	20.6	1.3	1.2	35.1	39.8	136.9	19.9
Oceania	25.4	5.2	1.3	1.2	9.7	11.0	36.4	5.3
Total	487.7	100.0	111.6	100.0	88.1	100.0	687.4	100.0

Source: World Energy Conference, 1980

TABLE 2

Additional Coal Resources

Region	Bituminous Coal and Anthracite 10⁹tce	%	Subbituminous Coal 10⁹tce	%	Lignite 10⁹tce	%	Total 19⁹tce	%
Africa	144.4	2.3	0.8	0.0	0.0	0.0	145.2	1.5
America	1,181.2	19.2	1,334.1	44.5	406.8	48.0	2,922.1	29.2
Asia	1,423.2	23.1	2.2	0.1	19.5	2.3	1,444.9	14.5
Soviet Union	2,480.0	40.2	1,570.9	52.5	381.5	45.0	4,432.4	44.3
Europe	429.5	7.0	1.1	0.0	12.1	1.4	442.7	4.4
Oceania	503.1	8.2	82.3	2.8	28.4	3.3	613.8	6.1
Total	6,161.4	100.0	2,991.4	100.0	848.3	100.0	10,001.1	100.0

Source: World Energy Conference, 1980

Conversion factors for coal equivalents (tce)
Bituminous/anthracite = 1.00
Subbituminous coal   = 0.78
Lignite   = 0.30 Turkey, German, Democratic Republic, German Federal Republic, Poland, Australia
                     = 0.33 All-African countries, all Asian countries except Turkey, Soviet Union, France, Greece, Hungary, Italy, Romania, Spain
                     = 0.50 Canada, United States, Albania, Austria, Bulgaria, Portugal, New Zealand, Yogoslavia
                     = 0.60 Czechoslovakia

at 687 · 10⁹tce and the resources at 10,001 · 10⁹tce. The increase is not primarily due to new discoveries but inclusion of resources at greater depths and in narrower seams (Tables 1 and 2).

Coal reserves are widely distributed in the world. The United States, Soviet Union and the People's Republic of China have the largest proved reserves and additional resources followed by Great Britain, Poland, Australia, Republic of South Africa, and the Federal Republic of Germany (Grathwohl, 1982). The coal reserves are sufficiently larger to permit a considerable increase in production. Based on present production and available reserves, the world has at least a 190 year supply of bituminous/anthracite coal. When total coal resources are included, the number of years' supply is greatly extended.

Petroleum

The petroleum reserves are being exploited most rapidly in the world and will be the first fossil fuel to be exhausted. This is particularly important for no alternative to petroleum is presently available. Because petroleum is so important to the world, a number of attempts have been made to estimate the

TABLE 3
World Petroleum Reserves 1979

Region	Proved Recoverable Reserves[1]		Estimated Additional Recoverable Resources[2]		Ultimate Recovery[3]	
	10^6t[4]	%	10^6t	%	10^6t	%
Africa	8,040	9	34,000	16	42,040	14
Anglo America	4,480	5	24,000	11	28,480	10
Latin America	7,770	9	12,000	6	19,770	7
Far East/Pacific	2,390	3	12,000	6	14,390	5
Middle East	51,050	57	52,000	24	103,050	34
Western Europe	2,710	3	10,000	5	12,710	4
Socialist Bloc	12,700	14	64,000	30	76,700	25
Antarctic	—	—	4,000	2	4,000	1
Total	89,140	100	212,000	100	301,140	100

(1) Proved recoverable reserves are those known to be in place and recoverable under present economic and technological conditions
(2) Reserves in addition to proved reserves that can be secured under economic and technological conditions with reasonable confidence
(3) Proved recoverable reserves and additional recoverable resources
(4) One ton of oil is equivalent to about 7.33 barrels of crude oil
Source: World Energy Conference, 1980

total available and recoverable petroleum resources in the world (Masters, 1985). Over the years the reserves have continually been raised. In 1920 it was estimated that the recoverable world reserves were only $5.9 \cdot 10^9$t. By 1979 about $54 \cdot 10^9$t petroleum had been produced and the remaining proved recoverable reserves and additional recoverable resources were estimated by the World Energy Conference to be about $302 \cdot 10^9$t (Kerr, 1981). Of this amount, the proved recoverable reserves were approximately $89 \cdot 10^9$t and the additional recoverable resources were around $212 \cdot 10^9$t. (Table 3).

The total amount of "oil in place" has been estimated by the Federal Institute for Geosciences and Natural Resources, Hannover, to be $725 \cdot 10^9$t. Other estimates indicate that the total "oil in place" could be as much as $1,000 \cdot 10^9$t.

It is likely that these estimates are low. Exploration for petroleum is at an initial stage in many regions of the world and the likelihood of new major discoveries is excellent. A recent study by the Exxon Corporation indicates that there are 60 countries in which oil exploration will occur in the future. Exploration will also continue in such countries as the Soviet Union, Saudi Arabia and the United States where large production exists today.

The proved recoverable reserves are extremely unevenly distributed over the world (Billo, 1979). The Middle East from Egypt to Iran has about 57 percent of the world's proved recoverable reserves of oil, with about 26 percent located in Saudi Arabia alone. The importance of Saudi Arabia in supplying the world oil in the immediate future cannot be overstated. The Soviet Union has about 11 percent of the proved recoverable reserves, but has the largest additional recoverable resources. The United States has had more than twice the cumulative production of any other country and still has an estimated 4.2 percent of the proved recoverable reserves in the world.

In addition to conventional oil, there is an enormous potential supply of non-conventional oil. These oil sources include offshore deposits at depths beyond 6,000 feet, polar deposits, enhanced secondary and tertiary recovery, oil shales, oil sands, heavy oil deposits and synthetic oil from coal. In the prediction of future production the 1980 World Energy Conference indicated that available reserves would permit a considerable increase in production. The International Institute for Applied Systems Analysis in 1980 in an optimistic estimate indicated that with appropriate technology efforts, world petroleum production could increase to about $5 \cdot 10^9$t in 1990. Annual production would then fall gradually to $2.5 \cdot 10^9$t by 2065. This estimate was based on reserves of $300 \cdot 10^9$t of oil at production costs of no more than $20 per barrel in 1976 dollars.

All estimates of future production have serious limitations due to a continually changing reserve picture. There is little doubt that the ultimate recoverable reserves of the Middle East will give these countries the largest production. In contrast, the lifetime of ultimate recoverable reserves of the United States is predicted to be much shorter.

Natural Gas

Natural Gas is emerging as an important substitute for oil. Within the past 10 years, world reserves of natural gas have more than doubled. The World Energy Conference of 1980 gave the proved recoverable reserves to be $74 \cdot 10^{12}$m^3, and the ultimate cumulative production falls in the range between 250 and 300 $\cdot 10^{12}$m^3 (Table 4). The total energy content of natural gas reserves is approaching that of petroleum, and as the search continues will probably surpass that of petroleum.

There are three major reasons for speculating that new natural gas deposits will be found. First, exploration in the past focussed on petroleum deposits. Now companies are interested in discovering only natural gas deposits. Second, natural gas can be created in the earth and survive under conditions where oil cannot be formed or is unstable. For example, natural gas can be formed from all types of organic material while petroleum must contain what was originally lipid rich organic matter, and finally, geophysical prospecting methods continue to improve.

Natural gas has been found in about 50 countries. Of the reserves, the Soviet Union has the largest deposits followed by Iran, United States, Algeria, Saudi Arabia, Venezuela, and Canada. The proved reserves of the Soviet Union have nearly doubled in the past ten years and are considerably larger than that of

TABLE 4

World Natural Gas Reserves 1979

Region	Proved Recoverable Resources		Estimated Additional Recoverable Resources		Ultimate Recovery	
	$10^{12}m^3$	%	$10^{12}m^3$	%	$10^{12}m^3$	%
Africa	7.3	10.0	26	13.6	33.3	12.5
Anglo America	7.5	10.1	42	21.9	49.5	18.6
Latin America	4.7	6.3	10	5.2	14.7	5.6
Far East/Pacific	3.3	4.6	10	5.2	13.3	5.0
Middle East	20.5	27.3	30	15.6	50.5	19.0
Western Europe	3.9	5.3	6	3.1	9.9	3.7
Socialist Bloc	26.9	26.4	64	33.3	90.9	34.1
Antarctic	—	—	4	2.1	4.0	1.5
Total	74.1	100.0	192	100.0	266.1	100.0

Source: World Energy Conference, 1980

the oil reserves of Saudi Arabia. In recent years the natural gas reserves of western Europe have increased considerably with the Netherlands and the United Kingdom having the largest amounts.

In addition to conventional gas, there is enormous potential in nonconventional gas sources in coal beds, shales, tight formations, geopressured gas, coal conversion and biomass conversion. Gas (methane) occurs naturally in coal beds. The World Energy Conference estimated that coal beds alone contain about 75 to 92 · 10^9tce of gas. Gas in geopressured zones is by far the largest source of unconventional gas. The geopressured zones are characterized by thick sedimentary deposits containing water trapped at higher than normal pressure. The trapped water contains the methane deposits. Such areas as the Gulf of Mexico, Siberian coastal basins, Indo-China Sea, Yellow Sea, Sea of Japan and the North Sea contain geopressured gas. For the United States only the WEC estimated reserves of unconventional gas as follows: gas in coal beds, 11 to 30 · 10^9tce; gas in shale, 19 to 22 · 10^9tce; gas in tight formations, 22 · 10^9tce and geopressured gas 1700 · 10^9tce.

Because of the large reserves, given appropriate technologies and production costs, annual production of conventional natural gas could increase greatly. From a level of 1.8 · 10^9tce in 1970 the level could rise to 5 · 10^9tce (about 3.8 · 10^9m^3) in 2000. From this point, production will likely fall gradually to 3.8 · 10^9tce (about 2.8 · $10^{12}m^3$) until about 2065. The development of nonconventional gas will extend this date many decades.

Fuel Reserves for Nuclear Fission

The initial exploitation of nuclear energy was based on fission of the uranium nucleus U^{235}. Natural uranium is composed of 99.274 percent U^{238}, 0.720 percent U^{235} and 0.006 percent U^{234}. When the nonfissionable U^{238} absorbs fast neutrons emitted in the fission process, it is converted to the fissionable isotope of plutonium Pu^{239}, which is not present naturally in the earth. A reactor has also been developed that uses thorium to produce energy.

It is difficult to determine precisely the reserves of uranium and thorium, for no uniform international classification has been devised. The World Energy Conference of 1980 estimated the uranium reserves in the western world (excluding the Soviet Union, People's Republic of China and Eastern Europe) in the cost classes up to $50/lb of U_3O_8 to be 2,590,700t of uranium and the estimated additional resources to be 2,552,500t. The Federal Institute for Geosciences and Natural Resources in Hannover, West Germany estimated that the socialist countries have 150,000 to 300,000t of uranium reserves and additional resources of 1,115,000 to 1,630,000t. In addition to these uranium reserves and resources, there is an additional 5,350,000 to 6,300,000t of possible and speculative potential uranium in the world. There is thus a potential of 13,720,200t of uranium. There is, however, a nearly inexhaustible supply of uranium in the oceans. The concentration is very low and mining costs will be high so that extraction will not occur until well into the twenty-first century.

The major uranium reserves are concentrated in a few countries. The United States has the largest assured reserves and estimated additional resources followed by Canada, Soviet Union, German Democratic Republic, Republic of South Africa, Australia, Sweden, Niger, Namibia and the People's Republic of China (Table 5).

Thorium, like uranium, is found in the earth. Its average abundance in the earth is 12ppm, which is about three times that of uranium. Because demand for thorium has been low, little exploitation has occurred. At the present time, the estimated reserves and additional resources of the world are 3,893,500t. The reserves are widely scattered. The North American reserves are about two-thirds in the United States and one-third in Canada; the South American reserves are found exclusively in Brazil; the European reserves in Denmark (Greenland) and Norway; the Asian reserves in India and the African reserves are in Egypt. The Soviet Union has the largest reserves in the Communist countries.

Fuel Reserves for Nuclear Fusion

Although the underlying processes have been known for decades, no laboratory has been able to build a fusion reactor. The first fusion bomb was exploded on November 11, 1952 in the United States, but the extreme temperatures and pressures required to detonate it were achieved by the explosion of a fission devise. Although it is uncertain when a fusion reactor will become commercially feasible, some estimates of lithium, deuterium and beryllium, the fuels for a fusion reactor have been made.

TABLE 5
Uranium Reserves in the Western World[1][2]

Region	Reasonable Assured Resources		Estimated Additional Resources		Total	
	10^3tU	%	10^3tU	%	10^3tU	%
North America	943.0	36.4	1,886.0	73.9	2,829.0	55.0
Africa (south of the Sahara)	745.4	28.8	264.9	10.4	1,010.3	19.6
Western Europe	432.3	16.3	113.6	4.4	536.9	10.4
Australia/Japan	306.7	11.8	53.0	2.1	359.7	7.0
Latin America	109.9	4.2	136.7	5.4	246.6	4.8
Middle East/North Africa	32.3	1.2	74.6	2.9	106.9	2.1
South Asia	29.8	1.2	23.7	0.9	53.5	1.1
East Asia	0.3	0.1	n.a.	—	0.3	—
Western World	2,590.7	100.0	2,552.5	100.0	5,143.2	100.0

(1) Data for Socialist Bloc countries not available
(2) Resource estimates based on recovery costs of $50 per pound, U_3O_8 (1977 Dollars)
Source: World Energy Conference, 1980

In 1978 the Max-Planck-Institut fur Plasmaphysik estimated the economic recoverable reserves of lithium to be $1.4 \cdot 10^6$t, representing an energy potential of 35 to $123 \cdot 10^3$q depending upon the type of reactor. The indicated and inferred reserves are $5.2 \cdot 10^6$t which corresponds to 128 to $455 \cdot 10^3$q. The total resources are estimated to be $1.2 \cdot 10^8$t (0.29 to $1.0 \cdot 10^7$q). Since reserves are large and consumption small, there has been no need to prospect for new lithium deposits. Lithium deposits are widely scattered. The major producers are the United States, Zimbabwe, Soviet Union, Canada and China. Lithium could also be obtained from the oceans, if there was a demand.

Deuterium is available in practically unlimited quantities in the oceans, in the form of D_2O or HDO. Natural oceanic water contains 16.68 ppm deuterium. It is estimated the oceans contain $4.6 \cdot 10^{13}$t of deuterium providing an energy source of $15 \cdot 10^{12}$q.

Most countries have access to the oceans so a supply of lithium and deuterium for nuclear fusion is readily available. When fusion becomes available, a safe and unlimited supply of energy will be available to the world.

In some types of reactors beryllium is the required breeder material. Estimates indicate that beryllium is available in sufficient quantities to satisfy all demands. The average abundance of beryllium in the earth's crust is 6 ppm. A beryllium reactor will consume about 1/11 the amount of lithium consumed.

WORLD TRENDS AND PATTERNS OF ENERGY CONSUMPTION

There has been a steady and rapid growth of energy consumption in recent decades. In 1965 the world's energy consumption was nearly 4,000 million metric tons of oil equivalent (one toe is about equal to 7.33 barrels of oil). By 1984 this figure had risen to 7,201 Mtoe (Table 6).

There are, however, marked differences in the trends of energy consumption among the different countries of the world. In the western industrialized countries (Western Europe, Anglo America, Japan, Oceania) energy consumption has risen dramatically. For example, between 1965 and 1972 energy consumption in Anglo America rose from about 1,475 Mtoe to nearly 2,000 Mtoe and in Western Europe from 825 to over 1,200 Mtoe. With the increase in the cost of petroleum, consumption of energy has been nearly stable since 1972.

Energy consumption trends contrast sharply in the Socialist countries with trends in the Western World. Energy consumption in the Soviet Union has been steadily rising from about 600 Mtoe in 1965 to just over 1,300 Mtoe in 1984. The Eastern European countries have experienced trends similar to the Soviet Union due to political ties. Because of the stagnation of the Soviet oil industry increases of energy consumption in the Socialist nations are likely to be small in the future.

In the developing countries the consumption of energy has risen in spite of the dramatic increases in oil costs. By 1984 it had reached a total of 1,690 Mtoe.

TABLE 6
Commercial Energy Consumption by Region 1984
(Million metric tons of oil equivalent, Mtoe)

Region	Oil	Natural Gas	Coal	Hydro-electric	Nuclear	Total	Percent
Anglo America	791.4	595.8	466.2	154.6	101.3	2,019.3	28.0
Western Europe (1)	591.0	190.1	256.7	107.0	104.6	1,249.4	17.4
Oceania (2)	35.5	14.4	34.6	9.3	—	93.8	1.3
Japan	214.6	33.1	64.0	19.8	30.6	362.1	5.0
Soviet Union	447.8	439.4	357.0	53.0	25.0	1,322.2	18.4
Eastern Europe (3)	98.7	77.0	274.2	7.2	7.5	464.6	6.5
Developing Countries (4)	688.5	150.1	726.9	134.5	13.2	1,690.2	23.4
World Total	2,844.5	1,409.9	2,179.6	485.4	282.2	7,201.6	100.0
Percent	39.5	19.6	30.3	6.7	3.9	100.0	

(1) Western Europe includes Austria, Belgium, Cyprus, Denmark, Federal Republic of Germany, Finland, France, Gibraltar, Greece, Ireland, Republic of Ireland, Italy, Luxemburg, Malta, Netherlands, Norway, Portugal, Spain, Sweden, Switzerland, Turkey, United Kingdom, Yugoslavia.

(2) Oceania includes Australia, New Zealand, Papua, New Guinea, Southwest Pacific Islands.

(3) Eastern Europe includes Bulgaria, Czechoslovakia, German Democratic Republic, Hungary, Poland, Romania.

(4) Developing Countries include all Latin America, Africa, Southeast Asia, and Middle Eastern Countries.

Source: British Petroleum Statistical Review of World Energy, London, 1985

The increased energy consumption has been a severe economic burden on these countries. In 1981, low and middle-income oil-importing countries were spending 61 and 37 percent, respectively, of their export earnings on oil imports. Between 1973 and 1981 the international debt of these countries increased by a factor of six and now totals over one trillion dollars. As the economy of these countries grows, the problem of energy becomes increasingly complex particularly with the depletion of their principal domestic source of energy-fuelwood.

Coal

In the 1970s, after the rise in oil prices, many countries in Western Europe and the United States began to investigate the possibility of substituting coal for oil in their domestic economies. Between 1973 and 1984, however, coal's share of world energy consumption had increased little from 28.2 to 30.3 percent. In 1984 coal provided 2,179 Mtoe of the world's energy consumption.

Of all the fossil fuels, the consumption of coal is most highly geographically concentrated (Griffith, 1979). China has become the world's largest producer of coal with a consumption in 1984 of 466 Mtoe. Anglo America also consumes 466 Mtoe of which the United States consumes well over 90 percent of the total. The third major consumer of coal is the Soviet Union, totaling 357 Mtoe in 1984. These three nations account for about 60 percent of the world's coal consumption. The importance of coal in the world's energy budget varies greatly between nations. In the Soviet Union, coal's contribution to the energy budget declined from 36 percent of the total in 1973 to 27 percent in 1984. In contrast, coal consumption in China during this decade increased by about 60 percent and coal now provides about 80 percent of China's energy. In the United States, the importance of coal in the energy budget changed little and provides less than one quarter of the nation's energy needs.

In Western Europe, consumption of coal in 1984 was 256 Mtoe. Coal has gradually declined in importance providing only 20 percent of the energy consumed, being replaced by petroleum and nuclear energy. Although Japan has little domestic coal production, coal provides 64 Mtoe of an energy budget of 362 Mtoe or about 18 percent of the total. In the developing nations of the world, coal provides about 726 Mtoe of their energy needs, or about 24 percent of the total. In Oceania coal provides about 43 percent of the energy consumed.

Petroleum

In 1984, oil provided 2,844 Mtoe or 39.5 percent of the energy consumed in the world. This is a decline, however, from the 47.4 percent share which oil held in 1973. This decrease in importance was due, in part, to the economic recessions following the oil price rises since 1973, partly to the improved efficiency with which oil is used, and partly due to substitution of other fuels for oil.

TABLE 7
Petroleum Production and Consumption in Selected Countries 1980

Country	Production (10^6t)	Consumption (10^6t)
Soviet Union	603	438
Saudi Arabia	493	19
United States	484	800
Iraq	130	5
Venezuela	116	17
Mexico	106	47
China, P.R.	106	92
Nigeria	102	4
Libya	86	4
Canada	80	88
United Kingdom	80	83
Indonesia	79	20
Iran	74	25
Kuwait	71	5
United Arab Emirates	65	n.a.

Source: World Energy Conference, 1980
B P Statistical Review of the World Oil Industry 1980, London 1981

There is wide disparity between production and consumption centers of petroleum (Table 7) (Ivanhoe, 1984). In most of the industrial nations of the world such as the United States, West Germany, France and Japan, consumption far exceeds production. The gap between consumption and production in the western industrialized nations has to be filled by petroleum from such areas as the Middle East, Venezuela, North Africa and Nigeria where production is high and consumption low. In contrast, the Soviet Union and China have a near balance between production and consumption (Flower, 1978).

There are wide variations of dependence on oil to supply the nation's energy in the world. Japan with 59 percent of its energy requirements from oil had, in 1984, the greatest dependence on oil in the industrialized world (Table 6). In contrast, only 14 percent of the energy consumed in China was from petroleum. Western Europe continued to secure 47 percent of its energy from petroleum and Anglo America had 39 percent of its energy consumption from petroleum. There were considerable differences of petroleum consumption in the Socialist countries. The Soviet Union secured about one-third of its energy from petroleum but the Soviet dominated Eastern European countries secured only one-fifth of their energy from petroleum.

Natural Gas

Natural gas provided about 19.6 percent of the world's energy consumption in 1984 slightly up from 18 percent in 1973, making it the third largest source of energy. This was about 1,409 Mtoe of natural gas. While the global picture

TABLE 8

Natural Gas Production and Consumption in Selected Countries 1980

Country	Production ($10^9 m^3$)	Consumption ($10^9 m^3$)
United States	577	531
Soviet Union	433	354
China, P.R.	98	13
Netherlands	88	35
Canada	74	53
United Kingdom	35	44
Norway	33	n.a.
Romania	33	n.a.
Mexico	27	n.a.
Algeria	23	n.a.

Source: B P Statistical Review of the World Oil Industry, London, 1981

of natural gas consumption was nearly stable, significant regional changes are occurring. Between 1973 and 1984 the United States natural gas consumption declined from 562 to 458 Mtoe. Nevertheless natural gas still accounted for about 25 percent of U.S. energy consumption (Table 8).

The trend in the other western industrial nations and the Soviet Union was directly opposite to that of the United States. Consumption of natural gas increased by about 90 percent between 1973 and 1984 from 502 to 951 Mtoe. The growth of natural gas consumption between 1973 and 1984 was particularly large in the Soviet Union when it rose from 198 to 439 Mtoe. The Soviet Union replaced the United States as the world's largest producer in 1983. In Western Europe, consumption rose from 129 to 190 Mtoe between 1973 and 1984. Natural gas has become an important substitute for oil in Western Europe.

Natural gas has been a traditional source of energy in a number of countries of the Third World. Utilization has been growing and in 1984 the Third World consumed 150 Mtoe of natural gas or about 9 percent of its energy needs. The lack of an infrastructure and exploration has delayed the use of natural gas for many of the Third World countries in Asia and Africa where huge natural gas resources exist. When developed, natural gas will become an important source of energy in these regions.

As with other fuels, the dependence on natural gas varies greatly from region to region. The Soviet Union has the greatest dependence with natural gas supplying about one-third of the nation's energy requirements. In the United States about 25 percent of the energy consumed comes from natural gas. In China, which has very limited domestic production, natural gas supplies only about 1.8 percent of the nation's energy.

Nuclear Power

The development of nuclear power has been spectacular in recent years. Electrical output has risen from 62.4 Mtoe in 1974 to 282.2 Mtoe in 1984. With the massive OPEC oil price increases in 1973 and 1979, the industrial nations tried to decrease their heavy dependence on imported oil. With a shortage of indigenous alternative fuel resources in many industrial nations, the need for development of nuclear power to provide electricity appeared a desirable alternative.

In 1984 nuclear power accounted for 3.9 percent of the world primary energy consumption. Ninety percent of nuclear energy production is concentrated in the highly industrialized nations with about 35.9 percent in Canada and the United States and 37.1 percent in Western Europe. The number of operable reactors has grown from 80 in 1970 to 345 in 1984. The United States has the largest number with 86, with an installed capacity of 71.2 GWe (gigawatts of electricity), followed by the Soviet Union with 46 and installed capacity of 24.1 GWe, France 41 and 33.3 GWe, United Kingdom 38 and 10.7 GWe, and Japan with 31 reactors with installed capacity of 21.8 GWe.

The dependence on nuclear power to generate electricity varies greatly in the world. France, with the greatest dependence, secured 64.8 percent of its electricity from nuclear power in 1986. In 1973 th Messmer-Pompidou government announced that it was embarking on a program to supply by means of nuclear power 70 percent of the electricity and 30 percent of the total energy needs of the nation by 1985. At that time nuclear power was considered the cheapest and best alternative to imported oil. Nuclear plants are widely dispersed in France and until recently opposition has been local and regional focusing on particular sites rather than the total program. Other countries with major dependence upon nuclear power for electricity include Belgium 59.8 percent, Sweden 42.0, Finland 38.2, Switzerland 34.3, Bulgaria 31.6, West Germany 30.0, Japan 23.0, United Kingdom 19.3, United States 16.0 and the Soviet Union 11.0 percent. Nuclear reactors have been built in a few Third World countries such as India, Pakistan, Brazil and Argentina, but dependence on them for the nation's electricity remains small.

The future of nuclear power to produce electricity is uncertain and is expected to slow down in most countries, owing to a combination of events in recent years, including increased capital costs, lower forecast demand for electricity, and environmental opposition. The principal environmental concerns are reactor safety following the Three Mile Accident in 1979 and Chernobyl in 1986. While human error was a major factor in both accidents, a symposium in Vienna in August 1986 concluded that there were design shortcomings in the Chernobyl graphite reactor that played a major role in the disaster and that projected modifications to the outdated Chernobyl-type plants would not make them safe.

Although the United States has the largest installed capacity of nuclear power, the future is particularly bleak. Capital costs per kilowatt of installed capacity have increased by a factor of four in real terms for plants constructed since 1983 compared with a typical plant completed in 1971. In addition, construction time has doubled and reactor performance has been mediocre. No new orders have been placed since 1978 in the United States and 13 plants ordered between 1975 and 1978 have been cancelled.

The disposal of spent fuel inventories remains a critical problem in the world. Between 1970 and 1984 the cumulative metric tons of heavy metal has increased from 6,045 to 56,654 in the Capitalist world only. No estimates are available for the Socialist countries. Of the 56, 654 metric tons 60.4 percent are located in Western Europe and 31.3 percent are in Anglo America. No safe depository has yet been found for the spent nuclear fuels that will continue to emit radiation long into the future. This may be the ultimate problem in the use of the nuclear reactor to produce electricity.

RENEWABLE ENERGY SOURCES

Renewable energy has played an insignificant role in supplying the energy of the world since the beginning of the Industrial Revolution in the 18th century. It has been the inanimate energy resources that have made possible the progress of human society in the past several hundred years. Of the renewable energy resources, hydroelectric power is the most important in the industrial world. In the Third World the use of biomass provides a fundamental source of energy in providing heat for cooking and heating homes. Only recently has attention been directed to the potential of solar and wind energy as a major source of power in the future. Geothermal energy has been utilized locally in a few places and tidal energy remains for the future.

Hydroelectic Power

The world's hydroelectric power has grown rapidly in recent years from 331.5 Mtoe in 1973 to 485.4 Mtoe in 1984. Its share of total world energy consumption rising from 5.6 to 6.7 percent. North America had the largest consumption with 154 Mtoe. This was about 7.6 percent of the region's energy consumption. North America and Western Europe consumed about 54 percent of the world's hydropower (Table 6).

In the Developing Countries, the consumption of hydro-electricity doubled from 57 to 134.5 Mtoe between 1973 and 1984, increasing their share of world consumption from 17.2 to 27.7 percent. In 1980 hydropower accounted for 40.5 percent of the electricity production in Developing Countries. This figure is expected to rise but the rate of growth will depend on cost factors, the growth of market potential, and the environmental and social difficulties.

Hydropower is poorly developed in most Socialist nations. In the Soviet Union only 53 Mtoe of hydropower were consumed in 1984. This was only 4 percent of the nation's energy consumption. Similarly, China had only 23.5 Mtoe of hydropower or 4 percent of its energy needs. The least dependence on hydropower was in Eastern Europe where only 7.2 Mtoe were consumed, or 1.5 percent of energy needs.

Biomass

The consumption of wood and other biomass materials had essentially ceased prior to the oil crisis of the early 1970s. Since then, the use of biomass fuels has grown in countries with wooded regions. In such industrialized countries as the United States, Canada, Sweden, and Norway, wood use for residential heating nearly doubled between 1975 and 1984. The United States and Canada now obtain from 3 to 4 percent of their primary energy from forest products.

As the price of petroleum has risen, the use of wood to provide the energy requirements in the Third World has increased in importance. Statistics of energy consumption in these countries usually does not include biomass fuels. A large proportion of this fuel does not enter commercial transactions. Nevertheless, surveys of residential fuel use have shown that biomass fuels, particularly fuelwood, are key sources of energy in developing countries.

In a survey by the world Food and Agricultural Organization, it was estimated that the world consumption of wood in 1983 totaled 1.63 billion cubic meters, or about 6 percent of the world's energy consumption. Within the Third World, the use of biomass fuels varies greatly from about 10 percent of energy requirements in the more advanced industrialized nations of Latin America to more than 90 percent in the least developed African nations. In a survey of 95 countries by FAO, it was found that fuelwood played a significant role in all of them, and 21 countries depended on fuelwood for more than 75 percent of their energy needs.

Within Third World countries, the use of fuelwood usually varies from rural to urban areas. In poor rural areas, fuelwood may be the only available source of energy. When wood is scarce, the quantity consumed decreases drastically. While wood is a renewable resource, its production requires decades. Because consumption of wood exceeds production in many areas of the world, wood has become so scarce that people must turn to other sources of energy, such as crop residues and dung. The greatest use of biomass residues occurs in relatively treeless areas such as China, India and Bangladesh. In some rural areas of these countries biomass residue provides about 90 percent of household energy.

Solar Energy

The importance of solar heat has been recognized throughout time by helping to heat and light buildings. The world's solar energy base is immense. Solar energy arrives at the earth's surface at an average rate of 4.76 kwh per square meter per day, so that in the course of a year, one square kilometer receives about 1.7 billion kwh. The United States consumes annually a total of about

24 trillion kwh. Accordingly, about 18,800 square miles, used for solar conversion at about 30 percent efficiency, could meet the United States energy requirements.

The basic question is, why has utilization of solar energy been so slow to develop? A wide variety of barriers have hindered development (Sawyer, 1986). The inherent constraints include (1) the large space requirements for energy collection, (2) the storage requirements and costs due to the intermittent supply of solar energy, (3) the great regional variations in supply, (4) private and public investment decisions are biased because of high initial cost, and (5) the long term gamble regarding energy prices, reliability, future innovations and other monetary needs. There are also such federal policy constraints as (1) tax policies, (2) exclusion of social costs because competing fuels exclude such social costs as air, water, land, pollution impacts, national security risks, climate changes, and health costs, (3) subsidies for the development of competing energy resources, particularly nuclear power and (4) the marketing and other promotion programs such as rural electrification programs, limits on utility liability on nuclear accidents and similar programs that subsidize centralized technologies. The market constraints include (1) the co-evaluation of solar energy with established energy industries with decades of technology research and market infrastructure developments, (2) the profit margins are too low to fund large scale research and development and (3) the low energy density of solar energy in market areas requiring a high energy density are best met by traditional energy sources. Finally there are a number of institutional restraints such as (1) resistance to change of established practices by the building profession, (2) lack of information reduces market appeal and prevents full assessment of solar energy potential, and (3) the legal uncertainties that assure access to sunlight for solar energy.

Of the uses contemplated for solar energy, heating for houses and other structures appears most feasible. Space heating requires low temperatures, which solar collectors can provide. Housing and solar energy are both widely distributed so the need to transport electricity is eliminated. Solar heating had the overwhelming share of the approximately 0.02 to 0.03 quads of energy secured from solar sources in 1984.

The future of solar energy in the production of electricity will depend on the solution of a number of complex problems. Because the sun's radiation is widely dispersed, a large area of the earth's surface must be covered with solar collectors in order to harvest sufficient energy for economic viability. No estimates have been made on the environmental impact of covering several thousand square miles with solar collectors. Because the sun does not shine continuously, a system is needed to store energy when it is available and to tap energy at other times. The battery is the most reliable method of energy storage but to have batteries in sufficient numbers to store massive quantities of electricity is still not feasible.

Solar energy will not provide a significant share of the world's energy needs in the near future. The use of solar energy for home heating and cooling appears most promising and should be well established by the year 2020. The commercial use of energy may begin by supplying electricity during peak demands. It is possible that within the next century solar energy will assume a greater role in supplying the world's energy needs, but even then it will be only one of a number of energy sources.

Wind Power

Wind has been used as an energy resource for several thousand years. Nevertheless it continues to play an insignificant role in the world's energy consumption. When the energy crisis of the early 1970s renewed interest in using wind to generate electricity, the industry was technologically backward. There was little meteorological information on such aspects as average wind velocities and local patterns of wind currents. Over the past decade, wind turbine technology has been advanced and wind power resource assessments begun.

Only crude estimates are available as to the potential wind power in the world, but it is tremendous. In 1961, the World Meteorological Organization selected some 350 sites around the world to measure wind speeds from air currents 160 to 360 feet above the ground. From these and more recent observations, it is estimated that some 20 billion kilowatts of wind power are available at favorable sites around the world.

Within the past decade, technical developments have proceeded in two directions (Gipe, 1985). The federally sponsored programs have focused on large machines with the goal of designing large multi-megawatt, technically sophisticated turbines for utility construction and operation. These systems were designed to produce large quantities of electricity. In contrast, the private developers have emphasized smaller, simpler turbine designs primarily for the single family market. As a response to providing an alternative source of power, wind power use has increased. There were approximately 12,000 wind turbines with 100 megawatts of generating capacity in the nation in 1982. California has made the greatest strides in developing wind power. By 1985 the installed capacity was about 1,000 megawatts. On some windy days over one percent of California's electricity is provided by wind. California has developed more wind power than the rest of the world combined.

The technical problems of operating wind mills have been solved and they are now economically feasible. It has been proved that wind power can supplement conventional sources of power, and in the United States, any excess power produced is purchased by electric companies. Nevertheless, the development of large wind machine networks producing large quantities of electricity is far in the future. Today wind power plays no role in the energy consumption of the world. The rate of wind power development will depend on its place in the total energy economy.

CONCLUSION

The energy for the world will be supplied in the foreseeable future by fossil fuels and nuclear power. The proved reserves and available resources of the traditional fuels will be sufficient to supply the world's energy demands for many decades. The environmental problems of supplying the world's energy needs in the foreseeable future are thus known. These include such aspects as restoration of mined lands, control of air and water pollution, and development of better safety standards for nuclear power plants. In order to provide a wholesome environment requires not only technological competence but also necessary economic and political commitments on a global scale.

SELECTED REFERENCES

Agnew, A.F., 1977, "Coal Resources and Production," in *Project Independence: U.S. and World Energy Outlook Through 1990,* Washington, D.C.: Congressional Research Service, Library of Congress. U.S. Government Printing Office, pp. 208-263.

Aoki, H., 1980, "International Cooperation in World Coal Development," *11th World Energy Conference,* Munich, 461-479.

Banks, Ferdinand E., 1983, *Resources and Energy: An Economic Analysis,* Lexington, MA: Lexington Books, 332 pp.

_____, 1980, *The Political Economy of Oil,* Lexington, MA: Lexington Books, 270 pp.

Bauer, L., G.B. Fettweis and W. Fiala, 1977, "Classification Schemes and their Importance for the Assessment of Energy Supplies," *10th World Energy Conference* (Division 1, 1.1-2): 1-20.

Billo, S.M., 1979, "Future Petroleum Resources," *The Oil and Gas Journal,* 77 (January 1): 98-103.

Chadwick, M. and N. Lindman, 1982, *Environmental Implications of Expanded Coal Utilization,* Oxford: Pergamon Press, 283 pp.

Dasgupta, P. and G. Heal, 1979, *Economic Theory and Exhaustible Resources,* Cambridge: Cambridge University Press, 501 pp.

DeSouza, G.R., 1981, *Energy Policy and Forecasting,* Lexington, MA; Lexington Books, 218 pp.

Deudney, D. and C. Falvin, 1984, *Renewable Energy: The Power to Choose,* Washington, DC: Worldwatch Institute, 448 pp.

Dunkerley, J. and W. Ramsay, 1982, "Energy and the Oil-Importing Developing Countries," *Science,* 216: 590-595.

Fettweis, G.B., 1979, *World Coal Resources: Methods of Assessments and Results,* Essen, West Germany: Glückauf, 415 pp.

Flower, A.R., 1978, "World Oil Production," *Scientific American,* 238 (March): 42-49.
Georgescu-Roegen, N., 1976, *Energy and Economic Myths,* New York: Pergamon, 300 pp.
Gipe, P., 1985, "An Overview of the U.S. Wind Industry," *Alternate Sources of Energy,* 75 (September/October): 25-27.
Govett, G.I.S. and M.H. Govett, 1974, "The Concept and Measurement of Mineral Reserves and Resources," *Resource Policy,* 1: 46-55.
Grathwohl, Manfred, 1982, *World Energy Supply: Resources, Technologies, Perspectives,* New York: Walter deGruyter, 439 pp.
Grenon, M., 1978, "On Fossil Fuel Reserves and Resources," International Institute for Applied Systems Analysis, Luxemburg/Vienna, RM-78-35.
Griffith, E.D. and A.W. Clarke, 1979, "World Coal Production," *Scientific American,* 240 (1): 28-37.
Hayes, E.T., 1979, "Energy Resources Available to the United States, 1985 to 2000, *Science,* 203: 233-239.
Ivanhoe, L., 1984, "World Crude Output Reserves by Region," *The Oil and Gas Journal,* (Dec. 24): 65-68.
Kerr, R.A., 1981, "How Much Oil? It Depends on Whom You Ask," *Science,* 212: 427-429.
Kuri, Mahmood A., 1979, *The World Energy Picture,* New York: Vantage Press, 256 pp.
Leach, G. et al, 1986, *Energy and Growth: A Comparison of 13 Industrial and Developing Countries,* Stoneham, MA: Butterworth, 208 pp.
Masters, C.D., 1985, *World Petroleum Reserves-A Perspective,* Reston, VA: U.S. Geological Survey, 85-248.
McKelvey, V.E., 1972, "Mineral Resources Estimates and Public Policy," *American Scientist,* 60: 32-40.
Odell, P.R. and K.E. Rosing, eds., 1983, *The Future of Oil: World Resources and Use,* London: Kogan Page, 385 pp.
Peterson, F.M. and A.C. Fisher, 1977, "The Exploitation of Extractive Resources," *Economic Journal*, 87: 681-721.
Sassin, W., 1980, "Energy," *Scientific American,* 243 (3): 106-110.
Sawyer, Stephen W., 1986, *Renewable Energy: Progress, Prospects,* Washington, DC: Association of American Geographers, 102 pp.
Wilson, D., 1983, *The Demand for Energy in the Soviet Union,* London: Croom Helm, 310 pp.
British Petroleum (BP), June 1985, *BP Statistical Review of World Energy,* London, pp. 1-32.
Food and Agricultural Organization (FAO), 1983, *Fuelwood Supplies in Developing Countries,* Rome: FAO, 125 pp.
_____, 1983, *Wood for Energy,* Rome: FAO.

———, 1985, *Yearbook of Forest Products, 1983,* Rome: FAO, 408 pp.
Office of Technological Assessment, 1984, *Nuclear Power in an Age of Uncertainty,* Washington, DC: Office of Technological Assessment.
World Energy Conference, 1983, *Energy 2000-2020: World Prospects and Regional Stresses,* London: Graham and Trotman.
———, 1980, *Survey of Energy Resources, 1980,* for 11th World Energy Conference, prepared by Hannover, West Germany: Federal Institute for Geosciences and Natural Resources.
World Energy Outlook, 1982, Paris: International Energy Agency, 473 pp.

Environmental Consequences of Energy Production: Problems and Prospects. Edited by S. K. Majumdar, F. J. Brenner and E. W. Miller. © 1987, The Pennsylvania Academy of Science.

Chapter Two
INTERNATIONAL ASPECTS OF THE MITIGATION OF THE ENVIRONMENTAL EFFECTS OF ENERGY PRODUCTION

HEINZ G. PFEIFFER
Manager, Technology & Energy Assessment
Pennsylvania Power and Light Company
Two North Ninth Street
Allentown, PA 18101

It is hard to draw a boundary between those impacts on the environment which are specifically related to energy and those which are a general consequence of the concentration of human activities. This survey of international activities on energy and the environment will be restricted to those effects which commonly operate across international boundaries and will not include such problems as the airborne chemical problems at Bhopal, the Rhine spill at Basel or the ozone depletion by chlorofluorocarbons. The major international energy-related concerns are: 1) carbon dioxide and the possible warming of the earth; 2) sulfur oxides, nitrogen oxides and photochemical oxidant effects on lakes, streams, forests and structures; 3) nuclear radioactive releases from weapons testing and power plants as they affect human health; and 4) the disposal of radioactive waste as a possible threat for the future groundwater supplies or oceans.

The per capita energy production of a country correlates with its per capita income to some degree. When only electrical production is considered, the correlation with economic level is very close. It is not surprising, therefore, that the strong movements for control of the environmental effects of energy production have been in the most prosperous countries. These countries are in North America, western Europe, and the western Pacific. The same countries have also had the resources to divert to environmental improvements. In most cases, the environmental movement has gotten its strength from the public at large rather than from the government. This is very apparent when the tardy environmental efforts of those countries with centrally planned economies are considered. Thus, there are three ingredients for a high level of environmental control in a country: a high level of energy production, availability of resources, and public participation. Almost all countries have adequate regulations for the control of pollution that has a direct and immediate effect on human health, but even there enforcement varies greatly.

Much of the pollution caused by high levels of energy production crosses political boundaries and leads to international tensions, particularly in western Europe. Scandinavia, which has been the leader in Europe on emission control, now sees that most of the remaining airborne pollution comes from England and the European continent. The costs of reduction in emissions from stationary sources has been high, but even more difficult is the control of automobile emissions where unleaded gasoline is necessary for the operation of a catalytic converter. With the free movement among all of the countries in western Europe, automobile emission control can only be considered if the countries reach a consensus. At present, non-leaded gasoline is unavailable in most of Europe.

Radioactive materials that result from nuclear fission provide a disposal problem. The high level of technology and surveillance required has concentrated the reprocessing and disposal in a selected few of the countries with numerous nuclear power plants. As this has improved control of the waste, it has compounded the transportation problem of nuclear wastes.

The theme of the 1986 World Energy Conference in Cannes, France was "Energy, Economy and Environment". The environmental papers were largely aimed at technical solutions to specific problems, with very little regard for the global effects of energy production. With the large representation from the underdeveloped countries, it is understandable that energy production was a larger concern than the environment.

In the industrialized nations, the serious episodes of air toxicity that produced short-term health effects have been eliminated; longer-term effects have been hard to determine and seem to be the result of synergistic effects of several factors including personal habits and individual sensitivity. In much of the world any long-term effects of air quality are hidden in the high mortality rates related to public health and nutrition. Internationally, the reduction of emissions from stationary sources has been led by Sweden, with Japan and the United States playing a role. The United States has been a leader in the reduction of emissions from motor vehicles. The Swedish work is in five problem areas: deposition, forests and soils, ground-water and corrosion, surface water, and wildlife. The deposition problem is centered on aerosols and very closely parallels the work in the United States, with a somewhat larger emphasis on oxides of nitrogen and road vehicles. Sweden was also instrumental in the establishment of an 80-station, 23-country, long-range transport monitoring network in Europe. Sulfur and nitrogen oxides and sulfate-bearing particulates are being tracked; and any precipitation is assayed for pH, ammonia, nitrate, and chloride. To date, the sulfur components have been decreasing and the oxides of nitrogen have been increasing. This is what would be expected as large stationary sources are more tightly controlled and vehicular traffic continues to increase with no attempts to control automobile emissions.

Japan has been a consistent leader in the control of sulfur emissions from large stationary sources. Most of the scrubbing techniques presently used were first applied there on a large scale and the necessary technology developed. Other countries with rapid industrial development are starting experimental work to anticipate problems they foresee, but in most cases, the antipollution efforts are not being applied on a commercial scale.

Extensive work on forest damage has been done in the Federal Republic of Germany as well as Sweden. The research has come to the same conclusion as much of the American work—that a complex group of stresses have led to the observed decline, with no one factor dominant. The stresses have been identified as extreme climatic conditions, airborne pollutants, soil acidification, and nitrogen saturation. The airborne pollutants, photochemical oxidants, ammonia, and nitrogen oxides cause leaf and needle damage; the soil acidification leads to heavy metal release and nutrient deficiency; and the nitrogen saturation causes plant physiology problems which result in decreased resistance to frost.

Some indications of reduction of alkalinity in groundwater in Sweden have been found, but an overall trend has not been established. On the other hand, the surface waters have serious problems. About 18,000 Swedish lakes have pH levels that are below values that are healthy for some living organisms. The surface water decline was the trigger that led to the extensive work on acidification, and large-scale mitigation efforts. By 1986, the addition of limestone had improved at least 4,000 lakes and has encouraged similar activities in other countries including the United States. Striking changes have been observed in the heavy metal content of the fish taken from the treated lakes. It has long been known that acid waters release heavy metals from rock and soil, and the metals stay in solution or are incorporated into the food chain. Some fish high on the food chain concentrate these metals in their tissues.

Wildlife research is in its early stages and is being concentrated on the effects of pollutants on the food chain for the higher animals such as birds, amphibians, and mammals. Forage flora and insect population are being monitored in the program.

Both formal and informal cooperation is increasing in Europe. The major agencies are the Nordic Council of Ministers, the Economic Commission for Europe, the European Economic Community, and Eurotrac. A large number of bilateral agreements between various countries also exist. In late 1986, England announced a program to reduce sulfur emissions from some of the coal-burning power plants. This initiative was stimulated by the evidence of long-distance effects (into Scandinavia) of the sulfur emissions. In Helsinki during 1985, the environmental ministers of 21 countries signed a "Protocol" on sulfur emissions. It calls for a 30% reduction between 1980 and 1993. The Protocol itself does not take effect until 16 countries ratify it, and 14 countries did not sign. The table below shows the status of agreement at the end of the conference.

SIGNATORIES TO THE SULFUR PROTOCOL

Sweden	Switzerland
Norway	Austria
Demark	
Finland	Czechoslovakia
	Soviet Union
Belgium	The Ukraine
The Netherlands	East Germany
Luxembourg	Bulgaria
	Hungary
Liechtenstein	
	Italy
West Germany	Canada
France	

COUNTRIES NOT SIGNING

Greece	Iceland
Spain	Eire
Portugal	The Vatican State
Turkey	U.S.A.
Malta	United Kingdom
Cyprus	Poland
Rumania	
Yugoslavia	EEC

The United States and Canada reached an agreement during 1986 that will provide improved air quality for both countries. The agreement calls for the development of efficient and less expensive control technologies, joint research and demonstration of clean coal technologies. The total program of five billion dollars would be financed by the United States government and industry.

The international situation is very much in flux, especially in situations where manufacturing and energy producing operations seek locations where their pollution control costs are minimum. Where emissions from economically troubled countries have affected more prosperous neighbors, some sharing of pollution control costs has been proposed. It is certain that in the next decade international practices and law will undergo major changes.

Environmental Consequences of Energy Production: Problems and Prospects. Edited by
S. K. Majumdar, F. J. Brenner and E. W. Miller. © 1987, The Pennsylvania Academy of Science.

Chapter Three
INTERNATIONAL COOPERATION IN REDUCTION OF ENERGY-RELATED ATMOSPHERIC POLLUTION

JOHN W. LANDIS
Senior Vice-President
Stone & Webster Engineering Corporation
245 Summer Street, P.O. Box 2325
Boston, MA 02107

INTRODUCTION

Ever since acid deposition (commonly called "acid rain") was identified as a serious atmospheric-pollution problem some thirty years ago, it has been the subject of widespread and often acrimonious debate. The debate has involved different factions in the various countries involved, but generally has pitted the operators of large combustors of coal, oil or wood against government agencies, environmental groups and public-interest organizations.

Conclusions drawn as a result of these exchanges have varied greatly from country to country and even more from faction to faction. In the United States, for example, the analyses and recommendations of the National Academy of Sciences are quite different from those of the Edison Electric Institute. Because this type of pollution cannot be confined to the originators' territories, however, it should be studied and combatted by joint action of the major countries involved, the great majority of which are located in the Northern Hemisphere.

Several proposals for such joint action surfaced about a decade ago as a result of intergovernmental conferences and protocols, but the actual and perceived negative effects on strong internal parties prevented effective implementation. Shortly thereafter, several privately-sponsored international organizations began to examine the problem. Among these organizations was the World Energy Conference.

The World Energy Conference (WEC) is a non-political association of energy experts from over 80 nations, including most of the countries that, like the U.S.S.R., have centrally controlled economies. Reasonably free of governmental constraints, it provides an open forum for debate of energy issues that span international boundaries and serves as a medium for development of basic data and consensus recommendations for worldwide use.

The WEC was established in the early 1920s and its first congress was held in London, where it is headquartered, in 1924. Congresses have been held periodically—except during World War II—since that time.

At the congress in Munich in 1980, which was attended by over 5,000 representatives of 81 nations, discussions of atmospheric pollution by energy facilities cropped up in a number of the sessions on energy resources, energy production and usage, and evolving energy technologies. Recognizing the need for more concentrated thought and effort on this subject, the United States National Committee shortly thereafter decided to establish a Special Committee on Cost/Benefit Analysis of Environmental Regulations to tackle one important aspect of the atmospheric-pollution problem.

After a series of meetings in 1982 and 1983, culminating in a major conference in the Washington, DC area, the leadership of the Special Committee, realizing that the work of the Committee would be more meaningful if broadened to include other aspects of environmental pollution, requested and received permission to address a broad range of issues, including acid deposition, and to identify possible remedial measures. The name of the Special Committee was thus appropriately changed to Protection of the Environment.

Following a particularly stimulating colloquium on acid deposition sponsored by the Special Committee in February 1984, Mr. W. Kenneth Davis, Chairman of the United States National Committee, proposed to the WEC International Executive Council that a WEC Working Group on Acid Deposition, with representation from all affected areas of the globe, be established to collect and analyze data on this phenomenon and to evaluate and correlate recommendations for remedial action received from concerned nations.[1]

The Working Group was organized late in 1984 under the chairmanship of the author of this chapter, who had served as chairman of the forerunner groups within the United States National Committee.

The first meeting of the Working Group was held in Washington, DC in May 1985. Initial discussions led to an enlargement of the scope of the Group's activities to include all types of atmospheric pollution resulting from the production of energy. Accordingly, the name of the Group was subsequently changed to Working Group on Energy-Related Atmospheric Pollution.

The main reason for this action was that the member-country representatives who attended the meeting unanimously agreed that it would be virtually impossible to separate a study of the causes and effects of acid deposition from examination of the concurrent and closely linked causes and effects of other types of energy-related atmospheric pollution.

---

[1]Reference A attached is a transcript of Mr. Davis's resolution.

A charter for the Working Group was roughed out during the meeting, revised by subsequent correspondence, and officially adopted by the Group at its second meeting in Sofia, Bulgaria in September 1985.

In view of its importance to the remainder of the chapter, this original charter is quoted here in its entirety:

"Energy-related atmospheric pollution is a highly complex technical topic that embraces many scientific and engineering disciplines. Although substantial agreement has been reached on certain fundamental data, much additional research needs to be done. This research will contribute to the development of an acceptable consensus on remedial action.

Many countries have already set targets and embarked on remedial programmes to reduce emissions of energy-related air pollutants. Nevertheless, a worldwide consensus on remedial action would probably be facilitated by better understanding of the physical and biological processes involved in the formation, transportation, conversion, and injurious action of the major pollutants. This goal can best be achieved by unreserved and effectual cooperation among the many nations involved. The World Energy Conference is an appropriate medium for such cooperation.

The Programme Committee of the World Energy Conference has therefore established a Working Group to promote a candid interchange of ideas among representatives of the international community interested in this subject. Activities of the Working Group will consist of the following:

1. Review of the current state of understanding of acid deposition and other types of energy-related atmospheric pollution and their effects
2. Identification of critical issues
3. Evaluation of recommendations received from affected nations
4. Preparation of consensus statements or alternative decision options for submission to the World Energy Conference
5. Participation in appropriate symposia such as the 13th Congress of the World Energy Conference."

The individuals who attended the first meeting of the Working Group are listed in Reference B. They represented twelve countries and one agency of the United Nations: Austria, Canada, Denmark, Federal Republic of Germany, France, Great Britain, Italy, Japan, Mexico, Sweden, United States of America, Yugoslavia and UNIPEDE. During the four months between the first and second meetings of the Group, four additional countries joined the project: Australia, German Democratic Republic, Poland and South Africa. Since that time, six other countries have expressed strong interest in becoming either full-fledged or "corresponding" participants. The WEC Secretary-General is investigating the feasibility of enlarging the Group to this extent.

CRITICAL ISSUES

In the four-month interval between the initial meeting of the Working Group in Washington and the second meeting of the Group in Sofia, the twenty-five individuals then involved in the Group's activities tackled the problem of selection of critical issues to be addressed. After hundreds of communications by telephone and mail, a select list was compiled for consideration in Sofia.

The Sofia deliberations resulted in an accord that near-term effort should be concentrated on:

1. Extraction of important data from national research programs
2. Validation of research results
3. Dissemination of pertinent information and data
4. Development of alternative energy options
5. Identification of mechanisms and processes resulting in adverse environmental effects
6. Characterization of transport phenomena.

It was then decided that the most efficient way to address these issues would be to divide the Working Group into three subcommittees, each having specific and basically independent or non-overlapping assignments. The breakdown that was agreed to and the appointments that were made were:

1. Sources and Physical/Chemical Processes—Mr. Elio Marchesi (Italy), Chairman
   (Because of a transfer to a new position in ENEL, Mr. Marchesi has recently resigned and been replaced by Dr. Edmondo Ioannilli.)
2. Effects—Mr. John Roden (Sweden), Chairman
3. Alternative Decision Analysis—Dr. Peter Chester (United Kingdom), Chairman

Several fundamental guidelines were drafted and approved in principle by the Working Group at this second meeting. Those pertaining to the charge to the three subcommittees were:

1. Energy-related atmospheric pollution includes emissions from stationary combustion systems, nuclear power plants, waste-disposal processes and facilities, the transportation sector, and certain end uses of electricity.
2. A number of aspects of this broad subject are not understood well enough at present to permit the Working Group to draw definitive conclusions.
3. The areas that are well understood include the effects of acidification of surface water, the effects of large doses of radioactivity, the health impacts of a number of heavy metals, and the damage done by excessive dust.

4. The areas that are not well understood include the causes of forest decline, the climatic impacts of the so-called "greenhouse gases", the health impacts of chemicals in coal ash, and the effects of small doses of radioactivity.
5. The initial conclusions of the Working Group should be ready for peer review by the end of 1988 and crystallized sufficiently for presentation to the 14th Congress in Montreal, Canada in the fall of 1989.

SUBCOMMITTEE I ACTIVITIES

An important endeavor of Subcommitee I will be an assessment of the amounts of relevant pollutants presently emitted from the major natural and man-made sources and the development of projections of these emissions for the next 10 to 20 years, taking existing and probable future regulations into account. More important than absolute emission amounts, however, are how, where and in what quantities the emitted and derived pollutants reach the ground and how the deposition patterns match the morphological, geological and biological features of the receptor sites. The key question is how to efficiently decrease deposition in *sensitive* areas, not how to decrease it in general. If global deposition were the main problem, correction would be extremely expensive, because natural emissions of sulphur and nitrogen oxides, for instance, are about 200 percent of man-made emissions.

Several factors greatly influence the patterns of concentration/deposition of man-made atmospheric pollutants. They are:
a) emission parameters (height, temperature, velocity, etc.);
b) meteorological conditions (wind direction and speed, stability, etc.);
c) climate (average temperature, average relative humidity, etc.);
d) presence of natural pollutants (hydrocarbons, photooxidants, ammonia, etc.).

A review of recent findings on the influence of these factors should help to answer some crucial questions about the relationship between emission and deposition of important pollutants in typical situations.

Many attempts have been made to estimate the global emissions of materials likely to have direct or indirect environmental impact—more often than not with less than satisfactory accuracy. Estimates of natural emissions generally have had large margins of error and even measurements of man-made emissions in industrialized regions of the world frequently have been in error by a factor of two or more.

Information available to its members and consultants, together with an extensive literature survey, should permit Subcommittee I to improve on past performance, however. One of the Subcommittee's first tasks will be to prepare a gamut of emissions estimates.

After these estimates have been prepared, the existing and proposed limitations for both stationary and mobile sources of pollution will be examined and the future impacts evaluated in an effort to develop realistic emission trends.

In addition to emissions estimates, spatial and temporal patterns of concentration/deposition together with relevant features of the receiving environment are of utmost importance in predicting ecological consequences. Recognizing this fact, many organizations, both public and private, have carried on major sampling and analysis programs ever since the late fifties. These programs have spread to almost every region of the world.

Subcommittee I will review what is being done in these programs and what the results indicate and include pertinent conclusions in its final report, with emphasis on long-term trends. Possible connections between spatial and temporal patterns of concentration/deposition and man-made emissions will be outlined and discussed.

SUBCOMMITTEE II ACTIVITIES

The main objectives of Subcommittee II are to review the current state of knowledge of the effects of energy-related atmospheric pollution on the environment, humankind, and various materials and to draw meaningful conclusions regarding these effects. Its report will attempt to identify the important effects that have been conclusively attributed to specific pollutants. It will also present what is known about dose-response relationships and discuss five critical concerns. The concluding portion of the report will identify the issues, tabulate data on levels of deposition of different pollutants and summarize the research needs.

The major pollutants being investigated are:

a) sulphur dioxide and sulphates
b) nitrogen oxides and reaction products
c) hydrocarbons
d) oxidants
e) carbon oxides (CO and CO_2)
f) particulate matter
g) metals
h) freons
i) dioxins and furans
j) radionuclides

The critical concerns that will be discussed are:

a) effects on human health
b) effects on aquatic ecosystems
c) damage to forests and crops
d) deterioration of materials
e) effects on climate

SUBCOMMITTEE III ACTIVITIES

A number of decisions have already been made in industrialized countries on emissions policy and others are likely to be made quite soon. Also, there will certainly be improvements in environmental standards in the immediate future running parallel with expectations of improved quality of life and reflecting competitive forces. These together constitute an envelope for the scenarios regarding energy-related atmospheric pollution that can usefully be considered. Examples include new-source performance regulations for SO_2 and NO_x in the United States, the SO_2 and dust air-quality directive recently promulgated by the European Economic Community, German Federal and Lander legislation and practice on SO_2 and NO_x emissions, the undertakings of the members of the 30% Club, the UK policy goals for SO_2 and NO_x by the end of 1999, and current automobile emission controls. Other examples can be found in the rivalry for "a clean image" among competing energy sectors and hardware manufacturers.

Thus we have a realistic and pragmatic starting point for alternative decision analysis. Primary input will come from Subcommittee I in the form of projections of the way in which emissions of SO_2, NO_x, hydrocarbons and ozone will change with time over the next 10 to 20 years as a result of these regulations and influences. Taking this as an "envelope of inevitability", Subcommittee III will attempt to illuminate the major decisions still required within the envelope, bringing into sharp focus the missing scientific data and hypotheses necessary to support sound decision-making. In this it will draw on the findings of Subcommittee II and will provide a framework of decision-making requirements to help guide the work of both Subcommittee I and Subcommittee II.

Because of the implications for the energy market of more stringent environmental requirements and because of the likely operation of the law of diminishing returns, the next set of decisions will be more sophisticated than those already made. It will be necessary to consider the precise environmental benefits to be achieved at each stage and, in the process, to distinguish carefully among:

- the benefit to be derived from an increase in the *rate* of reduction of any pollutant

- the benefit to be derived from an increase in the *extent* of reduction of any pollutant
- the benefit of simultaneously controlling other pollutants
- the possible disbenefits—e.g., a possible increase of ozone as a result of NO_x reduction, a possible increase in plant pathogens as a result of excessive SO_2 reduction.
- the benefit to be expected from locally-applied corrective measures, either alone or in combination with specific emission reductions
- the irreducible contribution of natural emissions
- the actual contribution of factors other than pollution to the problem.

Some of the questions to be answered are already clear—
- Have we given equal weight to all reasonable causes of the damage?
- What is the implication of a non-pollution-related cause?
- In a particular situation, what is the relationship between emissions and deposition and between these and the rate of growth of damage or rate of recovery from damage?
- In a particular situation, what is the relationship between emissions/deposition and the size of the area affected?
- Have we considered all reasonable short-term local ameliorative measures?
- Is full recovery possible without local intervention, and how long would it take without intervention?
- What limits are set by natural levels of SO_2, NO_x, hydrocarbons and ozone?
- What limits are set by other factors—e.g., land use or management?

These World Energy Conference activities have shown, and will continue to show, the author believes, that reduction of atmospheric pollution not only can, but *must*, be based on international cooperation. Within the next decade, mankind will probably either find ways to prevent the use of the earth's gaseous blanket as an energy sewer or begin a prolonged, chaotic demise.

REFERENCE A

Resolution from the Member Committee of the United States
Presented to the Programme Committee, World Energy Conference
Algiers, August 1984

Acid depositon is one of the most prominent and controversial concerns facing electric utilities and other consumers of high-sulfur coal and oil in the world's industrialized nations. Basically an international public-policy issue, it has engendered much public debate and disagreement. Decision on acid deposition should be based on sound scientific and engineering facts and principles.

In an effort to provide a forum for the exchange of information and positions on this complex subject, the Member Committee of the United States has conducted seminars on acid deposition in which leading industry executives, government officials and independent scientists have participated.

Research reports on acid deposition illustrate that suspected causes and recommendations for remedies differ to a substantial degree. Yet these reports tend to agree that coal-fired electric-generating plants are one of the major contributors to the acid-deposition phenomenon. In order to arrive at maximum effectiveness of proposed remedial measures, more data are needed on the formation of acidic atmospheric contaminants, their transport mechanisms, their interaction with the affected environment, and various technical measures for amelioration of the effects.

Because of its political and social ramifications, this issue cannot be resolved by technological means alone. It is, therefore, important that the conflicting issues and proposed solutions be examined closely from all standpoints before restrictions on coal-burning plants are enacted.

RESOLUTION

In view of the factors cited above, the Member Committee of the United States recommends that the Programme Committee of the World Energy Conference establish a Working Group on Acid Deposition whose activities would encompass collection and analysis of data on acid deposition and evaluation and correlation of recommendations received from affected nations. Results of the work of the Working Group would be reported to appropriate bodies of the World Energy Conference on a timely basis.

REFERENCE B

Country or Organization	Name
Austria	Dr. G.E. Obermair
Canada	Dr. Alex Manson
Denmark	Mr. Paul Olsgaard
Federal Republic of Germany	Dr. Jochen Seeliger
France/UNIPEDE	Madame Pierrette Larivaille
Great Britain	Dr. Peter F. Chester
Italy	Mr. Elio Marchesi
Japan	Mr. Toshio Senshu
Mexico	Mr. Humberto Lopez Rubalcava
Sweden	Mr. Inge Pierre
	Mr. Hugo von Sydow
	(Representing Mr. John Roden)
United States of America	Mr. William O. Doub
	Mr. John W. Landis (Chairman)
	Mr. Gorman C. Smith
	Mr. John W. Wilson
World Energy Conference	Mr. Eric Ruttley
Yugoslavia	Dr. Jovan Mandic

Chapter Four
ABANDONED MINE AND COAL REFUSE FIRES: A BURNING PROBLEM

RAJA V. RAMANI

Professor of Mining Engineering and
Chairman, Mineral Engineering Management Section
The Pennsylvania State University
University Park, PA 16802

"... A fire being made by workmen not far from the place where they dug the coal, and left burning when they went away, by the small dust communicated itself to the body of the coals and set it on fire, and has now been burning almost a twelve month entirely underground, for the space of twenty yards or more along the face of the hill or rock, the way the vein of coal extends, the smoke ascending up through the chinks of the rocks. The earth in some places is so warm that we could hardly bear to stand upon it; at one place where the smoke came up we opened a hole in the earth till it was so hot as to burn paper thrown into it; the steam that came out was so strong of sulphur that we could scarce bear it. We found pieces of matter there, some of which appeared to be sulphur, others nitre, and some a mixture of both. If these strata be large in this mountain it may become a volcano. The smoke arising out of this mountain appears to be much greater in rainy weather than at any other times. The fire has already undermined some part of the mountain, so that great fragments of it, and trees with their roots are fallen down its face ..."

Rev. C. Beatty: Journal of a Two Months Tour
in America, 1768 [*The Pittsburgh Coal Bed,
its early history and development*
by H.N. Eavenson, A.I.M.E., New York, NY, 1938]

The combination of fuel, oxygen and heat can start a fire. Once started, as long as fuel and oxygen continue to be available, and cooling is not sufficient to quench it, the fire will persist. In coal mining districts, there is no dearth of fuel. Furthermore, the ambient atmospheric conditions provide sufficient quantities of oxygen. Under favorable conditions, the slow oxidation of coal (called spontaneous combustion) can be the source of heat. In other cases, a combination of natural or human-induced causes can start a fire. Unless detected early and controlled effectively, the threats to life and property from mine fires are immense. Their disastrous consequences on the health, safety, and economic welfare of the communities can be sudden and catastrophic or slow in developing but nevertheless devastating in the long run. There are occasional fire emergencies in active coal mines; the discussion in this chapter, however, is restricted to fires in abandoned mines and in coal refuse piles.

The purpose of this chapter is to outline (1) the conditions under which mine and refuse bank fires can originate, (2) the environmental hazards posed by these burning fuel resources, (3) the methods of fire control for both abandoned mine fires and refuse bank fires, and (4) the present requirements with regard to closure of mines and design and engineering of refuse banks.

The present magnitude of the fire problem in the coal regions of the United States is not easy to assess. Readily available relevant data is summarized in Table 1. In 1977, the U.S. Bureau of Mines recorded 261 known fires burning along virgin coal outcrop and abandoned underground mines in 16 of the coal bearing states [Johnson and Miller, 1979]. Of the 42 reported fires in Pennsylvania, 12 were in the Anthracite region, and the remaining in the bituminous field. Three-fourths of these fires were in the Western United States, including three in Alaska. The states of Montana, Colorado, Wyoming, Utah and North Dakota had 65, 47, 26, 17, and 15 fires, respectively.

A 1964 Bureau of Mines report (Stahl, 1964) estimated that there were 495 coal-mine refuse banks on fire in 15 coal producing states and that 50 percent of these fires were in late stages or almost burned out. By 1969, this number had dropped to 292 in 13 of the 26 coal producing states. Less than ten percent of the banks on fire were in the Western U.S. [McNay, 1971]. In fact, West Virginia and Pennsylvania had 132 and 74 banks on fire, respectively. The number of fires in the Anthracite region was 26. In view of the fact that coal refuse banks are more easily accessed and contain limited fuel, the ability to extinguish a fire or achieve complete burn-out is high. However, if the coal refuse banks are adjacent to an abandoned mine or an exposed coal outcrop, the chances of setting this coal on fire are ever present. Fortunately, considerable progress has been made to prevent fires from starting in abandoned mines or refuse banks. In the same vein, significant developments in fire-fighting technology are being reported to extinguish fires in the coal-producing regions of the United States.

TABLE 1

Fires in Abandoned Mines and Coal Refuse Piles

State	Abandoned Mines and Inactive Coal Deposits*	Coal Refuse Piles**
Alabama		6
Alaska	3	
Arizona	10	
Colorado	47	15
Illinois		4
Kentucky	5	27
Maryland	2	2
Montana	65	3
New Mexico	9	
North Dakota	15	
Ohio	7	6
Oklahoma		1
Pennsylvania		
Anthracite	12	26
Bituminous	30	48
South Dakota	2	
Texas	1	
Utah	17	4
Virginia		17
Washington	2	1
West Virginia	8	132
Wyoming	26	
Total+	261	292

* 1977 U.S. Bureau of Mines data; Johnson and Miller (1979).
** 1968 U.S. Bureau of Mines data; McNay (1971).
+ The absolute numbers will change from year to year due to reclamation activities. Recent data is not readily available. U.S. Office of Surface Mining is currently compiling information under the National Inventory of Abandoned Mine Land Problems.

ORIGIN OF MINE FIRES

The three requirements for the start of a fire are fuel, oxygen and heat. In coal mining districts, the coal outcrops, the exposed coal seams in surface mines, coal stockpiles, the carbonaceous material in the refuse piles and surface mine spoils all provide a ready source of fuel. Even today, the recovery of coal from seams mined by underground mining methods is only slightly more than fifty percent. In the earlier periods, not only was the recovery lower, the seams were also shallower. In seams that were outcropping, many entries were driven from the outcrop into the coal seam for access, haulage and ventilation. When these mines were abandoned, both abundant fuel and a ventilation system which could provide adequate oxygen supply were left behind. With a source of ignition, a fire may be started.

Coal refuse piles, also referred to as culm banks, refuse banks, gob piles and waste piles, are deposits of coal refuse resulting from the mining and processing of coal which are left accumulating on the surface near a coal mine or a preparation plant. Coal refuse piles contain coal, rock, carbonaceous shale, and pyrites. They also contain such materials as grease, oil containers, oil soaked rags, brattice cloth, timber used in mine support and discarded mine supplies and machine parts. The size consist of the refuse pile material may vary from large coarse particles to extremely fine particles (< 1mm). While the carbon content of some of the coal refuse piles may be as high as 45 percent, most average less than 10 percent. Over time, the coal and carbonaceous material may release trapped methane, if any. The combustible material in the refuse pile and the availability of oxygen from the atmosphere may start and sustain a fire when ignited.

Complete recovery of the coal from a seam and the elimination of refuse from a mined product are desirable but difficult objectives to achieve. The ability to prevent the access of oxygen to abandoned workings or to coal refuse piles is also not easily achieved or completely assured. Therefore, removing the ignition sources presents the most viable approach to preventing fires from starting.

Ignition Sources

From a study of several fires in coal refuse piles and abandoned mines, a number of sources which can start a fire has been identified (Magnuson, 1974). Some of the common ignition sources are human-created. When not properly guarded against, the abandoned mines, the depressions found in mining regions, the strip mine pits, and mine waste areas have been used for dumping of industrial waste and domestic refuse from neighboring populated areas. Burning of this trash has ignited the coal bed or the coal matter in the refuse pile. The warming fires, started by such diverse groups as campers, hunters, hikers, and people mining house coal on exposed outcrops or refuse piles, have led to mine and refuse bank fires.

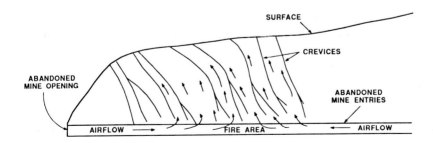

FIGURE 1. Section of a Mine Entry in a Shallow Mine Showing the Ability of a Fire to Establish its Own Ventilation Through the Cracks and Crevices to Surface.

Brush and forest fires, started by either lightning or other causes, can ignite an exposed outcrop or coal refuse piles. There is reason to believe that many fires in remote areas may have started in this manner. In some cases, coal refuse piles have been set on fire intentionally to manufacture "Red Dog" which is the residue of the burnt refuse, for use as construction material. Without proper control, these intentional ignitions have spread to other parts of the refuse piles. A burning refuse pile may itself be the source of ignition to exposed coal outcrops in abandoned coal mines and strip mines.

A mine, abandoned due to a fire during its active life or a portion of an operating mine sealed due to a fire, can be the source for a fire. If the abandonment or the sealing has not been done properly, the fire will gradually spread to other portions of the mine.

In many cases, a large portion of the heat generated from the fire dissipates into the surrounding materials which in an underground mine are the rock strata and in a refuse pile or surface mine can be refuse, or rock strata. In most cases, these materials store the heat increasing their temperature. Ignition of the coal or coaly material from the heat stored in the adjoining strata has been evidenced at a later date, i.e., the fire may reappear in an adjoining area. In a shallow mine, the fire may even break out to the surface through subsidence cracks and crevices (Figure 1).

Spontaneous Combustion

Spontaneous combustion of coal or coaly matter is a source of ignition which does not involve a flame or spark. This ignition source is a result of the heating and slow combustion of coal and coaly matter initiated by the oxidation of coal. The heat from the oxidation, when not dissipated to the surroundings, increases the temperature. This, in turn, increases (1) the rate of oxidation, (2) the amount of heat and (3) the rise in temperature, eventually igniting the coal or the coaly matter. The important factors are (1) the chemical and physical properties of the coal, and (2) circulation of sufficient air to keep the oxidation process going but not enough to remove the heat of oxidation. Poor quality coals, loose packing of coarse coals, oily rags, damp hay, etc., have been known to be liable to spontaneous combustion. Spontaneous combustion susceptibility is high for low rank coals and vice-versa. Absorption of water by coal also generates heat, called heat of wetting, which is high for low rank coals. However, a high inherent moisture content in the coal will ensure little or no absorption of moisture and will require much heat to raise its temperature. However, the situations are different for coal in storage, refuse piles, and transportation. Here, moisture absorption by dry coal may release enough heat to accelerate the oxidation rate. While the ignition temperature of coal is between 600° and 900°F, the oxidation of coal may advance to spontaneous combustion around 200°F.

Spontaneous combustion is not recognized as a likely source of ignition in abandoned bituminous mines [Magnuson, 1974]. However, nearly two-thirds of the 292 refuse pile fires, reported earlier, are believed to be due to spontaneous combustion (McNay, 1971). In any case, spontaneous combustion of coal and coaly matter in storage, coal refuse piles, strip mine spoils can lead to the ignition of coal outcrops and exposed coal seams with which they may be in contact.

The most effective defense against these fires is not creating conditions where oxygen and ignition conditions can occur at the same time, since fuel is always present. Eliminating the careless sources of ignition and precautions against spontaneous combustion will ensure that accidental ignitions will not occur. Proper abandonment procedures for mines and good engineering designs for coal refuse piles will prevent a fire from starting or spreading.

ENVIRONMENTAL HAZARDS

There are several environmental hazards which are common to fires in abandoned coal mines and coal refuse piles. There are also several unique features to each of the situations.

The most easily recognizable hazard is the damage to the atmospheric environment. Not only is the area a visual blight but the air is foul. Smoke, fine particulates, and high levels of carbon dioxide are ever present, impairing visibility and generally creating unsafe psychological and physiological conditions. Toxic gases such as carbon monoxide, sulfur dioxide, hydrogen sulfide, and oxides of nitrogen, may be present. There is also release of several hydrocarbons. All these gases have serious and potentially fatal effects on plant, animal and human lives. They corrode built up structures as well. An added danger from underground fires is from subsidence in burnt areas. The damage from subsidence may be local or extensive, affecting buildings and highways. The sinkholes pose an additional safety hazard as their existence is not readily apparent.

The loss of the coal resource, consumed by the fire, is obvious. However, greater amounts of coal may be lost due to the added difficulties and dangers created for their extraction. The magnitude of the damage caused to by fires to the land values and economic welfare of communities is apparent from the effects of the Centralia Mine Fire in the Anthracite region of Pennsylvania. The fire was discovered in 1962 and has continued to spread for the last 25 years. Coal was first mined in the Centralia area in 1842. Mining had continued under different owners from this early beginning. In may 1962, a fire was discovered burning in refuse material in a waste disposal area operated by the Borough of Centralia. The disposal area was an abandoned strip pit, originally mined in 1935.

Between 1962 and 1980, numerous efforts were made to control the Centralia

TABLE 2

Centralia Mine Fire Control Projects

Completion Year	Brief Project Description
1962	Excavation of burning material near IOOF Cemetery.
1963	Hydraulic flushing of mine voids near IOOF Cemetery.
1963	Partial trenching in eastern area.
1967	Exploratory boreholes, hydraulic flushing and backfilling near Borough.
1968	Reopen some flush boreholes.
1970	Hydraulic and pneumatic flushed underground barrier in western area.
1973	Hydraulic flushed underground barrier in eastern area, shallow excavation.
1978	Partial excavation and flushed barrier reinforcement in western area.
1980	Monitoring and assessment of fire control.

Source: Chaiken et al. (1983)

mine fire (Table 2). A Bureau of Mines report (Chaiken et al., 1983), summarizes these efforts and presented the status as of 1980:

"... *Today, Centralia is comprised of about 500 homes, four churches, public buildings, and businesses. The population is approximately 1,200. Of these, 310 structures and approximately 500 people are in the area currently or potentially affected by the Centralia mine fire.*

For the town's residents, the most immediate effect of the underground fire has been the combustion gases that have penetrated area homes. Gases were first discovered in 1969 when CO, excess CO_2, and O_2 deficiency were detected in several homes in the borough, and residents had to be temporarily evacuated. When gases again became a problem in 1979, monitors provided by the U.S. Department of the Interior were placed in 15 area homes. Analysis of bottle samples of air taken in the living area of homes has revealed CO_2 levels of less than 0.5 pct, while some basement floor areas have approached 1 pct CO_2. An exception is the basement floor of residence C, where CO_2 levels as high as 10 pct have been recorded; reading up to 1 to 2 ppm CO have been recorded.

Fumes and a lack of oxygen in areas homes have been an especially serious problem for area residents with chronic respiratory ailments. In an effort to carry sulfurous fumes above breathing space, pipes have been installed extending upward over the openings in the ground through which gases are escaping.

Another hazard presented by the fire is the danger of surface subsidence. As the fire consumes the coal below, surface subsidence can result, particularly where underground coal pillars supporting the roof are destroyed. Several incidences of subsidence have occurred in the fire area. In some cases, however, subsidence may simply be a result of the previous underground mining, and not the fire.

A third problem has been the concern about elevated ground temperatures as a results of the fire. A local gasoline station was closed in December 1979 due to fumes and concern that the high ground temperature made it unsafe to store gasoline in the underground tanks. . . ."

According to published reports, by 1986, all but about 40 of Centralia's 500-odd houses have been razed and more than 900 people have been relocated at government expense. The Centralia situation is yet another reminder of the traumatic effect of the adverse impacts of mine fires on a local economy and environment.

CONTROL OF FIRES

Since fuel, oxygen and ignition sources are necessary to start and sustain a fire, the measures to control fires depend on eliminating one or more of the three constituents. The most common method of controlling fires —- quenching it with water —- is not easily applied or universally successful in controlling abandoned mine fires due to several reasons. The two most important ones are (1) the inability to completely inundate a burning coal seam, and (2) the spread of fire above the general water table. In fact, no single method can be used to control or extinguish all underground fires. Each fire presents unique

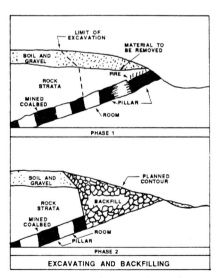

FIGURE 2. Mine Fire Control Techniques—Excavating and Backfilling.

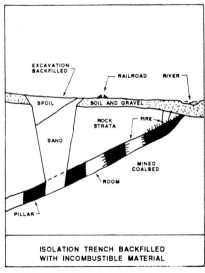

FIGURE 3. Mine Fire Control Techniques—Isolation.

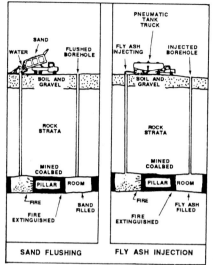

FIGURE 4. Mine Fire Control Techniques—Flushing.

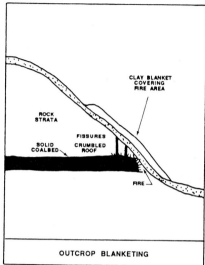

FIGURE 5. Mine Fire Control Techniques—Outcrop blanketing.

conditions of location, extent, and intensity [Magnuson, 1974]. Johnson and Miller (1979) report that four control methods have been used successfully either singly or in combination to fight abandoned coal mine and outcrop fires: (1) excavation and backfilling, (2) isolation barriers, (3) flushing and (4) surface sealing. Extended discussions of these methods and case histories can be found in Magnuson (1974) and Chaiken et al. (1983).

Excavation and backfilling (Figure 2) involves digging out the burning and heated material and cooling it with water or by spreading it on the ground. Once cooled, the material is disposed of in a safe manner either on site or in a disposal area.

The isolation barrier (Figure 3) is used to confine and isolate the fire from the main body of coal. A trench is excavated around the fire down to the bottom of the coalbed and backfilled with incombustible material to form a barrier. If properly isolated, the fire dies from lack of fuel when the combustible material in the confined area has been consumed.

Flushing (Figure 4) involves introducing pneumatically injected fly ash or incombustible solids mixed with water through boreholes to form a fire-containing barrier. Flushing is used where more direct methods cannot be employed because of excessive overburden depth or the presence of homes and other surface improvements. Flushing materials include sand, water-cooled slag, crushed slag, and crushed earth material consisting mainly of sand.

Surface sealing (Figure 5) is used to control underground fires by closing surface openings that permit ventilation of the fire. The seal is established by plowing the surface to a depth of several feet and blanketing with incombustible material to cut off airflow through fissures.

The cost of controlling or extinguishing underground and coal outcrop fires depends on geologic conditions, the depth of the fire, the extent of its progress, and the availability of fire control material. Chaiken et al. (1983) present an assessment of the relative effectiveness of the various control methods on the basis of a review of the 26 mine fire control projects at 17 abandoned mine sites. In five of these sites, 14 separate fire control projects were necessary before the fire was successfully controlled. At four sites the fires were extinguished by complete excavation. In the remaining 12 sites, only one attempt was made at each site to control the fire. The fires were successfully controlled at eleven of these sites. The following are some significant findings of the assessment:

(1) Only slightly more than 50 percent of the fire control projects were successful in controlling or extinguishing the fire.
(2) Excavation was the most successful control method (8 out of 10 trials).
(3) Flushing (with or without partial excavation) was successful in only about half of the attempts.
(4) Approximately $37 million was spent on successful projects, and approximately $6 million was spent on unsuccessful projects. On a per fire basis, success is five times more costly than failure.
(5) Fire projects started with insufficient funds required repeated efforts. The fire often spread and became more difficult and more costly to control. The history of the Centralia mine fire is typical (Table 2).

On the basis of this assessment, Chaiken et al. (1983) conclude that,

"... complete excavation ... has a high probability of success. Although flushing has occasionally been successful, ..., it requires extreme care in injection of slurry, and results must be monitored to insure the adequacy of the flushing operation. Surface sealing is considered only when the area to be sealed and the funds available are very limited. There is very little indication that underground tunnel barriers are successful in containing a fire, and may only serve to retard the fire and change the direction of propagation ..."

There are several methods to control coal refuse fires. These include (1) quenching the fire with water, (2) water-quenching the surface material and removing the quenched material, (3) smothering the fire by covering the refuse pile with clay or other inert material to keep the air out and (4) injecting fine noncombustible material into the refuse pile to decrease its porosity. Control of coal refuse fires presents difficulties of a far lesser magnitude than abandoned coal mine fires.

While most mine fire fighting methods are based on controlling oxygen or cooling the ignition source, a novel method of fire control under study is to accelerate the burning of the fuel under controlled conditions. This method called "Burnout Control," is being tested in field trails by the U.S. Bureau of Mines. If successful and economical, the method has the potential for safely harnessing the energy from the fire by accelerating the burnout of fuel isolated for consumption and exhaustion by the fire [Chaiken, 1983].

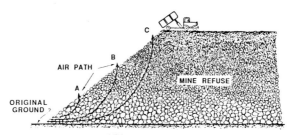

FIGURE 6. The past practice of end pumping

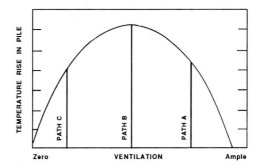

FIGURE 7. The permeability for ventilation and consequently, chances of temperature rise.

PRESENT STATUS

At the present time, the beneficial impacts of the increasingly stringent environmental legislation of the past three decades are beginning to be realized. Currently, there are Federal and State regulations concerning (1) abandonment of mines, (2) methods of coal refuse pile and other waste embankment construction and maintenance, and (3) reclamation of the affected areas.

Code of Federal Regulation, Title 30 (Mineral Resources) has extensive provisions for such things as location of refuse piles with respect to distances from coal seams, entrances to mine facilities, isolation from old refuse piles, etc. Minimum performance requirements are required with respect to compaction, sealing, drainage and stability during construction. Provisions to prevent burning, future impoundments of water, and major slope failures are to be identified before abandonment. The authority for these regulations is derived from the Coal Mine Safety and Health Act of 1969 and The Surface Mining Control and Reclamation Act of 1977. A common practice of the past to build a coal

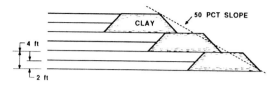

FIGURE 8. Compacted method of Mine Refuse Pile Construction.

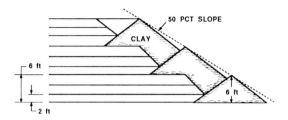

FIGURE 9. Uncompacted method of Mine Refuse Pile Construction.

mine refuse pile is shown in Figure 6. In this method of end dumping refuse, in addition to stability problems, the potential for heat build-up in Zone B is high (Figure 7) (Magnuson and Baker, 1974). The present requirements call for layering, compaction and facing with clay or other inert material (Figures 8 and 9).

With respect to abandonment of mines, surface mine reclamation plans call for returning the land to a condition desirable for future uses. This will involve sealing of coal outcrops, establishing vegetation on graded lands, and maintaining and monitoring water quality. Burial of toxic and coaly materials in the spoils, in a safe manner, is required. All these requirements will ensure that there is no fuel readily exposed or available for combustion. Abandonment of underground coal mines also requires sealing of unwanted openings to the coal seams and reclaiming and revegetating the lands affected on the surface. Examples of sealing the shafts, slopes and boreholes connecting a deep coal seam with the surface are shown in Figures 10, 11, 12 and 13 (Brezovec and Hedges, 1986). Under these conditions, access of oxygen to the abandoned workings is greatly reduced, if not eliminated. As the understanding of the causes of fire problems increased, the technology for their abatement also developed. Mine operators, working with Federal and State agencies, have responded with implementational schemes to ensure that current operations, when abandoned, do not create fire hazards.

With respect to fires in abandoned mines and coal refuse piles, Pennsylvania showed the way for the nation by accepting the Commonwealth's responsibility to clean up the environment in 1963. The monumental effort to eliminate the pollution to air, water and land resources from past mining practices was called "Operation Scarlift." The Scarlift program was directed at reclamation

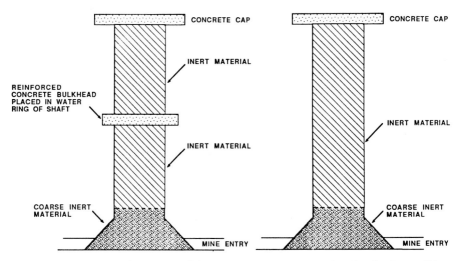

FIGURE 10. Shaft seal under watery conditions.

FIGURE 11. Shaft seal under dry conditions.

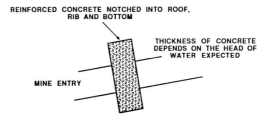

FIGURE 12. Hydraulic seal for mine entry or slope.

and restoration work to remove the scars of acid mine drainage, burning refuse piles, underground mine fires, surface subsidence and abandoned strip mines in Pennsylvania's coal mining regions. At the Federal level, with the enactment of the Surface Mining Control and Reclamation Act of 1977, an Abandoned Mine Reclamation Fund has been established for accumulating revenues designated for the reclamation of abandoned mined lands (30 CFR, Part 870). Under this regulation, the operators of surface and underground coal mines pay a reclamation fee on each ton of coal produced for sale, transfer or use. The monies from the fund will be available to the States for coal as well as non-coal reclamation activities for lands which were mined and left abandoned or inadequately reclaimed prior to 1977. A National Inventory of Abandoned Mine Land problems was created to help the Office of Surface Mining (OSM), participating states, and Indian tribes to locate and identify abandoned mine land problems and estimate the costs of reclamation. As of 1983, the States have

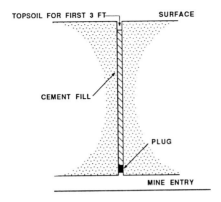

FIGURE 13. Typical seal for a borehole.

identified 3,960 problems, whereas OSM has determined that 1,373 of these have significant health, safety and general welfare problems. In the areas of surface and underground fires, the States have reported 149 and 104 problem areas whereas OSM has approved 37 and 11, respectively. Certainly, accelerated progress on controlling abandoned mine fires can be expected.

SUMMARY

The history of coal mining around the world is replete with stories of health and safety hazards from fires in abandoned mines and refuse piles. Equally well documented are the disastrous impacts on the social and economic welfare of the communities in the mining areas. The past practices for extraction, processing, transportation and waste disposal activities in the coal mining industry in the U.S. have all taken a tremendous toll on the environment. By the fifties, the increasing knowledge on the environmental problems and continually growing social responsibilities resulted in greater awareness to protect the environment. By the sixties, laws and regulations were being enacted by the states to control unregulated disposal of waste and abandonment of mines. By the seventies, these laws were being consolidated into national legislation for the protection of air, land and water resources. As a result of these actions, the chances of fire hazard in an abandoned mine or coal refuse pile are greatly reduced. Furthermore, the creation of the Abandoned Mine Reclamation Fund represents another positive step towards cleaning up the scars of the past. Vigilance, however, cannot be relaxed as the current efforts to fight the fires of the past show that they are, in many cases, accidentally started, and are tenacious. In addition to the technological advances in building mines and structures to prevent fires from initiating, the miners and general public must be educated in the safe role they must play to ensure that accidents don't happen.

REFERENCES

Brezovec, D. and Hedges, D.N., 1986, "With Careful Planning, a Mine Closes Economically and Efficiently," *Coal Age*, V. 91, n. 11, pp. 42-45.

Chaiken, R.F., 1983, Method for Controlled Burnout of Abandoned Coal Mines and Waste Banks, U.S. Patent N. 4,387,655.

Chaiken, R.F., Brennan, R.J., Heisey, B.S., Kim, A.G., Malenka, W.T. and Schimment, J.T., 1983, Problems in the Control of Anthracite Mine Fires: A Case Study of the Centralia Mine Fire (August 1980), U.S. Bureau of Mines Report of Investigations 8799, Washington, DC, 93 pp.

Code of Federal Regulations, 1986, Title 30, *Mineral Resources*, U.S. Government Printing Office, Washington, DC.

Johnson, W. and Miller, G.C., 1979, Abandoned Coal-Mined Lands: Nature, Extent and Cost of Reclamation, U.S. Bureau of Mines, Washington, DC, 29 pp.

Magnuson, M.O., 1974, Control of Fires in Abandoned Mines in the Eastern Bituminous Region of the United States, A supplement to Bulletin 590, U.S. Bureau of Mines Information Circular 8620, Washington, DC, 53 pp.

Magnuson, M.O. and Baker, E.C., 1974, State-of-the-Art in Extinguishing Refuse Pile Fires, *Proceedings First Symposium on Mine and Preparation Plant Refuse Disposal*, National Coal Association, Washington, DC. pp. 165-182.

McNay, L.M., 1971. Coal Refuse Fires, An Environmental Hazard, U.S. Bureau of Mines Information Circular 8515, Washington, DC, 50 pp.

National Academy of Sciences, 1986, *Abandoned Mined Lands: A Mid-Course Review of the National Reclamation Program for Coal*, National Academy Press, Washington, DC, pp. 47-58.

Stahl, R.W., 1964, Survey of Burning Coal-Mine Refuse Banks, U.S. Bureau of Mines Information Circular 8209, Washington, DC, 39 pp.

Environmental Consequences of Energy Production: Problems and Prospects. Edited by
S. K. Majumdar, F. J. Brenner and E. W. Miller. © 1987, The Pennsylvania Academy of Science.

Chapter Five
SURFACE SUBSIDENCE
LEE W. SAPERSTEIN
Professor of Mining Engineering
College of Earth and Mineral Sciences
The Pennsylvania State University
University Park, PA 16802

The action of removing solid substances from the earth's crust by processes such as mining, quarrying, tunneling, or other excavation causes a redistribution of stresses within that part of the crust surrounding the cavity. The resulting cavity is stable. Within appropriate limits of geometry and added artificial ground support, the additional stress redistribution causes only minimal and tolerable deflections. On the other hand, geometries leading to stresses that are greater than rock strength lead to cavity failures. In surface excavations, these are called slope failures or, colloquially, land slides, and underground they are called roof falls, roof failures, or roof caving. When the failure zone above an underground cavity is high enough to intersect the surface, the deflection of the surface, which is a manifestation of the subjacent failure, is called subsidence. Subsidence is surface movement; the question of whether or not this movement is damaging is an issue that is taken up separately.

Subsidence can occur above any cavity, including those formed remotely such as by pumping oil or groundwater from subterranean reservoirs. However, the broad, tabular nature of the U.S. coal deposits, lying primarily in relatively thin (three to ten feet thick), horizontal, and often wide-spread beds, means that subsidence is an environmental effect often associated with coal mining. However, the remarks that follow can be extended to the sub-surface extraction of any flat-lying bed.

These remarks are divided into general principles, problems of past mining, subsidence-associated damage, planned subsidence and modern mining, benefits of planned subsidence, and an introduction to subsidence control.

SUBSIDENCE IN GENERAL

Subsidence has been studied extensively and a large body of research literature has been published on it. Noteworthy are the several major reference works that have been published on the subject (National Coal Board, 1975, Kratzsch, 1983,

Bruhn and Turka, 1986). Although there are differing methods to predict subsidence and, then, to correlate structural damage with subsidence, there is general agreement on many of the general subsidence principles.

Types of Subsidence. Subsidence related to the unplanned failure of shallow, underground workings is called synonymously *pit, sinkhole,* or *chimney.* The name is derived from the shape of the failure zone, which is often temporarily stable and very steep sided. Pit-type subsidence events, caused by roof or pillar failures in old, abandoned, and very shallow (under 50 feet deep) room-and-pillar (partial extraction) coal mines, have been known to expose the old cavity to view from the surface. These unplanned subsidence events have caused much concern and, indeed, loss when they occur beneath developed regions.

Subsidence caused by planned longwall and full-extraction (pillar retreat) room-and-pillar mining is typified by the formation of a trough whose dimensions are linked to those of the mining zone. *Trough* subsidence is both planned and, in many cases, predictable. Many of the terms devised to describe subsidence are associated with trough-type subsidence. In an overall sense, a subsidence trough can be described by a mathematical model of the resulting surface and this facilitates prediction work.

Subsidence Geometry. The size of the mined zone, often called panel by miners, the depth to that zone, and the actual height of extraction all affect the development of surface subsidence. The width, W, of the mining zone may be but a few feet for an entry or tunnel and up to one thousand feet or more for a longwall panel. W is considered as a transverse measurement and is particularly significant for two-dimensional profiles. Most mine panels are rectangles, or aggregates of rectangles, so that the direction normal to W is the panel length, L, which is the longitudinal direction. L may be in the order of thousands of feet. The geological thickness of the seam controls, but is not synonymous with mining height, T. This is because the vicissitudes of mining may lead to extraction of more or less of the seam. T, in the United States, ranges from two to ten feet, with the occasional mine panel even thicker. The depth, D, to the seam of interest is another major parameter. U.S. coal seams that are underground mined are commonly between two hundred and one thousand feet deep, although depths to two thousand feet are not unknown. European producers have experience with coal mining depths in excess of three thousand feet.

Subsidence Site Parameters. The above geometric factors plus a number of derived factors constitute the subsidence site parameters. For example, the width of extraction, W, normalized by the depth, D, (W/D) gives a better understanding of the impact of mining on the surface than does either parameter by itself. The *subsidence factor* (S/T) is a normalization of the vertical subsidence, S, measured at the point of maximum subsidence (normally the panel center), by the extraction thickness, T. Subsidence factors can range from almost one (subsidence is practically equal to the extraction thickness), where the overburden is weak and the panel is supercritical in width (defined below), to an imperceptible number close to zero, where the overburden is strong and the W/D ratio is small. The *critical* width is that width of mine workings that will result in maximum possible subsidence, S_{max}, for that site. When critical width is achieved, the cantilevered support from strata at the panel's edge no longer affect the center of the opening. A sub-critical panel has S less than S_{max}, whereas a supercritical panel has a continued widening of the zone that is S_{max} in depth. In theory, a super-critical panel is a flat-bottomed trough. The *angle-of-draw* is a qualitative and descriptive term used to define the angle between the vertical and a line drawn from the edge of the underground support pillar to a point on the surface where subsidence is first measured. Positive is assigned to angles over the solid pillar and negative to those over the cavity. Inasmuch as the selection of precision for measuring subsidence can affect this line's location, as can other factors such as subsidence-related heave, it is best to think of angle-of-draw as a qualitative number. Angles of draw have been observed commonly in the range of 15 to 30 degrees, but numbers less than zero and greater than 30 have also been seen. The *time-lag* effect describes how long it takes for subsidence to develop fully. This may be almost instantaneous or it may be up to six months, depending on the consolidation nature of the rock.

Mechanical Parameters. Subsidence is a deformation of a solid mass due to the imposition of stresses on that mass; therefore parameters associated with solid-body mechanics are also used for subsidence. The *displacement* of a surface point is a measurement of the total movement that point undergoes. Displacement is frequently resolved into its horizontal and vertical components; strictly speaking, subsidence is only the vertical component of displacement. *Strain* is a relative term that represents the differential motion of one point with respect to another and divided by the length between the two points of interest. When points move toward each other, then the zone is said to be in *compression*; outward motions represent *tension*. The compression zone is often found toward the center of a panel and the tension zone at the edge, often just above the edge of the solid pillar.

SUBSIDENCE PROBLEMS OF PAST MINING

Wherever shallow underground coal mine workings exist, there is a potential for the sudden development of pit-type subsidence. One widely cited study (Miller and Johnson, 1979) suggests that all underground mining up to the time of the study resulted in more than seven million acres of undermined land, of which two million have already been affected by subsidence. Approximately 500,000 acres of the larger total are under urban areas and 80 percent of these urban lands are made up of population centers that grew in the later years of the 19th century.

Pennsylvania, of all the states, has, by far and large, the greatest amount of urban acreage (Zwartendyk, 1971), with the cities of the Wyoming Valley: Scranton, Wilkes-Barre, etc., affected by anthracite mining and the communities of the southwest, Pittsburgh as well as the myriad of coal-mining communities, affected by bituminous coal mining. However, substantial problems with abandoned coal-mine subsidence also exist in Illinois, Kentucky, Colorado, Wyoming, and in the many other States where coal was mined.

Plotting unplanned subsidence events reveals that it is difficult to predict the stability (or instability) of any particular site. However, it is safe to say that unplanned subsidence begins to occur about twenty years after mining ceases and is inevitable by the time fifty years have passed. To be explicit, not every abandoned room-and-pillar mine will cause pit-type subsidence; however, those shallower than some threshold depth will do so.

Inasmuch as the mines are abandoned and not readily accessible to observation, the mechanisms of failure can be conjectured only. Since pit subsidence occurs as much as several decades after mining, it is presumed that a gradual weakening of the pillars due to oxidation of the coal takes place until the pillar is no longer able to hold the overburden and a sudden failure occurs. Other, not necessarily exclusive, possibilities include the loss of foundation support to the pillars due to wetting of the underclays and weathering of shaly rock strata in the immediate roof. Sudden failures have been noted when mine fires consume the pillars or when large volumes of water suddenly enter the mine void, as happened in the floods of Hurricane Agnes, 1972.

When sink-hole or pit subsidence occurs, the manifestation on the surface is eminent and often damaging to any structure that is above. Unequal loss of foundation can cause structures to tip, damage to roads can make them impassable, and even failures that do not affect a structure can leave hazardous holes in the landscape. Even though subsidence does not affect a structure directly, the presence of nearby subsidence events can affect a structure indirectly by causing a general depression of property values.

Remediation of these circumstances is problematic and expensive. Often, in undeveloped land, the best approach is to remine the remaining resource by present-day surface mining methods, which include effective reclamation techniques. For developed, urban land, a number of backfilling techniques have been devised (Whaite and Allen, 1975) that are intended to stabilize the pillars and fill the void so as to prevent subsidence from occurring or, if it occurs, to lessen the subsidence factor into a tolerable range.

The National Research Council (National Research Council, 1986) concluded that remediation in urban areas was fraught with problems. The expense often exceeded the value of the property protected and the assurance of no further subsidence was incomplete. Instead, it was recommended that unplanned subsidence events be treated on a case-by-case basis and that subsidence insurance be considered as a way of spreading the cost of remediation over the affected parties. Pennsylvania, Illinois, and Colorado have, or contemplate having, such insurance programs. By managing the program, the State keeps the costs in an acceptable range for any individual home-owner.

Subsidence-Associated Damage

Subsidence-caused damage is of two broad types: land and structures. Inherently, the surface of the land is not damaged by lowering or tipping it a few feet. However, if the rearrangement of contours should cause drainage reversals or impoundments of shallow waters, then substantial loss of productivity could result and the land could be said to be damaged. Agricultural field drains can also be affected, either by changing their grade or, in the case of ceramic tile pipe, by cracking them. Typically, drainage problems are greater in terrain of limited relief such as in southern Illinois than in hilly country-side. Even in the latter, though, valley bottoms and flood plains can be affected by subsidence.

Soils, particularly those low in clay, have little or no tensile strength, so that they will contain large tensile cracks if the zone of tension at the edge of a subsidence trough is at all pronounced. This is the case where the angle of draw is small, implying a steep transition zone, or the amount of subsidence is large, or both. Fortunately, these cracks are easily repaired; indeed, weather-related processes will cause them to coalesce quickly.

A more substantial problem is that of damage to geologic strata that are water-bearing. These strata, called aquifers, are of economic and social value if the water is potable and consumed by humans. Although research on the effect of subsidence on groundwater flow is more limited than that on surface motions, there is some evidence to indicate that subsidence can dewater aquifers close to the mining zone in depth, which are the deeper strata relative to the surface, and actually enhance water flows in the shallower aquifers. To understand this seeming dichotomy, it is necessary to visualize what is happening to the rock in the strata above the mine panel. The immediate roof rock is falling into the mined-out zone and undergoes physical dislocation of both rotation and translation. The resulting rubble pile has an hydraulic conductivity almost as high as open channel flow and this conductivity extends vertically across former bedding planes. However, higher in the column of subsiding rock, the beds undergo only translation associated with bending and sagging. Here, former bedding planes remain relatively intact. In this upper zone, the aquifer undergoes some bending, which may increase its horizontal hydraulic conductivity, but the beds above and below the aquifer (aquicludes) continue to restrict vertical flow of groundwater. As a consequence, the aquifer may produce more water to a well, whereas an aquifer in the lower, rubble zone would drain its water into the mine.

In discussing potential water loss, then, the depth of the mine, including its position with respect to drainage, as well as the separation of the mine horizon from that of the aquifer need to be considered. The height of the rubble zone, and thus the potential for loss of water from intersected superjacent aquifers, can be estimated based on past experience with mining in a region. Upon abandonment, mines that are below drainage and, consequently, need to be pumped to remain dry will fill with water and may stop draining an aquifer (discharge) and reverse the action to that of refilling the aquifer (recharge). Shallow underground mines that drain naturally without pumping are difficult to seal and potentially will always be a discharge. Those aquifers that do not lose their water may or may not have the water quality affected. Quality changes depend upon the presence of soluble chemical species in the rock and whether or not they become exposed to the flowing groundwater.

Damage to a structure resting on the surface of the land relates, naturally enough, to the subsidence-induced strains that the structure undergoes. However, the nature of the structure, its construction method, its foundation, and the soil-structure interface will all modify the extent of damage caused by a given strain. The kind of physical damage that occurs depends upon the strains. Buckling is seen in the compression zone and separation in the tension zone. Where there are vertical strains (one point subsiding lower than its neighbor so that there is a differential motion), then tipping and cracking of a structure can occur. Vertical strains are expected in those parts of the subsidence profile that are on the edges of the panel. Because pit subsidence has such high strains, the damages observed in houses affected by subsidence from abandoned mines are often instantly apparent and severe.

The ideal location for a structure, with respect to a subsidence trough, is in the central part of a super-critical zone, where all subsidence is vertical and there are no differential motions (vertical strain). However, it should be remembered that a subsidence zone expands in step with the advance of the underground mining face. Consequently, a wave of differential motion will cross the land. This advancing wave may disrupt a structure, even though it is the center of a trough. Advance planning can do much to minimize the long-term damage caused by this wave. Utilities can be disconnected so that any cracking of water, gas, and sewer lines can be dealt with safely; masonry can be reinforced; and at the extreme, a house can be lifted off its foundation until such time as the land restabilizes and the foundation is repaired.

New construction that anticipates subsidence uses modular techniques, with each module being linked to others through expansion joints. Because of an ability to withstand tension, aluminum or wood siding is preferable to brick or stucco. Foundations should be placed on uniform material rather than having them partially on bedrock and partially on soil. Cross-bracing and tie downs help a structure to retain its integrity while being releveled. Repairs to a structure may be undertaken once the time-lag effects have finished, which is usually no more than six months after mining. However, adjacent panel mining can cause subsequent adjustment (Heasley and Saperstein, 1986c) so that final repairs should be done only after nearby mining is also completed.

PLANNED SUBSIDENCE AND MODERN MINING

The improving ability to predict subsidence movements in advance of mining means that the consequences of mining can be made internal to the overall mining scheme and not be left to chance or to future generations. Mines can be designed so that the physical lay-out of panels impinges on the fewest possible structures; those structures that remain at risk can be strengthened, moved, or otherwise arranged for compensation. The key, then, is an accurate and moderately precise method for predicting subsidence; fortuitously, many such models exist.

There are four broad classes of subsidence models used to predict subsidence.

a. Physical Models

Although seldom used for routine predictive work, a physical model is useful for studying the three-dimensional mass as it fails. Sand has been used to model discontinuous rock materials and, in this case, gravity caused failure when small foundational blocks were removed. Modeling clay and other cohesive substances have also been used to model the earth, in which case gravity has been mimicked by a sliding frictive base, a centrifuge, or externally applied loads.

b. Empirical Models

Data from existing subsided zones and obtained by standard surveying techniques have been fit to hyperbolic curves in order that an equation can be generated that describes best the overall population of data. This equation, sometimes called a *profile function*, is a two-dimensional idealized representation of the cross-section of a subsidence trough. Because of its ease of use, there are several different solutions programmed for use on a personal computer. The profile function is relatively accurate for slices taken through the panel at some distance from the ends and when these panels are in similar geological settings to those that provided the base-line data. In new sites, where no subsidence data exist, there are broad rules for creating profiles that take into account some geological differences. For instance, the slope of the curve may be changed to account for variations in percentage of hard rock in the overburden.

c. Influence Function

A somewhat more sophisticated technique for predicting subsidence than the profile function is the *influence function*. This is a computer-based technique that aggregates the subsidence influence (hence the name) of small, discrete extraction zones. Because of this approach, the influence function method can predict subsidence behavior at panel ends and in corners. There is an element of empiricism in this method as well, inasmuch as accuracy is enhanced if the system parameters are optimized by comparison with a known subsidence zone. SPASID, Subsidence Prediction and System IDentification, is an influence program written at Penn State for the Bureau of Mines (Kiusalaas and Albert, 1983) and used by the author (Heasley and Saperstein, 1986a). Once optimized, it is extremely accurate, precise, and moderately conservative of computer time. Even without previous subsidence data, SPASID can give moderately reliable predictions. Three-dimensional influence-function modeling represents, for this day, the best practicable method for subsidence prediction.

d. Analytical Models

Much has been attempted with numerical analysis models that follow the methods of classical mechanics. However, excessive computer time and unavailable values for *in-situ* behavior of geologic materials have kept these models from general use.

Progressive mining companies spend much effort in predicting subsidence from their active workings. The extraction method called longwall removes a regular and definable horizon of coal; hence the geometric site parameters are known precisely, which enhances the precision of the prediction. Although subsidence is usually inevitable with longwall panels whose W/D ratios are greater than 0.4, and supercriticality usually occurs above a ratio of 1.4, the uniformity of subsidence and the consistent behavior of strata are generally preferable to that which occurs above room-and-pillar extraction.

BENEFITS OF PLANNED SUBSIDENCE

It has been suggested already that some benefits accrue from planned subsidence and the longwall method of mining. These points are summarized below.

a. Longwall mining is safer, more productive and has lower overall costs than alternative underground methods of mining coal.
b. Longwall mining has a higher recovery factor, meaning that more complete recovery of the resource is effected and that less coal is abandoned or sterilized.
c. Longwall mining is uniform and predictable, whereas the evidence from old room-and-pillar mining suggests that subsidence above these workings may occur also but is not predictable.
d. Although subsidence from longwall mining is inevitable, it occurs within such time that the costs associated with subsidence can be borne by the mining company and not, in the future, by society at large.
e. Longwall mining can produce broad supercritical panels that minimize zones of large, damaging strains.

f. Damage to land and water is also predictable, although less precisely than with surface motions. If there is valid concern about water resource protection, mining may be deferred until technical reassurance of protection or alternative water supply is available.

SUBSIDENCE CONTROL

Through the years, subsidence-induced damage has created contention between coal-producing companies and owners of the surface. Because of severance of the mineral estate from the surface estate, often as much as a century before mining commences, these two are often unknown to each other. Although the author prefers to see this strife reduced by careful application of planned subsidence, along with appropriate restitution for any damage and sincere efforts on the coal producers' part to be good neighbors, this section would be deficient if it did not discuss techniques for subsidence control.

Reduced Extraction. Inasmuch as subsidence occurs only after coal is extracted, one approach to subsidence control is to prohibit mining beneath sensitive structures and surface features. This approach has been taken both legislatively and voluntarily. Before undertaking such a ban, a careful analysis should be made of the harm that could result to the structure if it undergoes movement as well as the cost associated with sterilizing the coal. Irreplaceable historic structures are candidates for total support as well as are high-density urban districts, where the cost of structural repair may well exceed the value of the coal mined.

A more moderate approach is to limit coal extraction to some presumably safe amount, often 50 percent, and to insist that the remaining coal be left in pillars of substance that will support the roof. Although, there is some evidence that sufficiently large pillars last for as long as we have maintained observations, there are still drawbacks to this method. The first drawback is that the room-and-pillar method is less productive, hence more expensive, than is longwall. The second is that the coal in the pillars is lost more certainly than even the coal in zones of total prohibition, for the latter may be available to some new technology of subsidence control whereas the former is degraded by exposure to air and water. The third is the uncertainty of success over extended times, if not even to perpetuity. Should there be failures of the support pillars in the distant future, then there will be a question of unresolved liability.

The size of support zones can be found by estimating rules, which are derived from W/D ratios and angles of draw. Preferably, the subsidence-protection zone should be calculated by use of a subsidence—prediction program (Heasley and Saperstein, 1986b). In this manner, the consequences of additional pillars and, thus, additional transition zones on the surface can be seen.

Back-Filling. There is an inherent logical attraction to the provision of subsidence support with artificial pillars. As will be seen, the attraction has been negated by some very real problems. Total face backfilling (sometimes called "stowing") has been developed beyond the prototype stage and has been used in Europe. Typically, coal refuse or waste is returned to the mined-out zone by pneumatic transport; occasionally, especially quarried sand has been used and emplaced hydraulically. The premise is that non-combustible and, in the case of the refuse, unwanted material substitutes for coal. The negative side is that the process is costly and slow, reducing productivity by over one half. The additional machinery and, in the case of pneumatic transport, dust and noise provide added hazards to the workers. Finally, experience shows that this backfill is compressible and, thus, while it reduces the S/T ratio, it does not prevent subsidence altogether. A recent study was undertaken to see if refuse could be strengthened (Carlson and Saperstein, 1986) by the use of additives. Portland Cement, in the range of seven or eight percent by weight, added strength, but it also added cost.

Legislative Control. The contentiousness of subsidence and any associated damage has led to many trips into the courtroom and, ultimately, legislation over subsidence. Although a limited number of States had independent subsidence legislation before 1977 (for example, Pennsylvania's "The Bituminous Mine Subsidence and Land Conversation Act of 1966, 52 P.S. 1406.1 *et seq.*"), all coal-mining states, by virtue of the authority of the "Surface Mining Control and Reclamation Act of 1977" (Public-Law 95-87, 30 U.S.C. 1201 *et seq.),* have either passed new legislation patterned after the federal law and its associated regulations or have adopted the federal regulations identically into their own laws (Bruhn and Turka, 1986).

In brief, the law requires operators "to adopt measures (that are technically sound and generally available) to prevent subsidence causing material damage," and to "maintain the value and reasonably foreseeable use of surface lands." The law also allows that the required measures should be economically and technologically feasible, should maximize mine stability, and should not restrict those circumstances where the "mining technology used requires planned subsidence in a predictable and controlled manner." The law distinguishes between material damage and value of the land and this has caused some interpretation problems. The previously referenced manual (Bruhn and Turka, 1986) has an extensive discussion of the federal regulations.

The Pennsylvania law, now amended to incorporate the federal requirements, also protects certain classes of structures and features that were in place at the date of original passage (April 27, 1966) and provides for owners of non-protected structures to purchase the coal needed to protect their structures from the owner of record, whether or not the coal owner wishes to sell. At present, this provision in the Pennsylvania law is being contested in the courts.

SUMMARY

In brief, complete extraction of flat-bedded deposits such as coal will result in subsidence. The extent of damage to land and structures caused by this subsidence will depend upon the width of the mined zone with relation to its depth, the thickness of the mined zone, the nature of overlying beds, and the nature of the structures or features themselves. Subsidence can be predicted and, for potentially damaging situations, steps can be taken before mining to mitigate damage. Total extraction mining, with planned subsidence, can have many benefits, not the least of which is the assumption of all environmental and damage-remediation costs by the coal producer.

REFERENCES

Bruhn, R.W. and R.J. Turka, 1986. *Guidance Manual on Subsidence Control,* Final Report by GAI Consultants, Inc. to OSMRE, USDI, 154 p.
Carlson, M.J. and L.W. Saperstein, 1987. "Efficient Use of Additives to Improve Pneumatically Emplaced Backfill Strength", Paper delivered to Annual Meeting, Society of Mining Engineers, Denver, CO, 9 p.
Heasley, K.A. and L.W. Saperstein, 1986a. "Practical Subsidence Prediction for the Operating Coal Mine," p. 54-67, in Peng. S.S. (ed.), *Proceedings 2nd Workshop on Surface Subsidence Due to Underground Mining,* West Virginia University, Morgantown, WV, 295 p.
Heasley, K.A. and L.W. Saperstein, 1986b. "An Investigation into the Use of Backfill Zones and Yielding Pillars for Subsidence Control," p. 19-27, in Singh, M.M. (ed.), *Mine Subsidence,* Society of Mining Engineers, Littleton, CO, 144 p.
Heasley, K.A. and L.W. Saperstein, 1986c. "Recent Insight into Longwall Strata Movements Deduced from Subsidence Analysis," Paper 86-331, Fall Meeting, Society of Mining Engineers, St. Louis, MO, 9 p.
Kiusalaas, J. and E.K. Albert, 1983. *SPASID: A Computer Program for Predicting Ground Movement Due to Underground Mining: User's Manual,* The Pennsylvania State University, University Park, PA, 201 p.
Kratzsch, H., 1983. *Mining Subsidence Engineering,* Springer-Verlag, Berlin, 543 p.
Miller, G.C. and W. Johnson, 1979. "Abandoned Coal-Mined Land: Nature Extent, and Cost of Reclamation," *Special Publication 6-791,* Bureau of Mines, USDI, Washington, D.C., 19 p. (available from NTIS as PB 299 535/AS).
National Coal Board, 1975. *Subsidence Engineers' Handbook,* Mining Department, National Coal Board, Hobart House, London, 111 p.

National Research Council, 1986. *Abandoned Mine Lands,* Committee on Abandoned Mine Lands, Board on Mineral and Energy Resources, Commission on Natural Resources, National Academy Press, Washington, D.C., 221 p.

Whaite, R.H. and A.S. Allen, 1975. "Pumped-Slurry Backfilling of Inaccessible Mine Workings for Subsidence Control," *Information Circular 8667,* Bureau of Mines, USDI, 83 p. plus append.

Zwartendyk, J. 1971. *Economic Aspects of Surface Subsidence Resulting from Underground Mineral Exploitation,* Ph.D. Thesis, The Pennsylvania State University, 411 p. (incl. 17 p. biblio.).

Environmental Consequences of Energy Production: Problems and Prospects. Edited by S. K. Majumdar, F. J. Brenner and E. W. Miller. © 1987, The Pennsylvania Academy of Science.

Chapter Six
RECLAMATION OF DEEP COAL MINING WASTES WITH PARTICULAR REFERENCE TO BRITAIN AND WESTERN EUROPE

MARTIN KENT

Department of Geographical Sciences
Plymouth Polytechnic
Drake Circus
Plymouth
Devon, England, PL4 8AA

INTRODUCTION

Derelict and despoiled land associated with deep coalmining represents a substantial environmental problem resulting from energy production in Britain and Western Europe. In particular, the spoil heaps found at most deep mining sites are highly intrusive visually and often dominate the landscapes of mining areas (Kent 1986). In addition, such despoiled areas and tips pose problems of erosion and instability, generation of acid minewaters and locally there may be spontaneous combustion of tipped materials. Environmental effects also extend well beyond the coal tips to problems of reworking of tipped material, subsidence, loss of productive land and the detrimental effects of transportation of both coal and waste. These problems are exacerbated because most coal mining regions have traditionally been areas of high population and many of these, particularly in Britain, have experienced overall economic decline in the past decade with both the urban infrastructure and the environment suffering as a consequence.

This chapter is concerned primarily with problems of coal spoil tipping and the reclamation of those tips once they are no longer active. In Britain and Western Europe, reclamation of despoiled land and colliery spoil heaps has been regarded as an important aspect of environmental management and landscape planning for over 20 yrs. Applied ecologists have studied problems of colliery spoil reclamation closely and a number of syntheses of their research findings now exist (Down and Stocks 1977, Bradshaw and Chadwick 1980, Univ. Newcastle-upon-Tyne 1971-72, Kent 1982). These studies demonstrate that colliery spoil is among the most difficult and expensive of derelict land types to reclaim (Kent 1982, Goodman 1974). The spoil comprises shales, mudstones, sandstones and the coal itself, which were originally deposited during the Carboniferous era some 280-345m yrs ago. The shales commonly represent 90 percent of all wastes and in consequence it is the properties of these shales which are most significant when re-vegetation and reclamation of spoil is to be undertaken. When present, iron pyrites (FeS_2) and more complex sulphides cause the greatest problems for long-term reclamation success.

The creation of a satisfactory soil and vegetation cover are the primary aims of most derelict land reclamation schemes. Such restoration decreases the problems of visual intrusion, reduces air and water pollution, stabilizes surfaces and offers a limited range of after uses, such as low quality grazing, recreation or forestry. Frequently, however, reclamation is largely 'cosmetic', with a complete vegetative cover and a stable soil environment representing the sole aim of a restoration project.

THE HISTORY AND SIZE OF THE PROBLEM

Taking Britain as a typical example, in the 19th century and the first two decades of this century, disposal of dirt and colliery waste was not a serious problem because the coal was sorted from the dirt at the coal face by the miner. Figure 1, however, shows that in the last 60 yrs, with increasing mechanization and with the richest seams being worked out, the ratio of dirt to coal has increased. This increase was particularly marked between 1960 and 1980 and although there is evidence to suggest that levels of spoil production have not continued to increase at a similar rate during the 1980's, the problems of colliery spoil disposal are now greater than ever before (Bradshaw 1982, Glover 1984). Recent estimates put spoil production in Britain at 50 - 55m tons of spoil per year, requiring some 200 ha of new land for tipping (Glover 1984).

Figure 1 also shows how coal production has varied over this century, with the substitution of alternative energy sources, notably oil, natural gas and nuclear power, causing a halving of coal production since the end of the second world war. In 1949, coal represented 91.4 percent of inland consumption of primary fuels in the United Kingdom, but by 1980, this had fallen to 36.9 percent (Flowers 1981). Similar patterns of change in energy production have occurred in other European countries, particularly France, Belgium and West Germany. These countries have also seen a slight rise in coal production during the late 1970's and early 1980's (Ghouzi 1982, Van der Haagen 1982, Petsch 1982). In the continental coalfields, amounts of spoil produced are roughly equivalent to the quantities of coal mined (Ghouzi 1982). As an example, according to Petsch (1982), in the Federal Republic of Germany in 1980, 87m tons of deep mined coal were produced, giving roughly the same quantity of spoil.

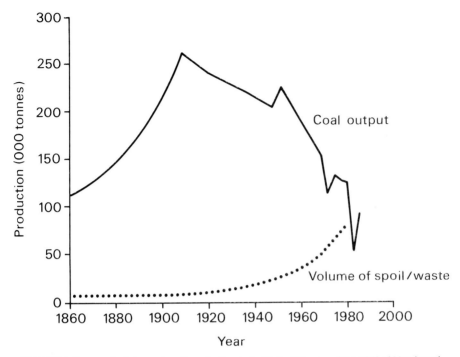

FIGURE 1. Deep coal mining and spoil production in the United Kingdom from 1860-1985; adapted from Bradshaw (1982).

NEW DEEP MINE DEVELOPMENTS

In the United Kingdom, the future of the coal mining industry has been the subject of two major studies. Firstly, the National Coal Board (NCB) published its Plan for Coal (1974) which was aimed principally at reversing the postwar decline in coal output, following the oil crises of the early 1970's. This report was mostly concerned with coal production and contained little on the environmental consequences of coal mining. However, in March 1978, a Commission on Energy and the Environment was set up with the terms of reference "to advise on the interaction between energy policy and the environment". The resulting report "Coal and the Environment," named after its chairman, Lord Flowers, was published in 1981 (Flowers 1981, North and Spooner 1982). The report examined "the longer-term environmental implications of future coal production, supply and use in the United Kingdom." The most frequently quoted conclusions on the environment were that "most of the potential environmental problems for increased coal production and use can be overcome by the more even application of current best practice." (Para 22.3) and "there are no insuperable environmental obstacles to the role of coal as currently envisaged in the U.K." (para 22.3).

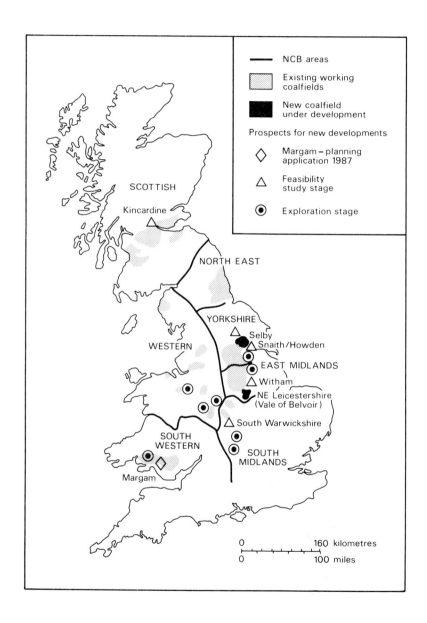

FIGURE 2. New coal developments and prospects in Britain; adapted from Flowers (1981).

At the center of the Flowers report was discussion of the environmental impacts of new deep mining developments in Britain. However, Flowers carefully avoided making any clear judgements as to the environmental suitability of any single new development. The largest sites are located to the east of the major coalfields of Yorkshire, Nottinghamshire and Leicestershire (Fig. 2). The most important of these have been near Selby in Yorkshire and in the Vale of Belvoir in North-east Leicestershire. Recently, investigations have also commenced into the feasibility of mining in South Warwickshire and a planning application for mining will be made in 1987. South Warwickshire has previously been a non-coalfield area. Evaluation has also been completed at Margam in South Wales (Fig. 2). Preliminary studies are also being made around Witham, in Leicestershire and in the Snaith-Howden area of Yorkshire. An important point is that the majority of these new developments are taking place in predominantly agricultural environments, rather than in areas of urbanization. This shift represents a major change in terms of environmental impact and requires new methods for environmental impact assessment (Selman 1986). The Selby coalfield is now operational and site work has commenced at the Asfordby mine in the Vale of Belvoir, but these and others have been subject to long delays because of local opposition, largely on environmental grounds, resulting in lengthy public planning enquiries.

North and Spooner (1982) point out that another important focus of major new NCB investment is in certain existing deep mines. For example, in the Yorkshire coalfield, Grimethorpe and Maltby Main, two of the largest collieries in the region, have both had new investment of well over £100m each in the late 1970's. The NCB have also reactivated Thorne mine at the very east of the Yorkshire coalfield, north-east of Doncaster, which had been mined between 1927 and 1959. Similar investment patterns are observable on other traditional coalfields with centralization and rationalization of existing mines. An advantage of these urban or old coalfield sites is that many of them are not under strict planning control, since under the General Development Order system, any mine which was in operation on July 1, 1948 is deemed to have planning permission for any further developments, both underground and within the existing pit-head site. These two different forms of investment and development have recently been given the names 'green-field sites' (agricultural) and 'black-field sites' (traditional coalfield areas) (North and Spooner 1982). The NCB has thus been able to compensate for the slowness of the planning process with respect to its new developments such as those at Selby and the Vale of Belvoir by its relative freedom to expand and invest at existing sites.

ALTERNATIVES TO RECLAMATION IN THE TREATMENT OF COAL WASTES

The large quantities of spoil generated in European coalfields each year means that most waste is tipped to await eventual reclamation. Nevertheless, several European coal mining countries have investigated the possibilities of using waste in a more positive manner and a range of different ideas have been proposed with varying degrees of success. In Britain, the NCB Minestone Executive was established in 1971 with the goal of pursuing "the application of the Board's waste products into useful engineering fields through extensive research and development of new technology." (Turnbull, p. 1.1) and a Symposium on the Reclamation Treatment and Utilization of Coal Mining Waste (Rainbow 1984) was held in Britain in 1984. The following classification of potential and actual uses for mining wastes is presented in Turnbull (1984):

Engineering Applications

Road embankments	Rail embankments
Canal, river and sea defences	Reservoirs
Impermeable linings	Land reclamation and land fill
Cement bound minestone	Reinforced minestone
Binding of toxic wastes	

Manufacturing Uses

Lightweight aggregates	Building blocks
Bricks	Cement

In addition to this, there is the possibility of back-stowage, but in most modern mines this is not an economic proposition at the present time. Modern longwall mining techniques, using power supports and shields, require the collapse of the roof behind the advancing coal face. Thus much possible backstowage space is lost, quite apart from the cost and practical difficulties of restowage itself in the confined space below ground.

THE NEED FOR RECLAMATION OF DEEP COAL MINING WASTES

Although substantial spoil heaps have been present in the coalfield landscapes of Britain and Western Europe since the last century, it was only during the 1960's that serious reclamation was undertaken for the first time. In Britain, an important distinction needs to be made between the tips and despoiled land created prior to 1970 and the spoil heaps created by the considerably improved tipping standards introduced thereafter. The basis of present tipping legislation in Britain was laid down in the Town and Country Planning Act 1947, which was subsequently amended in 1968 and 1971. Changes in tipping practice were also introduced by the NCB following the Aberfan disaster in South

Wales in 1966. Further important legislation followed in the Town and Country Planning (Minerals) Act 1981.

In the Federal Republic of Germany, as early as 1920, a law created an association to organize regional planning in the Ruhr and this was followed in 1962 by the formal establishment of a board for development planning to include both spoil disposal and land utilization policy (Schmidt 1984). The "Kommunalverband Ruhrgebiet" (KVR) is an administrative unit which determines environmental policy in the whole of the Ruhr area which produces around 80 percent of West Germany's deep mined coal. Its area includes 11 conurbations with over 100,000 population.

In Belgium, as far back as 1911, a law was passed compelling owners of mines or quarries to restore their sites and a royal enactment of 1967 enabled the state to provide loans to encourage reclamation work. Neither of these acts was particularly successful, because the laws provided no penalty if restoration was not completed. In 1978, a new law was passed for one specific area—the Walloon region, whereby generous grants were allocated by the state and a site could be taken over by the state if the owners failed to restore it. Even this had limited success. In 1981, there were still 400 tips in Belgium containing 800m tons of coal waste (Van der Haagen 1982).

In all Western European coal mining countries, for the period 1910-1950, large quantities of spoil were tipped mechanically by free-fall onto high-rise, conical or ridge heaps. By this process, pockets of air were trapped in the tips and spontaneous combustion of the colliery waste was commonplace. The smoke and fumes which resulted, particularly in wet weather, and the production of hydrogen sulphide, became a serious environmental problem. In Britain, a considerable number of these tips were reclaimed in the period 1965-80, but many still await treatment. The reclamation problems of these sites are very different to those of spoil heaps of more recent origin.

From 1970 onwards, spoil heaps were created by using earth-moving plant and heavy rollers to compact tipped spoil in a series of layers across a spoil site. This method also meant that tipping could be controlled carefully to reach exact contours. The advantages of this approach are considerable. The spoil has reduced permeability, increased density and shear strength, which virtually eliminates the risk of spontaneous combustion and greatly reduces the amount of pyritic oxidation which could lead to problems of acid mine drainage.

Increasingly, tipping has also involved the concept of 'progressive restoration', whereby only small areas are involved in active tipping at any one time and once tipping is completed and agreed contours have been reached, then the area is immediately restored. Such an approach is best suited to agricultural areas and is more difficult to implement in 'black-field' urbanized sites. Once an initial tipping area is established, valuable topsoil can be transported from a new tipping area to cover the tipped material in the previous area. This eliminates the double handling and storage of soil and thus improves reclamation cost and performance. It is also more effective if the area of land stripped for tipping is kept to an absolute minimum (one year's supply) (Selman 1986). The reclamation problems attached to 'progressive reclamation' are thus somewhat different to those of older tipped areas. Many older tips would never have been permitted to achieve their present size and height under present legislation.

Despite these recent improvements in tipping methods, the area of despoiled and derelict land associated with deep coal mining in Britain and Europe is still substantial and a large proportion of such land is so-called 'inherited dereliction' from earlier mining activity. In England, the 1982 Survey of Derelict Land lists 5081 ha of derelict coal spoil heaps of which 4679 ha are judged worthy of reclamation. This figure is still regarded as being grossly underestimated due largely to the General Development Order clause of the Town and Country Planning Act, which omits any site from the derelict land returns where there is still active working, even if it is on only a very small part of the total site area.

THE PROBLEMS AND PRACTICE OF COLLIERY SPOIL RECLAMATION

The many difficulties of reclaiming coal spoil heaps are now extensively documented, (Bradshaw and Chadwick 1980, Univ. Newcastle-upon-Tyne 1971-72, Kent 1982) and most problems of initial reclamation have been identified with differing solutions being offered, depending on the scale of the reclamation project, the finance available, the severity of toxicity problems in the raw shale and the availability or otherwise of top or sub-soil. Sites are usually first regraded, if they have not been already tipped to the limits of a specified landform, as with more recent tipping. In most cases and when 'progressive reclamation' is being carried out, top or sub-soil will be available and can be used to mask the spoil. If the spoil contains pyrites which is actively weathering, such masking becomes essential. On older sites, where there is no topsoil on site, importation of soil is very expensive. Unfortunately, many of these older sites of inherited dereliction often have severe acidity problems due to pyritic weathering and masking cannot be avoided. In some situations, it is possible to sow into bare spoil, but this is not as common as it might be and is likely to increase potential aftercare problems. With either bare or masked spoil, ameliorants are required, first lime and second varying proportions of macronutrients provided as fertilizers.

Major problems of reclamation may be summarized as follows (Kent 1982):

Physical Problems of Vegetation Establishment

Highly variable particle size distributions and degrees of stoniness

Compaction due to compression within tipped material
Steep slopes
Erosion and gulleying
Extremes of surface temperature, particularly in summer months
Highly variable soil moisture contents
Summer droughting problems
Spontaneous combustion
Compaction due to earthmoving machinery

Chemical Problems of Vegetation Establishment (Toxicities)

Pyrites and its weathering to produce sulphuric acid through oxidation and hydrolysis
Ankerite and carbonate content (acid neutralizing properties of spoil)
Chlorides and excessive concentrations of soluble salts
Secondary effects of low pH—phosphorus fixation and reduced availability of other ions e.g. potassium
High spatial variation in toxicities and occurrence of pyrites in spoil
Temporal variation in levels of acidity following weathering of pyrites
Alteration of shale properties following combustion
Toxic levels of heavy metals and trace elements, notably copper, zinc, nickel, molybdenum and selenium

Chemical Problems of Vegetation Establishment (Nutrient Status)

Low levels of available nitrogen in raw colliery shales and also often in masking top/sub-soil
No operational nitrogen cycle
Absence of legumes to promote nitrification
Absence of soil organism populations essential to successful establishment of nitrogen cycle e.g. *Nitrosomonas* and *Nitrobacter*
Serious deficiency or absence of available phosphorus
Absence of soil organisms for phosphorus cycle
Reduction in the availability of potassium, even though available in abundance in most colliery shales, because of low pH
Liming required to raise pH and to render potassium available
Calcium and magnesium highly deficient—can be corrected by liming
Rapid leaching of nutrient additions, particularly nitrates

Biological Problems of Plant Growth on Colliery Spoil

Total absence or very limited abundance of soil organism populations
Absence of organic matter in the upper horizons of the soil/spoil

Lack of earthworms to encourage soil fertility and organic turnover
Need for viable populations of the micro-organisms of the nitrogen and phosphorus cycles to become established

RECLAMATION PRACTICE

Figure 3 shows a typical sequence of actions in revegetating colliery spoil. The testing of spoil for pyritic material, both before and after earthmoving and at depth in the spoil is essential. After regrading, it is common to rip the surface to a depth of 20-50 cm prior to the application of lime and fertilizers, in order to reduce surface compaction. Amounts of lime required to neutralize acid spoil vary greatly (Bradshaw and Chadwick 1980). Many colliery spoils have been found to need over 100 tons of lime/ha, which is between two and four times the amounts recommended in earlier work (Costigan et al. 1981, Bradshaw 1981). Excessive amounts of lime may fix phosphorus, increase nitrate losses and affect performance of legumes sown to encourage nitrogen fixation. Different fertilizer treatments and hydroseeding have been experimented with on some sites but both slow release fertilizers and hydroseeding are generally too expensive for serious consideration (Bradshaw 1981, Doubleday and Jones 1977). Lists of common grasses and legumes planted in reclamation schemes in Britain are available in Bradshaw and Chadwick 1980, Kent 1982, and Richardson 1977, as are others for trees and shrubs.

Reclamation practice during the 1970's and 1980's has been criticized and limitations acknowledged (Lindley 1984). Foremost amongst these have been the problems of earthmoving, which frequently absorbs around 70-90 percent of reclamation costs (Chadwick and Lindeman 1982). Often the desirable shape or contours cannot be achieved within the finance or the land area available. The resulting steeper slopes then preclude most land uses other than grazing. Also there are problems with stability and load-bearing capacity. Most building construction is thus precluded. The lack of diversity in both landforms and planting designs has been criticized by several authors including Kent (1980, 1982) and Lindley (1984), but improvements will only be forthcoming if more finance is made available for individual schemes.

Finding positive after uses for more than a handful of sites in each colliery region is difficult. There are psychological problems on the part of developers (Chadwick and Lindeman 1982). If the developer knows that the site is a former coal spoil heap, he is unlikely to risk his investment capital. Many examples of light industrial development such as warehousing are suitable, but the premises are usually constructed by local government using central government finance rather than by private individuals. These problems are at the heart of 'Category A' schemes mentioned below.

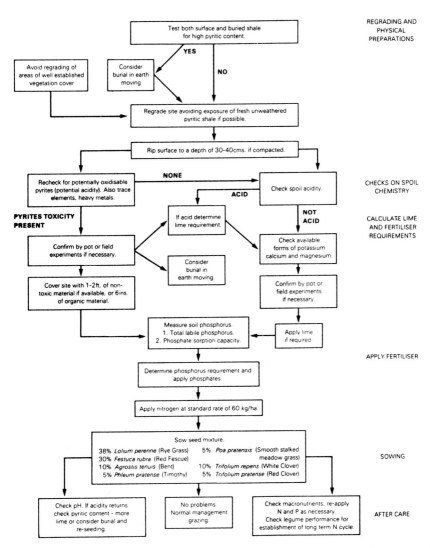

FIGURE 3. A flow diagram of a typical sequence of actions in the reclamation of colliery spoil; Kent (1982).

RECLAMATION COSTS AND FINANCE

Recent estimates for the cost of reclaiming colliery spoil heaps in Britain vary considerably. An average figure for 1984, quoted by Doubleday (1984) is £25,000/ha, which is probably nearer to £30,000/ha at the present time. In Britain, the main agencies of implementation are the local authorities, although they rarely finance their own reclamation projects apart from general environment improvement schemes. It is central government which has the principal say on how much reclamation is completed, since most work is done with 100 percent government grants. The Countryside Commission also offer 50 percent grants, particularly if recreation is to form part of the after use. However, it is often impossible for the local authorities to find the other 50 percent from their own funds. Attempts have been made to obtain European Community Grants but these have not been successful.

In Britain, since 1981, there has been a change in the way many central government grants for reclamation have been financed. A new class of 'Category A' schemes has been introduced which are projects jointly financed by public grant and a private developer. The aim of the government has been to try and encourage more examples of positive after use, rather than simply 'cosmetic' treatment. Unfortunately 'Category A' development is rarely suited to the derelict lands of the coalfields, because private developers do not wish to risk their finance in such areas. At the very most, 'Category A' finance has assisted with 10 percent of projects in recent years (Hartley 1984). Since 'Category A' schemes have priority, the availability of normal grants (presumably now 'Category B') have been reduced.

AFTERCARE AND MAINTENANCE

Until recently, long-term maintenance and aftercare received little attention, compared with the amount of research completed and money spent on the process of reclamation itself. Regression and die-back of vegetation on restored sites is commonplace (Kent 1982, Lindley 1984). Where species have been sown directly into bare spoil in combination with ameliorants, this is usually due to renewed weathering of pyritic material, generation of acidity and associated effects on nutrient status. One of the penalties of extensive earthmoving in a reclamation scheme is that fresh unweathered pyritic spoil is brought to the surface and can then cause serious problems for plant growth in the longer term if it starts to weather. Both Kent (1982) and Lindley (1984) have commented on this and point out that many older sites have areas of relatively mature vegetation cover and well-weathered spoil prior to regrading. Providing these areas are not a safety hazard, it may be possible to either keep them from being destroyed in earthmoving or attempt to remove some of the weathered spoil, which although still poor in nutrients, is of a less toxic nature. Such spoil could then be used for masking. Longer-term problems and the extent of renewed weathering of fresh shales may thus be reduced. A survey method for the evaluation of spoil heaps prior to reclamation has been described by Kent (1980). Even where masking layers of top/sub-soil have been used, toxicity problems can reoccur. Ensuring an even cover of masking material is very difficult and localized shallow areas can allow weathering spoil underneath to come into contact with plant roots. Also roots themselves can tap down into the underlying spoil if masking depth is insufficient. Water movement and weathering can be concentrated at the contact zone between spoil and the top/sub-soil. If this becomes severe, failure and slumping of the masking material on steeper regraded slopes can occur.

Doubleday (1984) emphasizes these and other problems of long-term management. He states (21.2):

"Most of the normal pressures operating on reclaimed land tend to restrict the site's capacity to achieve the designer's objective. Trespass and vandalism cause damage to fences, drainage systems, water supplies and other fixtures and fittings. On reclaimed pit heaps, weathering of the colliery shale usually reduces any fertility built up at the time of reclamation and may even produce toxic components which can kill off the new vegetation. Water logging, erosion, loss of vegetation and inability to sustain the after use for which the site was reclaimed are all symptoms of reclaimed land inadequately managed."

By far the greatest number of schemes are reclaimed by the establishment of a grass cover, followed by grazing or use as public open space. The goal then becomes sound management of a grass sward. Foister (1977) argued that grazing is one of the best means of encouraging bottom growth and root extension in grasses. However, sometimes the extra pressure of grazing may be too much for a sward which is already under stress from its rooting medium.

With grass swards, establishment and management of legumes is a particularly important and difficult aftercare problem. The use of nitrogen fertilizers is expensive and is only viable as a short-term solution. There are also problems due to the ease of leaching of nitrate fertilizers. However, although the sowing of legumes is vital for a functioning nitrogen cycle, the difficulties of establishment can be great. Jefferies (1981) and Gemmell (1981) have studied the effects of different liming treatments on legume success. In heavily limed spoil, *Trifolium repens* (White clover), for example, exhibits temporary but extreme inhibition of growth by surrounding grasses, which respond to treatment faster and are more competitive. Steady decline in clover cover with increasing time following reclamation is common for the same reasons and maintaining a balanced clover component in a reclamation sward is one of the most taxing jobs for a site manager. Palmer and Chadwick (1985) reinforce these ideas on the problems of legumes, having studied accumulation of nitrogen under fertilized plots with no legumes as opposed to unfertilized plots with legumes. Accumulation rates were significantly higher under the legume plots after 7 years.

Soil physical conditions can also cause aftercare problems. Rimmer (1982) studied 12 reclaimed spoil heaps of varying age and showed that all suffered to varying degrees from problems of water relations with summer droughting, winter waterlogging and high spatial variations of particle sizes across tips. High stoniness even following reclamation, caused difficulties. These results are important, particularly where grazing is to be practiced, since the trampling and poaching of animals can exacerbate these physical problems even further. Soils accumulate organic matter but only very slowly (Doubleday 1984, Rimmer 1982). Attempts to speed up the rate of organic matter accumulation are virtually impossible over large areas.

Doubleday (1984) also discusses problems of reseeding, since from an agricultural standpoint, the grass cover of most heaps can be seen as a ley. Reseeding can be required as frequently as every 3-5 yrs, although this can be extended to 8-10 yrs if a top/sub-soil mask has been applied.

Management of public open space is similar, except that the tolerance of the sown sward to trampling and recreational activity will in general be lower on colliery spoil. Trees, shrubs and woodlands planted within reclamation schemes are usually prone to moisture stress and can suffer from toxicity in the longer term if roots tap weathering pyritic spoil at depth. Vandalism is a serious problem. Nevertheless, attempts at tree planting are valuable in that they will relieve the all too frequent monotonous green sward.

Several authors (Bradshaw and Chadwick 1980, Kent 1980, 1982, Doubleday 1984) stress the importance of site monitoring. Only regular on-site inspections will convey useful information on aftercare problems and specific problems may require repeated chemical measurement and small-scale field experiments.

RECLAMATION PRACTICE IN WESTERN EUROPE

The problems and techniques of reclamation in France, Belgium and West Germany are very similar to those in Britain and the reclamation agencies have benefitted greatly from the conclusions of research published in both Britain and the United States. Similar problems to those in Britain also exist over financing of reclamation projects and the relative roles of public and private funding. Despite paying considerable attention to both legal and financial aspects of environmental pollution in the past decade, the European Community appear not to have addressed themselves to the specific problems of coal spoil reclamation (Johnson 1983).

EVALUATION OF RECLAMATION PERFORMANCE

As Tomlinson (1984) states, in recent years, attitudes towards reclamation have changed, with the emphasis increasingly being placed on the quality of reclamation rather than whether reclamation occurs. In Britain, the Minerals Act, 1981 has introduced new legislation with reference to aftercare provisions and the setting of reclamation standards. In the U.S.A. reclamation bonding is increasingly applied, whereby in order to gain planning permission for mining, a developer has to commit himself to restoration of the site to pre-defined standards within a specified time after the end of mining, with financial penalties if the standards are not met. The great problem is, of course, to set the necessary reclamation standards. At present, most schemes are judged subjectively by an acknowledged expert. More objective and consistent standards are required. These problems and criteria for standards are discussed in Tomlinson (1984) and Street (1985). An agency to oversee monitoring of standards is also necessary.

CONCLUSION

There is no question that significant environmental consequences result from the mining of deep coal in Britain and Western Europe. Equally, although most of the practical and ecological problems of restoration and aftercare have been solved, both the quantity and quality of reclamation in all countries could be greater if sufficient finance were available. The perennial problem of who should pay for reclamation is as relevant as ever. In the end, it is the consumer who pays the price for derelict and despoiled land and its reclamation, whether this is done through an increase in the price of coal or by increased central taxation. If for political and/or economic reasons the money is not forthcoming by either route, then the local people in coal mining areas simply have to live with and accept the environmental consequences. The evidence is that all too often, they do.

ACKNOWLEDGMENTS

The author would like to thank Dr. Chris Down, of the Department of Mineral Resources Engineering, Imperial College, University of London for constructive comment on the manuscript and Brian Rogers and Jenny Wyatt (Plymouth Polytechnic) for drawing the diagrams.

LITERATURE CITED

Bradshaw, A.D. 1981. Nitrogen and the maintenance of productivity, pp. 41-44. In: Land Decade Education Council (Eds.) *The Productivity of Restored Land*. Land Decade Education Council, London.

Bradshaw, M. 1982. Environmental problems of coal production in the United Kingdom with particular reference to the Yorkshire coalfield, pp. 47-61. In: Down C.G. (Ed.) *Proceedings of the European Conference on Coal and the Environment. Min. Environ.* 4, 112 pp.

Bradshaw, A.D. and M.J. Chadwick, 1980. *The Restoration of Land. The Ecology and Reclamation of Derelict and Degraded Land*. Studies in Ecology, Volume 6. Blackwell Scientific, Oxford, 317 pp.

Chadwick, M.J. and N. Lindeman. 1982. *Environmental Implications of Ex-*

panded Coal Utilization. Pergamon Press, Oxford, 283 pp.

Costigan, P.A., A.D. Bradshaw and R.P. Gemmell, 1981. The reclamation of acidic colliery spoil: I. Acid production potential. *J. Appl. Ecol.* 18: 865-78.

Department of the Environment. 1982. *Survey of Derelict Land in England 1982.* Department of the Environment, London.

Doubleday, G.P. 1984. The management of reclaimed land, Paper 21. In: Rainbow, A.K.M. (Ed.) *Proceedings of the Symposium on the Reclamation Treatment and Utilization of Coal Mining Wastes.* National Coal Board Minestone Executive, London, 772 pp.

Doubleday, G.P. and M.A. Jones. 1977. Soils of reclamation, pp. 85-124. In: B. Hackett (Ed.) *Landscape Reclamation Practice.* IPC Press, Guildford, 235 pp.

Down, C.G. and J. Stocks. 1977. *Environmental Impact of Mining.* Applied Science Publishers, London, 371 pp.

Foister, J. 1977. Development and maintenance organizations, pp. 205-18. In: Hackett, B. (Ed.) *Landscape Reclamation Practice.* IPC Press, Guildford, 235 pp.

Gemmell, R.P. 1981. The reclamation of acidic colliery spoil: II. The use of lime wastes. *J. Appl. Ecol.* 18, 879-87.

Ghouzi, D. 1982. The case of the French Nord/Pas de Calais coalfield, pp. 67-74. In: Down, C.G. (Ed.) *Proceedings of the European Conference on Coal and the Environment. Min. Environ.* 4, 112 pp.

Glover, H.G. 1984. Environmental effects of coal mining waste utilization, Paper 17. In: Rainbow, A.K.M. (Ed.) *Proceedings of a Symposium on the Reclamation Treatment and Utilization of Coal Mining Wastes.* National Coal Board Minestone Executive, London, 772 pp.

Goodman, G.T. 1974. Ecology and problems of rehabilitating wastes from mineral extraction. *Proc. R. Soc. Lond. A.* 339: 373-87.

Hartley, D. 1984. Reclamation—a waste of valuable resources? Paper 22. In: Rainbow, A.K.M. (Ed.) *Proceedings of the Symposium on the Reclamation Treatment and Utilization of Coal Mining Wastes.* National Coal Board Minestone Executive, London, 772 pp.

Jefferies, R.A. 1981. Limestone amendments and the establishment of legumes on pyritic colliery spoil. *Environ. Pollut. A.* 26: 167-72.

Johnson, S.P. 1983. *The Pollution Control Policy of the European Communities.* 2nd Ed., Graham and Trotman, London, 244 pp.

Kent, M. 1980. Regional assessment of plant growth problems for colliery spoil reclamation. I. Introduction and site survey. *Min. Environ.* 2: 165-75.

Kent, M. 1982. Plant growth problems in colliery spoil reclamation—a review. *Appl. Geogr.* 2: 83-107.

Kent, M. 1986. Visibility analysis of mining and waste tipping sites—a review. *Landscape Plann.* 13: 101-110.

Lindley, T. 1984. Landscape and land use of reclamation sites, Paper 19. In:

Rainbow, A.K.M. (Ed.) *Proceedings of the Symposium on the Reclamation Treatment and Utilization of Coal Mining Wastes.* National Coal Board Minestone Executive, London, 772 pp.

Lord Flowers (Chairman) 1981. *Commission on Energy and the Environment Coal Study.* H.M.S.O., London, 257 pp.

North, J. and D. Spooner, 1982. A future for coal. *Town Country Plann.* 51: 93-98.

Palmer, J.P. and M.J. Chadwick. 1985. Factors affecting the accumulation of nitrogen in colliery spoil. *J. Appl. Ecol.* 22: 249-57.

Petsch, G. 1982. Environmental problems of coal production in the Federal Republic of Germany with particular reference to the Ruhr, pp. 75-80. In: Down, C.G. (Ed.) *Proceedings of the European Conference on Coal and the Environment. Min. Environ.* 4, 112 pp.

Rainbow, A.K.M. (Ed.) 1984. *Proceedings of a Symposium on the Reclamation Treatment and Utilization of Coal Mining Wastes.* National Coal Board Minestone Executive, London, 772 pp.

Richardson, J.A. 1977. High performance plant species in reclamation. pp. 148-172. In: Hackett, B. (Ed.) *Landscape Reclamation Practice.* IPC Press, Guildford, 235 pp.

Rimmer, D.L. 1982. Soil physical conditions on reclaimed colliery spoil heaps. *J. Soil. Sci.* 33: 567-79.

Schmidt, H.G. 1984. The activities of a regional administrative union of municipalities of a coal mining district in the field of reclamation, treatment and utilization of mining wastes, Paper 2. In: Rainbow, A.K.M. (Ed.) *Proceedings of a Symposium on the Reclamation Treatment and Utilization of Coal Mining Wastes.* National Coal Board Minestone Executive, London, 772 pp.

Selman, P.H. 1986. Coal mining and agriculture: a study in environmental impact assessment. *J. Environ. Mgmt.* 22: 157-86.

Street, E.A. 1985. Evaluation procedures for restored land. *Environ. Geochem. and Health* 7: 56-63.

Tomlinson, P. 1984. Evaluating the success of land reclamation schemes. *Landscape Plann.* 11: 187-203.

Turnbull, D. 1984. The role of the Minestone Executive in British mining and civil engineering, Paper 1. In: Rainbow, A.K.M. (Ed.) *Proceedings of a Symposium on the Reclamation Treatment and Utilization of Coal Mining Wastes.* National Coal Board Minestone Executive, London, 772 pp.

University of Newcastle-upon-Tyne (Ed.) 1971/2. *Landscape Reclamation.* 2 vols. IPC Press, Guildford.

Van der Haagen, M. 1982. The coal problem in Belgium: matters related to the economy and the environment, pp. 62-66. In: Down, C.G. (Ed.) *Proceedings of the European Conference on Coal and the Environment. Min. Environ.* 4, 112 pp.

Environmental Consequences of Energy Production: Problems and Prospects. Edited by S. K. Majumdar, F. J. Brenner and E. W. Miller. © 1987, The Pennsylvania Academy of Science.

Chapter Seven
ENVIRONMENTAL PLANNING FOR SURFACE MINING OF COAL
RAJA V. RAMANI
Professor of Mining Engineering
and
Chairman, Mineral Engineering Management Section
The Pennsylvania State University
University Park, PA 16802

The demonstrated reserve base of U.S. coal, as of 1982, is 4.83 trillion tons. Nearly 3.25 trillion tons are potentially mineable by underground methods. The other 1.58 trillion tons are potentially mineable by surface methods. In the short span of 20 years, from 1967 to 1987, the contributions of surface and underground methods to the total production have reversed. The ratio of underground to surface coal in 1967 was of the order of 2:1. In 1987, it is 1:1.5. In the same period, the coal production increased from 551 million tons to over 875 million tons. Clearly, the surface mining of coal is projected to play an important role in the energy picture of the U.S.

As compared to any other industrial activity, a surface mining operation is much more disruptive to the environment, particularly to both the land surface and the subsurface. Reclamation and rehabilitation of the land is difficult and costly. Understanding the environmental consequences of surface mining and reduction of the potential negative impacts are important.

The purpose of this chapter is to (1) outline briefly the methods of surface coal mining; (2) discuss the environmental impacts of the mining activities; and (3) present a detailed procedure for the environmental site planning for lands with surface mineable coal reserves. The contents of this chapter summarize, among others, some of the results of research pursued at the Pennsylvania State University during the period 1967-1986 and supported by U.S. Office of Coal Research, U.S. Environmental Protection Agency, U.S. Office of Surface Mining, U.S. Bureau of Mines and Pennsylvania Department of Environmental Resources. Extended discussions and detailed developments with regard to environmental planning of surface mines can be found in Stefanko, Ramani, Ferko (1973); Ramani and Clar (1978); Ramani et al., (1980); Clar and Ramani

(1983); Ramani and Sweigard (1983a, 1983b, 1983c, 1983d) and Sweigard (1984). This chapter will purposefully sidestep discussions on the legal requirements, particularly the Surface Mining Control and Reclamation Act of 1977, as their implications are discussed in Chapter 35 of this book.

SURFACE MINING

Surface Mining is a very broad term and refers to the removal of soil and strata (bedrock) over a coal seam to remove the seam itself. Since coal seams occur under diverse geological and topographical settings, several methods of removing the soil and strata, and coal seam have evolved, each most suitable for the encountered conditions.

Surface mining of coal is conducted in a relatively simple sequence of unit operations which includes: (1) preparing the surface by removing vegetation and topsoil, (2) drilling for rock fragmentation, (3) blasting with explosives, (4) removing overburden by loading and hauling or direct casting to expose the coal, (5) removing the coal by loading and hauling, (6) reclaiming the mined area including (a) backfilling the mined-void with the blasted overburden material (spoil), (b) regrading and spreading of topsoil, (c) stabilizing the surface with temporary and quick growing vegetative cover, and (d) ensuring permanent stabilization and productive use over a longer period of time. All unit operations may not be necessary in a particular operation. Where the soil and bedrock are weak, there may be no need for drilling and blasting. In the direct casting method of overburden removal, much of the backfilling is accomplished during the overburden removal process itself.

Surface mining methods have been broadly classified as: (1) contour mining, (2) area mining, (3) open-pit mining (or bench mining), (4) quarrying, (5) auger mining, and (6) dredging. Quarrying and dredging are not commonly employed for surface mining of coal. Auger mining is usually associated with contour mining.

In the surface mining industry, experimentation with variations to the basic methods is the rule. Such experimentations are facilitated by the wide choice of equipment combinations and their modes of operations (Ramani and Grim, 1978). Detailed discussions of mining methods, equipment combinations and reclamation plans are provided in Stefanko et al. (1973); Skelly and Loy Engineers and Consultants (1975); Chironis (1978); and Ramani et al., (1980).

Contour mining is practiced where coal seams occur in rolling or mountainous terrains. Basically, the method consists of removing the overburden above the coal seam by starting at the outcrop and proceeding along the hillside (or the contour). Additional cuts may be taken into the hillside until the economic limit for surface mining is reached. At this time, the coal remaining under the knob of the hill may be removed by auger mining or underground mining. A schematic of conventional contour mining is shown in Figure 1. This method is no longer

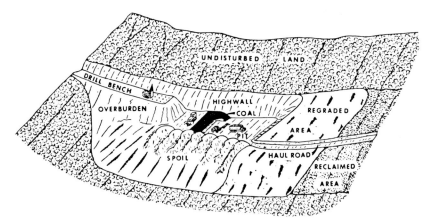

FIGURE 1. Conventional Contour Mining Method in Steep Hilly Terrain (Notice the spoil in the downslope side).

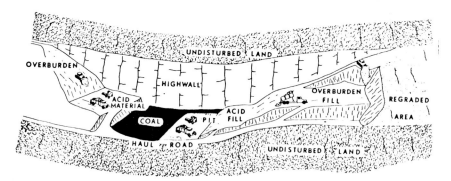

FIGURE 2. Contour Mining Method with Truck Haulback Technique (Spoil is stored within the previous cut).

practiced due to severe environmental problems arising from stability and erosion of the spoil. A modern variation of a contour method, employing truck haulback of the overburden material is shown in Figure 2. In this method, the amount of loose material pushed on the downslope side is greatly reduced and the toxic material (acid fill) is selectively buried. An extended discussion of contour mining methods can be found in Grim and Hill (1974). A summary discussion is provided in Ramani and Grim (1978).

Area mining is practiced in lands with flat topography where the coal seam is also relatively flat. This method is also commonly referred to as "strip mining." In this method, a trench, called a box cut, is made through the overburden by shovels, draglines, scrapers etc. to expose the coal seam. The overburden material, called spoil, is placed on the adjacent unmined land. After the coal is removed, a second trench (or cut) is started immediately adjacent and parallel to the first cut. The spoil from this cut is placed in the void created by the first

Environmental Planning for Surface Mining of Coal 81

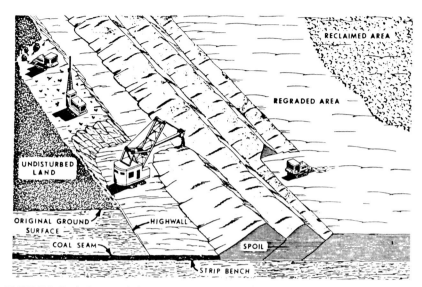

FIGURE 3. Typical Area Mining Method with a Stripping Shovel.

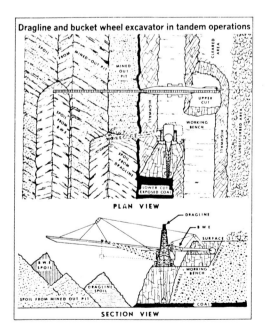

FIGURE 4. Area Mining Method with Bucket Wheel Excavator Removing Loose and Soft Top Material and Dragline Removing the Hard Overburden.

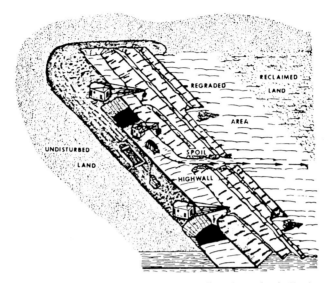

FIGURE 5. Area Mining Method Employing Two Draglines Operating in Tandem.

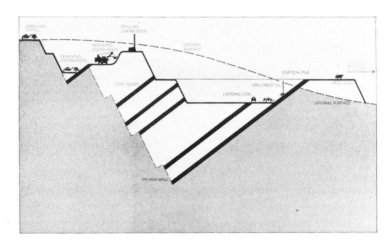

FIGURE 6. An Open-pit Coal Mining Method for Steeply Pitching Multiple Seam Deposit.

cut. Cut by cut, the mining progresses towards the property boundary (Figure 3). The spoils in the cut are graded, seeded, stabilized and restored for return to surface uses. Schematics of large area mining operations employing draglines, and wheel excavators in combinations for overburden removal are shown in Figures 4 and 5.

Applications of open-pit mining for coal removal are necessary under two specific occurrences of coal. In one case, the coal seams may dip so steeply that direct overcasting (as in area mining) is not feasible (Figure 6). In the other case,

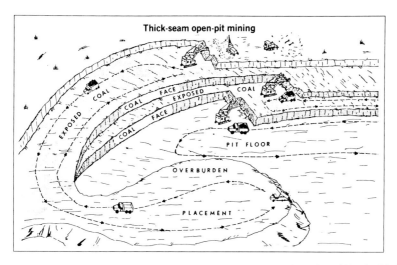

FIGURE 7. Thick Coal Seam Open Pit Mining Employing Loading Shovels and Trucks (Notice two benches in coal and one bench in overburden).

coal or overburden or both may be very thick and benches are required in coal and overburden for their removal (figure 7).

The environmental impacts of surface mining have become more pronounced in the U.S. in recent years for the following reasons: (1) the high visibility and the encroachment of surface mining operations into populated areas in the east; (2) many operations are involved with handling greater volumes of overburden due to the ever increasing ratio of overburden to coal thickness; (3) equipment size and capability have increased allowing deeper deposits to be surface mined; (4) economies of scale allow larger operations to benefit; and (5) thick western U.S. coal reserves have become more significant to the U.S. coal production. The mines are located in semi-arid to arid areas which did not have an extensive history of surface coal mining. In 1987, the top 30 producing mines are surface mines, mostly in the west. Collectively, they account for over 25 percent of the total U.S. coal production.

The increasing knowledge of the environmental consequences of surface mining is resulting in the development of several extraction and reclamation techniques aimed at controlling adverse environmental impacts and increasing the land values after mining.

ENVIRONMENTAL IMPACTS

Any activity, be it natural or human-induced, disturbs the existing environmental equilibrium. In time, an equilibrium may again be achieved though it may not be identical to the one that was disturbed. The surface mining of coal affects both the natural and cultural components of the environments.

The natural factors include geomorphology, climate and soil. The cultural factors are a combination of demographic, economic and market conditions.

The major phases in the life cycle of a mine are: (1) exploration, (2) development, (3) extraction and (4) abandonment or (return to other productivity uses). The search for the coal seam necessarily involves clearing ground cover, building access roads, and transporting heavy equipment. The actual processes of exploration such as drilling, test pitting and other drivages are usually temporary in nature till the feasibility of a mining operation is established. Surface mine locations are predetermined by the disposition of the coal and overburden relationships. Once opening a mine appears feasible, the major development activities of establishing construction facilities, building roads, erecting equipment, and building coal processing and coal load-out facilities all occur simultaneously over a relatively short period of 3 to 4 years. The extraction phase may last for 30 to 50 years for large operations. The surface areas affected by the mine itself may be small as in large open-pit mines or large as in area surface mines. However, if not properly planned, the affected area may be much

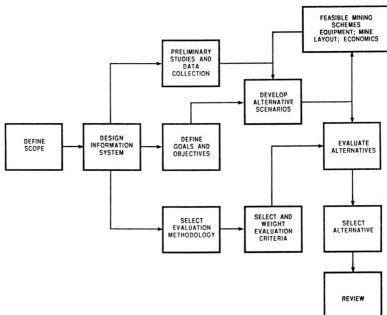

FIGURE 8. A Macro Flow Diagram Showing the Planning Process for Surface Coal Mine Planning for Improved Reclamation and Land Use Planning.

Environmental Planning for Surface Mining of Coal   85

TABLE 1
Potential Major Effects of Surface Coal Mining on Land, Water, Air and Other Resources.

Surface Coal Mining Unit Operations	Land and Soil									Water							Air			Wildlife								Other	
	Soil Erosion	Overburden Swelling	Toxic Strata	Soil Inversion	Soil Stability	Landslides	Spoil Piles	Oil Spills	Coal Spills	Aquifer Effects	Runoff Alteration	Sediments	Toxic Substances	Groundwater	Industrial Effluents	Sludge	Fumes & Emissions	Dust	Noise	Habitat Altered	Species Diversity	Aquatic Life	Animal Essentials	Accident/Deaths	Soil Organisms	Vegetation Potential	Wildlife Disturbed	Aesthetic	Dangerous Material
1. Exploration	x	x							x	x	x			x			x	x	x										
2. Development	x						x	x	x		x	x	x	x			x	x	x								x	x	
3. Drilling										x	x						x	x	x								x		
4. Blasting			x			x				x	x						x	x	x								x		
5. Overburden Removal			x	x	x	x	x	x									x	x	x	x	x	x	x	x	x				
6. Haulage	x			x	x	x		x	x			x					x	x	x	x	x			x					
7. Top Soil or Other Soil Storage	x		x				x							x											x				
8. Coal Beneficiation								x	x	x	x	x			x	x		x	x							x	x	x	x
9. Coal Loadout	x								x	x	x						x	x	x					x		x	x	x	x

Source: Ramani et al., (1980).

larger. For example, the surface topography and soil conditions which existed before mining commenced may be destroyed. Even if attempts are made to preserve the topography, surface mining results in a complete restructuring of the subterranean environment. The potential productivity of the soil for plant growth may be greatly reduced after mining. The soils that are disturbed may be toxic and become chemically active, causing water pollution. Huge rocks and boulders may occur in the graded spoil making it difficult for the smooth passage of farm machinery. Unconsolidated spoil heaps and preparation plant waste piles may affect drainage patterns and the natural processes of erosion and sedimentation may be accelerated. When not properly guarded against dumping of waste and refuse, these sites become potential health and safety hazards. The surface mining effects on air, water and land are not mutually exclusive. Air pollution may lead to water pollution and both may degrade the land, creating dangerous health and safety conditions and interfering with community development. A summary of the major environmental impacts due to various surface mining unit operations on land, water, air and wildlife resources is presented in Table 1. Though this listing is not exhaustive, the potential for long-term environmental damage is clear.

Fortunately, the growing environmental consciousness in the industrial, social and legislative sectors has led to the development of methods of restoration through which, in many cases, surface-mined lands are satisfactorily reclaimed within 2 to 3 years of mining. In fact, surface-mined lands, abandoned open-pits, orphaned spoil and waste piles are now being shaped, graded, seeded, stabilized and returned to productive uses (Ramani et al., 1980).

PREMINING PLANNING FOR ENVIRONMENTAL CONTROL

Mining operations are, at one and the same time, development and use of land. In addition to consuming land in a real sense, they have a finite life. While profitable extraction of the coal resources is a major objective, the other objectives are to ensure, as far as practicable, that no non-renewable uses of the land are pre-empted and no renewable uses are destroyed. In essence, the search to preserve or even enhance the value of mined lands for future uses must be a guide for the long-term environmental planning. To achieve this long-term goal, the planning steps are to (1) make an inventory of the premining conditions, (2) evaluate and decide on the postmining requirements of the region, consistent with the needs and desires of the affected groups, (3) analyze alternative mining and reclamation schemes to achieve the requirements optimally, and (4) develop an acceptable integrated mining and reclamation scheme most suitable under the technical, social and economic conditions.

The planning process described hereafter has been developed from studies of surface coal mining operations in the United States (Ramani and Sweigard, 1983a,b,c,d). A general outline of the procedure is shown in Figure 8. This procedure is an adaptation from the regional planning sources (Chapin, 1979) with modifications for environmental site planning for surface coal mines. Since mining operations can vary greatly in size, and must adapt to different regional conditions, it is difficult to make specific site planning recommendations that are applicable in all circumstances.

Define Scope

Scope definition must include determination of the specific output of the planning process. It also entails the organization of personnel and the assignment of tasks to execute the planning process. As a minimum, the output from the planning process should satisfy the Reclamation Plan Requirements given in Section 508 of PL 95-87. This includes a description of the premining capability, the proposed use of the reclaimed land, a discussion of the utility and capacity of the reclaimed land to support a variety of alternative uses, and a description of how the postmining land use will be achieved. If a change in land use is proposed, the output must also satisfy the requirements listed in Section 515(c)(3), which constitute a justification for the proposed postmining land use. The actual output will contain both text and drawings. The required detail may vary from case to case.

In some regions of the U.S., various components of the reclamation plan may have more impact on postmining land use. These components should be identified during the initial step of the planning process. For example, in the Midwest, where prime farmland may constitute a large percentage of the area to be mined, topsoil handling is a very important consideration. Likewise, in Appalachian regions, where mountaintop removal mining is practiced, the disposal of excess spoil is significant as this may figure into the postmining land use plan.

Initially, a company may need to spend a considerable amount of time in scope definition since it involves policy decisions relating to land management. Once the scope has been defined for a particular case, however, subsequent cases should only require minor modifications.

Design Information System

The site planning process depends heavily on the collection, analysis, storage, and retrieval of many data types. The amount of data required for environmental site planning is generally large and quite specific in nature. The categories of data include environmental background data (permit application requirements), additional cultural data, cost data (planning, reclamation, and mine closure) and operating factors such as equipment availability. The major functions that must be performed on the data are storage, updating, and retrieval.

Although a simple filing system will suffice for smaller operations, many larger companies may recognize a need for a computerized information system. Companies that presently use computers for accounting purposes may easily adopt a standard format for storing environmental and other data. The geographic location of a mine does not enter into the design of an information system, as such. However, larger surface mine operators, such as those located in the Midwest and Western U.S., may have more resources and more reasons to employ a sophisticated information system. To be of most use, site-specific environmental data should be referenced to a common coordinate system. A further step in the direction of management information systems could allow site planners to make inquiries of the data base concerning land use suitabilities of specific areas thus expediting the site planning process. The value of such a complex system would have to be weighed against the number of land use plans that a given company must generate in the course of a year.

Define Goals and Objectives

The definition of local land use goals and objectives is basically a function of the public planners. This is an area where the mine planners should interact with local and regional planners to ensure that the proposed postmining land use plan is compatible with the overall or comprehensive plan of the area. Reclamation plan regulations, in fact, require the mine operator to discuss the relationship of the proposed plan to existing land use policies and plans. In general, these goals and objectives address such issues as accommodating the residential needs of population growth, protecting natural and cultural resources, and providing needed public facilities and services. In addition to satisfying local goals and objectives, the mining company may wish to establish other goals for itself, such as improving the value of the land or promoting good public relations.

Ideally, comprehensive planning should precede site planning and provide input to the site planning process. It is not uncommon, however, for site planning to be conducted in an area that has no comprehensive plan. In the absence of formalized goals and objectives by a planning agency, the mine planners can either rely solely on company goals or make a limited survey of goals by contacting neighboring landowners and local elected officials (Clar, 1982; Clar and Ramani, 1983).

Preliminary Studies and Data Collection

The preliminary studies and data collection phase of the planning process should provide the necessary background information for the formulation and evaluation of mining and postmining land use plans. This step will encompass

TABLE 2

Information Needs on Ecosystem Factors for Surface Mine Planning and Land Use.

I. SOCIOECONOMIC INVENTORIES
   - (a) Land demands for mining and associated activities, and availability, encroachment into critical areas, other resources affected and foregone benefits.
   - (b) Past, current and future land use plans for the area-industry, agriculture, wilderness, etc.
   - (c) Transportation networks
   - (d) Recreation networks
   - (e) Archeological and historical interests
   - (f) Demographic and population characteristics
     - (i) employment patterns
     - (ii) tax base
     - (iii) land value alterations and their impacts on local government
     - (iv) attitudes of public, political and educational institutions in the community/region.

II. GEOLOGIC ANALYSIS
   - (a) Topography
   - (b) Overburden Characteristics
     - (i) stratigraphy
     - (ii) physiography
     - (iii) geomorphology
     - (iv) chemical nature
   - (c) Soil and Toplayer Characteristics
     - (i) physical
     - (ii) chemical
     - (iii) limitations for vegetation and future land uses
     - (iv) limitations for engineering structures

III. COAL CHARACTERIZATION
   - (a) Coal seam attitude
   - (b) Coal seam thickness
   - (c) Coal seam analysis
     - (i) proximate analyses
     - (ii) ultimate analyses
     - (iii) washability analyses

IV. TERRESTRIAL ECOLOGY
   - (a) Natural vegetation, characterization, identification of survival needs
   - (b) Crops
   - (c) Game animals
   - (d) Resident and migratory birds
   - (e) Rare and endangered species

V. AQUATIC ECOLOGY
   - (a) Aquatic animals—fish, waterbirds; resident and migratory
   - (b) Aquatic plants
   - (c) Characterization, use and survival needs of aquatic life system

VI. WATER RESOURCES
   - (a) Surface hydrology
     - (i) watershed considerations
     - (ii) flood plan delineation
     - (iii) surface drainage patterns
     - (iv) amount and quality of run-offs
   - (b) Groundwater hydrology
     - (i) groundwater table
     - (ii) aquifers
     - (iii) amount and quality of groundwater flows
     - (iv) recharge potential

VII. CLIMATOLOGY
    (a) Precipitation
    (b) Wind—airflow patterns, turbulence
    (c) Humidity
    (d) Temperature
    (e) Climate type
    (f) Growing season

VIII. OTHERS
    (a) Noise
    (b) Particulates
    (c) Aesthetics

Source: Ramani et al., (1980).

the environmental baseline data required for a permit application. Baseline studies, as a minimum, include a premining inventory of the ecosystem factors listed in Table 2. The listing is not complete but points to the recognition that any attempt to extract coal resources involves a change in the physical, biological, social, physiosocial, biosocial, and psychosocial factors. Almost any economic, demographic or environmental information can be useful. In fact, Pugliese et al., (1979) have identified at least 135 environmental characteristics which can have an impact on post-mining land use planning.

Socioeconomic Inventories: The mine planners should collect information on and analyze aspects of the local economy, population trends, transportation networks and demographic characteristics. Such information is useful to develop and analyze alternative mining and postmining land uses.

Physical and Chemical Attributes: Physical and chemical attributes of the premining environment are factors in several areas of design and planning, particularly in the engineering design of excavation, materials handling and ground support systems. They are also necessary to assess water quality control provisions and to determine the potential future land uses.

Geology: Geological aspects of the coal occurrence (e.g., attitude, depth, thickness, stratigraphy, etc.) is the basis for establishing alternative mining and reclamation schemes including slope stability. The nature of the overburden, especially those components as sulfur, having potential environmental impacts must be studied. The structural and stratigraphic evaluations which affect the hydrological conditions are critical for developing effective mining and reclamation plans.

Soils: Soils must be considered to ensure their proper removal, placement and postmining dispostion. Topsoil and subsoil have certain advantages over other materials, including higher organic matter, better moisture-holding capacity, natural seed course, and soil microorganisms. The removal and storage of topsoil and replacement of it on top of the graded soil, in many instances leads to better revegetation potential, which in turn leads to better erosion control.

The mining operation also provides an opportunity to modify the soil strata. A few of the soil modification and reconditioning methods include treatment with lime or limestone, fertilizers, and other additions as mulch, fly ash or sewage sludge, to adjust pH, nutrient and physical conditions.

Hydrology: Hydrology, both surface and underground, together with the physical and chemical characteristics of overburden and coal, is important for any reclamation program planning. A knowledge of water movement and quality in the mine area and of modifications during and after mining is essential. In all cases, erosion and sediment production are immediate concerns in the critical postmining, prerevegetation time period, although their control during storms is a long-term problem that requires detailed planning. Also, despite selective backfilling, designed contouring, and revegetation, it may require years until a new deep water flow-storage system is reestablished. During this interval, some change in water quality may be inevitable.

Topography and Climate: Topography and climate influence the selection of the mining method and mining equipment. Severe terrains also influence the methods of revegetation, and limit alternatives for future land use. Extremes in climate affect the selection and operation of equipment. Seasonal fluctuations determine desirable planting periods. Heavy rains can cause erosion complicating water quality control, sediment formation and the establishment of vegetation.

Certain environmental or cultural factors may require special attention due to specific regional conditions. For example, in Appalachia, the amount of pyrite in the overburden is particularly important. So is the amount of limestone, which helps to neutralize acid drainage. In the Northern Great Plains, the presence of sodium in the soil is a key factor that affects vegetation and, ultimately, land use. Where mines are located near urban centers, greater attention should be given to population trends and local economic growth as part of the preliminary studies and data collection phase.

Develop Alternative Scenarios

Although the development of alternative scenarios is shown as a single step in Figure 8, it is actually an iterative process that is closely related to the evaluation of alternatives of both mining and land use plans and is addressed in more detail later. The development of alternative scenarios for land use planning begins with broad land use classifications such as agricultural, forestry, wildlife, recreational, residential, commercial, industrial, institutional, and water resources. As the evaluation process continues, unsuitable land uses are eliminated and the remaining uses are combined to form the desired number of alternative scenarios. One alternative that should always be considered is that of restoring the mined area to its premining land use.

The selection of feasible mining schemes (mining method as well as equipment and unit operations) is greatly influenced by the thickness and altitude of the coal seam, the thickness of the overburden and the surface topography. A broad and general classification of surface mining districts in the U.S. with applicable surface mining methods is presented in Table 3.

Various types of alternative postmining land uses have been observed throughout the U.S. In Indiana and Illinois, residential and recreational uses are common, particularly in association with final cut lakes. Improved wildlife habitat has been made possible through the reclamation of surface-mined land in the Northern Great Plains. Mountaintop removal sites have been used in West Virginia for industrial and institutional purposes. Agricultural use of reclaimed land is one of the most common uses throughout the U.S. but particularly so in the Midwest. In this region, where the average farm size is larger than in the eastern U.S., improved crop management can be realized through the consolidation of row crop areas.

Some design skill is required at this stage to properly formulate alternatives from the acceptable land uses. A site plan rarely consists of a single land use but often combines several uses along with the specifics of grading, planting, traffic circulation, utilities, and site furnishings where required. Design methods

TABLE 3

Applicable Mine Methods in the Various Regions

	Mining Methods
REGION I West Virginia, Virginia, Tennessee, Kentucky	Steep-sloped Contour Mining Methods; Mountain top removal; Head-of-Hollow Valley Fill Methods; Haulback methods with front-end loaders, loading shovels and trucks.
REGION II Pennsylvania, Ohio, Maryland, Alabama	Conventional Contour Methods; Block-cut Methods, and dragline and shovel area methods. Draglines, front-end loaders, loading shovels and trucks are common.
REGION III Ohio, Kentucky, Illinois, Indiana, Missouri, Kansas, Oklahoma, Texas, North Dakota	Large area strip mines with draglines, shovels, bucket wheel excavators, multi-seam operation with large equipment in tandem.
REGION IV North Dakota, Montana, Wyoming, Colorado, New Mexico, Arizona	Large area strip mines, open-pit mining, steeply pitching multi-seam operations. Draglines, loading shovels and trucks are common.

Source: Chironis (1978)
Skelly and Loy Engineers and Consultants (1975)

such as overlay maps (McHarg, 1969) may be used for the development of alternative scenarios. Also, general site planning standards (Dechira and Koppelman, 1978; Jensen, 1967) which relate land use potential to such characteristics as distance from population centers and steepness of slope should be applied.

Select Evaluation Methodology

Evaluation techniques can be divided into three categories: economic analysis, environmental impact analysis, and social impact analysis. Economic analysis of alternative scenarios can range from informal discussions of potential land values to detailed engineering cost analysis. Detailed cost analysis must account for all of the project economic impacts of land use planning and the postmining land use plan. A discounted cash flow analysis as well as a fuzzy set analysis can be performed (Sweigard, 1984). In evaluating environmental and social impacts, the level of effort expended should be proportional to the size of the operation and the potential for creating land use benefits.

Select and Weigh Evaluation Criteria

Evaluation criteria should be selected and weighed, based on the evaluation methodologies chosen. A combination of criteria that reflects the economic, environmental, and social aspects of the plan is desirable. Specific regional natural or cultural characteristics that may have been identified in the Preliminary Studies and Data Collection stage should be considered in the selection of evaluation criteria. These would include, for example, overburden characteristics where acid production is a potential land use constraint or population growth where high intensity land uses are being considered. Economic criteria may include a minimum rate of return or a minimum benefit-cost ratio (Sweigard and Ramani, 1987a). Environmental criteria may be directed toward not only meeting various performance standards, but also the minimization of negative environmental impacts. Social impact criteria are more difficult to establish since social impacts are difficult to quantify. Various techniques such as cost-effectiveness analysis have been developed which attempt to subjectively quantify social impacts. Minimum standards can be set based on one of these techniques (Sweigard and Ramani, 1987b).

Evaluate Alternatives

In this phase of the process, the land use alternatives are subjected to the evaluation methodologies discussed previously. The necessity of having an evaluation process that is usable under a wide variety of mining and environmental conditions is satisfied by the four-stage process described more fully in the following section entitled, "Expanded Evaluation Process."

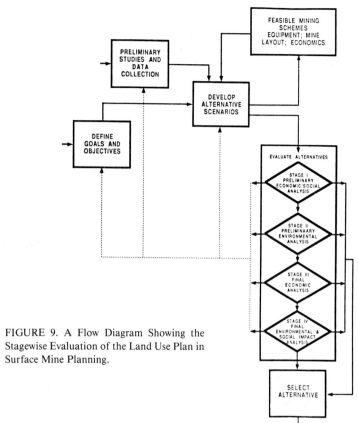

FIGURE 9. A Flow Diagram Showing the Stagewise Evaluation of the Land Use Plan in Surface Mine Planning.

Select Alternative

The result of the previous step will be a ranking of the feasible land use alternatives. The highest ranking alternative would be recommended. There may, however, be other factors relating to company policy that influence the final decision. In actual practice, the difference in relative attractiveness between the alternatives may be quite small. Also, before a final land use selection can be made, the relative importance of the economic, environmental, and social impacts must be weighed (Sweigard and Ramani, 1987b). Since one of the primary objectives of a business under the free market system is fair return on invested capital, land use schemes that would decrease the profitability would not generally be accepted unless they clearly demonstrated some compensating benefit to the company.

Review

The final step in the procedure is a review by company management and any public land use authority that would have to approve the postmining land use. In addition to the review made immediately after the plan has been selected, there should be periodic review by the mine planners and company management, during mining, to ensure that land reclamation takes place so that the land use plan is still workable.

EXPANDED EVALUATION PROCESS

The model developed for evaluating postmining land use alternatives is illustrated in Figure 9. This model expands the Evaluate Alternatives step of the general site planning model shown in Figure 8. This expansion provides for a four-stage evaluation process and a feedback loop to earlier steps in the planning model.

The four-stage evaluation methodology is analogous to the procedure used in mineral exploration (Bailly, 1968). Mineral exploration investigations proceed from an initial regional appraisal, in the first stage, to a detailed three-dimensional site evaluation, in the final stage. There are decision points built into the exploration model at the conclusion of each stage where it is necessary to evaluate the data already available and decide either to terminate the process or continue into the next stage.

This type of methodology can also be applied to the evaluation of postmining land use alternatives. The goal of this process is to design a postmining land use plan that is well suited to the environmental, social, and economic conditions at the mine site. As the mine planner proceeds with the evaluation, a decision must be made after each stage to either select a final land use plan, enter the next stage, or return to an earlier step in the planning process. The dashed lines in Figure 9 feeding back to earlier planning steps indicate that this entire process can be iterative in nature. In practice, the time of the planning steps will overlap. Therefore, it is best to begin the evaluation procedure with a comprehensive list of general land use classes (e.g., agricultural, forestry, wildlife, residential, recreational, industrial, commercial, etc.) rather than detailed site plans. A preliminary consideration of social and economic conditions will generally eliminate a number of potential land uses. As the process continues, alternative scenarios may need to be reformulated, additional data may be required, or it is possible that the land use goals may need to be redefined if the initial goals prove unattainable.

In the simplest case, data collection will consist mainly of gathering the required data for the permit application. A number of general land uses would be considered, including the premining uses. A preliminary consideration of the economic and social conditions may indicate that there is no justification for further evaluation of alternative land uses and the postmining land use plan will be designed, based on the premining uses.

The various stages of the evaluation process can also be viewed as a sieve. Typically, Stage 1 would have a short duration and be limited to an overview of such economic and social characteristics as land ownership, current real estate values, current land use patterns, population growth, and development patterns. Consultation with realtors, developers, and local planners may be helpful at this stage.

Those land use alternatives that initially appear to be economically and socially acceptable may then be subjected to a preliminary environmental analysis in Stage 2. This analysis should also be relatively short and general in nature. It would entail reviewing a checklist of natural environmental factors to determine if there are any physical conditions that would eliminate a given land use from further consideration. For example, poor agricultural soil characteristics may limit agricultural use to pasture rather than row crops, or an abundance of acid-producing material in the overburden would likely eliminate any recreational use that would require a water impoundment.

Some refinement in the alternative land use plans would likely be necessary after the completion of the first two stages. The final economic analysis conducted in Stage 3 can take a number of forms. The simplest analysis would estimate the resale value of the land upon completion of reclamation based on real estate transactions in the area. A more complex alternative would be to consider the final land value or income from retained land (e.g., agricultural production) in the overall cash flow of the project.

Stage 4 of the evaluation procedure would employ analytical techniques that have been developed by environmental scientists and social scientists to differentiate between alternative plans. This stage would only be reached when two or more land use plan alternatives demonstrate significant potential for economically improved land use. A possible method for evaluating the environmental impact of various alternatives is the matrix method developed by Leopold et al., (1971). Runyan (1977) has identified 12 methods for evaluating the social impact of alternative plans. He has also qualitatively rated these methods with respect to their simplicity and usefulness. The final environmental and social impact analyses may also be combined and handled by multicriteria decision-making techniques such as forced decision analysis (Fasal, 1965) or fuzzy sets in a structured hierarchy (Saaty, 1978; Sweigard, 1984).

ORGANIZATION FOR SURFACE MINE PLANNING

The size of the company will likely determine, to a great extent, how a mine plan and a postmining land use plan are selected. Small companies may not employ a staff of engineers and planners. Therefore, consultants are contracted to perform all necessary environmental planning including preparation of permit applications, reclamation plans, revegetation plans, and erosion and sediment control plans. Since the consultant's performance is judged upon successful completion and approval of the permit application, postmining land use planning may not be highlighted.

Regardless of where the planning is done, either in-house or by consultants, recent public policy enactments have caused an expansion of mine planning into areas that were not routinely addressed in the past. The land use potential evaluation requirements are a prime example of this expanded role. Expertise in many different disciplines is essential to the overall mining planning effort which includes planning for extraction, reclamation, and postmining land use. A listing of specialization areas required in each phase of the mine planning process is given in Table 4 along with the specific land use implications.

For larger mining companies that perform their own premining environmental planning, there are basically two organizational alternatives. Each has its own merit and can be adapted to various circumstances. The first alternative integrates the mined-land planning function into a company's engineering department. Within the engineering department, it would operate as a cost center and be allocated a portion of the overall engineering budget. This allocation would be based upon a specified cost per ton for postmining land use planning. The advantages of this approach are constant interaction with engineers responsible for overall mine planning and a strict accounting of all planning costs. It is important that the mine supervisory personnel have input into the planning process and that the resulting postmining land use plans are adequately communicated to them since they will supervise the actual reclamation operation.

The other general alternative for postmining land use planning involves the formation of an organization charged with overall land management. This organization could take the form of a land management department or it could be a separate subsidiary. The organization would be given responsibility for acquiring land, obtaining permits, developing postmining land use plans, and either managing or disposing of reclaimed land. Effective communications with the operations and engineerng departments would be required to insure that land would always be available for continuing in light of mining practices and equipment availability. The budget for this organization would have to cover the purchase of land and mineral rights, leases, payments for loss of agricultural production, and the premining planning function. Income from the sale of reclaimed land would be considered as a positive cash flow and the portion of the overall budget allocated to postmining land use planning could be fixed, by a formula, to income from land disposal and reductions in reclamation costs, lost production costs, and leasing costs.

SUMMARY

Public policy has succeeded in expanding the role of environmental planning with regard to surface mining operations. This is particularly true for the

TABLE 4

Areas of Technical Expertise Essential in the Mined-Land Planning Process.

Mine Planning Phase	Planning Activities	Areas of Specialization
Legal Requirements Analysis	Identification of regulatory constraints related to land use.	*Land Use Planner Attorney or Paralegal Specialist
Land and Reserve Acquisition	Prepare land use/land cover maps.	Land Use Planner Photogrammetrist/ Cartographer Plant Biologist
	Prepare land ownership map.	Photogrammetrist/ Cartographer Surveyor
Market Development	Check market potential of site.	Geographer Transportation Engineer Land Use Planner
Financial Evaluation	Check if land development potential of the site will justify reclamation to a higher, more costly land use.	Engineering Economist Land Use Planner Real Estate Specialist Fiscal Planner
Coal Beneficiation Studies and Plant Design	Determine the impact of waste disposal on the postmining uses of land.	Mineral Processing Engineer Environmental Engineer Agronomist Geologist Hydrogeologist
Environmental Impact Studies	Evaluate the impact mining will have on the site with respect to capability and productivity.	Mining Engineer Environmental Engineer Agronomist Geologist Hydrogeologist Terrestrial Ecologist Plant Biologist Agricultural Engineer Archeologist Land Use Planner Social Scientist
Preliminary Mine Planning	Preliminary identification of postmining land uses.	Mining Engineer Land use Planner Agronomist Engineering Economist
Permits Acquisition	Land use information and postmining land use plan.	Land Use Planner Environmental Engineer Agronomist
Administrative Detail Analysis	Submittal/approval of final land use plan.	
Detailed Mine Planning	Detailed land use plan design.	Land Use Planner (specifically Landscape Architect) Mining Engineer Environmental Engineer Agricultural Engineer Agronomist Hydrogeologist Plant Biologist Engineering Economist

* Refers to someone trained in regional planning, landscape architecture, or site design.
Source: Ramani and Sweigard (1983a).

land use planning aspects of surface-mined land. However, it is not yet apparent whether these changes will actually yield more productive or beneficial postmining land uses. There are many complex issues involved that cannot be resolved by simply collecting more data. Some positive changes in mine-land planning may only take place as attitudes and perceptions change, and this may require many years to accomplish.

Flexibility is required on the part of regulatory authorities in the review of alternative postmining land use plans. Although the objective of recent public policy has been to encourage more productive and beneficial uses for reclaimed land, many surface mine operators justifiably fear that proposed land use changes may delay approval of surface mining permits. As long as this impression persists, there is little chance that more productive postmining land uses will result.

Both the surface mining industry and the public land use planning agencies must recognize that the value of a parcel of land stems from a unique combination of renewable and nonrenewable resources. Although mineral resources are nonrenewable, the land still has value after minerals have been removed due to its other resources. These resources include not only the land's natural characteristics but also the value attributed to it because of its geographical location or other cultural significance. Planning for the use of these lands should have the objective that, as far as practical, no nonrenewable uses are pre-empted and no renewable resources are permanently destroyed.

REFERENCES

Bailly, P.A., 1968. "Exploration Methods and Requirements," *Surface Mining,* E.P. Pfleider, ed., AIME, New York, NY, pp. 19-42.

Chapin, F.S., Jr., 1979. *Urban Land Use Planning,* 3rd edition, University of Illinois Press, Urbana, IL, p. 82.

Chironis, N.P., 1978. "Regional Aspects Affects Planning of Surface Mining Operations," *Coal Age Operating Handbook of Coal Surface Mining and Reclamation,* Ed. N.P. Chironis, McGraw Hill, New York, NY, pp. 3-21.

Clar, M.L., 1982. "An analysis of Requirements and Guidelines for Surface Mine Land Planning," M.S. Thesis, The Pennsylvania State University, University Park, PA, pp. 30-105.

Clar, M.L., and R.V. Ramani, 1986:"User's Manual for Premining Planning of Eastern Surface Coal Mining, Volume 6: Mine Land Planning," EPA-600/7-83-051, NTIS PB 83-262907, EPA, Cincinnati, OH, 201 pp.

Dechiara, J., and L.E. Koppelman, 1978. *Site Planning Standards,* McGraw-Hill, New York, NY, 351 pp.

Grim, E.C., and R.D. Hill, 1974. Environmental Protection in the Surface Mining of Coal, EPA Publication No. 670-2-74-093, EPA, Cincinnati, OH.

Imhoff, E.A., W.J. Kockelman, J.T. O'Connor, and J.R. LeFevers, 1978. "Integrated Mined-Area Reclamation and Land-Use Planning; Vol. 2: Methods and Criteria for Land Use and Resources Planning in Surface Mines Areas," Argonne National Laboratory, Argonne, IL. 56 pp.

Jackson, D., 1978. "Multi-seam Surface Mining in the West," *Coal Age Operating Handbook of Coal Surface Mining and Reclamation,* Ed. N. P. Chironis, McGraw Hill, New York, NY, pp. 130-136.

Jensen, D.R., 1967. *Selecting Land Use for Sand and Gravel Sites,* National Sand and Gravel Assoc., SilverSpring, MD.

Leopold, L.B., 1971. "A Procedure for Evaluating Environmental Impact," U.S. Geological Survey Circular 645, Government Printing Office, Washington, D.C.

McHarg, I.L., 1969. *Design with Nature,* Natural History Press, Garden City, NY, 197 pp.

Paone, J., J.L. Morning, and L. Giorgetti, 1974. Land Utilization and Reclamation in the Mining Industry, 1930-71, U.S. Bureau of Mines, Information Circular 8642.

Preston, J., E. Strauss, and T. Friz, 1974. "Model Mineral Reservation and Mine Zoning Ordinance," Wisconsin Geological and Natural History Survey IC 24, Madison, WI, 43 pp.

Pugliese, J.M., D.E. Swanson, W.H. Englemann, and R.R. Bur, 1979. "Quarrying Near Urban Areas: An Aid to Premine Planning," USBM Information Circular 8804, Washington, D.C.

Ramani, R.V., and E.C. Grim, 1978. Surface Mining—A Review of Practices and Progress in Land Disturbance Control, *Reclamation of Drastically Disturbed Lands,* American Society of Agronomy, Madison, WI, pp. 241-270.

Ramani, R.V. and M.C. Clar, 1978. "User's Manual for Premining Planning Eastern Surface Coal Mining. Volume 1: Executive Summary," EPA 600/7-78-180, NTIS: PB-287-086AS, EPA, Cincinnati, OH, 81 pp.

Ramani, R.V., C.J. Bise, C.W. Murray, and L.W. Saperstein. 1980. "User's Manual for Premining Planning Eastern Surface Coal Mining. Volume 2: Surface Mine Engineering," EPA 600/7-80-175, NTIS: PB-81-109-415, EPA, Cincinnati, OH, 349 pp.

Ramani, R.V., and R.J. Sweigard, 1983a. "Development of a Procedure for Land Use Potential Evaluation for Surface-Mined Land—Final Report," Office of Mineral Institutes, U.S. Bureau of Mines, NTIS: PB-83-253401, Washington, D.C. 114 pp.

Ramani, R.V., and R.J. Sweigard, 1983b. "Development of a Procedure for Land Use Potential Evaluation for Surface-Mined Land—Appendix I: Eastern U.S. Surface Mine Case Study," Office of Mineral Institutes, U.S. Bureau of Mines, NTIS: PB-83-264325, Washington, D.C. 87 pp.

Ramani, R.V., and R.J. Sweigard, 1983c. "Development of a Procedure for land Use Potential Evaluation for Surface-Mined Land—Appendix II: Central

U.S. Surface Mine Case Study," Office of Mineral Institutes, U.S. Bureau of Mines, NTIS: PB-83-264333, Washington, D.C. 135 pp.

Ramani, R.V. and R.J. Sweigard, 1983d. "Development of a Procedure for Land Use Potential Evaluation for Surface-Mined Land—Appendix III: Western U.S. Surface Mine Case Study," Office of Mineral Institutes, U.S. Bureau of Mines, NTIS: PB-83-264341, Washington, D.C. 106 pp.

Ramani, R.V. and R.J. Sweigard, 1984. "Impacts of Land Use Planning on Mineral Resources," *Mining Engineering*, Vol. 36, No. 4, pp. 362-369.

Runyan, D. 1977. "Tools for Community-Managed Impact Assessment," *Journal of American Institute of Planners,* April, pp. 125-135.

Saaty, T.L., 1978. "Exploring the Interface Between Hierarchies, Multiple Objectives and Fuzzy Sets," *Fuzzy Sets and Systems,* Vol. 1, No. 1, North-Holland Publishing Co., Amsterdam, pp. 57-68.

Skelly and Loy Engineers and Consultants, 1975. Economic Engineering Analysis of U.S. Surface Coal Mines and Effective Land Reclamation, *U.S. Bureau of Mines, OFR 74-75,* NTIS: PB-245315, Washington, D.C. 619 pp.

Stefanko, R., R.V. Ramani, and M.R. Ferko, 1973. "An analysis of Strip Mining methods and Equipment Selection," U.S. Office of Coal Research, NTIS: 223778, Washington, D.C. 144 pp.

Sweigard, R.J., 1984. "A Procedure for Environmental Site Planning and Postmining Land Use Planning for Surface-Minable Land: Development, Analysis, and Application," Ph.D. Thesis, The Pennsylvania State University, University Park, PA, p. 301.

Sweigard, R.J. and R.V. Ramani, 1986. "A Regional Comparison of Postmining Land use Practices," *Mining Engineering,* Vol. 38, No. 9, pp. 897-904.

Sweigard, R.J., and R.V. Ramani, 1987a. "Economic Aspects of Land Use Planning," *SME Transactions,* Vol. 286, Society of Mining Engineers, Inc., (In Press).

Sweigard, R.J., and R.V. Ramani, 1987b. "Evaluation of Post Mining Land Use Plans using Fuzzy Set Analysis," *SME Transactions,* Vol. 286, Society of Mining Engineers, Inc., (In Press).

Chapter Eight
COAL REFUSE PILES: PROBLEMS AND PROSPECTS
RAJA V. RAMANI
Professor of Mining Engineering and Chairman
Mineral Engineering Management Section
The Pennsylvania State University
University Park, PA 16802

Coal, as it occurs in nature, is a heterogeneous mixture of organic and inorganic minerals. The organic minerals are generally rich in carbon. Since coal is mined for its heating value, if there is a high proportion of materials which affect this important property, the Run-of-Mine (ROM) coal is processed before utilization. The steps in the processing of coal may vary from simple screening for sizing purposes to crushing, screening, and washing to remove the inorganic materials. The surface accumulations of the reject material from the processing of the raw or ROM coal are known as coal refuse piles. Synonyms include coal waste pile, refuse bank, culm bank, gob pile and slate bank.

The purpose of this chapter is to (1) outline the current status of coal refuse generation and disposal; (2) review the characteristics of coal refuse piles; (3) discuss the environmental impacts from surface disposal of refuse; (4) present the potential uses of the refuse pile material; and (5) summarize the procedures for safe disposal of coal mine refuse.

COAL REFUSE PRODUCTION

The problems of handling, storage and utilization of coal refuse is international in scope. Coal mining regions around the world have an unenviable heritage of derelict land, spotted with abandoned and burning refuse piles and waste impoundments, from past and current mining activities (Figures 1 and 2). In view of the long history of mining in the anthracite region of Pennsylvania, dating back to the late 1700's, the anthracite refuse pile accumulations are monumental. While absolute numbers can be misleading, according to one estimate there is over one billion tons of anthracite refuse in more than 800 piles.

These piles occupy 12,000 acres of surface area (Spicer and Luckie, 1970). Thousands of large active and abandoned waste piles are also found in other U.S. coal-mining areas. It is estimated that there are 3,000 to 5,000 refuse piles and impoundments in the eastern coal fields alone containing 3 billion tons of refuse (National Academy of Sciences, 1975). The Bureau of Mines has estimated that about 174,000 acres used for the disposal of underground mine and coal processing waste remains unreclaimed (Johnson and Miller, 1979).

The production of waste is an inevitable consequence of any coal mining operation, only the volume of waste is large when the grade of coal is poorer. The comparison of coal preparation statistics for the years 1973 and 1983 is revealing (Table 1). In 1973, there were 382 coal preparation plants in operation. They handled a total of 398 million raw (ROM) tons, producing 109 million tons of refuse, amounting to 27% reject. The resulting 289 tons of clean coal represented approximately 49 percent of the total clean coal production that year (592 million tons). The ratio of waste produced to total ROM can be estimated approximately at 16 percent (701 million tons ROM; 109 million tons refuse). In 1983, there were 873 plants which were processing 666 million raw tons and producing 376 million tons of waste, an increase of 250 percent over that generated in 1973.

The problem of coal refuse disposal has grown to enormous proportions as the volume of refuse to be disposed has greatly increased due to a number of reasons. The most important reason is the growing need to process ROM coal, not only in greater volumes but also to a greater extent than ever before, to meet

FIGURE 1. Photograph showing an anthracite coal refuse bank.

FIGURE 2. Photograph showing a bituminous coal refuse pile.

TABLE 1

U.S. Coal Production/Preparation Statistics, 1973 and 1983 (millions tons).

	1973	1983
Total Production, clean tons	592	886
No. of Preparation Plants	382	873
Raw Coal Processed, tons	398	666
Clean Coal Yield, tons	289	290
Refuse, tons	109	376
Reject, percent	27.4	56.5
Processed Production to Total Production, percent	49	33
Refuse Production to ROM Production, percent	15.6	29.8

Source: King (1977) and Grape (1986), National Coal Association, Washington, DC.

the sulfur-dioxide emissions standards from power plants. Other contributory factors include (1) the generally lower quality of the coal seams, resulting from the depletion of higher grade coal seams, (2) greater amount of dilution and finer size consist resulting from highly mechanized and non-discriminating mining systems, (3) increased demands placed on available lands from competing and more desirable uses and (4) growing recognition of the health, safety and general welfare hazards arising from coal refuse piles.

Unless properly constructed and maintained, coal refuse piles can cause water pollution, catch fire, and be a general threat to health and safety of all living

FIGURE 3 and 4.
Photographs showing a refuse pile before and after it was reclaimed.

things. In the U.S., since the passage of the Coal Mine Safety and Health Act of 1969 and the Surface Mining Control and Reclamation Act of 1977, considerable engineering requirements have been placed on the design, construction and monitoring of mine refuse piles and embankments. As a result, the health and safety problems from new coal refuse piles are greatly reduced. However, the problems posed by abandoned refuse piles will need to be addressed. Reclamation efforts are also successful (Figure 3 and 4). However, finding alternative safe disposal methods for new and old refuse will continue to challenge the ingenuity of man.

COAL REFUSE PILE CHARACTERISTICS

Is it difficult to generalize the characteristics of the materials in the refuse piles. They are very dependent on the roof, floor and seam conditions, and the methods used in the mining and processing of coal. For example, while anthracite refuse is a generic term, it would include breaker refuse, silt, mine refuse and tunnel rock. Breaker refuse can further be broken down into culm, the refuse of old dry breakers and rock, refuse from modern wet breakers. Culm banks contain appreciable amounts of coal. Furthermore, as the refuse material ages in the pile, its characteristics also change.

In the coal processing plant, two types of refuse are generated: coarse waste and fine waste. The need for removing greater amounts of sulfur from the ROM coal has required that the coal be ground to very fine size-consist (28 mesh) for processing. A comparison of the size analysis of samples from a number of West Virginia refuse piles with that from refuse piles samples from Britain is shown in Figure 5 (Moulton et al., 1974). Since the object of preparation is to yield a cleaner product of higher Btu, the refuse is a concentrate of partings, roof rocks, pyrites and other inorganic material that form a part of the ROM

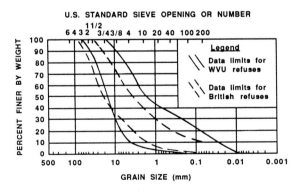

FIGURE 5. Limits of particle size distribution curves for samples from refuse piles in North Central, North Western West Virginia and British coal mines.

coal. The coal refuse pile material also contains alumina, silica, germanium, gallium and many other trace elements. The following information on some important physical and chemical characteristics of coal may be useful as a guide:

specific gravity: 1.4 to 2.7, average of 2.0
size: fine dust to 6″; more fines with aging
moisture content: 10-20%
carbon: 5-35%, average 10%
Btu per lb: 3,000-8,000, usually higher with age
sulfur: 5-10%, lower with age

The engineering properties such as permeability, shear strength, compressive strength, and cohesive strength are important factors in determining refuse pile stability, and the potential use of the refuse material as a construction material.

Some coal refuse piles contain appreciable amounts of commercially saleable coal. For example, in 1985, the production from anthracite culm banks accounted for nearly 20 percent of the 5.2 million tons of U.S. anthracite production.

ENVIRONMENTAL IMPACTS

The adverse impacts from abandoned coal refuse piles include air pollution, water pollution, health and safety hazards, and psychological effects. Many abandoned coal refuse piles throughout the U.S. are burning creating health and safety hazards, inhibiting economic growth, and contributing to the general destruction of surrounding communities (see Chapter 4).

The air pollution from refuse piles can arise from the release of methane in the refuse and windblown dust and particulate matter. In addition to being aesthetically undesirable, the dust and particle matter are harmful to vegetation. A more serious air pollution problem can arise when a refuse pile starts burning due to accidental or intentional ignition. The burning material will emit toxic and noxious fumes which are harmful to living things and to built-up inanimate structures as well. The potential for explosions from ignitions of pockets of methane in the refuse is also present.

A variety of water pollution problems can arise from coal refuse piles. The most common are physical pollution such as increase in turbidity and suspended solids, and chemical pollution due to leachates. When turbidity is increased, light penetration decreases with resulting lowering of aquatic food production, and of size and quality of the aquatic habitat. The refuse material is usually high in acid forming materials such as pyrites. These react with oxygen and water forming sulfuric acid. The water draining from the piles can also contain elevated levels of other ions such as iron, aluminum and manganese. This chemical pollution may reach levels which are toxic to aquatic life. The streambeds and other structures are often coated with a yellow precipitate of ferric hydroxide. The extent of pollution may be so much as to affect drinking water sources and aquifers. The ultimate impact of water pollution from coal refuse piles is a lowering in the quality of life of all living things. The potential for greater harm in terms of lasting effects on the health of the communities is ever present.

The safety hazards from abandoned coal refuse piles arise from their poor design and construction. End dumping of refuse was a common practice (Figure 6). This practice creates piles susceptible to not only instability but also spontaneous combustion. In some cases, coal refuse piles are built on unstable foundations such as supercharged aquifers and clay bed. In mining areas, the subsidence from past mining may extend to areas covered by the refuse banks creating movements in the pile foundation. Refuse pile instability is the result of many factors such as angle of slope, stratification and slope of the individual layers, the segregation of fine and coarse particles, the compaction and the degree of saturation with water. Some of the refuse piles in the Appalachian region are on steep slopes and have been exposed to high precipitation for several years. The stability of these piles is questionable. The manner in which piles may start moving are shown in Figure 7 (Glover, 1971).

The magnitude of the sudden and tragic consequences of these hazards was brought to world attention in two realizations, one on either side of the Atlantic (National Academy of Sciences, 1975):

> "...a flowslide of 140,000 cubic yards of waste from a 200-foot waste pile at Aberfan, South Wales in 1966, and a failure of a waste impoundment at Middle Fork, Buffalo Creek, West Virginia, in 1972 when 650,000 cubic yards of water and 220,000 cubic yards of material were transported downstream. The cause of both slides was improper disposal of coal mine waste. In each case the physical cause was saturation of the waste by water which reduced its stability to such an extent that the material flowed as a liquid. The geologic cause of the Aberfan disaster was the disposal of waste over a spring

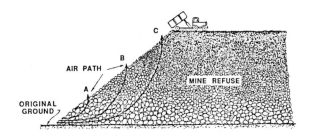

FIGURE 6. Past practice of end dumping showing refuse segregation by size.

of water. The climatic cause of the Buffalo Creek disaster was a rainstorm which deposited 3.7 inches of rain in 72 hours (a 2-3 year frequency storm). Both incidents were avoidable and unexcusable..."

In the Aberfan disaster 144 men, women and children were killed. In addition, 29 children and 6 adults were injured, 16 houses were damaged and 60 houses had to be evacuated. In the Buffalo Creek disaster, the confirmed deaths numbered over 115; over 500 homes were destroyed and about 4,000 persons

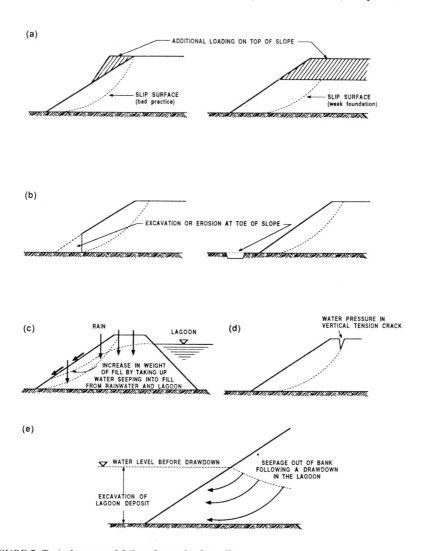

FIGURE 7. Typical causes of failure for coal refuse piles.

were left homeless. The destruction and damage to other property, public structures and conveniences was also extensive. Threats to life and property from abandoned refuse piles and impoundments must be recognized.

Abandoned coal refuse piles have unnatural and unsightly physical attributes. They are aesthetically displeasing. They are generally very steep, dark grey in color, and devoid of vegetation. In many cases, the preparation plants are located near waterways and rail transportation networks, and the refuse disposal areas in adjacent areas. Such locations ensure adequate water for cleaning purposes and result in cost savings for transportation of both clean coal and refuse. Unfortunately, with abandonment of mining, the refuse piles inhibit the land development for other uses. In some cases, such as in Illinois and West Virginia, they occupy prime farmland and valuable flat bottomland, respectively. There is an increased public awareness of the need to improve the refuse pile appearances from an aesthetic point of view and of greater need to preserve the value of land for post-mining uses. In the location and design of modern refuse design facilities, these factors are given their due importance.

USES OF COAL REFUSE

Finding uses for the material in the coal refuse piles should be a major objective in planning new mines. If suitable uses can be found, in addition to decreasing the disposal and several associated problems, economic benefits can accrue.

The list of potential uses for the refuse in coal piles is large. Among the several uses tried in a laboratory, pilot-plant or economic scale include the following:

(1) low grade fuel
(2) processing for the recovery of a higher grade fuel
(3) landfill and backfill material
(4) highway construction material
(5) manufacture of building blocks and bricks
(6) "Red dog" production by burning the refuse
(7) raw material for the mineral wool industry
(8) soilless media for growing plants
(9) transporting and stowing in underground mine voids to reduce subsidence
(10) recovery of minerals and trace elements

As a first step, the suitability for a particular use must be determined. The evaluation of the physical and chemical properties of the refuse material in the raw state is necessary. Depending on the properties required for the selected use, the methods of processing the refuse need to be developed. The economics of the refuse utilization program must be established through an evaluation of the costs and benefits.

Among the uses listed above, beneficiation of the refuse pile material for generating a recycled fuel is a viable option, particularly for old refuse piles. The practice is quite widespread in the Anthracite region of Pennsylvania. Though not as common, bituminous coal refuse piles have also been utilized from time to time for secondary recovery of the coal. The waste impoundments for the fines from coal preparation plants are also potential feed sources for fuel briquetting and pelletizing plants (Maneval, 1977). The attractiveness for reworking coal refuse piles and fines impoundments will increase with the correct combination of such conditions as (1) the price of competitive fuels, (2) the technology for using high ash fuels, (3) the costs of mining and processing the coal refuse materials, and (4) incentives (e.g. permits) from local, state, and federal agencies to carry out these operations in an expeditious manner.

The use of the refuse material to satisfy the needs for construction fill material in nearby localities is prevalent. However, only a small fraction of the refuse can be used this way. Other uses such as lightweight aggregate, anti-skid materials and bricks are even less common. In any case, construction projects in localities where there are abandoned coal refuse piles nearby must consider using as much of this material as possible.

The disposal of coal waste in mined-cavities underground is technically feasible and has many attractive features. It has the potential to eliminate all the adverse environmental impacts of surface disposal, and at the same time, to reduce subsidence (Figures 8 and 9). Before this alternative is recommended, several issues must be resolved. Firstly, the resource value in the refuse may be lost forever. Secondly, unless the waste is properly emplaced, the potential for contamination of ground water resources cannot be ruled out. Additional health and safety concerns arising from disposal operations must be evaluated. Finally, careful evaluation of the economics of transporting and stowing refuse in underground mines is necessary (National Academy of Sciences, 1975).

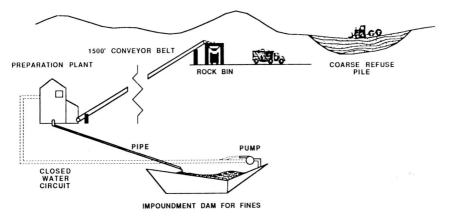

FIGURE 8. Surface disposal of coarse and fine refuse from coal preparation plants.

Old and abandoned coal refuse piles present potential for new uses through reclamation. Reclamation process for these piles is essentially the same as that for any surface mining operation. The steep slopes have to be made gentle; the toxic material needs to be excavated and encapsulated; the pile needs to be stabilized, and sterilized against spontaneous combustion; ground cover has to be established first to prevent erosion and sedimentation; and trees have to be planted later for aesthetic beauty. These reclamation activities can be directed to increase the land values for recreation, housing development and light commercial industry.

There exists enormous amounts of refuse in abandoned piles to be completely utilized for productive purposes. The amount of new refuse generated is far in excess of the needs for various uses outlined before. Therefore, surface disposal of coal refuse will continue to be the practice at least in the near future.

PRESENT STATUS

The 1972 Buffalo Creek disaster led to numerous investigations by government, industry, and consultants to assess the status of coal refuse piles and impoundments in the U.S. The need to develop improved design and engineering manuals for coal refuse disposal facilities became apparent. These studies traced the source of stability and other environmental problems (fires, water pollution etc) to the manner in which the early refuse piles were built up. The process of end-dumping waste led to segregation of coarse particles in the toe of the pile and fines, near the top. This loose pile becomes unstable when saturated

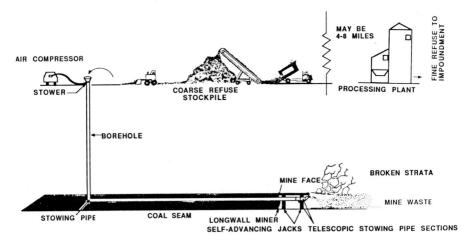

FIGURE 9. Underground disposal of coarse refuse from coal preparation plants.

by heavy rain, ground water and impounded water. Furthermore, conditions for spontaneous combustion are created. To eliminate these problems, several guidelines were developed for coal refuse disposal facilities (E. D'Appolonia Consulting Engineers, Inc., 1974). The guidelines are simple in concept and include the following:

(1) a geotechnical evaluation of the disposal site for ground water, run-off water and foundation suitability;
(2) a detailed evaluation of the physical and chemical properties of the refuse, both when fresh, ashed and weathered;
(3) suitable design analysis and evaluation of foundation, disposal facility configurations, including detailed engineering of the selected design;
(4) layering and compaction of the coal refuse material to eliminate particle segregation, reduce air and water flow, and increase stability;
(5) control of surface and underground water through ditches and subdrains to reduce acid formation and erosion;
(6) monitoring of pile stability and other conditions through measurement of surface movements, internal movements, ground water conditions, water quality, pile pore pressure, temperature, vibrations, etc.;
(7) vegetation of the pile areas where disposal has ceased;
(8) general maintenance through routine inspections for health, safety and environmental conditions;
(9) specific inspection and maintenance after unusual events such as severe frosts and heavy rains; and
(10) long-term maintenance for aesthetic and land value considerations.

With the enactment of the Coal Mine Safety and Health Act of 1969, considerable emphasis has been placed on the safe construction and regular monitoring of refuse piles and impoundments. The provisions with regard to these two surface facilitates have been modified and strengthened since that time (CFR 30, Part 77, Section 77.214 to 77.217). The passage of the Surface Mining Control and Reclamation Act of 1977 also brought along with it requirements on refuse piles and waste disposal, their design, construction, reclamation and abandonment (e.g., CFR 30, Parts 780, 784, 816 and 817). An important feature of the 1977 Surface Mining Act was the collection of a reclamation fee per ton of coal sold from the current operators of coal mines to reclaim abandoned mined land (e.g., CFR 30, Sub-chapter R, Part 870, 872, 874, etc.). Under these provisions, funds are available to state government to reclaim abandoned refuse piles and embankments which pose imminent and serious threats to health, safety, general welfare and environment. As such, it is reasonable to expect that new refuse piles and embankment will be constructed with adequate safeguards against air and water pollution, and health and safety hazards, and the number of old problem piles reclaimed will continue to increase.

REFERENCES

E. D'Appolonia Consulting Engineers, Inc., 1974. Engineering and Design Manual: Coal Refuse Facilities, Pittsburgh, PA, 744 pp.

Glover, H.G., 1971. Coal Mine Refuse Disposal in Great Britain, Special Research Report Number SR-81, Coal Research Section, The Pennsylvania State University, University Park, PA, 1971, 33 pp.

Grape, P.A., Coal Data 1986 Edition, National Coal Association, Washington, DC, 95 pp.

Johnson, W., and G.C. Miller. 1979. Abandoned Coal-Mined Lands: Nature, Extent and Cost of Reclamation, U.S. Bureau of Mines, Washington, DC, 29 pp.

King, L., Coal Data 1977 Edition, National Coal Association, Washington, DC, 90 pp.

Maneval, D.R., 1977. Recent Developments in Reprocessing of Refuse for a Second Yield of Coal, Proceedings Third Symposium on Coal Preparation, National Coal Association, Washington, DC, pp. 152-162.

Moulton, L.K., D.A. Anderson, R.K. Seals and S.M. Hussain, 1974. Coal Refuse: An Engineering Material, Proceedings First Symposium on Mine and Preparation Plant Refuse Disposal, National Cost Association, Washington, DC, pp. 1-25.

National Academy of Sciences, 1975. Underground Disposal of Coal Mine Wastes, National Academy Press, Washington, DC, 172 pp.

Spicer, T.S. and P.T. Luckie, 1970. Operation Anthracite Refuse, Proceedings of the Second Mineral Waste Utilization Symposium, IIT Research Institute, Chicago, IL, 1970, pp. 195-204.

Chapter Nine
ALTERNATIVE RECLAMATION STRATEGIES FOR MINED LANDS
FRED J. BRENNER[1] and RICHARD P. STEINER[2]

[1]Biology Department
Grove City College
Grove City, PA 16127

and

[2]Richard P. Steiner
Department of Mathematical Sciences
University of Akron
Akron, Ohio 44325

Coal has remained a vital energy source in the United States and throughout the world for over two centuries. In the United States, coal was first mined in 1701 along the James River in Virginia. Many of the early operations could be classified as strip, open pit or open cast mines. Exposed coal seams along river banks were exposed with hand tools and the coal mined by pick and shovel. Mechanized mining began around 1825 when mule drawn scrapers were introduced into the coal fields. The amount of coal produced by strip mining has increased continually since 1877 when the Otis steam shovel was first introduced into the coal fields near Pittsburgh, Kansas and today strip mining accounts for over 50 percent of the total United States coal production. The primary reasons for the increase in strip mining are : (1) greater output/man day, (2) lower operating costs, (3) greater recovery rate, and (4) reduction of lost time due to accidents. Strip mining will continue to increase in the United States and elsewhere in the world at the expense of deep mine operations as larger equipment and new techniques are developed to mine the world's coal reserves. Today, coal as deep as 60m (200 ft.) is routinely included in the calculation of minable reserves. The amount of overburden that may be economically profitable per meter of coal has increased from 6:1 in 1946 to 20-30:1 today depending on the quality of the coal.

Prior to and during World War II, little if any regulations concerned with strip mine reclamation existed in the Unite States or elsewhere in the world. Without adequate regulations, strip mining often caused the destruction of land resources, pollution of streams, contamination of ground water aquifers, and other environmental problems. For over 50 years, the lack of concern by government and industry alike resulted in thousands of hectares of unreclaimed mine lands throughout the world.

In the United States, the first surface mining law was passed by the West Virginia legislature in 1937, which provided the basic requirements for land reclamation. In the 1940's and 1950's, numerous laws were enacted by state governments which provided not only reclamation guidelines but also provided for bonding of lands to be mined in an effort to ensure adequate reclamation. Until 1977 when the Federal Surface Mining Control and Reclamation Act (SMCRA) was passed by Congress, considerable variation existed among the coal producing states as to how mined lands were to be reclaimed. SMCRA not only brought about more uniform regulations for the reclamation and bonding of mined lands but also provided for more public input into the permitting process. The enforcement of these regulations is the responsibility of various state agencies once the state meets the primacy requirements set forth by the Office of Surface Mining (OSM). The federal government provides the states with funds to help offset the cost of these additional responsibilities.

With regard to reclamation, current regulations require that the land be returned to its approximate original contour (AOC) or other acceptable alternatives, that spoil horizons and top soil be separated, and that the land be returned to a productivity that is equal to or exceeds its productivity prior to mining. Federal regulations also require that land designated as prime farmland by the USDA Soil Conservation Service be returned to agricultural production. These regulations also encourage the revegetation of mined lands with native species and recommend that wildlife be given prime consideration in reclamation planning, but in reality this does not occur universally throughout the industry. In general, the federal act has resulted in an improvement in postmine land use practices. However, if regulations are too restrictive, they discourage the development of innovative techniques. For this reason, Section 711 of SMCRA allows for a variance from standards on an experimental bases, thereby encouraging alternative mining and reclamation practices. In this chapter, land use alternatives and reclamation practices will be discussed both in light of current regulations as well as development of diverse and stable ecosystems following mining.

SITE PREPARATION

Returning strip mined sites to approximate original contour (AOC) may result in long, gradual slopes and excessive compaction that may accelerate erosion thereby making revegetation difficult. In terms of the amount of erosive soil loss, the length of a slope may have more importance than the degree of the slope (Brenner 1948a,b, 1985; Doyle 1976). Doyle (1976) reported that doubling the length of slope increases soil loss approximately 2.6 times. This suggest that interrupting long reclaimed slopes with the use of water diversions and terracing may be useful in combating erosion on reclaimed lands (Brenner 1984a, B, 1985). In this section we describe models which illustrate potential effects of such soil conservation measures.

Perhaps the most commonly used soil erosion model in the United States is the Universal Soil Loss Equation (USLE) (USDA 1983). This multiplicative model estimates average annual soil loss and has the form

$$A = RKLSCP \qquad [1]$$

where

A = *Soil loss per unit area* in metric tons per hectare (t ha^{-1})
R = *a rainfall and runoff factor* (t m ha^{-1} cm hr^{-1} x 10^{-2})
K = *a soil erodability factor* (t ha^{-1} (t m ha^{-1} cm hr^{-1} x 10^{-2})$^{-1}$
L = *slope length factor* (unit-free; see below)
S = *slope steepness factor* (unit-free; see below)
C = *a cover and management factor* (unit-free; see below)
P = *a support practice factor* (unit-free; see below)

Values of the rainfall and runoff factor (R) may be obtained from the isoerodent map of Wischmeir and Smith (1973). A table of R values for different areas of the country is also available (USDA 1983). To convert the map and the table values to the metric units given above, the map or table R value must be multiplied by 1.735 (Wischmeier and Smith 1973).

Soil erodability factors (K) are tabled for numerous soil series (USDA 1983). The soil erodability factor for surface mine spoils has been reported (Brenner 1984, 1985) as K = 0.020 t ha^{-1} (t m ha^{-1} cm hr^{-1} x 10^{-2})$^{-1}$. Stein *et al* (1983) reported that soil erodability tended to be higher in test plot field calculations for reclaimed soils (K_{calc}) than values predicted (K_{pred}) by a monograph developed by Wischmeier *et al* (1971). We calculated the median difference $K_{calc} - K_{pred}$ as +0.098 (two-tailed sign test, n = 14, p = 0.013). The differences ranged from −0.020 to +0.647 K_{calc} values and from 0.500 to 1.167t ha^{-1} (t m ha^{-1} cm hr^{-1} x 10^{-2})$^{-1}$ for the reclaimed soils at the three Indiana sites studied by Stein *et al* (1983).

The slope length factor (L) is the ratio of soil loss from the reclaimed slope length to the loss from a 22 m length under identical conditions. This factor is defined as

$$L = (\lambda/22)^m \qquad [2]$$

where

λ = slope length in meters
m = 0.5 for slopes 5% or more; 0.4 for slopes 3.5 to 4.5; 0.3 for slopes 1 to 3%; and 0.2 for slopes less than 1%.

The slope steepness factor (S) is the ratio of soil loss from the degree of slope in the reclaimed site to a 9% slope under identical conditions. Wischmeier and Smith (1973) defined this factor as

$$S = 65.41 \sin^2 \theta + 4.56 \sin \theta + 0.065 \quad [3]$$

where θ = angle of slope. If x is defined as the *percent* slope, then

$$\theta = \tan^{-1} (x/100) \quad [4]$$

Both Equations [2] and [3] were developed from cropland data on slopes approximately 9 to 91 m in length and ranging from 3 to 18% slope.

Hahn *et al* (1984) developed a linear slope steepness factor for use in Equation [1]. Their study was based on 22 test plots of reclaimed mine soils in Indiana with slopes ranging from approximately 5 to 16%. Their S factor is

$$S = 0.114 \, X - 0.023$$
$$= 11.4 \tan \theta - 0.023 \text{ where X and } \theta \text{ are as above} \quad [5]$$

An application of this S factor is illustrated in Table 1 (see footnotes c and d).

Factors C and P are also ratios of site soil loss to that from an area tilled continuous fallow and from straight row tilling lengthwise on the slope, respectively (Wischmeier and Smith 1973). The mining and regrading processes effectively destroy the vegetative cover and soil root zone such that C = 1 for bare reclaimed soil (Wischmeier and Smith 1973). Straw or hay mulch applied to bare soil on 6 - 10% slopes no longer than 70 m, at a rate of 4.5 t ha^{-1}, can reduce C to as little as 0.06, provided the mulch is anchored to the soil (Wischmeier and Smith 1973). Regrading to AOC with no support practices such as terracing, contouring (for agricultural land), or water diversions result in value of P = 1 (Brenner 1984 a, 1985).

Brenner (1984 a) suggested that soil loss (T) on reclaimed lands should be maintained at about 5.4 t ha^{-1} or less. In the western Pennsylvania counties where surface coal mining is prevalent (eg. Mercer, Butler), the rainfall erosion index is R = 216.9 t m ha^{-1} cm hr^{-1} x 10^{-2} (Wischmeier and Smith 1973). Using the reported soil erodability factor (Brenner 1984 a, 1985) for surface mine spoils of K = 0.220 t ha^{-1} (t m ha^{-1} cm hr^{-1} x 10^{-2})$^{-1}$, consider a 300 m slope of bare mine spoil in Mercer County, Pennsylvania, returned to AOC with a uniform 7% gradient. The slope length factor, by applying Equation [2] would be L = 3.69. Application of Equation [3] yields a slope steepness factor of S = 0.702. The USLE (Equation [1]) estimates a mean annual soil loss of 123.8 t ha^{-1}, well above the recommended value of 5.4 t ha^{-1}. For slopes longer than 137.2 m it has been suggested (USDA 1983) that a better estimate of soil loss is to (a) compute A using half the actual slope length for λ in Equation [2], then (b) multiply the resulting A value by 1.5 to get the final adjusted mean annual soil loss. When this is done for the above example, an estimated mean annual soil loss of 131.1 t ha^{-1} is obtained. (See Table 1 footnotes a and b).

TABLE 1

Examples of mean annual soil loss estimated by the Universal Soil Loss Equation for some hypothetical uniform slopes. For all slopes $R = 216.9$ t m ha^{-1} cm hr^{-1} x 10^{-2}, $K = 0.220$ t ha^{-1} (t m ha^{-1} cm hr^{-1} x $10^{-2})^{-1}$, and $P = 1$

Length (λ)	Slope X	θ	L	S	Soil Loss (A, t ha^{-1})			
					$C = 1$	$C = 0.331$	$C = 0.141$	$C = 0.004$
300 m	7%	4.0°	3.69	0.702	123.8	41.0	17.5	0.5
			(2.61)[a]		(131.1)[b]	(43.4)[b]	(18.5)[b]	(0.5)[b]
90 m	7%	4.0°	2.02	0.702	67.8	22.4	9.6	0.3
90 m	15%	8.5°	2.02	2.181	210.2	69.6	29.6	0.8
				(1.687)[c]	(162.6)[d]	(53.8)[d]	(22.9)[d]	(0.7)[d]

[a] This is L based on 0.5 = 150 m
[b] These are the soil losses for A values (based on L = 2.61) multiplied by 1.5.
[c] This is the slope steepness factor of Hahn et al (1985), S = 0.114 X − 0.023.
[d] These are soil losses based on S = 1.687.

Terracing produces a site composed of a series of differing slope gradients. The USLE assumes a uniform gradient along the entire length of the slope. Irregular slopes may be handled by dividing them into segments (indexed by i) of a fairly uniform slope and computing L£i and S£i values separately for each segment (Wischmeier and Smith 1973). It is assumed that the change in gradient from segment to segment is not sufficient to cause soil depostion. Then, if the segments are of equal length, LS for the entire slope may be computed as:

$$LS = \sum_{i=1}^{N} L_i S_i f_i \quad [6]$$

where $f = \dfrac{i^{m+1} + (i-1)^{m+1}}{N^{m+1}}$

i = segment sequence index (segment 1 is at slope top)
N = total number of segments
m is defined as in Equation [2]

The product LS is called the topographic factor by Wischmeier and Smith (1973).

Use of Equation [6] is necessary in the calculation of mean annual soil loss for an entire reclaimed site when terracing (or other gradient irregularity) is present and runoff flows from the backslope to the frontslope within a terrace, and from terrace to terrace.

When inverted terraces with water diversion channels or underground outlets between the backslopes and frontslopes are constructed, use of Equation [6] becomes unnecessary. For the remainder of this section the work "terrace" shall refer to this type of structure. Frontslopes and backslopes of all terraces may be considered independently. Mean annual soil loss (A_i) may be computed using the USLE separately for each segment and the A_i values summed to yield total soil loss for the site. Spacing of such terraces may be accomplished by rearranging Equation [1], the USLE as follows and replacing A with T, a soil loss tolerance, (Brenner 1984 a, b, Wischmeier and Smith 1973).

$$LS = \frac{T}{RKCP} \qquad [7]$$

Wischmeier and Smith (1973) discuss how the LS factor may be used to obtain guides for the horizontal spacing and vertical spacing of terraces. Their discussion centers on points particularly pertinent if the intended use of the reclaimed land is agricultural.

In general, the use of Equation [7] in the planning of inverted terraces with diversions to maintain a specified soil loss tolerance (T) is quite flexible. Once LS is calculated, a series of adjacent terraces may be constructed with frontslopes and backslopes having any convenient gradient and length combinations that yield an LS value less than or equal to that calculated initially by Equation [7]. For example, applying Equation [7] to the hypothetical 300 m, 7% slope in Mercer County, PA with T = 5.4 t ha^{-1} yields LS = 0.113. Combining Equations [2] and [3] shows LS in terms of slope length (λ) and slope gradient (θ)

$$LS = (\lambda/22)^m \; 65.41 \sin^2 \theta + 4.56 \sin \theta + 0.065 \qquad [8]$$

So,

$$0.113 = (\lambda/22)^{0.5} \; 65.41 \sin^2 \theta + 4.56 \sin \theta + 0.065 \qquad [9]$$

since m = 0.5 for a 7% slope. Thus by selecting a desired length, Equation [9] may be solved for the necessary gradient, θ. Conversely, by selecting θ the required gradient (λ) may be found. LS values for some selected C $\leq$ 1 are given in Table 2.

Vegetative cover can dramatically reduce soil loss. This is modeled in the USLE through the cover and management factor, C. Establishment of vegetation on surface mines is directly related to soil organic content and soil moisture (Brenner and Goughler 1983). Reclamation efforts should strive to enhance these characteristics early in the initial process (Brenner *et al* 1985). Brenner and

TABLE 2

Topographic factors (LS) for various C values and a soil loss tolerance T = 5.4 t ha^{-1}. For all calculations R = 216.9 t m ha^{-1} cm hr^{-1} x 10^{-2}, K = 0.220 t ha^{-1} (t m ha^{-1} cm hr^{-1} x 10^{-2})$^{-1}$, and P = 1.

C	LS
1.000	0.113
0.331	0.342
0.141	0.803
0.004	28.3

Goughler (1983) have shown that sorghum (*Sorghum vulgare*) appears to be an excellent cover because it provides shade and mulch, promoting moisture conservation and increasing organic matter. An average increase of 12 g per 100 g soil was reported in the second year. In the same study, oats (*Avena sativia*) did not show an increase in soil organic matter. Sorghum, then, tends to stabilize the site while providing suitable habitat for the invasion of native species (Brenner *et al*, 1984).

Next we consider some cover and management factors (C) that may be useful when considering soil loss in light of revegetation efforts. These C values must be considered approximate, as they were developed for agricultural cropping systems, and not reclaimed mine soils. For a legume and grass mixture C = 0.004 (1) (Brenner 1984 a, 1985), Wischmeier and Smith (1973), suggest developing C values for sorghum based on the same soil loss ratios as corn. Two management sequences are presented for initial vegetative reclamation involving sorghum. Both assume spring plowing with conventional tillage and are actually derived for corn in the coal district of western Pennsylvania (ecological area C—17) (USDA 1983). Some examples of soil losses with these C values are given in Table 1.

Reclamation practices suggested by the USLE and it applications should only be considered as guides for designing strategies aimed at reducing soil losses and enhancing revegetation of surface mined sites. Other soil factors affecting these goals and the total ecological reclamation are lack of bonding between replaced spoil horizons, top soil replacement during very wet or dry conditions, and possible changes in the biological and chemical properties of top soil due to lengthy storage (Brenner 1984 a,b).

There is no simple formula for reconstructing soil that applies to all mined lands (Jansen 1981). Soils are complex entities whose compositions vary with climate and previous land use management systems. Each site will have a unique set of materials available for soil reconstruction. Soil characteristics that may be desirable for one land use, crop land, tree production, etc. may not be applicable to another land use category. A soil must be established that is suitable for the intended land use. Usually, surface material will be the best available, but on some sites deeper horizons are as good or better than the A horizon, depending on the intended land use. In all cases, soil material must be replaced in a manner that will avoid excessive compaction. And finally, after reconstruction, the soil should be managed to maximize the rate of soil improvement over time that yields the minimum productivity for the intended land use.

Although current regulations require that mine sites be returned to approximate original contour (AOC) and that soil horizons be separated and replaced in their order of removal, these procedures do not ensure adequate reclamation on mined lands. Several authors (Brenner 1984 a, 1985, Stein *et al* 1983, Hahn *et al* 1985) proposed the application of the soil loss equation $A = R K L S C P$ and the terrace spacing equation $V = X S + Y$ to predicate soil loss and number of diversions or terraces necessary to reduce soil loss on mined lands. Stein *et al* (1983) found that the soil loss equation overestimated erosion on steeper slopes. These authors also found that soil loss was greater on slopes constructed on A horizon slopes than it was on those constructed of B horizon soils.

The cost of reclaiming mined lands under current regulations has been estimated at between $9,000 and $16,000/ha with the majority of the cost involved in backfilling and recontouring the site (Brenner 1984 a,b, 1985). These costs may be excessive, however, since backfilling and recontouring are part of the continuous mine operation and special equipment is generally not purchased for this phase of the mine operation. For this reason, as well as site to site variations, it is difficult to obtain an exact cost analysis of each phase involved in mine reclamation.

REVEGETATION ACCORDING TO LAND USE CATEGORIES

Reclaimed mined lands may be divided into four broad land use categories: (1) agroecosystems, (2) natural grasslands, (3) forest lands, and (4) wetlands (Brenner 1985). It is possible and even desirable to have a combination of several land use categories on any given mine site. The degree of management required to maintain these systems will depend on the ultimate land use objective for the site (Brenner 1984 a, 1985).

Agroecosystems
As stated previously, mined lands that were classified as prime farmlands must be returned to a state where production is equal to or exceeds the pre-mined condition. The term agricultural land has been used loosely for years to refer to lands that are highly productive and well suited to agricultural use (Jansen 1981). Moreover, the concept of what qualifies as prime farmland varies from one geographical area to another or from one agency to another. In SMCRC (PL 95-87), prime farmland must meet minimum soil property conditions (federal Register 1978) but these conditions pertain to a variety of different soil types which may be subject to interpretations depending on local conditions.

In the midwest, studies have shown that through the use of soil conditioners, forage crops and irrigation (Dunbar *et al* 1982) yields can be obtained on mined lands as good or better than before mining. Corn planted on reclaimed mines the same year as soil reconstruction averaged 700-893.5 Kg/ha which was between 76 and 97 percent of the yield from pre-mine soils under extensive management but exceeded the yield from soils under a basic management level (Kleinman and Layton 1979). Based on a yield of 892.5 Kg/ha, the net income was $955/ha. At an estimated reclamation cost of $9000-$16,000/ha, it could take between 9.5 and 16.7 years for the agricultural income to equal or exceed the reclamation cost.

As with row crops, where overburden conditions are favorable, productivity of pasture and forage on mined lands can be equivalent or exceed that of unusual soils (Kleinman and Taylor 1979). On unfertilized 100 year old mine spoils in the Appalachians, the nutritive quality, species diversity, and biomass were superior to those of unmined sites (Smith et al 1971). In the midwest, when fertilizer was applied to forage lands, the total forage production the first year on mined lands was twice that of native soils the second year. Again in the midwest, there was not a significant difference in the weight gain of cattle grazing on mined and unmined sites, respectively (Grant and Tang 1958).

Agriculture is a valuable land use alternative for mined lands, however, when the average income from farming of $738/ha is taken into consideration, it will take between 12-22 years for agriculture income to equal or exceed the cost of reclamation (Brenner 1984 a, 1985). Another economic consideration is that mountain top removal and valley fill mining in mountainous regions has created extensive nearly level areas which did not exist before mining, producing new potential for agricultural or alternative land uses.

Grassland Ecosystems

Natural grassland communities occur in many areas of the coal mining regions of the world. These ecosystems are in various succession stages of climax communities and hence, the revegetation regime may be designed to duplicate as closely as possible the successional stage that existed on the site prior to mining. This involves careful consideration and selection of species used to revegetate the site, seeding rates, as well as a management plan that insures that the system remains in the desired successional stage. We must remember, however, that succession is a natural occurence in all ecosystems and mined lands will eventually return to the climax stage for that region unless they are managed accordingly.

Unfortunately, in the United States and possibly elsewhere, exotic rather than native species are often used in the reclamation of mined lands in the eastern United States, only 6 are native to the region (Brenner 1984 b). The usual practice is to seed a mixture of two or three species of grasses (rye, *Lolium perenne*; fescue, *Festuca elatior*) and legumes (clovers *Trifolium* spp, birdsfoot trefoil, *Lotus corniculatus*) with annual rye, oats or winter wheat as a nurse crop. These areas, therefore, often lack the diversity and stability that existed prior to mining and hence, should be classified as an immature ecosystem characterized by high productivity and low diversity.

It appears therefore, that the succession of both native and exotic species is an important component in the development of diverse and stable grassland communities on mined lands. Although some opportunistic species that commonly invade mine sites, such as ox-eye daisy (*Chrysanthemum leucanthemum*) and foxtail millet (*Scleria italica*) were introduced into the United States from Eurasia, localized populations or demes may be genetically distinct and hence, might be considered as part of the native flora (Brenner 1984 b). Natural succession increases the diversity of plant communities, of the 28 species found on mine sites in Pennsylvania, 82 percent were volunteers and of these 80 percent were native to the region (Brenner 1984a, Brenner *et al*, 1984). Based on these results, it appears that by either planting native species or by the encouragement of natural succession that diverse and stable plant communities will be established in a shorter time frame than might occur if only grass and legume mixtures were used to revegetate mined lands.

On the premise that succession is a phenomenon that occurs on all disturbed lands, it may be possible to manage succession to enhance the reclamation process. If the ultimate objective is to create a diverse and stable ecosystem, then the initial reclamation should be designed to stabilize the site while at the same time encouraging natural succession. On test plots in Pennsylvania, the number of volunteer plants (12141) was 50 percent greater when sorghum (*Sorghum vulgare*) was used rather than rye or oats as a nurse crop (Brenner and Gougher, 1985). Natural succession may also be encouraged by the development of terraces and ledges along the outer slopes which will hold seeds and fertilizer during the initial seeding as well as retaining moisture and organic matter, thereby providing a seed bed for invading species. This procedure, however, may require a modification of the required practice of regrading to AOC (Brenner 1984 a,b, 1985).

In addition to providing a greater diversity and stability to mined lands, volunteer species are also used to a greater extent as food and cover for wildlife than are the exotic species used in reclamation. The development of natural grasslands in forested areas has been shown to be beneficial to species such as elk (*Cervus canadensis*) as well as increasing the amount of nesting habitat for bobolinks (*Dolichonyx oryzivorus*), upland Sandpipers (*Bartramia longicauda*) and other grassland species (Brenner, 1986).

Forested Ecosystems

In the United States, the earliest reclamation efforts involved tree planting but little and conflicting data exists on timber growth on mined lands primarily because these lands have not been consistently managed or monitored (Ashby, 1978; Brissler, *et. al.*, 1984; Byrnes, *et. al.*, 1984). Studies in the eastern and midwestern United States have shown that the growth rates of both conifers and decidious species are equal to or exceed those on unmined sites. On the other hand, the growth of red pine (*Pinus resinosa*) on acid soil in Pennsylvania was significantly less than that on an undisturbed site (Aharrah and Hartman, 1973).

As with grasslands, natural succession is an important factor in increasing the diversity of forest communities on mined lands. Of the 34 species identified on 82 reclaimed mines in Pennsylvania, 22 were native species that volunteered on the sites (Brenner *et. al.*, 1984 b, Brenner 1983). Opportunistic species such as aspen (*Populis spp.*), red maple (*Acer rubrum*) and black cherry (*Prunus serotina*) invade these sites within 1-3 years followed by oaks (*Quercus spp.*), hickories (*Carya spp.*) and other hardwoods. Volunteer species also provide food and cover for wildlife including the ruffed grouse (*Bonasa umbellus*) (Brenner and Michalski, 1984), wild turkey (*Meleagris gallopavo*), white tailed deer (*Odocoileus virginianus*) (Brenner, *et. al.*, 1984) and numerous passerine species (Brenner and Kelly, 1981; Brooks, *et. al.,* 1985).

Current reclamation practices, as stipulated by the regulations, are detrimental to the establishment of forest ecosystems on mined lands (Ashby, 1978). Recontouring to AOC often results in soil compaction reducing permeability and water holding capacity, thereby making the site less favorable for tree establishment. Dense herbaceous cover has also been shown to be detrimental to tree growth and survival on mined lands. The survival and growth of tree seedlings can, however, be increased by selective herbicide treatment prior to and after planting (Byrnes, *et. al.,* 1984). Treatment may either be as alternating grass and tree strips or a spot treatment of individual seedlings. This procedure allows for the stabilization of the site with grasses and legumes, followed by secondary reclamation with trees and shrubs. Direct seeding of pines, hardwoods, and shrubs either simultaneously with grasses and legumes or as a secondary reclamation has been shown to be an effective method of tree establishment on mined lands (Richards and Graves, 1984). Planting trees and shrubs along contours, diversions and terraces as hedgerows between grass strips has been suggested as a way of enhancing mined lands for wildlife (Brenner, 1984 b, 1985).

It is difficult to compare the economic return between forest, grassland and agroecosystems since it may be 20-30 years before an economic return can be realized from forest lands. The cost of preparing the site for tree planting, however, may be significantly less than when returning the site to agricultural production because less contouring may be required. Pulpwood production from reclaimed strip sites has been estimated at approximately $378/ha but these sites had poor soil conditions and were not managed for either timber or pulpwood (Misiolek and Bailey, 1980). The establishment of forested ecosystems on mined lands should be based on long term ecological benefits rather than short term economic gain.

Wetland ecosystems

Shallow and deep water wetlands have been developed on mined lands for over five decades. These ecosystems have been shown to be productive both in terms of carbon fixation (Brenner, *et.al.,* 1985) and as a potential fishery source. Wildlife habitat especially for waterfowl on these wetlands may be improved by the installation of artificial nesting structures and/or islands and the plant-

TABLE 3

Wetland species recommended for planting on mitigated wetlands in the eastern and midwestern United States.

1. Wet Soils
    A. Herbaceous Species
        Lovegrass—*Eragrostis reptans*
        Orchard Grass—*Dactylis glomerta*
        Fescue—*Festuca elatior*
    B. Trees and Shrubs
        Lespedeza—*Lespedeza* spp.
        Dogwoods—*Cornus* spp.
        Smooth Alder—*Alnus rugosa*
        Red Maple—*Acer rubrum*
        Pin Oak—*Quercus palustris*
        Red Oak—*Quercus rubra*
2. Intermediate Zone—periodic flooding
    A. Herbaceous Species
        Millet—*Echinocloa* spp.
        Switch Grass—*Panicum virgatum*
        Cord Grass—*Spartina pectinata*
        Nut Sedge—*Cyperus* spp.
        Reed Canary Grass—*Phalaris arundinacea*
    B. Trees and Shrubs
        River Birch—*Betula nigra*
        Pin Oak—*Quercus palustris*
        Alder—*Alnus rugosa*
        Buttonbush—*Cephalanthus occidentalis*
3. Shallow Water Zone
    A. Maximum Depth—15 cm (6 inches)
        Herbaceous Species
            Millet—*Echinochloa* spp.
            Rice Cutgrass—*Leersia* spp.
            Spike Rush—*Eleocharis* spp.
            American Lotus—*Nelumbo lutea*
        Trees and Shrubs
            Button bush—*Cephalanthus occidentalis*
            Bald Cypress—*Taxodium distichum*
    B. Maximum Depth—25 cm (10 inches)
        Herbaceous Species
            Burreed—*Spargariaceae* spp.
            Smartweed—*Polygonum* spp.
            Spike Rush—*Eleocharis* spp.
            Cattail—*Typha latifolia*
            Arrowhead—*Sagittaria* spp.
            Hardstem Bulrush—*Scirpus acutus*
        Trees and Shrubs
            Bald Cypress—*Taxodium distichum*
    C. Maximum Depth—45 cm (18 inches)
        Herbaceous Species
            Arrowhead—*Sagittaris* spp.

                Wild Rice—*Zizania aquatica*
                Hardstem Bulrush—*Scirpus acutus*
                Pickerel Weed—*Pontederia* spp.
    4. Deep Water Zone
        A. Depth 0.3-0.6 m (1-2 ft.)
                Arrowhead—*Sagittaria* spp.
                Yellowwater Lilly—*Nuphar* spp.
        B. Depth 0.3-1 m (2-3 ft.)
                Sago Pondweed—*Potamogeton pectinatus*
                Wild Celery—*Vallisneria americana*
        C. Depth 1-2 m (3-6 ft.)
                Coontail—*Ceratophyilum demersum*
                Elodea—*Anacharis canadensis*
                Muskgrass—*Characeae* spp.
                Pondweed—*Potamogeton* spp.
                Water Millfoil—*Myriphyllum* spp.

ing of wildlife food and cover species (Brenner, 1973; Brenner and Mondok, 1978) along the shores and adjacent upland areas. In addition to the development of wetlands during reclamation, if sedimentation and treatment ponds are allowed to remain in place after mining; the area will be enhanced for wildlife.

At several mine sites in the midwest, slurry ponds have been reclaimed as productive wildlife areas. (Nawrot and Yarch, 1982; Thompson, 1984). Slurry consists of a mixture of water and the fines that remain after the coal has been cleaned; this mixture is then discharged into a depression for storage. Because of its pyritic sulfur content and acid producing potential, slurry is considered a potentially toxic substance by OSM and must be disposed of in a prescribed manner. Current regulations require that slurry be covered with at least 1.2m (4 ft.) of non-toxic material and revegetated. This procedure is not only costly to the mine operator but also generally results in land that has limited potential for secondary reclamation.

As a alternative to the prescribed method of slurry pond reclamation, researchers at Southern Illinois University and AMAX Coal Company are developing viable wetlands on slurry disposal sites. The company received an Experimental Practices Permit under Section 711 of SMCRA which allows them to develop an alternative practice for slurry pond reclamation. Briefly, the procedure involves the determination of highly toxic materials and covering them with top soil according to OSM regulations; then the remainder of the ponds are revegetated with wetland species. Species are selected and planted according to the water depth and available moisture in each zone (Table 3). This alternative to slurry pond reclamation enabled the development of a viable wetland with the pH of both the slurry and water returning to acceptable pH ranges. Reclaimed slurry ponds are currently being used by migrating and nesting Canada Geese (*Branata canadensis*), blue and green-winged teal (*Anas discors*,

A. carolinensis), great blue herons (*Andea herodias*) and a variety of shore birds. The deep water portions are supporting a viable sport fishery (Thompson, 1984). In addition to developing a valuable wildlife habitat, the company was able to reduce their reclamation costs by approximately 80 percent (Thompson, 1984). These procedures should be given serious consideration by regulatory agencies and the coal industry as an alternative to slurry pond reclamation.

In addition to their value as fish and wildlife habitat, wetlands allow for the recharge of ground water, act as nutrient and sediment traps and provide for future water supplies and recreational resources (Brenner, 1984 b; Rosso, 1980). Wetlands of all types should be encouraged by regulatory agencies and the regulations changed whenever necessary to permit their construction.

LITERATURE CITED

Aharrah, E.C. and R.T. Hartman. 1973. Survival and Growth of Red Pine on Coal Mine Spoil and Undevastated Soil in Western Pennsylvania. pp. 429-443. R.J. Hutnik and G. Davis (eds).*Ecology and Reclamation of Devastated Land*. Gordon and Breach, N.Y. Vol.

Ashby, W.C., C. Kolar, M.L. Guerke, C.F. Pursell, and J. Ashby. 1978. Our Reclamation Future: The Missing Bet on Trees. Document No. 78/04 Illinois Institute for Environmental Quality, Chicago.

Brenner, F.J. 1973. Evaluation of Abandoned Strip Mines as Fish and Wildlife Habitats. *Trans. N.E. Wildlife Conf.* 30: 205-229.

_____. 1974. Ecology and Productivity of Strip-mine Areas in Mercer County, Pennsylvania. Research Technical Completion Report. A-0290 PA. Institute for Research on Land and Water Resources. The Pennsylvania State University. 70 pp.

_____. 1983. Environmental Aspects of Coal Production in Pennsylvania: A Guide to Reclamation. pp. 405-422. In. S.K. Majumdar and E.W. Miller (eds). *Pennsylvania Coal: Resources Technology and Utilization*. Penn. Acad. Sci. 594 pp.

_____. 1984a. Management Strategies and Ecological Approaches to Facilities Reclamation on Mined Lands. *Strategies for Environmentally Sound Development in Mining and Energy Industries*.

_____. 1984b. Restoration of Natural Ecosystems on Surface Mined Lands in Northeastern United States. pp. 211-225. In T. N. Vezoroglu (ed). *The Biosphere: Problems and Solutions*. Elsevier Science Publishers. Amsterdam.

_____. 1985. Land Reclamation After Strip Coal Mining in the United States. Mining Magazine. September pp. 211-219.

_____. 1986. Habitat Preservation and Development for Rare and Endangered Species Management. pp. 37-52. In S.K. Majumdar, F.J. Brenner, A.F. Rhoads (eds). *Endangered and Threatened Species Programs in*

Pennsylvania and Other States: Causes, Issues and Management. Penn. Acad. Sci. 519 pp.

_____, M.J. Masaus, and W. Granger. 1977. Selectivity of Browse Species by White-tailed Deer on Strip Mined Lands in Mercer County, Pennsylvania. Proceed. Penn. Acad. Sci. 51: 105-108.

_____, and J.J. Mondok. 1978. Waterfowl Nesting Rafts Designed for Fluctuating Water Levels. *J. Wildl. Manage.* 43: 978-982.

_____, and J. Kelly. 1981. Characteristics of Bird Communities on Surface Mined Lands in Pennsylvania. *Environmental Management.* 5: 441-449.

_____, and Goughler. 1983. Role of Nurse Crops in Encouraging Establishment of Natives on Reclaimed Land: An Evaluation (Pennsylvania). Restoration and Management Notes. 1: 31-32.

_____, and S. Michalski, III. 1984. Evaluation of Surface Coal Mines as Ruffed Grouse Habitat. Better Reclamation With Trees Conference. 4: 87-100.

_____, M. Warner and J. Pike. 1984. Ecosystem Development and Natural Succession in Surface Coal Mine Reclamation. *Minerals and Environment.* 6: 10-22.

_____, W. Snyder, J.F. Schalles, J.P. Miller, and C. Miller. 1985. Primary Productivity of Deep Water Habitats on Reclaimed Mined Lands. pp. 199-209. In R.P. Brooks, D.E. Samuel, J.B. Hill (eds). *Wetlands and Water Management on Mined Lands*, Pennsylvania State University. 393 pp.

Brooks, R.P., J.B. Hill, F.J. Brenner and S. Capets. 1985. Wildlife Use of Coal Mines in Western Pennsylvania. pp. 327-352. In R.P. Brooks, D.E. Samuel, J.B. Hill (eds). *Wetlands and Water Management on Mined Lands*. Pennsylvania State University. 393 pp.

Bussler, B.H., W.R. Byrnes, P.E. Pope and W.R. Chaney. 1984. Properties of Mine Soil Reclaimed For Forest Land Use. *Soil Science Soc Amer.* 4: 178-184.

Byrnes, W.R., P.E. Pope, W.R. Chaney, B.H. Bussler, and C.R. Anderson. 1984. Ground Cover Control Essential for Successful Establishment of Hardwood Seedlings on Reclaimed Mined Lands. Better Reclamation with Trees Conference. 4: 22-32.

Doyle, W.S. 1976. Stripmining of Coal: Environmental Solutions. Noyes, Park Ridge, N.J. pp. 43-73.

Dunker, R.E., I.J. Jansen, and M.D. Thorne. 1982. Corn Response to Irrigation on Surface Mine Lands in Illinois. Agronomy Journal. 74: 411-414.

Federal Register. 1978. Prime and Unique Farmlands. Dec. 657.5 Jan. 31.

Grandt, A.F. and A.F. Lang. 1958. Reclaiming Illinois Strip Coal Land with Legumes and Grasses. Agricultural Experiment Station Bull. 628 University Illinois, Urbana.

Hahn, D.T., W.C. Moldenhauer and C.B. Roth. 1985. Slope Gradient Effect on Erosion on Reclaimed Soil. Trans. ASAE 28: 805-808.

Jansen, I.J. 1981. Reconstructing Soils After Surface Mining of Prime Agricultural Land. Mining Engineering March 1981. 3 pp.

Kleinman, L.H. and D.E. Layton. 1979. Reclamation Techniques and Vegetation Response of Lecker-Coal. *Proceed. Symposium Surface Coal Mining and reclamation.* McGraw-Hill Mining Information Service. New York. pp. 255-590.

Misiolek, W.S. and J.E. Bailey. 1980. Analysis of Bituminous Coal Production Costs in the Warrior Coal Field: Economic Analysis of the 1977 Reclamation Law. University of Alabama.

Nawrot, J.R. and S.C. Yarch. 1982. Slurry Discharge Management for Wetland Soil Development. pp. 11-18. In *Symposium on Surface Hydrology, Sedimentology and Reclamation.* University of Kentucky. Lexington, Kentucky.

Richards, T.W. and R. H. Graves. 1980. Preliminary Experimentation in Mechanically Planting Large Seeded Tree Species. pp. 413-415. In *Symposium on Surface Mining Hydrology, Sedimentology and Reclamation.* University of Kentucky, Lexington, Kentucky.

Rosso, W.A. 1980. Reclaim Your Lands With Lakes. Coal Age Magazine. Aug. 1980. pp. 76-77.

Stein, O.R., C.B. Rother., W.C. Moldenhouser and D.T. Hahn. 1983. Erodability of Selected Indiana Reclaimed Strip Mine Soils. pp. 101-106. *Symposium on Surface Mining, Hydrology, Sedimentology, and Reclamation.* University of Kentucky, Lexington, Kentucky.

Smith, R.M., E.H. Tyron and E.H. Tyner. 1971. Soil Development on Mine Spoil. Bull. 6047. West Virginia Agricultural Experiment Station. Morgantown, West Virginia.

Thompson, C.S. 1984. Experimental Practices in Surface Coal Mining—Creating Wetland Habitat. National Wetlands Newsletter. Mar.-April 1984.

United States Department of Agriculture. 1983. Soil Conservation Technical Guide. U.S. Government Printing Office. Washington, D.C.

Wischmeier, W.H., C.B. Johnson and B.V. Cross. 1971. A Soil Erodability Nomograph for Farmland and Construction Sites. *J. Soil and Water Conservation.* 26: 189-193.

Smith, D.D. 1973. Predicting Rainfall Erosion Losses—A Guide to Conservation Planning. USDA Agriculture Handbook No. 537.

Environmental Consequences of Energy Production: Problems and Prospects. Edited by S. K. Majumdar, F. J. Brenner and E. W. Miller. © 1987, The Pennsylvania Academy of Science.

Chapter Ten
ACID MINE DRAINAGE
L. BARRY PHELPS
Associate Professor of Mining Engineering
College of Earth and Mineral Sciences
The Pennsylvania State University
University Park, PA 16802

All mining today must undergo a permitting process prior to the commencement of operations. It is during this process that an evaluation is made of the potential for degradation of surface and ground water. Premining sampling and monitoring of existing waters is required to establish a base line to make an assessment of any changes.

Thus water quality which is affected by the mining process can be monitored. Changes in quality are made up of both short- and long-term effects. Short-term effects can generally be placed into two categories, those associated with physical parameters and generally limited to total suspended solids and those associated with chemical parameters. Long-term effects are generally limited to those associated with the chemical parameters. Short-term effects deal with the life of the mine; long-term effects extend to time beyond the mine life. Short-term effects are dealt with through the use of sedimentation and treatment ponds. Both of these structures should be able to be removed following the completion of mining. Long-term effects are dealt with through a knowledge of the acid/base accounting of the overburden, selective placement of toxic material, alkaline addition and spoil water management. It is the short- and long-term chemical effects of mining dealing with the formation of acid mine drainage (AMD) that is the subject of this chapter.

FORMATION AND CONTROL MECHANISM

Acid mine drainage is one of the components of coal mine drainage. A suggested range of major, minor and trace values for post coal mining is shown in Table 1. It is produced by the oxidation of pyrite (FeS_2). Pyrite is present in varying quantities in most coal, the adjacent rock strata and sometimes present in the rock strata overlying the coal (overburden) and in the rock strata between coal seams (interburden).

Pyrite, when it is exposed to water (H_2O) and air (O_2), reacts to form ferrous sulfate ($FeSO_4$) and sulfuric acid (H_2SO_4).

$$2FeS_2 + 2H_2O + 7O_2 \rightarrow 2FeSO_4 + 2H_2SO_4 \qquad (1)$$

In this environment, the ferrous sulfate reacts with the formed sulfuric acid and additional air to form ferric sulfate ($Fe_2(SO_4)_3$) and water.

TABLE 1

Ranges of Component Concentrations in Appalachian Coal Mine Drainage (Lovell, 1982).

Major Components	Suggested Ranges (mg/l)
Hydrogen ions	
pH	1.4 - 7.0
acidity, as $CaCO_3$	0 - 45,000
alkalinity, as $CaCO_3$	0 - 400
Soluble iron-total	1 - 10,000
Fe^{+2}	1 - 10,000
Fe^{+3}	1 - 1,000
Sulfate	1 - 20,000
Aluminum	1 - 2,000
Calcium	1 - 500
Magnesium	1 - 200
Total dissolved solids	5 - 10,000
Suspended solids	to 3,000
ferric oxyhydroxides, clays, silica and organics as: coal fines, sewage particles, plant components as algae.	
Minor Components	
Sodium	to 100
Potassium	to 50
Silica	to 100
Chloride	to 200
Trace Components	
Mn	to 50
Ni	to 5
V	to 2
Zn	to 10
Sr,Ba,Ti	to 5
Cd,Be,Cu,Ag,Co,Pb,Cr	<1
Hg	<0.05

$$4FeSO_4 + O_2 + 2H_2SO_4 \rightarrow 2Fe_2(SO_4)_3 + 2H_2O \qquad (2)$$

The two formed products then hydrolyze to form ferric hydroxide ($Fe(OH)_3$) and sulfuric acid.

$$Fe_2(SO_4)_3 + 6H_2O \rightarrow 2Fe(OH)_3 + 3H_2SO_4 \qquad (3)$$

Ferric hydroxide is an insoluble precipitate known as "yellow boy." It is the reddish-brown coating noted on exposed coal and stream bottoms affected by AMD. Finally, the formed ferric sulfate reacts with more pyrite and water to form greatly increased quantities of ferrous sulfate and sulfuric acid.

$$FeS_2 + 7Fe_2(SO_4) + H_2O \rightarrow 15FeSO_4 + 8H_2SO_4 \qquad (4)$$

Of the two pyrite reactions, this second reaction is considered to be much faster than the first (Lovell, 1982).

The interaction between various parameters relating to chemical, biological and geological factors in the formation of AMD is extremely complex (Kim et al., 1982). However, the proceeding set of chemical equations can be used to determine the stoichiometric relationships in the formation of AMD.

The procedure for preventing AMD is to keep the pyrite liberated during the mining process in an alkaline environment. The principal material used for longer term control is limestone, principally calcium carbonate ($CaCO_3$),

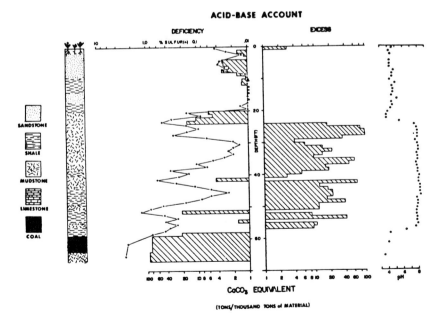

FIGURE 1. Acid-base account and rock type of the overburden above an Upper Freeport coal seam (Smith et al., 1976).

because it is the most readily available rock which can be used as a neutralizing agent. This is shown by the following reaction:

$$CaCO_3 + H_2SO_4 \rightarrow CaSO_4 + H_2O + CO_2 \qquad (5)$$

The products of formation are calcium sulfate, a naturally occurring precipitate anhydrite, water and carbon dioxide (CO_2) gas.

With rising pH, the ferrous iron is removed by aeration in treatment ponds to change the iron to insoluble ferric hydroxides according to the formula:

$$Fe_2(SO_4)_3 + 3CaCO_3 + 3H_2O \rightarrow 2Fe(OH)_3 + 3CaSO_4 + 3CO_2 \qquad (6)$$

For short-term control and rapid neutralization, lime (CaO), the product of limestone, when heated, is used because it is more reactive due to its chemical state and particle size. This latter characteristic is a result of the limestone disintegrating during heating and its effect is illustrated by an example. A cube

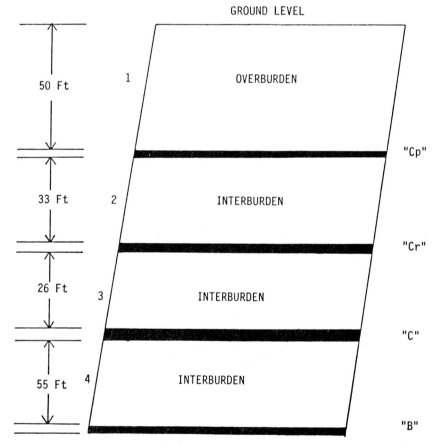

FIGURE 2. Mining base case showing a column section of four coal seams to be mined (Pfaff, 1987).

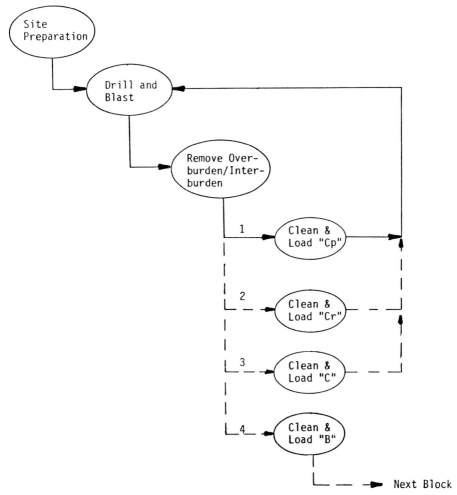

FIGURE 3. Sequence of operational procedures for the mining base case with no toxic materials (Pfaff, 1987).

with an edge of one centimeter has a surface area of six square centimeters. If this cube is powdered such as happens when limestone is heated to form lime, cubes, each with, for example, a side of 0.01 millimeters are formed. Thus, the total surface area available for reactions becomes 6000 square centimeters and the number of particles increases from one to one million.

OVERBURDEN CHARACTERIZATION

Where the potential exists for spoil to generate AMD, a determination of

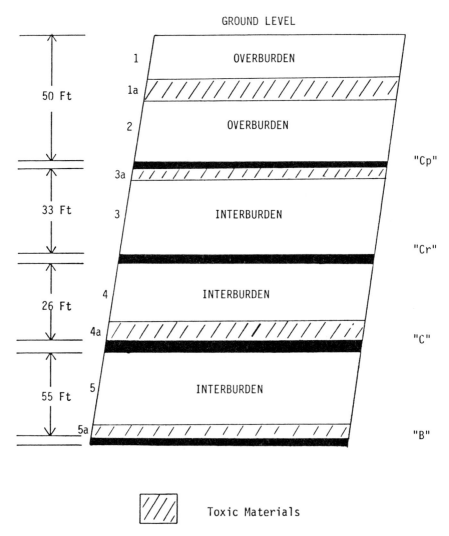

FIGURE 4. Mining case showing a column section of four coal seams with four zones of toxic material (Pfaff, 1987).

the potential must be provided during the permitting process. The process requires an acid/base accounting of the overburden and interburden. This requires two basic measurements, total or pyritic sulfur and the neutralization potential of calcium carbonate equivalent of bases present in the rock.

The materials, either rock or coal, are considered toxic where a pH of the material is less than four or where a net potential deficiency of five or more tons of calcium carbonate equivalent per 1000 tons of rock is encountered. The acid/base accounting method is generally used to make this determination. It

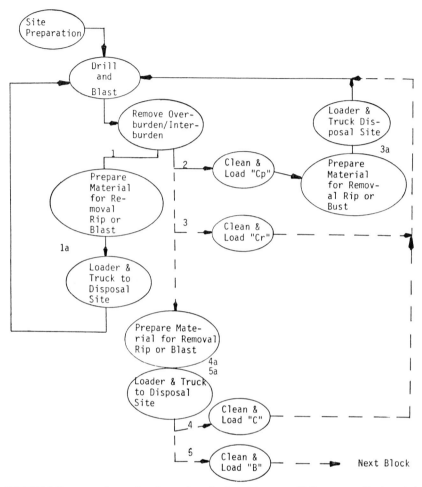

FIGURE 5. Sequence of operational procedures for the mining case with four zones of toxic material (Pfaff, 1987).

must be done on samples obtained in development drilling, that is prior to the commencement of mining.

An example of a drill hole above the Upper Freeport coal seam illustrates this (see Figure 1). There are net deficiencies in two zones of overburden rock. These occur where there is a net deficiency of more than five tons per thousand tons of calcium carbonate equivalent. The zones occur in intervals of 20 ft to 24 ft and 51 ft to 52 ft. The only other zone which contains a deficiency is the coal seam and underlying rock. Although there are other zones with net deficiencies, they are not considered toxic. Base rich zones occur extensively throughout the hole. Notice that in two of the deficient zones, the pH was also less than four, but it was not in the zone from 51 ft to 52 ft.

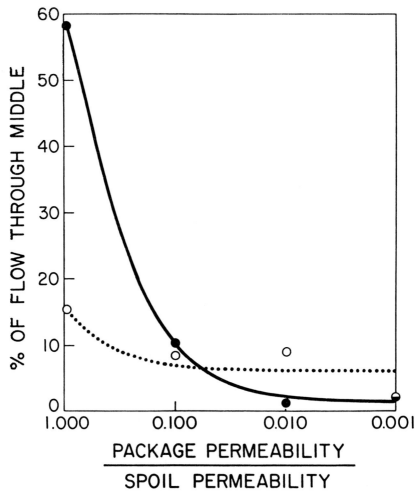

FIGURE 6. Graphs of the percentage of the total flow passing through the middle of a selective placement zone, against the package/spoil permeability ratio. Barrier permeability is 1/1000 that of the surrounding spoil (Phelps et al., 1983).

The neutralization potential is calculated for each zone in the drill hole by determining the calcium carbonate equivalent. When balanced against acidity from the total measurement, a net acid/base account can be made. Also, in a situation such as shown in Figure 1, placing a base such as limestone or lime on the pit floor prior to backfilling to neutralize the potential acid formation from this zone is standard current operating procedure. In addition to the common use of limestone and lime, new dolomitic-based by products have been found to be effective neutralizers (Heunisch, 1987).

It can readily be seen that the selection of a sufficient number of drill holes to characterize the overburden of a mine site is essential (Rodgers, 1984). Proper selection will define the extent of potentially toxic zones as well as indicate where potential calcarious material may be found.

MINING PROCESS

During the mining process, the potentially toxic layers of overburden and interburden rock that have been delineated in the overburden analysis procedure must be selectively handled. The process adds to the cost of mining where these strata exist. A base line case for examination of the unit operations for a four seam mine and how each block of material is removed is shown in Figure 2 and Figure 3. If potentially toxic zones are found in the rock adjacent to the coal or in the overburden rock as the example shown in Figure 4, then the mining process has become considerably more complex as shown in Figure 5. For this case, potentially additional and/or different equipment is necessary to achieve the same level of production as the base case shown in Figure 2. These additional costs of mining must be considered in the economic evaluation of the property prior to permitting.

Where potential acid producing zones have been identified, a selective handling procedure may be required. Such a procedure should, at a minimum, provide for a toxic spoil location which is not subject to alternate wetting and drying cycles due to a fluctuating water table. Where possible, it should be buried below the water table to limit oxygen exposure. It may be placed in a "package" of shaped, compacted, low permeable material (Phelps et al., 1981; Phelps et al., 1983). Such a selective placement zone can provide significant isolation to air and water contact (see Figure 6). These zones should be maintained alkaline and therefore be mixed with some inexpensive form of lime for short-term rapid control and limestone for longer term control.

CONCLUSIONS

Acid mine drainage can be a significant problem in mining if premining planning does not consider potentially toxic material placement and handling. Good engineering practice as well as the process for permitting a mine today requires an analysis for determining the potential for AMD. If a potential for acid production exists, special handling of the toxic material and alkaline addition can be engineered to provide the protection to the environment required. Research into the complex formation of AMD is continuing, however, with current knowledge mining can be conducted in an environmentally sound manner and in such a way that maximum resource recovery can be obtained from the energy resources available.

REFERENCES

Heunisch, G.W., 1987. "Lime Substitutes for the Treatment of Acid Mine Drainage," *Mining Engineering*, Vol. 39, No. 1, pp. 33-36.

Kim, A.G., B.S. Heisey, R.L.P. Kleinmann, and M. Duel, 1982. Acid Mine Drainage: Control and Abatement Research, U.S. Bureau of Mines Information Circular 8905.

Lovell, H.L., 1982. "Coal Mine Drainage in the Unites States - An Overview," 11th Conference, International Association on Water Pollution Research, Pretoria, South Africa, April 5-6.

Pfaff, M., 1987. "An Analysis of the Cost Estimation and Planning Procedures for a Pennsylvania Surface Mine: A Case Study," unpublished Master of Engineering Paper in Mineral Engineering Management, The Pennsylvania State University, University Park, PA (in progress).

Phelps, L.B., L.W. Saperstein and J.C. Mills, 1983. Reclaimed Surface Mine Flow Definition Model with High Density Zones, Final Report to the Pennsylvania Science and Engineering Foundation on Grant Agreement No. PSEF 21-24, Harrisburg, PA.

Phelps, L.B., L.W. Saperstein, W.B. Wells and D. Yeung, 1981. Burial of Potential Toxic Surface Mine Spoil, NTIS PB-83-236051, Springfield, VA.

Rodgers, K.W., 1984. "Overburden Analysis for Surface Mining on the Upper Freeport Coal Seam in Armstrong County, Pennsylvania," Proceedings of the Symposium on Overburden Analysis as applied to Surface Mining, Clarion University, Clarion, PA, November, pp. 38-58.

Smith, R.M., A.A. Sobek, T. Arkle, J.C. Sencindiver and J.R. Freeman, 1976. Extensive Overburden Potential for Soil and Water Quality, U.S. Environmental Protection Agency, EPA-600/2-76-184, Cincinnati, OH.

Sobek, A.A., W.A. Schuller, J.R. Freeman and R. M. Smith, 1978. Field and Laboratory Methods Applicable to Overburdens and Minesoils, U.S. Environmental Protection Agency EPA-600/2-78-054, Cincinnati, OH.

Environmental Consequences of Energy Production: Problems and Prospects. Edited by S. K. Majumdar, F. J. Brenner and E. W. Miller. © 1987, The Pennsylvania Academy of Science.

Chapter Eleven
OIL AND NATURAL GAS DRILLING AND TRANSPORTATION— ENVIRONMENTAL PROBLEMS AND CONTROL

MICHAEL A. ADEWUMI
and
TURGAY ERTEKIN

Petroleum and Natural Gas Engineering Section
Department of Mineral Engineering
College of Earth and Mineral Sciences
The Pennsylvania State University
University Park, PA 16802

The importance of 'mineral energy' or simply energy in an industrial economy cannot be overemphasized. Whereas the less developed economy derives the bulk part of its production energy requirements from human and animal physical energy, the industrial economy derives hers from fuel such as oil, natural gas, coal, etc. It is an indisputable fact that oil and natural gas take the lion share of the energy sources in this regard. In spite of all the talks in recent years of evolving alternative energy sources to replace these commodities, it is widely believed that oil and natural gas will continue to play a dominant role in the energy consumption pattern of the industrialized nations in particular and the energy-hungry world in general.

Tracing back the evolutionary history of petroleum as a source of energy, it is amazing how within a short period of time, and with such a humble, almost mythical beginning, this commodity has risen to a position of pre-eminence in the total energy consumption picture of the world. Even though the modern usage of petroleum transcends several industries including medical, plastic, agricultural, etc., its use as energy source has received the greatest attention. Perhaps the main reason for this seemingly monopolitic emphasis on energy end-use of petroleum stems from lack of public education on one hand and issues-of-the-time dogma on the other. Nevertheless, this commodity, petroleum, serves as a major raw material in the manufacturing sector and this has added to the position of indispensability that it enjoys today.

The modern petroleum industry, as it is known today, is a direct offshoot of the celebrated Drake's 69-ft deep well of 1859 in Titusville, Pennsylvania. However, the true history of petroleum usage dates back to the time of ancient civilizations as far back perhaps as 3000 B.C. History has it that during those times, asphaltic bitumen had been used in hydraulics, ship building and for medicinal purposes. Whereas the modern petroleum industry has several distinct phases, namely exploration, drilling, production and marketing, the ancient one confined its activities to natural seepages and its end-use was rather limited in scope. It was not purely an accident of history that the modern petroleum industry was born in the 19th century, especially if one recalls some of the major world events that rocked the entire world in the preceding century. The essence of the industrial revolution of the mid-18th century was to change forever the ability and capability of human beings to create and recreate machinery which transcends the imagination of its creators. This industrial revolution, coupled with the earlier French and American political revolutions on one hand and also interspersed by such major inventions as the steam engine constitute a sequence of events which changed the shape of man's destiny. This, in part, invariably fuels the desire of man to search for better life in the true meaning of industrialization and its accompanying consumptive propensity. This invariably led to the search for a source of cheap light that could provide effective illumination both in the plant and at home. This led to the invention of a process for distilling kerosene, first from coal and later from petroleum; the design of the oil lantern, earlier invented, was improved considerably to such an extent that its usage became safe and widespread. This, of course, led to skyrocketing demand for crude oil and resultant hike in price. This set the stage for the development of the petroleum industry into the organized one which it is today.

Whereas coal continued to enjoy a position of pre-eminence in the energy scene, with petroleum contented with its second position, it did not take too long for the 'black gold' to overtake coal at the end of the second world war. It is pertinent to explore historically the major developmental process of drilling, production and transportation technology of oil. This should help to foster a better understanding of where the industry is today relative to where it was and where it could be in the future with respect to the technology or the lack of it.

With such a humble and almost accidental beginning, it is noteworthy what strides have been made in the areas of oil well drilling and completion, oil production and transportation. While initially, petroleum had been recognized and obtained solely from the surface findings where this rather strangely dark liquid seeps out of the soil, many of the water wells drilled had petroleum afloat the water. In many instances, it became a nuisance since it was regarded as a pollutant, even though the full implication of this term was not recognized in its usage at the time but nevertheless the idea of oil as a source of pollution was unceremonially born. When the value of oil was later recognized and the search for it began in earnest, it was natural for its explorers to look beneath the surface, and this they did by drilling wells. At this point, of course, the perception of oil as being a pollutant had changed and it is ironical that oil became the real thing while the water became a nuisance. The famous Lucas gusher truly marked the beginning of several important developments that culminate in the present day activities in the petroleum industry. It is so much so that the drilling of Lucas gusher has been called the most important event in the history of the petroleum industry since Drake's well. It was the first well drilled using rotary drilling rig, even though this was in the crudest form. It was also the first recorded experience of a blowout, that is an uncontrolled gushing out of oil and gas during drilling operation. It also marked the first attempt at using a system of valves to control the flow from the well, a system now referred to as Christmas tree. By the time this first blowout was curtailed, a lake of about 800,000 barrels of oil had formed that spread three-quarters of a mile, an obvious pollution hazard. Unfortunately at that time, this kind of situation was regarded as good and it was not until years later when the technology of oil production had developed significantly that the Christmas tree and a system of valves and pipes were used to contain and direct the flow of this valuable commodity.

There is little gain in saying that the technology of exploring and drilling for oil, producing and storing it safely, and transporting it has grown rather rapidly since Drake's well. This rapid development has removed some of the safety laxity associated with its transportation and production as well as allay many of the fears that may be associated with it. Needless to say that this technological advancement has helped bring the much needed energy to the needy world at a reasonable cost, both in terms of dollars and values.

BASIC MECHANICS OF OIL AND GAS DRILLING

Drilling is the process of making wells. The well serves as the only means of reaching the petroleum bearing formations whose depth from the surface could range from a few hundred feet to several thousand feet. It is the avenue through which the petroleum is produced as open mining of crude oil is impracticable for most cases and hence production through the well is the rule rather than an exception. Drilling, as one can imagine, is a very capital-intensive operation and it requires large supplies of equipment and personnel. With large number of wells that must be drilled to effectively exploit the oil deposit, the drilling industry employs thousands and makes use of several thousand tons of equipment.

Why all these wells? Does one have to drill so many wells to produce so much petroleum? A quick analogy with water wells would clarify and answer these questions. If one were to be in the desert where there are pockets of underground oases scattered all over, the only means of confirming where these oases exist would be to drill wells. It is conceivable that many of these wells will be "dry" while others will find producible quantity of water. The situation is even more uncertain in the case of oil fields where the hydrocarbon-bearing formation may exist several thousand feet below the surface. Despite the giant strides made in the geological and geophysical exploration techniques, there is as yet no perfect predictive technique that is capable of positively identifying, with 100% certainty, where petroleum could be found and in what quantity, other than by making physical connections to it via the wells. In view of this, after geological and geophysical prospecting has suggested the possible existence of petroleum in certain places and horizon, wells must be drilled to confirm if this is so or not. Furthermore, through these wells, tests could be performed to determine the extent and producibility of this formation. This is particularly necessary to justify further investment in the field development. The information generated from the first successful well drilled in a field is, more often that not, insufficient to justify further development and possibly embark on production planning, hence many more wells are drilled to map out the extent of the reservoir and generate more test data. It is, therefore, quite clear that only a rather small fraction of the total number of wells drilled actually serve as production wells. Besides those wells on which is an array of valves called the "Christmas tree", many more wells are drilled and subsequently plugged. Those include exploration and wildcat wells. This is the main reason why the drilling industry is such a big one. A lion share of the capital outlay for oil production is actually consumed in drilling. The drilling industry employs thousands of personnel ranging from roustabouts to drillers, tool pushers, drilling supervisors, engineers and drilling program managers. These personnel man an assorted array of machinery, tools and chemicals all of which make up the drilling rig.

THE EVOLUTIONARY HISTORY OF DRILLING TECHNOLOGY

Drilling is usually the first visible sign of the oil prospecting even though this activity must have been preceded by both geological and geophysical prospecting. It is also the first of the oil development activities that arouse the suspicion of the public as to possible damage to the environment, both real and imagined. In order to understand the root causes of this legitimate, but sometimes unwarranted concern, it is pertinent to examine the historical perspective of drilling technology from its rather crude beginning to its present day sophistication. In many respects, one would see at a glance what great strides have been made in a rather short period of time and also, it would put into better perspective, the considerable improvement in technology now available. The latter should allay some of the fears about the environmental impact of drilling technology, especially when coupled with the high standard the industry observes.

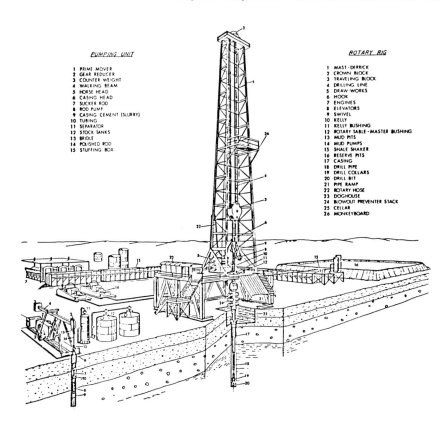

FIGURE 1. Schematic Diagram of a Rotary Rig Showing the Towering Derrick and the Pumping Unit. [Courtesy: Stanley B. Supon (original source unknown)].

Drilling wells for water and brine is as old as human civilizations; at the earliest stages, wells were dug by hand using the manually percussive technique. In this case, the digger uses a chisel-like tool which he strikes on the ground repeatedly to break the rock into pieces. Soon there are enough broken pieces of rock as to impede further progress, hence they have to be bailed out and thereafter digging resumes. Slow as it is, this procedure has been used by man from time immemorial and it was this that motivated the earlier semi-mechanized version, generally referred to as the percussion technique. In this system, a chisel-like tool, later and still called the "bit", is suspended at the end of a rope. A reciprocating up-and-down motion is mechanically imparted upon the bit which repeatedly pounds the hole, chips the rock continually and by this process deepens the hole. This was the first logical step developed from the traditional digging by hand. When the debris start getting in the way, a container is used to bail out the rock material and drilling is subsequently resumed. As the well gets deeper, the wall may cave in causing more unwanted debris and in order to prevent this, some hollow pipe (wooden pipe or bamboo had been used in early days) is run in the hole to hold the wall back from caving in. This pipe is known as casing, even though steel pipes cemented to the wall of the well are used today.

With the industrial revolution came the era of mechanization. This greatly improved the percussion drilling and gave birth to what then became known as cable tool drilling technique. In essence, steam engine replaced the manual labor required to impart reciprocating motions on the bit through the spring pole. In addition, it supplied the necessary power needed to hoist the drilling and bailing tools out of the well. The now familiar symbol of a drilling rig on site, the tall tower known as the derrick was born to accommodate and help support the drilling, bailing and fishing tools before they are run into the hole. It also helps to support the several lengths of casing pipes needed to prevent the hole from caving in. This system ushered in a new era of deeper and safer drilling of wells. Even though this system was limited in applicability for drilling medium hard rocks, it nevertheless made a tremendous impact on the development of the oil industry. This technique was used in the famous Drake well in Titusville, Pennsylvania. The use of cable tool rig in oil well drilling reached its zenith in the period of time spanning the first third of this century. In that period, steel replaced timber in the construction of derricks, larger bits of up to 24 inches in size became available, steelpipes were used for casing and perhaps most significant development of all, casing cementation was successfully established. All these made it possible to drill deeper wells, and such wells could be adequately completed in such a way that oil flows to the surface through a coordinated system of valves, called the well head assembly or the Christmas tree.

With ever increasing demand for oil, drilling activities became more competitive and it was imperative to drill faster and deeper. The technological advancement brought about by the industrial revolution offered an opportunity which the cable tool technique could not fully exploit. The idea of rotary drilling, which hitherto had not been popular, was explored more deeply. It became obvious that rotary drilling was what the technology needed to meet the contemporary challenges posed.

In a complete departure from the up-and-down percussive motion involved when cable tool method is used in drilling a well, rotary drilling as its name implies, involves rotary motion of the drill string which consists of several joints of a hollow pipe. Coupled to the end of the drill string is the drill bit which does the actual cutting of the rock. Several types of drill bit have been designed for use in specific rock environment and the technology involved in bit design has grown to a level of sophistication involving today the use of high technology. No purpose will be served here going to its details. A major component of the rotary rig operation is the continuous circulation of fluid in-and-out of the well bore, primarily to remove the cuttings and carry them to the surface. This fluid, called drilling mud, is usually a concoction of mixture of fluids and solid chemicals. The fluids and solids utilized vary and are determined by the nature of the formations being drilled and the geo-pressure profile envisaged and/or encountered. The drilling mud is truly the lifeblood of rotary drilling. Apart from its primary function as the solid cuttings conveying medium, it lubricates and cools the drill bit and the drill string; it helps maintain the integrity of the well bore, controls formation pressure thus averting possible blowouts. The simplest drilling mud would be a mixture of water and ordinary earth whereas its sophistication in terms of constituents chemicals knows no bounds. At the very crudest stages of rotary drilling, ordinary water is used which mixes with the dirt and the end product being recirculated is like mud, hence the generic name, drilling mud.

IMPACT OF DRILLING ON THE ENVIRONMENT

From the environmental point of view, there are several ways and areas in which drilling activities could impact adversely on our environment unless adequate precautions are taken promptly. First of all, it could be contested that drilling activities open up the wilderness since apart from the actual drilling site, roads must be constructed for the purpose of moving the heavy equipment involved, noise pollution could disturb the rather serene atmosphere of the wilderness, fuel waste could contaminate the atmosphere and so on and so forth. The list could be endless. However, from a pragmatic point of view, all these are essential to human survival and besides, wherever human beings live, the face of that environment must as a matter of necessity be changed immaterial of whether drilling takes place or not. After all, the dependence of human survivability on the environment is a matter of fact.

More realistically examined, there are two basic primary areas in which drilling can encroach upon the peace and harmony of the environment. Happily though, both are avoidable by way of technology and human probity. The first borders on the use of drilling mud which unavoidably is a concoction of chemicals (after all, water is a chemical by scientific definition), some of which if not properly disposed of could be harmful to plants and animals alike. The other is through accidents, particularly those that may result in a blowout, which is essentially uncontrolled flow of oil and gas out of hole. The formation pressure must be surpressed and closely monitored throughout the entire drilling operation. Preventive measures are rigorously taken by way of using blowout preventers which must be routinely tested at intervals to ensure its functionability when needed. Nevertheless these blowouts have occurred in the past and they could result in massive oil spill, the effect of which needs no telling. The technology of drilling has improved so much since the era of spindle top gusher when the blowout was regarded as the primary sign of oil producibility and was even looked forward to. Not only are blowouts comparatively rare today, where they occur, procedures are in place for containing it and cleaning up the damaged area and restoring the environment.

While there is no denying that drilling for oil could impact adversely on the environment, one can argue that the technology of drilling has advanced so greatly that it rises to eradicate these impacts. In its present stage it has reduced the probability of those adverse impacts considerably and judging from novel hardware invented and put to work in this field each year, one cannot but foresee an era of perfect safety in drilling. The industry itself continues to work very assiduously with the relevant government agencies to preserve the environment.

One other fact which is often overlooked is the array of standards observed for the design of pieces of drilling equipment on one hand and their operations on the other. Such standards are created by professional bodies and organizations such as the American Petroleum Institute (API), American Society of Testing Materials (ASTM) to mention just a few. Subject to all of these regulations, the equipment manufacturers have been able to produce safer pieces of equipment and the operators operate them more cautiously and safely. These development have reduced the probability of equipment failure, thus making the drilling process safer with respect to the environment. While the impacts of the drilling process on the environment are probable, they are minimally so.

In the area of emerging technology which could help to alleviate the possible harmful effects of mud chemicals on the environment, it is pertinent to mention air drilling. Essentially, here air is substituted for drilling mud as the medium of lifting solid particles out of the hole. At the present moment, air drilling technique is used in drilling certain types of wells and in certain kinds of formations. Shallow wells and hard rock formations such as obtained in many areas of Pennsylvania and New York are good candidates for air drilling and it is indeed being used quite extensively in those areas. One of the bottlenecks in its widespread use is the lack of understanding of the well bore hydraulics involved. It is expected that as research progresses into this emerging technology, it will gain wider acceptability within the industry and have a wider use even in those situations not currently thought possible. Air drilling eliminates the chemical effect of the traditional mud. Other reasons for its possible attractiveness is the fact that it achieves faster drilling rate and at a lower overall cost. If this technology can be enhanced such as to be usable in other situations, it definitely would lessen some of the environmental problem associated with the conventional mud usage.

OIL AND GAS PRODUCTION AND TRANSPORTATION

The primary incentive for the existence of the oil companies is to produce oil and/or gas. Without the hope of producing oil, there would not be any need for drilling either. Thus it is understandable while oil and gas production gains quite a lot more visibility than the other operations which constitute the oil industry.

After the well is drilled and if it is successful, it is prepared for production. Essentially, the primary purpose of any completion program is to create the necessary avenue for the oil and/or gas to flow to the surface in a controlled manner and at a pre-determined rate within certain technical limits. Even though, oil production is desired, its flow must be controlled, channelled and directed so as to optimize the efficiency of the production facility on one hand, reducing possibility of pollution on the other and ultimately to maximize recovery of capital investment. If this seems like a complex task, that is because it is. However, the technology of petroleum production has advanced to such a stage that this entire operation is almost routine today. Of course, research continues to improve upon the system especially in the area of automation and safety.

The well completion program depends uniquely on the type of well, the mode of production, the available downstream processing facilities, etc. However, in general, production casing must be run and cemented. The primary function of the production casing is to ensure isolation of the formation fluids from the well bore especially the unwanted ones and also to control the flow of the desired ones. Secondarily, it provides the anchor or foundation to which the well head assembly is bolted. The well head assembly or Christmas tree is a system of control valves that controls and directs the flow of fluids from the well as desired by design. It is thus the primary safety device that prevents what is called the 'gusher phenomenon' of the spindle-top well.

Within the well bore itself, the oil and gas flows through several joints of pipes that run the entire depth of the well from the pay zone to the well head assembly. It is called the production tubing. In order to direct the production through this tubing, other equipment, mainly valves and packers, are installed. The latter is a means of isolating one zone from the other. Many producing fields could have more than one productive layer. For technical reasons, it is usually desired to produce each zone independently of the others even though one well can be used to produce from two or more zones except that in this case, the well would be completed by more than one production tubing in the well. Since the primary objective of this discussion is to examine the possible impact of these activities and/or processes on the environment, no useful purpose will be served describing the technical details involved. In general, two broad categories of production systems are discerned based on the source of production energy. One is the primary production system where all energy needed for moving the oil from the formation to the surface is supplied by the reservoir and the other is the case where this movement is surface induced or supplemented by a mechanical device, usually a pump. The only visible difference between these two systems is the fact that a pump (e.g. beam pump) could be seen around the well, perpetually (or so it seems) performing a reciprocating motion. In the one with a primary production system, all that is visible is the Christmas tree probably with hissing sound indicative of flow through constrictions.

It is pertinent at this point to indicate that the design of all components of the well completion gadget such as valves, tubings, casings, well head assembly, etc. are designed by proven standard set by API and ASTM. Even the completion procedure such as the cementation, and so on, follow certain standard practices developed over several years of testing, experience and practice. It can be stated, therefore, that within the limits of human error, the system is generally fail-safe. In addition, especially for offshore wells, additional safety valves are installed downhole to avoid loss of petroleum and possible pollution in case of accidental damage to the well head assembly such as a ship colliding with the well head. These valves are called subsurface safety valves. They are designed and set to close if the pressure drop across them is abnormally high which is indicative of uncontrolled flow at the well head. All these combined makes oil spill from the well head or completion assembly a rarity.

The next thing after the oil is produced is that it has to be transported. First of all, for effective exploitation of any particular field, a certain number of production wells must be drilled and these wells must produce in most instances simultaneously. The determination of the number of wells, their spacing and flow rates require sophisticated reservoir engineering analysis which, more often than not, involves sophisticated mathematics. Our main concern here, however, is what to do with the oil and gas produced by these various wells. They are gathered or collected through a network of pipelines called the gathering lines which essentially direct the flows into a common pot called the flow station or production platform for further processing and storage. Processing at this stage is usually limited to purification of the oil and the gas. Through array of vessels, some of them separators, heater treaters, surge tanks and other bank of tanks, one obtains the crude oil in the state in which the refinery wants it and the gas in the state in which the pipeline company wants it.

In essence, the oil and gas are produced at one place, processed for preliminary removal of unwanted constituents at some other place and transported to the refinery or somewhere else for its end use. All these involve transportation of oil and gas and there lies some potential for spills due to accidents or material failure or human error. A detailed look at some of the means of transporting these commodities will expose some of those possibilities.

Apart from its primary use as source of energy, petroleum is used as a raw material in several industries ranging from pharmaceutical to plastic and synthetic fiber. For all these and several industries, petroleum, the raw material must be available in reliable quantity and at the right time. Of course, in its traditional role as the primary energy source for home and industries—heating, power, electricity, and so on, it must be made available where it is needed. All these brings to bear the fact that reliable means of transporting this all-important commodity must be available.

The commonest means of transporting this commodity include pipeline, trucks, train and ships generally called oil tankers. In several cases, natural gas in its liquefied form is transported using LNG tankers which ply oceans and seas. In fact, this is the primary method by which many European countries including France and Germany obtain their natural gas energy from the faraway Middle Eastern countries. It is pertinent to examine some potentials for pollution that are inherent in these means of transporting petroleum. For comparative analysis, each of these major means of transportation will be examined separately.

Petroleum via Pipelines: History has it that the Chinese, perhaps more than 2000 years ago, transported natural gas from shallow wells through bamboo poles. At that time, this natural gas was said to be burnt to evaporate sea water to produce salt (Springborn, 1970). Also, several centuries later, William A. Hart who has been described as the "father of natural gas" in the United States, made natural gas pipelines out of hollow logs in the 1820's. From these humble beginnings, the technology of oil and gas pipelining has grown to a position of pre-eminence in petroleum transportation. For instance the construction of the alaska pipeline in such a very unfriendly environment represents a major proof of the level of pipeline technology. Today, pipelines are constructed from iron and steel. As a matter of fact, the pipeline industry consumes a lion share of the pipes produced in this country and is a multi-billion dollar industry involving several thousand workers. Furthermore, the day-to-day energy need of millions of Americans is met using network of pipelines to bring this commodity to their doorstep.

In spite of these advances, accidents can and do occur either due to human error, or faulty equipment or equipment malfunctioning. However, the rate at which these occur is very low that its level is acceptable just as automobile accidents do occur and yet we ride cars. The technology of pipeline design is so information-based that computer-control is now a common place. Several safety checks are usually and routinely installed on the pipeline to detect even the minutest leaks and definitely any breaks. These devices are computer-controlled so that flow in the affected pipeline is immediately shut down to prevent further waste and curb possible pollution. At the entrance to any major city, a city gate station is usually installed whose purpose is to further clean and meter the incoming gas, step down its pressure and odorize the gas, all being additional safety measures. Except for the possible fire it may cause, natural gas does not usually constitute a pollutant since it can diffuse and disperse very rapidly through the atmosphere such that its local concentration would be negligibly low. Oil spill due to pipe breaks on the other hand can have adverse pollution effect on land, animals and plants since it is in liquid form under atmospheric conditions. If such an oil pipeline traverses a body of water such as a lake, sea, etc., its resultant pollution effect would be much harder to control since the wave helps to spread it fast. Even in this worst-scenario cases, established and proven procedures are available for the clean-up of the pollutant.

In a nutshell, one can assert that pipelines have become an economically attractive, safe and convenient means of transporting petroleum and petroleum products. Even though, cases of water-wells pollution due to leaks from nearby pipelines are reported but these are exceptions rather than a rule.

Transportation of Petroleum by Tankers: Transportation of crude oil by water has played a very significant role in the development of the oil industry. There is no gain-saying in the fact that further growth of the "upstream" oil activities including exploration, drilling and production would have ben seriously hampered without the corresponding growth in the "downstream" activities, namely refinery, distribution and consumption. The indispensable connecting link between these seemingly different sets of activities is moving the "upstream" product to the "downstream" input as feedstock. While the Drake well was producing oil from the rather obscure location of Titusville, the crude oil had to be transported initially by boats via Oil Creek to Oil City. It was then transferred to larger vessels for shipment to Pittsburgh which at the time was a major refinery and transportation center for the oil industry. Judging by the accounts of these early activities, there is no doubt that transportation by water played a key role in the movement of crude petroleum and the refined products between the oil field and the refinery on one hand and between the refinery and the consumer centers on the other. In the continental United States, for instance, rivers, streams and creeks quickly formed a natural transportation network for petroleum and petroleum products. They were transported in barges, passenger boats and rafts to centralized shipping posts from where steamboats took over the continued transportation.

As industrial development of nations scattered over the earth and separated by oceans and seas are *interdependent,* it is natural to perceive that sooner or later, petroleum would have to be transported across the seas and oceans. The first reported overseas shipment of crude oil was said to be in 1861 when three thousand barrels were shipped from the United States to London. It quickly dawned on the ship builders that putting crude in barrels and then shipped aboard ships was rather cumbersome. They quickly rose to this challenge by retrofitting a vessel, "S.S. Charles" of Antwerp, Belgium for bulk shipment of crude oil. Even though this first attempt could only be used to ship about 800 tons of crude oil, it proved the viability of this approach. It was not too long after that when the era of tanker transportation of oil was ushered in.

From this rather humble and uncertain beginnings, the oil tankers have grown to massive sizes. Today, crude oil is transported by VLCCs (very large crude carriers) whose capacity varies between 160,000-320,000 DWT (dead weight tonnage) and ULCCs (ultra large crude carriers) with as much as 540,000 DWT capacity. These super-giant tankers are seemingly oversized, exceeding 1000 ft in length and capable of carrying 4 million barrels of crude oil. As these tankers grow in size, so also is the size and complexity of the loading and receiving terminals. In the United States, for instance, only one port has deep enough water to accommodate these super tankers. Complex moving buoys are designed and installed to handle the loading and unloading of these tankers. they ply every ocean and sea, transporting crude oil from one end of the earth to another and their routes are as complex as these monstrous-looking ships. In effect, one can say that tanker routes have replaced the notorious "slave routes" of the past. It is ironical that while large ships were reminiscent of the slave triangular trade of old, tankers are now indicative of crude oil trade between the remotest corners of the world to the industrial centers of Europe and North America.

While the design, construction and operations of these super oil tankers may seem like an insurmountable challenge, even more challenging is the technology of transporting natural gas over the seas. While some long distance under-the-sea pipelines have been built in recent years (e.g. the Algeria-Europe pipeline), the most common method of transporting gas across the seas is by LNG (Liquefied Natural Gas) tankers. As the name implies, LNG is the liquefied form of natural gas. Liquefaction of natural gas reduces its volume by a ratio of 630 to 1, thus enabling more gas to be stored and transported in a relatively small tank. However, the LNG has certain unique characteristics which makes its storage very challenging. These characteristics particularly affect the design of the tankers and docking terminals. LNG has extremely low temperature of about 259 F below zero and a density of about half that of water, meaning that it will float on water. Moreover, at ambient conditions, it evaporates very rapidly, thus expanding to about 630 times the liquid volume. When it is in this vapor state and still cold, it is heavier than air and hence hugs the earth's surface in case of spill. This means that it will remain on the earth's surface for a substantial period of time before it dissipates up into the atmosphere. It will do the latter on attaining a temperature of about 100 F when it becomes lighter than air. While it is not poisonous per se, it may cause asphyxiation due to absence of oxygen. Also alarming is the fact that in the vapor state, a concentration of 5 to 15% natural gas is flammable, thus posing a fire hazard.

It is good to know that this odorless and colorless liquid, LNG, which looks like water except that it is extremely cold, is relatively safe. In bulk form, it will not burn or explode, however prolonged contact with skin may cause freeze burns. Because of these unique characteristics, design of LNG storage and handling equipment requires a great deal more technology than for crude oil. For instance, many materials including carbon steel are brittle and easily crack when subjected to the LNG temperatures. In spite of this heavy technology demand to its design, construction and operation, use of LNG tankers to transport natural gas has become almost a routine operation. This technology continues to improve and LNG tanker routes now span the globe.

The technology of LNG tankers has grown very rapidly in the last 30 years and the fleet of LNG tankers is estimated at about 100. These tankers are bulk cargo ships which require very unique design and material to handle very low temperatures. Most LNG tankers vary in size from about 40,000 to about 165,000 cubic meters. However, the LNG industry standard is in the range 125,000 to 130,000 cubic meter capacity tankers. This size tanker has enough capacity to carry sufficient LNG to heat a city of 100,000 population for one month [OTA (1977)]. Even though such LNG tankers are only one-half to one-fourth the total size of the crude oil super tankers, VLCC, they appear to be more visible on the ocean. That is because of their design which makes them a shallow draft vessel, about 36 feet with unusually large amount of freeboard, sometimes rising about 50 feet above the water surface. Capacity is usually added to the LNG tanker by increasing its length rather than depth.

The LNG tanker is usually a high-powered, high-speed ship with speed in the order of 20 knots. The design and construction of an LNG tanker is rather complex and capital-intensive, requiring more than $150 million investment. In addition, LNG receiving terminals must be built to handle this cold liquid. One can safely make the claim that LNG tankers have become a very viable and safe transportation means for the much needed natural gas from its faraway points of production overseas to the consuming places in the United States and elsewhere.

Other Means of Transporting Petroleum: Even though pipelines and tanker ships provide the means of transporting the bulk of crude oil and natural gas, especially in terms of both quantity and distance involved, nevertheless, other means are not less common. These include railroad tank cars, barges and tank trucks. Which of these methods is used is determined by several factors including distance, product and alternative means and, of course, safety. For instance, given the network of roads criss-crossing the United States and other developed nations of the world, tank trucks provide the most flexible means of transporting petroleum liquid products such as gasoline, kerosene, etc. The basic constraint, with this means of transportation, is that it can only involve relatively small quantities and over shorter distances. On the contrary, pipelines are economical and can transverse longer distances but are rather inflexible as to route and destinations. By similar token, railroad cars can handle intermediate distances and sizeable quantities. Barges, as a means of transportation of goods including petroleum, dates back to the colonial days in the United States and other parts of the world. This is to be expected with the tremendous mileage of inland waterways naturally existing. The use of inland waterways facilitates distribution of petroleum and petroleum products since, traditionally, towns and cities are built along these waterways. Of course, none of these compete favorably with the major overseas means of transportation—the tankers.

By way of comparisons, very close to 100% of the over-the-seas transportation of petroleum is done by way of tankers and ships whereas on land, the story is different. Very close to 50% of the oil transported in the continental

United States is done by pipelines. There are more than 250,000 miles of pipelines transporting crude oil and petroleum products in the United States. These network of pipelines convey more than one billion tons annually. Moreover, natural gas is transported by a network of pipelines totalling more than one million miles. More than 160,000 tank trucks are involved in the United States for the haulage of about 30% of oil while tankers and barges run a close third of hauling about 22 percent. Only about two percent of oil is transported by the rails in the United States. It, therefore, needs no telling that while tankers and ships (numbering a staggering 5,200) move the bulk of the petroleum and natural gas in international trade, the pipeline is still king for domestic transportation.

Safety Considerations in Modes of Transportation: In general, all the modes of petroleum transportation have to comply to certain safety standards set by the industry and the relevant regulatory bodies. Several notable organizations involved in setting such standards and design codes include the American Society of Testing Materials (ASTM), American Petroleum Institute (API), etc. These design codes are usually stringent and hence they ensure safe operation of the end product when in use under the prescribed conditions. Additional safety precautions are usually taken to limit loss of fluid and subsequent possible pollution and fire hazards in case of accidents.

For instance, on pipelines are installed remote sensing devices to detect leaks and activate the closure of automatic safety valves. Similar devices are installed in the compressor or pump stations to effect shut-down of these in case of emergency. Today, all these processes are computer controlled and this makes them very reliable and also drastically reduces the time lag between occurrence and correction of problems. The testimonies of the Senate and House of Representatives Committees that recommended the passage of the Pipeline Safety Act attest to the excellent safety record the industry enjoys. This was as early as 1970.

Similarly, the construction and regulation of oil and LNG tankers are subject to strict design and safety codes that must be observed. Under the various Safety Act for Ports and Waterways, the Coast Guard is entrusted with the responsibility of establishing and enforcing design and construction standards for the tankers carrying the U.S. flag and foreign ones entering the 3-mile territorial waters of the United States. This is usually achieved through letters of compliance for foreign vessels and certificates of inspection for U.S. vessels. In either case, both are subject to the same federal regulations. The receiving and loading terminals also have to meet similarly stringent design and safety codes.

Of course, road and rail transportation modes have to comply to even higher standards since they pass through large communities where human beings could be affected by any safety lax. In general, safety considerations are uppermost in the development of these design codes and they have served the industry and the public as well as the environment quite adequately. The advent of powerful computers even makes the thought of technical failures more remote. One thing that cannot be totally eliminated is the human factor which permeates any industry. Human judgement and failure could render any safely designed equipment useless to the extent that the public suspicion of such system can be aroused from time to time. This is one of the reasons why the petroleum and natural gas industries invest heavily in training and re-training of their personnel to sharpen their skills and improve productivity.

In spite of all the sound technical base, the ingenious computer control and excellent human training, accidents have occurred and will occur either due to human error, unforeseen circumstances and also, but very rarely, equipment failure. In these cases, plans must be made for prompt actions that need to be taken to avert catastrophy and curb damage to humans, animals, and the environment. The next section attempts to examine possible curative measures with regards to the impact of oil spills on the environment.

CURATIVE MEASURES FOR OIL SPILLS

So long as there is no such thing as "perfect technology" and "perfect operator," accidents are inevitable, failures and disasters will continue to occur from time to time. These are eventualities we will live with as long as industrialization and economic exploitation of our planet remains a human endeavor. The risk of these eventualities will continue to be reduced by various technological advances and as the appropriate level of our awareness of the environment and our dependence on it grows. By and large, pollution including oil spills are man-made, be it the consequence of negligence, lack of discipline, indolence or design errors. Industrialization and economic development are built around energy consumption. It is estimated that the world energy consumption has tripled in the thirty years spanning 1950 to 1980. During the same period of time, the oil consumption has increased six-fold. Energy expenses do constitute a sizeable percentage of the Gross National Product (GNP) of most industrialized nations. Oil forms such a central focal point of economic development that the constant tension between the super powers arise in part because of their struggle to ensure access to the oil and gas deposits around the world.

In spite of all the talk about alternative energy sources, oil and gas are likely to remain the main energy source for continued industrial growth and a major raw material for the chemical, petrochemical and the pharmaceutical industries. While production of oil will continue on-shore, there is likely to be a major shift in oil production activities to an off-shore environment. This means an another additional source to possible sea pollution by oil, the traditional source being through transportation. Recognizing the tremendous use the human beings and animals make of the seas for food, raw materials, recreation, transportation, etc., every effort must be made to prevent pollution of these seas. When polluted, we must promptly clean it up. This forms the main theme of this section.

The indepth technical treatment of this problem, namely oil spill cleanup, cannot be dealt in the few pages allowed here. Besides, books have been written on the subject, periodic conferences are held where lucid discussions of the issues involved, ranging from technical to economic and legal aspects, are undertaken. Two of such books are those by Marshall Sittig and A. Brenel, both of them published by Noyes Data Corporation. Another one is edited by Cox et al. and published by the American Petroleum Institute. Proceedings of the Oil Spill Biennial Conferences sponsored jointly by the American Petroleum Institute (API), Environmental Protection Agency (EPA), and the United States Coast Guard (USGC) contain an excellent pool of research results and case studies of this problem. The most recent one held in 1985 provides up-to-date information on this subject. They both deal with the four important areas of Oil Spill Prevention, Behavior, Control, and Cleanup. Similar ones are also held under the auspices of the Association Europeene Oceanique (UEROCEAN). Since the information on this subject is readily available, we will not dabble indepth into it but rather summarize the main points concerning how the spills arise from the oil production and transportation activities and mention briefly the common curative measures for removing the spilled oil. Having dealt with the various safety measures inbuilt into the various equipment involved in petroleum production and transportation, which are basically preventive, one should deal with the post-spill curative measures. In this respect, of utmost importance is the pre-spill contingency planning especially for high risk coastal regions, the control of the spill and cleanup [Stacy (1985)].

In the early 1960's, the major sources of sea pollution by oil were said to be (1) sludge from oil tankers discharged on the beach, (2) disposal of oil-containing wastewater at sea, and (3) accidental rupture of oil tankers [Brenel (1981)]. While the last may not be within the control of the shipmaster, the first two are. This may be one of the reasons why the public took a radically uncompromising look at the effect of the oil operations on their much-valued environment. The decade 1970's witnessed a tremendous upsurge in the public consciousness for the environment, especially in North America. While other industries can be tied to one pollution problem specifically such as noise, water pollution, etc., the oil industry, because of its array of activities upstream and downstream making it a multi-faceted business, rendered it vulnerable to attack on all sides. Among the environmental problems that become associated with oil and gas exploitation activities are subsidence, water pollution, air pollution, noise pollution, visual pollution and other ecological changes. While the other forms of pollution, listed above, can usually be associated with oil and gas production/processing activities, subsidence could not for a long time. Subsidence is a phenomenon by which earth surface depression is caused by the fact of oil and gas production from the area beneath it. Examples of this problem are the Wilmington field located under the heavily industrialized port of Long Beach in California and Lake Maracaibo in Venezuela. Between 1938 and 1945, more than 4 ft of surface subsidence had been recorded in Long Beach, by which time some of the surface facilities were threatened [Allen (1972)]. Measures taken included water-injection programs to stabilize the earth from further subsidence by replacing the produced fluids. This is just one example of how the public awareness and industry-government cooperation and prompt action could alleviate these pollution problems.

The key to pollution control due to oil spill, as in the case of other environmental problems, is prompt action. Such action cannot be taken unless a contingency plan is already in force to envisage such problems and delineate clear path to handling them. Such contingency plan helped to avert the threat of a major oil spill pollution in the Arabian Gulf area in 1980 [Brown and James (1985)]. In this case, the industrial initiative between Bahrain Petroleum Company and other operating companies in the Gulf area established a mutual aid organization as a regional oil spill cleanup cooperative venture. This was set up in response to the Torrey Canyon catastrophe which occurred in early 1970's and which caught international attention.

There are certain aspects of this contingency plan that deserves elaboration. First of all, information and the plan for disseminating it promptly are very important [Cox (1980), Collins (1981)]. Targeted areas with some possibility of spill, especially those closer to the coastlines, must be identified prior to any disaster. Detailed geological map of such areas must be prepared especially highlighting areas that may need special attention. For instance, areas that may be home or resting ground to endangered species or where the use of chemicals must be restricted. An up-to-date information on the currents and tide movement and motion in such areas must be on hand. This is particularly important to develop a prior knowledge of the possible direction of oil slick movement and its possible speed. This would enable experts to quickly develop the control strategy. While it is recognized that these pieces of information cannot be acquired by a single organization, cooperation is very important. Instead of creating adversarial relationships between oil prospectors, environmental regulatory organizations, either private or governmental, these bodies should work as partners in progress to develop this bank of data and ultimately a contingency plan to combat and control possible pollution problems. In some cases, such partnership may have to be internationalized where it could involve intergovernmental cooperation. Part of this contingency plan could be a pool of funds, material and personnel to be used in such emergencies. Funding could be through some taxes levied on the oil prospecting companies, shipping companies supplemented by the appropriate government bodies.

Granted that the contingency plan is on hand, the next aspects of spill control is execution of the plan itself. Of course, each oil spill situation will present a new challenge in terms of control and cleanup but nevertheless, a general execution plan should be on hand. The mode of disseminating information on the spill, about the environment and about the solution technique must be clearly spelled out so as to avoid confusion after-the-fact. When spill occurs, panic may occur among the public who may perceive both the real and imaginary effects. This is normal especially since in the final analysis, the people are the 'guardian angel' of their environment. Initial emotions can lead to confusion and haphazard solution rather than a well-thought, planned execution program. The latter is the only acceptable method of alleviating this problem and it can only be done if developed and allowed to mature before the fact.

What is there to execute? Three broad categories of spill spread control mechanism, cleanup processes and ultimate effect on the environment can be delineated.

Spill Spread Control Mechanism: One of the very essence of oil spill response strategy is its confinement in as small an area as possible. This prevents spread especially to environmentally vulnerable areas and also aids the ultimate spilled oil recovery. This process can be quite difficult, recalling that many such open sea spills take place due to bad weather in the first place. A most popular piece of equipment used in this operation is the containment bloom. Blooms are floating devices that form a closed embarkment around the oil slick. Various types available include fence blooms, skirt blooms, sorbent blooms, improvised blooms. All these fall under two general categories of containment and deflection blooms. Invariably, any of these blooms will fail and allow oil to escape over or underneath it under high wave and swift current conditions. As their names connote, while the containment blooms are used to contain the slick, the deflection blooms are used to deflect its motion in a predetermined direction, say towards a calmer sea whence spread can be minimized before cleanup. In either case, the ultimate use of the bloom is to facilitate later cleanup. The equivalent of blooms on land would be the permanent embarkments built around oil and LNG storage tanks to arrest spill in case of accidents. The onshore situation is much easier since there are no waves to dislocate these embarkments.

Cleanup Processes: Even if the containment and/or diversion of the oil slick is one hundred percent effective, the spilled oil must be eventually removed and the polluted area(s) cleaned up as soon as possible. The basic response path is always such as to minimize the overall environmental damage. This path must also take cognizance of practical and safety considerations. In other words, unrealistic goals should not be set. Several response options are available for the cleanup exercise. Broadly, they can be classified into three major categories, namely, mechanical recovery, chemical treatment and natural removal of oil from the environment. None of these options is a panacea for getting rid of the spilled oil but their prudent and intelligent use will minimize the impact of oil spill on the environment. On many occasions, several of the options may have to be used simultaneously.

Skimmers are mechanical devices for recovering oil from the water surface. This can usually be done once the spilled oil is contained. Among these devices are equipment used to vacuum the oil off the water surface, such as floating weirs and suction heads. They require the use of skimmer pumps, rotating down and belt skimmers which absorb the oil. This equipment in turn deposits the oil in a vessel such as a sump tank, barge, etc., for eventual disposal at the dumpsite.

It must be mentioned that by their manner of operation, this equipment takes in a large amount of water with oil, which may combine with the latter to form emulsion. Eventually, this emulsion must be broken down to separate oil from water and this is a different problem altogether. Nevertheless, oil skimmers can be very effective in recovering the spilled oil if weather conditions are favorable. Absorbents are equally effective in many cases at skimming the oil. These are materials designed to float on water and absorb the spilled oil by capillary action. They are primarily effective for small- to medium-sized spills and for thin oil film left after initial skimming had been completed.

The idea of injecting chemicals into the environment is in itself inexpedient unless its effect on the environment is negligible compared to the alternative option of the oil pollution. There are certain other cases where this option is even preferable to the mechanical options. For instance, where winds, currents and waves cause an indispensed oil to have a dramatic impact on a sensitive shore line, or even where enough time cannot be afforded for mechanical salvage of the oil. The most commonly used chemical agent in combatting spill is the dispersant. Dispersants are chemical agents which help to disperse the oil in water. Basically they break down the oil so that it mixes with water, thus permitting eventual biodegradation of oil. Their use is quite controversial since the public perceive every kind of chemical to be harmful. While this may be true, one has to perceive it from the point of view of 'give and take'. If the use of dispersants constitutes the only option that minimizes damage to the environment, then by all means, they should be used.

The key to safe use of dispersants is advance planning, involving testing in the laboratory environment and controlled testing at sea. In other words, research must be conducted as to the possible effect of dispersants on the ecological disposition of the environment. This invariably leads to standards that are set by the appropriate regulatory agency of the government such as the Environmental Protection Agency (EPA). The team controlling the spill must exercise good judgement and prudence in their decision process as to the use or non-use of dispersants in a particular situation. In addition, information should be available as to what areas dispersants may or may not be acceptable and such information must be factored into the decision process.

Natural recovery process of a cleanup of the shoreline of an oil spill is probably the oldest technique [Bell (1981)]. Today, this method has been replaced by "un-natural" recovery techniques discussed hitherto. However, researches have shown that there are indeed some situations when this old technique is supreme. Again, a systematic decision-making process must be undertaken before deciding on this route. Such questions as to whether any "un-natural" clean-up exercise might cause more harm to the shoreline than leaving oil to natural recovery; Is the spilled oil of low toxicity? Is the short-term presence of the oil acceptable on that sea shore ecologically and otherwise? etc. A reasonable checklist of these questions is found on page 78 of Brenel's book [Brenel (1981)].

Whichever of these various techniques used will depend, to a large extent, on several factors, paramount among them are ecological and ecosystem preservation demands, standards set by regulatory arm(s) of governments, weather conditions and, of course, feasibility. Each situation will always present a new challenge but how well that challenge is met depends on the level of readiness which is intricately tied to a good contingency plan.

CONCLUSION

It needs no telling today that energy is an essential part of our daily life. Leaving aside the 'natural' energy stored within our bodies and/or produced by physiological processes, we consume a great deal more 'external' energy to perform our day to day functions. At the flicker of a switch or button, there comes light; at the turn of an ignition key and a shift of a few "sticks" and pedals, automobiles, trains, ships, airplanes, etc., move us and our goods around. These are only a few examples of our dependence on 'external' energy which is the prime mover behind all these actions/processes. It used to be said that to sustain life, air, water and food are needed. In the modern society, with the rapid population growth, energy may well be the fourth realm of life sustainability.

BIBLIOGRAPHY

Allen, D.R. (1972), "Environmental Aspects of Oil Producing Operations— Long Beach, California," Journal of Petroleum Technology, pp. 125-131.

American Petroleum Institute (1981). *1981 Oil Spill Conference.*

American Petroleum Institute (1985). *1985 Oil Spill Conference.*

Bell, V.A. (1981), "Protection Strategies for Vulnerable Coastal Features," *1981 Oil Spill Conference,* API, pp. 510-507.

Berger, B.D. and Anderson, K.E. (1981), *Modern Petroleum,* PennWell Publishing Company, Tulsa, Oklahoma.

Brenel, A. (ed.) (1981), *Oil Spill Cleanup and Protection Techniques for Shorelines and Marshlands,* Noyes Data Corporation, Park Ridge, New Jersey.

Brown, D.J.S. and James, J.D. (1985), "Major Oil Spill Response Coordination in the Combat of Spills in Bahrain Waters," Journal of Petroleum Technology, pp. 131-116.

Collins, J.W. (1981), "Information from Oil and Gas Industry Essential to Protect Coastal Environment from Oilwell Spills," Journal of Petroleum Technology, pp. 1005-1007.

Congress of the United States—OTA (1977). *Transportation of Liquefied Natural Gas*, U.S. Government Printing Office, Washington, D.C.

Cox, G.V. (ed.) (1980), *Oil Spill Studies: Strategies and Techniques.* American Petroleum Institute.

Deslauriers, P.C. (1979), "The Bouchard No. 65 Oil Spill in the Ice-Covered Water of Buzzards Bay," Journal of Petroleum Technology, pp. 1092-1100.

Duerden, F.C. and Suiss, J.J. (1981), "Kurdistan—An Unusual Spill Successfully Handled," *1981 Oil Spill Conference,* American Petroleum Institute.

Eurocean (1981), *Petroleum and the Marine Environment,* Proceedings of PETROMAR 80, Graham and Trotman Ltd., 14 Clifford Street, London W1X 1RD.

Gaines, T.H. (1970), "Oil Pollution Control Efforts—Santa Barbara, California," Journal of Petroleum Technology, pp. 1511-1514.

Gatlin, C. (1960), *Petroleum Engineering — Drilling and Well Completions.* Prentice-Hall, Inc., Englewood Cliffs, New Jersey.

Giuliano, F.A. (1985), *Introduction to Oil and Gas Technology,* (2nd Edition), Scientific Software-Intercomp, Denver, Colorado.

Kleij, A.M. and Gubbens, J.M. (1985), "Case History of a South Holland Oil Spill—Organization and Cooperation," *1985 Oil Spill Conference,* American Petroleum Institute, pp. 293-297.

Littletown, J.H. (1986), "Regulations Complicate Offshore Mud Disposal," Petroleum Engineer International (March), pp. 53-56.

Lyle, S. (1972), "The Petroleum Industry As It Affects Marine and Estmarine Ecology," Journal of Petroleum Technology, pp. 385-392.

Oil and Gas Journal (1979), "More Regs Seen No Solution to Spill Problems," April 9, pp. 222-223.

Ibid (1984), "House Votes to Boost Taxes for Superfund," August 20, pp. 84, 86.

Petroleum Engineering International (1981), "Aerial Reconnaissance of Spills at Sea," october, pp. 10-12.

Ibid (1982), "Factors Influencing Containment of Spills," January, pp. 14-16.

Ibid (1982), "Effective Distribution of Spill Dispersants," November, pp. 12, 17, 20, 22, and 24.

Sittig, M. (1978), *Petroleum Transportation and Production—Oil Spill and Waste Treatment.* Noyes Data Corporation, Park Ridge, New Jersey.

Springborn, H.W. (1970), *The Story of Natural Gas Energy.* American Gas Association, Inc., New York, New York.

Stacey, M.L. (1985), "Marine Pollution Contingency Planning—Recent Changes in United Kingdom Organization," *1985 Oil Spill Conference,* American Petroleum Institute, pp. 89-91.

Straughan, D. (1972), "Factors Causing Environmental Changes After an Oil Spill," Journal of Petroleum Technology, pp. 250-254.

Welker, A.J. (1985). *The Oil and Gas Book.* SciData Publishing, Tulsa, Oklahoma.

Chapter Twelve
THE INFLUENCE OF OIL DRILLING OPERATIONS AND CRUDE OIL ON THE BIOLOGICAL COMMUNITY

E.C. MASTELLER
Professor of Biology
Division of Science, Engineering and Technology
The Pennsylvania State University
The Behrend College
Station Road
Erie, PA 16563

INTRODUCTION

Oil in the environment produces a variety of physical, chemical, geological, and biological processes. The extent of oil's impact on the ecosystem is influenced by the type of oil, volume, hydrography, climatic or seasonal changes, previous exposure of the area to oil, and indigenous biota. For millions of years natural seeps and erosion have caused petroleum hydrocarbons to enter the biosphere. Oil and oil seeps were described in biblical times, and hydrocarbon-using microorganisms have been known since 1895 when Miyoshi described the growth of the fungus, *Botrytis cinerea* on paraffin (Cooney 1984).

Since 1945 the increasing use of petroleum largely for energy has led to the introduction of large amounts of crude and refined petroleum materials into the biosphere (Cooney 1984). Sources of petroleum in the environment include natural seeps, spillage during production, transportation, refining, condensation from exhausts of gasoline and diesel engines, and runoff from roads and fueling areas. Relatively little effort has been devoted to studying the fate of petroleum hydrocarbons in freshwater ecosystems, particularly small ecosystems receiving chronic inputs of hydrocarbon contaminants (Cooney 1984). Most studies of terrestrial habitats have failed to assess the effects on soil invertebrates.

Chronic oil-field contamination can be caused by construction, broken oil pipes, overflowing separator tanks, and pumping accidents. Jaffe and Gemperlein (1985), reporting a 1984 check by the EPA in Warren County, Pennsylvania, found that 21 of 43 oil operations did not have spill control plans. Because of the potential for crude oil to flow directly into the environment, this paper will deal primarily with the effect of crude oil on organisms. In other oil related activities, Van Gundy (1969) reported that waste brines produced during drilling were toxic to aquatic life, both osmotically and physically. Then, too, numerous drilling additives have been used, but their toxicity has not been measured adequately.

CHARACTERISTICS OF PETROLEUM

Crude petroleum is a complex mixture of 100s of hydrocarbon components. These can be classified as straight chain and branched paraffins, cycloparaffins, aromatic hydrocarbons, and asphaltenes (Bartha 1977). Low boiling hydrocarbons have high contact toxicity, but because of their volatility, they have a less lasting effect. Benzene, toluene, and xylene are often present in these lighter fractions. Aromatics and lighter fractions may go into solution or evaporate, while heavier hydrocarbons may adhere to particulate matter and form tar balls. Snow and Rosenberg (1975) reported that the lighter fractions (n-alkanes < 9 carbons) disappeared from a lake within 2 hours following a crude oil spill. The processes of solubilization, volatilization, photochemical changes, emulsification, and microbial attack are collectively described as the "weathering" of oil. Skorvic, which was reported by Vestal et al (1984) found weathered oil less toxic than fresh oil. Bartha (1986) described the behavior of oil spilled on land and water as very different. On soil there is rapid vertical infiltration. On water surfaces oil spreads out (1 g of petroleum will cover 1-10 square meters) favoring evaporative and photo-oxidative losses. Hydrocarbons reach the freshwater ecosystem either directly or by rainfall and runoff. Where water-air interface organic compounds accumulate; water-in-oil emulsions are termed chocolate mousse (Cooney 1984). Anoxic conditions that prevail in oil- and water-saturated sediments are very unfavorable to biodegradation and result in extended persistence.

DRILLING PROCEDURES

At drilling sites various materials and chemicals are found at higher than normal levels in the environment and may have detrimental effects on the ecosystem. Well drilling often employs a rotary system requiring circulation of a drilling fluid or mud to remove cuttings from the hole. Drill cuttings brought to the surface are screened from the drilling fluid by shakers on the drilling platform. They, along with rig washing compounds, waste lubricants and oils, coolants and domestic waste are directed to a sump or injection well along with formation gas, oil and brine. During drilling, "muds" are used to remove and transport cuttings from the bottom of the hole to the surface, lubricate and cool the bit and drillstring, wall the borehole with an impermeable cake, release sand and cuttings at the surface in pits or sumps and transmit hydraulic power to the bit (Berger and Anderson 1981). Bentonite, barium sulphate and potassium chloride are common in the drilling fluids (Falk and Lawrence 1973). Drilling muds contain barite, hematite, ground iron ore, bentonite, and clay. High pH is present, as well as bacteriacides, detergents, and surfactants. Bacteriacides are paraformaldehyde, caustic soda, lime, and dowcide. Viscosifiers are bentonite, sodium carboxymethyl cellulose, attapulgite clays, and sub-bentonites. Barite, lead compounds, and iron oxides are used to control formation pressure, check caving, and facilitate the pulling of drill-pipe (Falk and Lawrence 1973).

Most oil wells in Pennsylvania are "stripper" wells which produce less than 10 barrels a day (Jaffe and Gemperlein 1985). Wells in Pennsylvania are usually 278-455 m (850-1500 ft) deep, shallow by most standards. Often "hydrofracking" and enhanced oil recovery (EOR) methods are used. The former uses high pressure to fracture the rocks and allows more oil to flow into the shaft as well as create fissures. These fissures may enter freshwater aquifers or abandoned wells (it is estimated that there are 200,000 abandoned wells in western Pennsylvania) (Jaffe and Gemperlein 1985). EOR is defined as the additional production of oil resulting from the introduction of artificial energy into the reservoir (Collins 1977) and includes waterflooding, gas injection, miscible liquified petroleum, condensing gas drive, carbon dioxide miscible process, caustic solution flooding, micellar solution flooding, polymer flooding, thermal recovery by hot fluid injection, and thermal recovery by in situ combustion. Alkaline biocides are added to the injection system to prevent bacteria build-up (Anderson and McElravy 1977, Albers et al. 1985). Primary sulfonates and alkylaryl sulfonates are the principal "primary" surfactants used in EOR, and most are supplemented by "cosurfactants" such as alcohol ethoxysulfate or an alkylphenol ethoxysulfate or by a "cosolvent" such as a low molecular weight alcohol or low molecular weight alcohol ethoxylate (Nelson 1982). Surfactants, micelles, or surface active agents (soaps) are used to reduce interfacial tension between water and oil. Sulfonate surfactants ($NaRSO_3$) where R is an alkyaryl group are frequently used. Polymers such as polyacrylamides and polysaccharides are used as mobility control buffers decreasing floodwater mobility.

Albers et al. (1985) found that Pennsylvania oil is often removed from sandstone formations typical of the Allegheny National Forest (Pennsylvania) area by high pressure injection of gas or water. To establish the wells, roads must be constructed, often leading to excessive sediment runoff. Oil field water is first brought to the surface and contains sodium, calcium, magnesium, strontium, barium, iron, manganese, sulfate, bicarbonate, chloride, bromide, and dissolved gases (Collins 1977). Relatively high concentrations of barium, strontium, and cadmium have been reported from Pennsylvania (Jaffe and Gemperlein 1985). In western Pennsylvania, Masteller (1980) found oil drilling operations increased levels of sodium, barium, potassium, and chlorides as well as alkalinity and conductivity. The chief constituents of Pennsylvania waste brines were chloride and sodium ions which were similar in composition to seawater (Van Gundy 1969). These brines contained more calcium and less magnesium than seawater. Calcium accounted for nearly 11 percent of the total solids. Brine-affected sites had values of 1,550 to 10,500 umhos/cm specific conductance (Van Gundy 1969). Oil field brine that is saturated with dissolved solids will precipitate salts such as NaCl when brought to the surface.

The oil-water mixture from the well is pumped into gravity separation tanks where the heavier water is drawn off the bottom of the tank and directed into settlement ponds. These ponds are drained from the bottom into the nearest stream or ravine. Albers *et al.* (1985) indicated that the most persistent source of oil contamination is the water discharge from settlement ponds. Water discharged from separators is usually high in salt (Crain 1969). Van Gundy (1969) found separator effluents to be approximately a 6:1 dilution of the connate water. Anderson (1974) reported salt concentrations in excess of 6,000 mg/l in streams affected by oil operations in Western Pennsylvania. Barton and Wallace (1979) stated that the disposal and storage of the residual, aqueous tailings which result from the refining oil and sand derived from open-pit mining constitute major environmental management problems. The tailings sludge was characterized as a thick slurry of inorganic particles ranging from clay to fine sand, mixed with globules of tar-like hydrocarbon material suspended in an aqueous solution of a variety of organic and inorganic compounds.

Gas drilling in Lake Erie was studied by Parker and Ferrante (1982). They had concerns about concentrations of suspended solids resulting from drill cuttings. Limits of 25 mg/l were recommended for suspended solids (measures in their study were 12.6 mg/l). Low impact was predicted if circulating fluid was ambient water and drilling fluid additives were limited.

GENERAL EFFECTS ON THE ECOSYSTEM

The effects of oil on the ecosystem include the following:

1. *Direct lethal toxicity* to organisms caused primarily by the low-boiling, water soluble aromatic fraction (benzene, toluene, xylene, and phenolic compounds) may cause death by coating, asphyxiation, or contact poisoning. Juvenile forms are most sensitive. The more volatile, "lighter fractions" are chiefly responsible for the high initial toxicity of fresh crude oil to aquatic organisms (Dickman 1971). Toluene is known to cause taste and odor problems (Buikema and Hendrick 1980). The short chain paraffins (< 10 carbons) are assumed to be toxic because of their solubility in water and interactions with lipoidal membranes. Polycyclic aromatic hydrocarbons interact with cells in two ways to cause toxic responses: they bind reversibly to lipophilic sites in the cell or their metabolites may bind covalently to cellular structures (Neff 1979). Bartha (1977) states that the

mechanism of toxicity is the dissolution of lipids which are an integral part of the cell membrane.
2. *Sub-lethal disruption of physiological* or *behavioral activities* may reduce many species' resistance to infection or stress. Lighter fractions of crude oil can affect yolk utilization, growth, hatching, oxygen consumption, feeding behavior and changes in cell permeability and histology (Buikema and Hendrick 1980). Aromatics may disrupt the orderly arrangement of grana in chloroplasts (Parker and Ferrante 1982).
3. Direct coating by oil *prevents light penetration* to plants and *increases temperature* by absorbing solar radiation. Coating of respiratory tissues is a mechanical impediment clogging vital processes of respiration and feeding (i.e., filter feeders).
4. *Incorporation of hydrocarbons into the food chain,* wherein a food loss in one section may be an energy crisis or short-term bonus to another, means support or destruction of a population (Cowell et al. 1979). Cowell et al. (1979) indicates, however, that oil does not bioaccumulate and biomagnify as do some of the persistent organics.
5. Changes in biological habitats, may occur through *alteration of substrate characteristics.* Each habitat type is in a different successional stage, so diversity and biomass may be at different levels—some are narrow niche specialists with low resilience and slow recovery while others may be mobile or highly resilient.
6. *Diversity changes and opportunistic species increase.* Numerous studies (Snow and Rosenberg 1975, Roeder *et al.* 1975, McCauley 1966, Krauss et al. 1973) have reported blue-green algae increases (potentially the source of increased nitrogen in such areas).
7. Oil at the air-water interface may act as a *physical barrier interfering with gas exchange.* Oxygen balance of streams may be affected by oils that exclude oxygen from the surface by forming a surface film, thereby hindering absorption of atmospheric oxygen. Surface films of oil are especially toxic to the hyponeuston (Hart and Fuller 1974). In fact, the basis of mosquito control dates back to about 1793 when whale oil or vegetable oil (50 yrs before petroleum was available) was used to control mosquitoes in rain barrels. Oil was the major mosquitocide until 1943 when DDT cut oil's use in half (Micks 1968). Oil may also absorb light wavelengths essential for photosynthesis.

Damage to the ecosystem is a result of the type of oil, volume spilled, and the length of contact (Roeder *et al.* 1975). Petroleum entering the freshwater ecosystem has two major effects on the indigenous microorganisms (Vestal *et al.* 1984). Oil may act as a carbon source to heterotrophic organisms, which in turn, through increased production, accelerate regeneration of nutrients for periphyton. Oil supplies inorganic nutrients which stimulate algal production

(Rosenberg and Wiens 1976). Oil may stimulate nitrogen-fixing bacteria, thus increasing the supply of nitrogen for periphyton. Oil may trap nutrients by sorptive processes as well as supply growth-stimulating compounds (Lock et al. 1981).

McCauley (1966) reported that petroleum effluents in a river killed plankton, the decomposition of which led to high nitrate-nitrogen values and to the release of soluble phosphate bacterial decomposition. Copeland and Dorris (1964) found that community metabolism in refinery holding ponds was higher than in most natural communities but lower than in sewage oxidation ponds. Dickman (1971) found primary productivity in an Inuvik Marsh was reduced tenfold from exposure to MacKenzie Valley Crude oil. Falk and Lawrence (1973) found suspended solids in drilling sumps were lethal to benthic organisms and toxic to spawning fish. Suspended solids from gas drilling could alter fish behavior, metabolism and reduction of feeding rates of indigenous phytoplankton and reduction of feeding rates and numbers of zooplankton (Parker and Ferrante 1982).

Hydrocarbon-degrading organisms represent opportunistic species in the oil-impacted habitat. Oil particles in water are decomposed by aerobic microbes including 100 species of bacteria, yeasts, and fungi (McCauley 1966). In another study, Bartha and Atlas (1977) list 37 genera of hydrocarbon-degrading microorganisms that have been isolated from the aquatic environment. Cairns and Buikema (1984) have reported numerous soil bacteria that degrade crude oil. When oil is ingested, it may pass through the organism bound to relatively dense fecal matter. Many organisms can degrade paraffinic and aromatic hydrocarbons and depurate them through feces and urine. Depuration of hydrocarbons is relatively rapid (one day to three weeks) (Neff 1979). Organisms exposed to chronic contamination or spills tend to depurate more slowly.

Another method of releasing oil into the environment is through benthic heterotrophic bacteria, which release dissolved oil in pelagic ecosystems by biochemical transformation, and these benthic consumers may transport oil to areas not initially affected (Cairns and Buikema 1984). Organisms with large surface-to-volume ratio were more susceptible to the effects of crude oil than others (Vestal *et al.* 1984). Biological effects are greatest and last the longest if oil reaches soft-bottom sediment, productive shallow waters, or low-energy shorelines (little wave action) (Cairns and Buikema 1984). Hoehn *et al.* (1974) reported that the effect of an oil spill did not affect the diversity but decreased total numbers indicating that each taxa was affected equally rather than selectively and that toxicity was due to the water soluble component of the oil. Van Gundy (1969) found numbers of individuals, numbers of species, and species diversity were negatively related to conductivity and the amount of dissolved solids. Masteller (1980) found numbers of individuals, numbers of species, and species diversity were related to the degree of oil contamination as did Anderson and McElravy (1977). Snow and Rosenberg (1975) found high mortality in the zoobenthos and pleuston in a lake the first two days after an oil spill.

In the case of oil sand tailings, a 60 percent reduction of the standing stock of benthic invertebrates was observed over a 4 week period. The stream bed was smothered by a fine sediment which covered breathing, feeding, and living surfaces and toxic components of the sludge were distributed on the benthos (Barton and Wallace 1979). Van Gundy (1969) reported that waste brines were apparently responsible for the abundance of approximately 35 species of macroinvertebrates in affected streams and for the presence of 18 species not collected in unaffected areas.

EFFECTS ON BIOLOGICAL SPECIES
MICROORGANISMS

Hydrocarbon-using microorganisms in pristine systems usually compose 10 percent or less (often less than 1 percent) of the total heterotrophs (Vestal *et al.* 1984). Experimental applications of oil result in increased numbers of bacteria (Cairns and Buikema 1984). The contribution of oil to the bacterial and fungal communities is obvious from the following research. Heitkamp and Johnson (1984) reported that chronic exposure to an oil field effluent stimulated the sediment microbial activity. Total numbers of microorganisms, especially those that degrade hydrocarbons, were elevated by oil-wastewater. This wastewater discharged into a river stimulated both electron transport system activity and carbon dioxide production in the sediments. The greatest increase was in ammonifiers, hexadecane degraders, protein hydrolyzers, and sulfate reducers. Sulfate reducing bacteria, such as the anaerobic *Desulfovibrio desulfricans,* may occur in oil field waters (Collins 1977). Iron bacteria*(Crenothrix, Gallionella,* and *Sphaerotilus*) sheath themselves in iron hydroxide and use the iron ions in the water. Aerobic slime formers, such as *Aerobacter, Bacillus, Escherichia, Flavobacterium,* and *Pseudomonas,* may also be present. Some of the opportunistic hydrocarbon bacteria that were found to be psychotrophic and used paraffinic, aromatic, and asphaltic compounds were *Pseudomonas, Brevibacterium, Spirillum, Xanthomonas, Alcaligenes,* and *Arthrobacter* (Cairns and Buikema 1984). They also report the following hydrocarbon degrading fungi from Northern Canada as follows: *Penicillium* spp., *Verticillium* spp., *Beauveria bassiana, Mortieriella* sp., *Phoma* sp., *Scolelcobasidium obovatum,* and *Tolypocladium inflatum.* Song *et al.* (1986) found hydrocarbon biodegradation maximal at a pH of 7.5-7.8 and n-hexadecane mineralization was 82 percent bacterial and 13 percent fungal.

Protozoan research at the species level is limited and most observations are general. Ciliates and flagellates are frequently associated with oil spills which Vestal *et al.* (1984) assumes is due to resilience. McCauley (1966) found *Trachelomonas* and *Vorticella* tolerant of oil pollution. Conversely, it has been observed that several aromatics cause swelling and eventual disruption of the plasma membrane (Neff 1979). Protozoans may accelerate the nutrient turnover increasing nitrates and phosphates which can limit factors in microbial degradation processes. Prokaryotic microorganisms convert aromatic hydrocarbons, by an initial dioxygenase attack, to trans-dihydrodiols that are further oxidized to dihydroxy products. Eukaryotic microorganisms using monooxygenases produce benzene 1, 2 oxide from benzene. In the presence of H_2O benzene 1, 2 oxide yields dihydroxy-dihydrobenzene, which is then oxidized to produce catechol, the key intermediate in biodegradation of aromatics (Bartha 1986).

ALGAE

The response of algae to crude oil appears to be species dependent. Krauss *et al.* (1973) reported that algal species differ markedly in their response to an oil spill, varying from inhibition of growth, to tolerance, to stimulation. They found that *Dinobryon sertularia* and *Peridinium* sp. had suppressed growth, while increased growth was recorded for *Anabaena flos-aqua, Ankistrodesmus convolutus,* and *Tabellaria fenestrata.* McCauley (1966) found the following algae tolerant of oil pollution: *Lyngbya, Oscillatoria* (Cyanophyceae); *Ankistrodesmus, Chlamydomonas, Closterium, Gonium* (Chlorophyta); *Scenedesmus, Asterionella, Cyclotella, Fragilaria, Meridion, Navicula, Tabellaria* (diatoms), and *Euglena* (McCauley 1966). Krauss *et al.* (1973) found Naphthalene (an aromatic hydrocarbon of crude oils) had a negative effect on the ability of *Chlamydomonas* cells to photosynthesize. They also indicate that short term toxicity derives from the rapid loss of volatile compounds. These compounds result in decreased growth and bicarbonate uptake of algal cells. Roeder *et al.* (1975) found that crude oil, whether experimentally spilled or applied directly to substrates, did not cause a decrease in periphyton biomass and on certain substrates it caused an increase. Cyanophyceae were always more abundant on oiled substrates than on non-oiled substrates. *Fremyella* sp. (coccoid blue-green) and three filamentous forms were present on oiled substrates along with large numbers of fungal filaments.

In addition to variations in species response to crude oil, seasonal variations also occur. In a study using Norman Crude oil (Roeder *et al.* 1975) diatom populations displayed seasonal periodicity. In June, *Achnanthes minutissima, Diatoma* sp. 1, *Synedra ulna,* and *Synedra* sp. 1 were prominent; in August, the 4 species mentioned here, as well as *Gomphonema angustatum* va. *producta* and *Cymbella ventricosa,* were abundant; and in September only *Achnantes minutissima* and *Cymbella affinis* remained on oiled plexiglass plates. When using styrofoam balls as substrate, they found *Achnanthes linearis* and *Cocconeis placentula* absent from oiled substrate, while *Cymbella affinis, Cymbella ventricosa, Synedra* sp. 1, and *Diatoma* sp. 1 were tolerant of oil. Oiled rocks supported a macroscopic algal community of *Closterium* sp., *Oedogonium* sp., *Bulbochaeta* sp., *Mougeotia* sp. with the red algae *Batrachospermum* sp. in small quantities. Rosenberg and Wiens (1976) found periphytic algal biomass increased on oil contaminated artificial substrates. This increase possibly may derive from a reduction in numbers or elimination of zoobenthic grazers by toxic fractions of the crude oil or from nutrients stimulating growth which could be supplied by the oil.

VASCULAR PLANTS

In terrestrial oil spills or pipe leaks the herbaceous vegetation is killed immediately. The mode of action of petroleum on plants is complex and involves both contact toxicity and indirect deleterious effects mediated by interactions of the petroleum with the abiotic and microbial components of the soil. Contact toxicity is primarily due to the low-boiling components of petroleum affecting tender portions of roots and shoots but may have little effect on the woody parts of trees and shrubs. Contact toxicity is due to the solvent effect on lipid membranes (Bossert and Bartha 1984). Trees and woody vegetation are destroyed slowly via their root-hairs and fine roots, which are subjected to an anoxic condition in the oiled subsoil. Conditions created by biodegradation of hydrocarbons by soil microorganisms exhausts the oxygen supply of the soil and root hairs may die due to lack of oxygen or toxic H_2S generated in the oxygen-depleted soil (Bartha 1977). These indirect effects, along with competition for nutrients by the oil-degrading microorganisms and decreases in soil air and moisture holding capacity, are deleterious to plants. Studies dealing with barley and carrots (Currier and Peoples 1954) reported that foliage odor was lost first, then dark areas appeared in the leaves followed by loss of turgor, wilting, and finally death. *Anacharis canadensis* (elodea) was killed by a half-saturated solution of benzene and cyclohexene within one hour.

TERRESTRIAL INVERTEBRATES

Crude oil devastates soil animals. Because of their higher lipid content and higher metabolic rates, soil invertebrates are likely to be more sensitive than plant roots. The higher-boiling and less-phytotoxic hydrocarbons may plug spiracles and breathing pores of microarthropods and interfere with respiration as has been suggested by the historic use of oil for mosquito larval control (Bossert and Bartha 1984). Nematodes have shown varied effects depending on the type of oil (Bossert and Bartha 1984).

ROTIFERS

McCauley (1966) found *Asplanchna, Keratella,* and *Polyarthra* tolerant of oil pollution.

PLATYHELMINTHES AND ANNELIDA

Several authors have noted the absence of planarians (Neel 1953, Lock 1981, McCauley 1966) in rivers receiving effluents from oil refineries in Wyoming, wastes from oil fields in New York, and a river in Massachusetts. Hirudinea (leeches) and *Tubifex* appear to be tolerant to refinery effluents (McCauley 1966). Van Gundy (1969) found *Tubifex* sp. tolerant of brine. But Barton and Wallace (1979) found burrowing and negatively phototropic organisms (i.e., Oligochaeta) significantly less abundant on oil sand than on rubble. Lock *et al.* (1981) found synthetic crude oil caused an increase in Naididae.

ARTHROPODA

CRUSTACEA

Daphnia middendorffiana and *Branchionecta paladosa* were eliminated within five days after an oil spill (Vestal *et al.* 1984). McCauley (1966) noted that *Gammarus* was conspicuously absent after an oil spill.

INSECTA

EPHEMEROPTERA

Rosenberg *et al.* (1980) found the following effects of crude oil on Ephemeroptera (mayflies) in Canada. *Heptagenia (flavescens* Walsh ?), *Stenonema vicarium* (Walker), *Ameletus* sp. 1, and *Baetis* spp. all decreased in response to oil, while *Pseudocloeon* sp. 1 and *Ephemerella aurivillii* Bengtsson increased. *Ephemerella bicolor* (Clemens ?) and *Ephemerella simplex* McDunnough appeared unaffected by oil contamination. These researchers believed that *H. (flavescens* ?) should be the best indicator of low level oil contamination because of its stable taxonomic position and ease of identification. In western Pennsylvania, Anderson and McElravy (1977) found the genera *Cinygmula* and *Epeorus* especially sensitive. Masteller (1980) found *Baetis* sp., *Baetis flavistriga* McDunnough, *Epeorus (Iron) fragilis* (Morgan), *Heptagenia pulla* (Clemens) and *Stenacron* sp. were the most sensitive to contamination.

Cloeon sp., *Paraleptophlebia debilis* (Walker), *Ephemerella invaria* (Walker) and *Ephemerella (Drunella)* sp. were somewhat tolerant. Barton and Wallace (1979) found that during high discharges, riffles were flooded to become pools and the rheophilic forms (i.e., *Baetis*) were then eliminated. Van Gundy (1969) reported that baetid mayflies appeared to be tolerant of brine.

ODONATA

Neel (1953) found oil to be lethal to dragonfly nymphs. McCauley (1966) found *Agrion* sp. absent from oiled areas and believed it could be used as an indicator organism. Snow and Rosenberg (1975) found that damselflies expired in a lake within 1-2 hours of a crude oil spill.

PLECOPTERA

Plecoptera were consistently lower in areas of oil sand exposure (Batron and Wallace 1979). Anderson and McElravy (Personnel Communication) found Plecoptera to be sensitive to oil drilling operations in western Pennsylvania. Masteller (1980) found the Peltoperlidae, Perlodidae, and Chloroperlidae to be the most sensitive to oil contamination in Pennsylvania. (Specifically *Peltoperla arcuata* Needham,*Isoperla* sp., *Hastaperla brevis* (Banks), *Hastaperla orpha* (Frison), *Sweltsa lateralis* (Banks), *Sweltsa naica* (Provancher), and *Sweltsa onkos* Ricker; some species of Nemouridae and Leuctridae were tolerant: *Amphinemura wui* (Claassen), *Amphinemura nigritta* (Provancher), *Leuctra ferruginea* (Walker) and *Leuctra tenella* (Provancher). Van Gundy (1969) found stoneflies were very rare in brine-affected areas.

HEMIPTERA

Hemiptera of the family Corixidae (*Sigara bicoloripennis, Sigara conocephala,* and *Sigara solensis*) were present in oil sand areas (Barton and Wallace 1979).

TRICHOPTERA

Barton and Wallace found Trichoptera consistently lower in areas of oil sand exposure (Barton and Wallace (1979). Masteller (1980) found Hydroptilidae (*Palaeagapetus celsus* (Ross) and *Hydroptila consimilis* (Morton) to be sensitive; and possibly Lepidostomatidae (*Lepidostoma griseum* (Banks). Hydropsychidae, Polycentropodidae, Psychomyiidae, Rhyacophilidae, and

Glossosomatidae all seem able to tolerate some contamination. Lock *et al.* (1981) found synthetic crude oil caused increases in Lepidostoma. In his collections from New York, Simpson (1980) found abdominal gills of Hydropsychidae with specks of a black tar-like substance on 18 of 40 specimens collected. In 6 other specimens, the gills were completely impregnated with this tar and the branches stuck together. All abnormal specimens were considerably smaller than the rest of the sample. Van Gundy (1969) found only *Hydropsyche bifida* able to tolerate brine affected areas.

COLEOPTERA

Snow and Rosenberg (1975) found that gyrinid beetles (*Gerris* sp.) expired within 1-2 hours after a crude oil spill on a lake.

DIPTERA

Of the Diptera, the Chironomidae seem to have been given the most attention in research related to oil contamination. Rosenberg and Wiens (1976) found the Orthocladiinae increased in species and individuals in oiled artificial substrates while Tanypodinae and Chironominae decreased. Herbivorous species were unaffected by crude oil. Rosenberg and Wiens (1976) evaluated 11 species as potential indicators of oil contamination: *Nilotanypus fimbriatus* (Walker), *Rheotanytarsus exiguus* (Johannsen), and *Polypedilum (Pentapedilum) tritum* (Walker) displayed a negative response to crude oil contamination; positive responses were shown for *Cricotopus bicinctus* (Meigen), *Cricotopus varipes* Coquillett, *Cryptochironomus fulvus* complex, *Polypedilum (Pentapedilum) fallax* (Johannsen), *Nanocladius (= Microcricotopus) alternantherae* Dendy and Subblette, *Eukiefferiella claripennis* (Lundbeck), *Eukiefferiella paucunca* Saether, and *Conchapelopia* sp. The authors indicated that reduction of *Nilotanypus fimbriatus* and increases of *Cricotopus bicinctus* and *C. varipes,* individually or as an assemblage, are the best indicators of oil or petroleum product contamination. *N. fimbriatus* numbers are depressed by the presence of petroleum, and *C. bicinctus* and *C. varipes* will be increased relative to control areas. From observations in western Pennsylvania, Masteller (1980) agrees that *N. fimbriatus* is a good indicator. He also found several species only in uncontaminated streams: *Zavrelimyia thrytica* (Sublette), *Gymnometriocnemus* sp., and *Smitti* sp. Other species that may prove to be good indicators are *Parachironomus* sp. and *Paratendipes albimanus* (Meigen). Barton and Wallace (1979) found several species of Chironmidae above oil sands and not in or below them; *Microtendipes* cf. *pedellus, Heterotrissocladius* cf. *marcidus, Nemocladius* cf. *rectinervis,* and *Synorthocladius.* McCauley (1966) found *Tendipes* was tolerant. Lock (1981) generalized that synthetic crude oil caused significant increases in Chironomidae. Micks (1968) reported that the primary lethal effect of oil on mosquito larvae was from the volatile gases in the oil. Masteller (1980) found that Simuliidae species did not occur in contaminated areas except for *Simulium tuberosum* (Lundstrom). Seven other species were found only in uncontaminated streams. Barton and Wallace (1979) found Tipulidae and Empididae accounting for 90 percent of the Diptera in oil sand areas.

MOLLUSCA

Mussels have a relatively slow rate of depuration of petroleum hydrocarbons. Bivalves bioconcentrate PHC (polycyclic hydrocarbons) from the water in both dissolved and particulate phases (Cairns and Buikema 1984). Loch and Gregory (1973) found *Psidium* sp., *Spherium* sp., and *amnicola* sp. unable to tolerate oil refinery effluents in a river. Van Gundy (1969) found that the gastropods, *Physa* sp. and *Lymnaea* sp. appear brine dependent.

FISH

The life history stage of the species of fish is highly relevant to its susceptibility to adverse effects of oil contamination. Year-class strength of recruitment of many species is determined by the survival through critical larval stages. Woodward *et al.* (1979) reported that Wyoming crude oil at a level of 520 ug/l concentration resulted in a 48 percent reduction in survival of cutthroat trout (*Salmo clarkii*). At levels between 450-520 ug/l, slower growth rates resulted and induced retinal and lens lesions. Swift et al (1969) found diesel fuel in concentrations of 350-1000 mg/l acutely toxic to rainbow trout. Gasoline was lethal at 100 mg/l and jet fuel at 500 mg/l. Falk and Lawrence (1973), using Lake chub (*Couesius plumbius*), rainbow trout (*Salmo gairdneri*), and ninespine stickleback (*Pungitius pungitius),* discovered that chlorides were the main active poisonous components at lower concentrations. They found 13 of 27 drilling fluids lethal at concentrations below 1000 mg/l. Sodium pyrophosphate and sodium polyphosphate were lethal to marine invertebrates and fish at concentrations of 500-7500 mg/l. Caustic soda, oil well cement, and white lime were toxic at 70-450 ppm. Peltex was lethal to trout at 1000 mg/l. Kelzam Polymer Al was lethal at 5 ppm but Kelzam XC Polymer was non-toxic to 500 mg/l.

Pessah *et al.* (1973) used Rainbow trout and fathead minnows (*Pimephales promelas* Rafinesque) to test acute toxicity of petroleum refinery effluent. They found zinc and cyanide in all six of the refineries studied. They referred to a study by McKee and Wolf that 0.1-4 mg/l of cyanide killed trout and toxicity was found to be greatly affected by synergism. Pessah *et al.* (1973) also found ammonia-nitrogen levels of 160 mg/l and phenols at 3.3 mg/l were lethal. The toxic actions of many chemicals were affected by temperature, pH, synergism, and the form in which they were presented. Fremling (1981) found that sunfish and waterfowl were sufficiently stressed by a No. 6 fuel oil spill to trigger a kill by the bacteria, *Flexibacter columnaris,* but there was no problem in the fish flesh.

BIRDS

Patton and Dieter (1980) reported that the two factors primarily responsible for the high mortality rate of birds exposed to oil were external oiling of waterfowl and ingestion of petroleum hydrocarbons. Ducks ingest significant amounts of oil while preening contaminated feathers but without deleterious effects. Gay *et al.* (1980) fed mallard ducks a number of different aromatic and aliphatic hydrocarbons after which the highest concentrations were found in the fat following analysis of the liver, kidney, fat, and brain tissue. Stickel and Dieter (1979) found that crude oil from South Louisiana and Prudhoe Bay reduced hatching of mallard ducks by 90 and 26 percent respectively, when applied to 8 day-old eggs. Ducks eating crayfish with napthalene had the most accumulation in the gall bladder and fat followed by uptake in blood, brain, liver, and kidney in that order (Tarshis and Rattner 1982). Hoffman (1979) applied crude oil and aliphatic hydrocarbons to mallard eggs. The aliphatic hydrocarbons (paraffin) had no effect on survival, but crude oil caused a reduction in crown-rump length, shorter and deformed bills, incomplete ossification of the phalanges, smaller liver lobes, and incomplete feather formation.

RECOMMENDATIONS

Falk and Lawrence (1973) recommended that waste fuels, lubricants, detergents, and other substances used during drilling operations should be contained and disposed of by means other than discharge into a sump. Flow into streams could be monitored by insect occurrence, as their life cycle is long enough to observe the effects of chronic contamination. Unused wells must be capped. Sedimentation from road construction should be controlled. Restricting the season of construction activities and runoff containment are necessary. The geology of the area needs to be surveyed critically for the impact of EOR and hydrofracking in regard to underground water tables. Monitoring wells should be used for determining water quality. Reestablishing plant cover in terrestrial environments requires biodegradation of oil on the surface through pH control, as well as fertilization and tillage. Reclamation of streams may occur if contaminants are eliminated.

LITERATURE CITED

Albers, P.H., A.A. Belisle, D.M. Swineford, and R.J. Hall. 1985. Environmental contamination in the oil fields of western Pennsylvania. *Oil and Petrochemical Pollution* 2:265-280.

Anderson, J.K. 1974. Current status of cooperative investigations of problems related to oil recovery operations in northwestern Pennsylvania. U.S. Fish and Wildlife Report, 7 pp.

Anderson, F.K. and E.P. McElravy. 1977. Effects on bottom fauna of chronic oil discharges into a northwest Pennsylvania stream. Personal communication.

Bartha, R. 1977. The effect of oil spills on trees. Journal of Arborculture 3:180.

Bartha, R. 1986. Biotechnology of petroleum pollutant biodegradation. Microb. Ecolo. 12:155-172.

Bartha, R. and R.M. Atlas. 1977. The microbiology of aquatic oil spills. *In:* d. Perlman (Ed.) Adv. Applied Microbiology, Acad. Press, N.Y., 22:225-266.

Barton, D.R. and R.R. Wallace. 1979. The effects of an experimental spillage of oil-sand tailings sludge on benthic invertebrates. Environ. Pollut. 18:305-312.

Barton, D.R. and R.R. Wallace. 1979. Effects of eroding oil sand and periodic flooding on benthic macroinvertebrate communities in a brown-water stream in Northeastern Alberta, Canada. 57:533-541.

Berger, B.D. and K.E. Anderson. 1981. Modern petroleum. Pennwell Books, Tulsa, OK, 255 pp.

Bossert I. and R. Bartha. 1984. The fate of petroleum in soil ecosystems. *In:* R.B. Atlas (Ed.) Petroleum Microbiology, Macmillan Pub., pp. 435-473.

Buikema, A.L. and A.C. Hendricks. 1980. Benzene, xylene, and toluene in aquatic ecosystems: a review. petro. Inst., Virginia Poly Inst. and State University, Blacksburg, VA, 69 pp.

Cairns, J., Jr. and A.L. Buikema, Jr. 1984. Restoration of habitats impacted by oil spills. Butterworth Pub., Boston, 182 pp.

Collins, A.G. 1977. Enhanced oil recovery injection waters. *In:* D.C. Wright, et al. (Ed.) *American Society for Testing and Materials.*

Cooney, J.J. 1984. The fate of petroleum pollutant in freshwater ecosystems. *In:* R.M. Atlas (Ed.) Petroleum Microbiology. Macmillan Pub., pp. 399-433.

Copeland, B.J. and T.C. Dorris. 1964. Community metabolism in ecosystems receiving oil refinery effluents. Limnol. Oceanogr. 9:431-447.

Cowell, E.B., G.V. Cox, and G.M. Dunnet. 1979. Applications of exosystem analysis to oil spill impact. Proceedings 1979 Oil Spill Conf., Los Angeles, Ca., 19-22 %, March 1979, pp. 517-519, Amer. Petroleum Inst., Washington, D.C.Pub. No. 4308, pp. 22, pub. research.

Crain, L.J. 1969. Ground-water pollution from natural gas and oil production in New York. Conservation Dept. Water Resources Commission, New York Rep. Invest. 5, pp. 14.

Currier, H.B. and S.A. Peoples. 1954. Phytotoxicity of hydrocarbons. Hilgardia 23(6):155-173.

Dickman, M. 1971. Preliminary notes on changes in the algal primary productivity following exposure to crude oil in the Canadian Arctic. *J. Fish Res. Board of Canada* 33:249-251.

Falk, M.R. and M.J. Lawrence. 1973. Acute toxicity of petrochemical drilling fluids components and wastes to fish. Environment Can. Fish. Mar., Tech. Rep. Serv. No: DEN T-73-1, Resource Management Branch, Fisheries and Marine Service, Operations Directorate, Department of the Environment, Freshwater Institute, Winnipeg, Manitoba, pp. 1-108.

Fremling, C.R. 1981. Impacts of a spill on No. 6 fuel oil on Lake Winona. Prev. Behavior Control Cleanup Conf., API Publ. 419-421:4334.

Gay, M.L., A.A. Belisle, and J.F. Patton. 1980. Quantification of petroleum-type hydrocarbons in avian tissue. *J. Chromatogr.* 187:153-160.

Hart, C.W., Jr. and S.L.H. Fuller. 1974. Pollution Ecology of Freshwater Invertebrates. Academic Press, N.Y., 389 pp.

Heitkamp, M.A. and B.T. Johnson. 1984. Impact on an oil field effluent on microbial activities in a Wyoming river. Can. J. Microbiol. 30:786-792.

Hoehn, R.C., J.R. Stauffer, M.T. Masnik, and C.H. Hocutt. 1974. Relationships between sediment oil concentrations and the macroinvertebrates present in a small stream following an oil spill. Environ. Lett. 7:345-352.

Hoffman, D.J. 1979. Embryotoxic and teratogenic effects of crude oil on mallard embryos on day one of development. Bull. Environ. Contam. Toxicol. 22:632-637.

Jaffe, M. and J. Gemperlein. 1985. Ravaged land, poisoned water: How the energy scramble fouled Pennsylvania. The Philadelphia Inquirer. Sunday, Feb. 3, 312:1,20,21,22.

Krauss, P., T.C. Hutchinson, C. Soto, J. Hellebust, and M. Griffiths. 1973. The toxicity of crude oil and its components to freshwater algae. Proceedings Jt. Conf. on Prevention and Control of Oil Spills, #4172:703-714.

Loch, J.S. and L.A. Gregory. 1973. A benthic survey of the Red River in the vicinity of an oil refinery. Technical Report Series No. CEN/T-73-11, Resource Management Branch, Fisheries and Marine Service, Operations Directorate, Department of the Environment, Central Region, pp. 1-26.

Lock, M., R.R. Wallace, and D.R. Barton. 1981. The effects of synthetic crude oil on microbial and macroinvertebrate benthic river communities—Part I: colonization of synthetic crude oil contaminated substrata. Environ. Poll. (Series A) 24:207-217.

Masteller, E.C. 1980. The impact of oil drilling operations on aquatic insects. U.S. EPA Report #68-03-2647. The Pennsylvania State University, The Behrend College, Erie, Pa., 63 pp.

McCauley, R.N. 1966. The biological effects of oil pollution in a river. Limnol. Oceanogr. 11:475-486.

Micks, D.W. 1968. Petroleum hydrocarbons in mosquito control. Soaps, Cosmetics and Chemical Specialties 44:80-88.

Neel, J.K. 1953. Certain limnological features of a polluted irrigation stream. Amer. Microscopical Soc. Trans. 72:119-135.

Neff, J.M. 1979. Polycyclic aromatic hydrocarbons in the aquatic environment. Applied Sci., London, 262 pp.

Nelson, R.C. 1982. Application of surfactants in the petroleum industry. J. Am. Oil Chem. Soc. 59(10):823A-826A.

Patton, J.F. and M.P. Dieter. 1980. Effects of petroleum hydrocarbons on hepatic function in the duck. Comp. Bioch. Physiol. 65:33-36.

Parker, J.I. and J.G. Ferrante. 1982. A survey of discharges from a natural gas drilling operation in lake Erie. Environ. Sci. Technol. 16:363-367.

Pessah, E., J.S. Loch, and J.C. MacLeod. 1973. Preliminary report on the acute toxicity of Canadian petroleum refinery effluents to fish. Technical Report No. 408. Fish. Res. Board Can., Freshwater Institute, Winnipeg, Manitoba, pp. 1-43.

Roeder, D.R., G.H. Crum, D.M. Rosenberg, and N.B. Snow. 1975. Effects of Norman Wells crude oil on periphyton in selected lakes and rivers in the Northwest Territories. Tech. Rep. 552 #69. Dept.Env. Fisheries and marine Service Research and Develop., 31 pp.

Rosenberg, D.M. and A.P. Wiens. 1976. Community and species responses of Chironomidae (Diptera) to contamination of fresh waters by crude oil and petroleum products, with special reference to the Trail River, Northwest Territories. *Fisheries Research Board of Canada* 33:1955-1963.

Rosenberg, D.M., A.P. Wiens, and J.F. Flannagan. 1980. Effects of crude oil contamination of Ephemeroptera in the Trail River, Northwest Territories, Canada. Adv. in Ephemeroptera Biology pp. 443-455.

Simpson, K.W. 1980. Abnormalities in the tracheal gills of aquatic insects collected from streams receiving chlorinated or crude oil wastes. Freshwater Biol. 10:581-583.

Strickel, L.F. and M.P. Dieter. 1979. Ecological and physiological/toxicological effects of petroleum on aquatic birds. U.S. Fish Wildl. Serv. Biol. Serv. Program. FWS/OBS-79/23:1-14.

Snow, N.B. and D.M. Rosenberg. 1975. Experimental oil spills on Mackenzie Delta Lakes. II. Effect of two types of crude oil on lakes 4C and 8. Technical Report No. 549, Department of the Environment, Fisheries and Marine Service, Research and Development Directorate, Winnipeg, Manitoba, pp. 1-19.

Swift, W.H., C.J. Touhill, W.L. Templeton, and D.P. Roseman. 1969. Oil spillage prevention, control, and restoration—state of the art and research needs. J. Water Pollut. Control Fed. 41:392-412.

Song, H.G., T.A. Pederson, and r. Bartha. 1986. Hydrocarbon mineralization in soil: relative bacterial and fungal contribution. Soil and Biol. Biochem. 18:109-111.

Tarshis, I.B. and B.A. Rattner. 1982. Accumulation of 14C-Naphthalene in the tissues of redhead ducks fed oil-contaminated crayfish. Arch. Environ. Contam. Toxicol. 11:155-159.

Van Gundy, J.J. 1969. The effects of oil field wastewaters upon the diversity and abundance of macroinvertebrates in a woodland stream. MA thesis. The Pennsylvania State University. 47 pp.

Vestal, R., J.J. Cooney, S. Crow, and J. Berger. 1984. The effects of hydrocarbons on aquatic microorganisms. *In:* R.M. Atlas (Ed.) Petroleum Microbiology. Macmillan Pub., pp. 476-505.

Woodward, D.F., P.M. Mehrle, Jr., and W.L. Mauck. 1979. Effects of a Wyoming crude oil on survival of cutthroat trout (*Salmo clarki*). Report to U.S. Fish and Wildlife Service, Columbia National Fisheries Research Laboratory, U.S. Department of Interior, Columbia, MO, pp. 1-16.

Environmental Consequences of Energy Production: Problems and Prospects. Edited by S. K. Majumdar, F. J. Brenner and E. W. Miller. © 1987, The Pennsylvania Academy of Science.

Chapter Thirteen
RECLAMATION OF URANIUM MINING AND MILLING DISTURBANCES
EUGENE E. FARMER[1] and GERALD E. SCHUMAN[2]

[1]Research Hydrologist
USDA Forest Service
Intermountain Research Station
Forestry Sciences Laboratory
Logan, Utah 84321
and
[2]Soil Scientist
USDA Agricultural Research Service
High Plains Grasslands Research Station
Cheyenne, Wyoming 82009

INTRODUCTION

Most of the uranium resources of the United States are located in the western part of the country. The states of New Mexico, Wyoming, Colorado, Utah, and Texas hold about 80 percent of the total probable uranium resources. However, uranium also occurs in the states of Washington, South Dakota, North Dakota, Idaho, California, Arizona, Nevada, Oregon, Pennsylvania, and Tennessee. In the western United States most of the uranium deposits are found in sandstones. However, a small amount of uranium has been obtained commercially from uraniferous lignite. Underground (deep) mining, surface mining, and solution mining methods have been used. In sandstone formations the uranium occurs as a coating on the grains of sand in the formation. The uranium content of these ores is on the order of 0.05 to 0.25 percent by weight. These low ore values dictate that large volumes of waste, both as overburden from open pit mining, and as mill tailings, are produced in the uranium recovery operation. Uranium also occurs in association with a variety of other trace elements, including arsenic, cobalt, copper, molybdenum, nickel, lead, selenium, vanadium, and zinc. Any of these metallic ions can occur in substantial quantities in mill tailings!

Since 1945 the history of uranium mining and milling in the United States has been a story of wide fluctuations in market prices and in mining and milling capacity. Immediately following World War II uranium changed almost overnight from a mineral commodity of minor commercial interest to a mineral vital to weapons production and national security. Following the Atomic Energy Act of 1946, the prospecting, mining, and milling of uranium was encouraged by the Atomic Energy Commission through price guarantees, bonuses, allowances, and ore-buying stations. During the early 1960's the number of operating mills producing concentrated uranium oxide, called yellowcake (U_3O_8), varied from 22 to 26. The annual production of yellowcake was about 15 thousand metric tons from the treatment of about 7 million metric tons of uranium ore. The late 1960's and the 1970's saw a sizeable reduction in the production of yellowcake because of an earlier over-supply, a leveling off of the military demand, and a failure of the nuclear electric power industry to create the anticipated commercial demand. The decline in the domestic production of yellowcake has continued through the early 1980's to the present. Today, there are five operating uranium mills in the United States: one in Wyoming, two in Utah, one in New Mexico, and one in Texas. Of these five mills, three are operating on a reduced schedule, as little as three days a month. A significant portion of the current United States production of uranium goes overseas to fulfill Japanese, French, and other European contracts.

There is still a sizeable reclamation job to be accomplished on old uranium wastes, both tailings impoundments and overburden embankments. Before the Uranium Mill Tailings Control Act of 1978 (PL 95-604), reclamation was frequently omitted altogether, or else done in a haphazard fashion. We do not know the total area of unreclaimed, radioactive, uranium overburden wastes in the western United States, but the area is large, probably several thousand hectares. Fortunately, these overburden wastes are almost entirely located in remote areas. Mill tailings are more difficult to reclaim than overburden, and tailings represent a more serious health hazard. There are approximately 25 million metric tons of unreclaimed uranium mill tailings, with variable health hazards, located in the United States.[2]

THE URANIUM MINING RECLAMATION PROBLEM

Reclamation Uncertainties

Since 1970 reclamation technology has been highly developed in this country. A growing environmental awareness and concern by the people, a more environmentally responsible mining industry, the passage of state and federal reclamation laws, combined with adequate reclamation research funding, has provided the United States with a technological reclamation base that is unmatched by any country. However, certain reclamation problems persist. Past reclamation efforts on uranium waste have, largely, been unsuccessful. These failures have been partially due to the operators' unrealistic attitudes regarding reclamation, the nature of the waste material itself, the adverse character of the climate in most uranium production areas, and the unusual demands of uranium reclamation. We will examine each of these factors: the waste materials, the climatic influences, and the unusual demands associated with reclamation of uranium wastes.

Uranium Mining and Milling Wastes

Wastes left over from mining and milling, i.e., the contaminated overburden spoils, acid-leach piles, and tailings, are radioactive and potentially hazardous. There are 14 naturally occurring isotopes of uranium; the most abundant of these is uranium-238. Therefore, uranium-238 is the isotope of greatest concern.[3] However, there are other important isotopes that can potentially be translocated upward by vegetation. These are uranium-234, thorium-230, radium-226, lead-210, and polonium-210.[4,5,6] Another important isotope, radon-222, is a radioactive noble gas that is released from tailings. Radon-222 has a half-life of 3.82 days and may be wind blown long distances, exposing human populations at very low levels.[7]

In addition to the considerations associated with radioisotopes, uranium wastes may pose a variety of chemical and physical difficulties to successful reclamation. Some waste materials have pH values greater than 9 or lower than 5.5. In the latter case, the wastes will be strongly acidic and will frequently exhibit high levels of soluble heavy metal ions in the spoil water. These ions are potentially toxic to vegetation and to receiving surface waters. If the pH is greater than 9, the materials will be strongly sodic and/or alkaline. Boron may be at high levels under these conditions. Boron is a potential toxicant to most vegetation. Salt concentrations may be too high for effective revegetation. Besides the chemical extremes that may or may not exist on a specific site, most spoils are sterile in terms of their microbiological component and are extremely infertile in terms of their plant nutrient content. Soil organic matter and cation exchange capacities are low, and the ability of the site to recycle plant nutrients will be marginal.

In addition to adverse spoil chemistry, uranium wastes often exhibit a variety of physical properties that hinder reclamation. Nearly all mill tailings will exhibit fine texture; sometimes 20 to 60 percent of the tailings materials will be in clay size particles. Conversely, overburden wastes will normally be coarse textured and droughty; but, a sandy-loam or gravelly-loam texture would not be unusual. Whenever wastes contain more than 33 percent clay or more than 66 percent sand, they will be difficult to revegetate. Waste materials often exhibit high bulk density values, i.e., values greater than 1.4 g/cm^2. High bulk densities are associated with inadequate plant root penetration, low water infiltration rates, and low soil-water holding capacities. Some waste materials develop tough surface crusts, often induced by sodic deflocculation of the soil particles, with consequent increases in soil cohesion. Finally, uranium mining and milling wastes may exhibit low values of water saturated hydraulic conductivity, which reduces water movement in the spoils and may restrict water use by vegetation.

Revegetation Under Arid Conditions

In those states where the greatest volumes of uranium wastes are located, viz., New Mexico, Wyoming, Utah, and Colorado, uranium deposits are typically located in arid or semiarid areas. These areas are also most likely to see future renewed interest in uranium mining and milling. Therefore, the potential for vegetative reclamation to stabilize spoils and to restore productivity to a site should be considered in light of our demonstrated ability under similar situations. Our track record is not good, but there have been successes.[8]

In the semiarid and arid portions of the western United States, rainfall is not only sparse but also highly variable from year to year. Annual precipitation amounts range from 150 to 300 mm at the lower elevations, to 500 mm at the higher elevations. About half of the precipitation falls during the summer season, July through October. The spring and fall months tend to be dry. The driest months are April, May, and June. Daytime high temperatures during the winter vary from about $-7°C$ to $7°C$. Summer high temperatures vary from about $22°C$ to $38°C$. Summer low temperatures are about $14°C$.[9] High wind velocities occur throughout the year. Climatic conditions in the more northerly, semiarid, areas are somewhat more favorable than those described. For instance, in parts of Wyoming and Colorado the precipitation distribution is more favorable; the spring season is often a wet period and winter precipitation may amount to half of the total. Nevertheless, the point remains the same: in the most likely uranium production areas, climatic influences deter successful vegetative reclamation.

What can reasonably be expected of revegetation under such adverse climatic conditions; what is successful revegetation? These questions tend to be site specific and are open to interpretation.[10] Assuming a conscientious effort, using the best technology, it seems reasonably certain that a stand of grasses, forbs, and shrubs can be established. But, in arid and semi-arid regions vegetation establishment is episodic, occurring only in years with favorable precipitation. Once the stand is established it is likely that it would be self-sustaining and "permanent" for at least 1000 years.[11] However, the permanency of vegetative stands under arid and semi-arid conditions is problematical. Vegetative stands are subject to numerous perturbations over a period of a 1000 years. For instance, drought, flood, erosion, grazing, insects, wildfire, and disease can all be expected to influence the ability of vegetative stands to sustain themselves. Establishing the nutrient cycle on revegetated lands is an absolute requirement if vegetative stands are to be permanent.

The importance of soil organic matter to nutrient cycling on reclaimed uranium disturbances was reported by Woods and Schuman.[12] Organic matter concentrations in revegetated uranium mined lands ranged from 1 to 21 g of organic matter per kg of soil. The authors suggest that sustainable plant nutrient cycles require more than 1 but less than 7 g of organic matter per kg of soil. Plant nitrogen concentrations and plant biomass were more closely related to mineralizable carbon and nitrogen than to soil organic matter. Long term plant fertilization would be required to maintain vegetal stands if levels of soil organic matter are insufficent to maintain the nutrient cycle.

Assuming the most likely condition of a self-sustaining stand of vegetation, some vegetative stand characteristics may be undesirable. From the viewpoint of isolating radioactive uranium wastes, perhaps the key stand characteristic is ground cover. It is likely that aerial ground cover will be substantially less than 25 percent. Under arid conditions, such low ground cover values are dictated by the competition for available water and nutrients. Low values for vegetal ground cover also imply that soil erosion from the site may exceed the rate of soil formation or deposition, and that covered uranium wastes might eventually be exposed. Therefore, it appears that if uranium wastes are to be isolated with a high degree of certainty, for long periods of time (1000 years), vegetative reclamation alone will not be adequate. The biological reclamation techniques should be supplemented with sound engineering practice.

Reclamation Goals for Uranium Wastes

Reclamation goals for nontoxic, nonhazardous mining disturbances can assume many different characterizations. There may be disagreements about objectives and techniques, post-mining land use, or the conditions that constitute successful reclamation;[13] these disagreements often involve value judgments. In any case, goals for reclamation normally emphasize the restoration of a level of site productivity that is commensurate with the post-mining land use objectives. On uranium reclamation sites the goals are quite different.

Uranium wastes from inactive mining or milling operations pose a health hazard.[14] The legitimate concerns for site productivity and post-disturbance land use are secondary to the need for isolating all radioactive waste materials and taking measures to ensure that they stay on-site. The concerns for isolating and immobilizing the material are codified in law and regulation: the Uranium Mill Tailings Radiation Control Act of 1978; reclamation rules in the Federal Register, volume 45, page 65521; EPA Proposed Standards for Inactive Uranium Processing Sites (40 CFR 192); regulatory guides issued by the Nuclear Regulatory Commission; and most recently the EPA Final Rule for Radon-222 Emissions from Licensed Uranium Mill Tailings.[15] These laws and regulations generally require the following treatment of uranium tailings:

> 1. Sufficient earth and/or rock cover should be placed over tailings to reduce radon-222 exhalation to specified radiation levels above natural background levels. The current specified level is 20 $pCi/m^2/s$.
> 2. A self-sustaining, permanent, vegetative cover should be established on earth covers. In this case "permanent" means lasting for 1000 years.
> 3. If it is unlikely that a full cover of self-sustaining vegetation can be established or maintained, rock can be used as the stabilizing cover.

4. A rock cover should reduce the tailings erosion potential to a negligible amount. Potential erosion of a rock cover, and specifications for rock (riprap) depth and size, are calculated from design elements of the Probable Maximum Flood for the area of interest. A rock cover shall normally last 1000 years, but under some circumstances may be designed for 200 years.

5. Post-reclamation topographic features, including embankments, should provide protection against wind and water erosion.[16]

The laws and regulations are aimed specifically at uranium mills and mill tailings. Where uranium overburden wastes pose a health hazard the same rules should apply, but that is not necessarily the case. Clearly, these rules call for a combination of biological and engineering techniques to contain uranium wastes; site productivity and future land uses have a low priority.

Potential Health Effects

The possibility of radon-222 gas exhalation by uranium spoils has already been mentioned. Due to their carcinogenic nature, radon-222 and its daughters, principally lead-210, have been of concern within uranium mines and near mill tailings for more than 20 years. Since radon-222 is a noble gas, it can be transported long distances downwind from uranium wastes. Human populations at considerable distances from uranium mill tailings have been studied, and dosage estimates have been made.[17] Open air inhalation of radon-222 probably is not a significant hazard for persons living more than 2 km downwind from uranium tailings. However, homes that are built on or near uranium milling or mining wastes may contain hazardous levels of radon-222.[18,19]

Radionuclides can also be released from uranium spoils by a variety of biological vectors.[20,21] The principal concern is the upward translocation of radionuclides by vegetation. Other concerns are that vegetal cover will encourage the establishment of populations of burrowing mammals and invertebrates. Such animal populations can lead to increased levels of radon gas exhalation, or to the release of radioactive wastes by bringing them to the surface. The scale and consequences of potential animal activity are not well documented. However, as previously mentioned, the uptake of radionuclides by vegetation is documented,[5,6,7,22] but only a small number of plant species have been tested.

Radionuclides, Forage Quality, Grazing, and Predation

The concern for radionuclides and toxic elements entering the food chain is directly related to the long term containment of uranium mill tailings and overburden spoils. Dreesen and Marple[23] reported that four-wing saltbush (*Atriplex canescens*) had elevated concentrations of molybdenum, selenium, radium-226, uranium-238, arsenic, and sodium when grown on tailings covered with 5 cm of soil. Alkali sacaton (*Sporobolus airoides*) also contained high levels of molybdenum, selenium, radium-226, and nickel when grown on tailings. In both plant species the levels of molybdenum and selenium were considered to be toxic to grazing animals. However, this was a greenhouse study and may not be representative of a field condition when the same plants are grown in tailings covered with a thicker layer of soil.

Stanley et al.[24] sampled western wheatgrass (*Agropyron smithii*) growing on untopsoiled, reclaimed, uranium spoil covered with 1-2 m of overburden materials. Wheatgrass samples were also collected from an adjacent rangeland site. Nutrient and metal concentrations in the forage were compared for the two sites; the soil pH was about 8 in both areas. None of the forage contained toxic levels of metallic ions.

To further evaluate the radiological risk to the food chain, Doerges and Kennington[25] sampled a variety of small animals, birds, and vegetation in the vicinity of a uranium mine in Wyoming. They collected raptorial birds, sage grouse, mourning doves, mammalian predators, mice, ground squirrels, several plant species and soil samples, both within and outside of the mining district. The soil and plant samples collected adjacent to the mine were significantly higher in radium-226 than samples collected away from the mine. Over 3000 samples of soft animal tissue were analyzed, but none of the samples showed alpha radiation. On the other hand, analysis of animal bone tissue did show measurable alpha activity. Herbivorous mammals showed significantly greater radioactivity than did carnivorous mammals. The same was true for herbivorous birds compared to carnivorous birds. These data suggest that grazing vegetation puts more radioactivity into the food chain than animal predation. This may be partially due to the fact that carnivores do not eat a large portion of animal bones, which is the body part that showed the greatest level of alpha radioactivity.

Schuman et. al.[26] conducted a grazing study on 49.2 ha of revegetated uranium spoils that were topsoiled and reseeded in 1974-1976. Four 12.3 ha pastures were established; two were grazed by 3 steers each, and two were grazed by 6 steers each. During the study period the plant basal ground cover increased. Steer gains between the two grazing intensities were not different from one another and were equal to or greater than gains obtained from nearby unmined rangeland sites. The authors concluded that these reclaimed lands are capable of supporting post-mining grazing without detrimental effects to the plant community. In 1980-1981, in conjunction with the grazing study, Doerges and Kennington[25] collected samples of vegetation from each pasture and slaughtered one steer from each pasture. In each year they also slaughtered a (control) steer that had not grazed on revegetated uranium spoils. The grass samples showed that stems and seeds contained higher levels of radioactivity than other plant parts. Grass samples from the reclaimed uranium spoils also showed greater levels of radioactivity than grass samples from the unmined rangeland. In 1980, bone samples from the steers grazing the revegetated uranium spoils showed lower levels of radioactivity than did the bone sample from the steer grazing the (control) unmined rangeland. In 1981 there were no differences in the levels of radioactivity in any of the bone samples. The authors suggest that these reclaimed uranium overburden spoils can be grazed without concern for the effects of radioactivity in the animal or its consumers.

RECLAMATION TECHNIQUES

A significant portion of the reclamation technology that is applicable to uranium reclamation was developed for coal mined lands. Given similar soils/spoils, landforms, and climate the coal reclamation methods and techniques are generally applicable. However, uranium and its radioactive daughters require additional considerations.

Revegetation
There is a voluminous literature on revegetation of mining disturbances in the western United States, although some of it is not applicable to uranium revegetation. However, it is not our intent to summarize the revegetation literature. Rather, we will suggest elements of revegetation that should not be overlooked.

As a general case, it can be asserted that the current reclamation technology is adequate for the revegetation of all uranium mined lands. However, some situations present considerable technical difficulties and increase reclamation costs. Primarily, these are phytotoxic soils/spoils, and extreme aridity. The radioactivity associated with uranium reclamation may present special considerations, such as the vegetal uptake of radionuclides.

It is important to develop a detailed reclamation plan well in advance of the actual revegetation operation. Part of the advance planing for uranium revegetation includes health considerations for humans.[27] Advance planning allows time to have the plan constructively reviewed by a variety of experts, to accumulate the needed plant materials, soil amendments, people, and reclamation equipment, and allows time for mandated regulatory review.

Prior to the actual revegetation operation, the mill tailings, overburden spoils, topdressing soils, and other disturbed soils should be extensively sampled to determine the radiological, chemical,[28,29] and physical properties of the soil-like materials. The possibility of long term changes in the measured properties of these soils/spoils must also be considered, especially if the materials are strongly acidic or alkaline. The availability of suitable topdressing soil should also be determined because this will affect the depth of the soil covering that will be used.[30,31] The optimum depth of soil covering is influenced by many factors: cost, availability, spoil characteristics, climate, potential plant rooting depth, and reclamation objectives. Soils and spoils usually need various amendments that will aid in vegetation establishment and growth.[32,33] Organic soil mulches are often useful; Slick and Curtis discuss comprehensive guides for their use.[34]

The selection of adapted plant species for revegetation is conditioned largely by whether primary importance is placed on the below-ground or above-ground biomass. There are also some differences in opinion concerning the utility of native versus introduced species. However, for any section of the country, there are many adapted plant species from which to choose.[35,36,37,38] Whatever species are selected, it is frequently useful to develop some early knowledge concerning the potential for vegetation production and ground cover. Packer et. al.[39] present some simple concepts concerning revegetation potentials for western mined lands. They developed a revegetation model based on annual precipitation, growing season length, soil potassium, soil salinity, soil pH, and seven different post-reclamation management strategies.

Vegetation establishment and growth in arid and semiarid regions can be enhanced by soil replacement, surface modification for soil water conservation, water harvesting, and minimal irrigation.[40,41] Irrigation makes it possible to establish vegetation even in drought years.

The details of vegetative reclamation tend to be somewhat site specific and are beyond the scope of this chapter. However, detailed revegetation information, for a variety of arid and semiarid conditions, is presented by Richardson,[42] Ferguson and Frischknecht,[43] Gifford et. al.,[44] and by Van Epps and McKell.[45] Revegetation of tailings ponds presents special difficulties due to salts and metallic ions in the tailings; radionuclides in uranium tailings also present special problems. An intensive sampling and evaluation of the factors limiting revegetation is essential. Practical experience and suggestions are given by Neilson and Peterson[46] and by Dean et. al.[47]

Various types of vegetation cover and vegetation with rock-mulch covers were evaluated by Beedlow.[48] He suggests that vegetation is necessary in order to prevent excess water from moving through the cover and into the tailings, where excess water could leach contaminants out of the tailings. He found no difference in soil water with or without a rock mulch. However, a rock mulch generally increased the amount of weeds, shrubs, and forbs in the stand and decreased the amount of grass cover. A rock mulch might limit the use of the area by some types of grazing animals.

Engineering Techniques

Numerous methods have been used to evaluate the effectiveness of radon barriers over spoil or tailings in order to meet regulatory requirements for radon exhalation. Asphalt emulsions were applied directly over tailings by Hartley et. al.[49] and Matthews.[50] The asphalt was covered by about 1 m of overburden and then soil. The radon flux was reduced as much as 99.9 percent. Barriers consisting of a coarse rock layer, a clay layer, another coarse rock layer, followed by overburden and soil for a total thickness of 1 to 1.5 meters, were evaluated by Gee et. al.[51] Such covers are intended to be stabilized by shallow rooted vegetation. Deep rooted species might threaten the integrity of the asphalt cover. However, shallow rooted species would be at a competitive disadvantage in arid climates.

Caliche slurries have also been sprayed over tailings to minimize erosion of the tailings and to immobilize radon within the tailings pond.[52] The caliche effectively precludes the establishment of vegetation. We consider this as a less desirable option than using vegetation to aid in the stabilization.

A variety of reclamation and control measures for mineral tailings, including radioactive tailings, was reported by Dean et. al.[47] They looked at rock mulches, chemical controls for dust, and vegetative stabilization, including uranium wastes.

Exhalation of radon-222 gas through various types of soil, rock, and asphalt covers by the diffusion process can be predicted for any combination or thickness of cover materials. Phillips and Bell present a finite element model of the radon diffusion process that gives exact solutions for any reasonable number of homogeneous cover layers.[53]

In some of the less arid portions of the United States there may be problems associated with the degradation or contamination of groundwater by radionuclides. While the protection of all groundwater systems is currently receiving increased study, the science of cleaning up contaminated groundwater is poorly developed; groundwater contamination is easier to avoid than to repair.[54] Houghton et al.[55] were able to prevent groundwater contamination from uranium mining in North Dakota by special handling of noxious materials and selective placement of such materials above the groundwater table. Clay caps and special topographic shaping were used to minimize the movement of water through buried radioactive materials.

CONCLUSIONS

Probably the most important points regarding the reclamation of uranium mining and milling wastes are the management implications of the current technology. We have also briefly stated some of the research areas that impose serious technical or cost limitations on our ability to reclaim lands disturbed by uranium mining and milling.

1. Adequate technology is available to reclaim virtually all disturbances associated with uranium mining or milling, but the costs may be high.
2. Perhaps the most important objective of reclaiming uranium disturbances

is the total containment of low-level, hazardous, radionuclides for the long term (1000 years). A favorable outcome is not assured in all cases.

3. Plant establishment and growth is often hindered by the physical and chemical nature of the wastes and by the climate.

4. It appears that with the current level of mining and milling, the existing regulatory controls are adequate to safeguard humans and other animals from radionuclides in the food chain.

5. Additional reclamation research is needed: (1) to assure the total containment of uranium tailings, (2) to quantify the influence of radionuclides in the food chain, (3) to improve the efficacy of revegetation technology for areas receiving less than 300 mm of annual precipitation, and (4) to determine the extent of groundwater contamination by radionuclides.

LITERATURE CITED

1. Barth, R.C., 1986. Reclamation technology for tailing impoundments, Part 1: Containment. Colorado School of Mines, *Mineral and Energy Resources,* 29(1):1-25.
2. U.S. Nuclear Regulatory Commission, 1980. Final generic environmental impact statement on uranium milling. Project M-25, Volume 1, Summary and Text, NUREG-0706, U.S. Nuclear Regulatory Commission, Washington, D.C., pp. 361.
3. Moffett, D., and M. Tellier, 1977. Uptake of radioisotopes by vegetation growing on uranium tailings. *Canadian Jour. of Soil Science*, 57(4):417-424.
4. Barth, R.C., 1986. Reclamation technology for tailing impoundments, Part 2: Revegetation. Colorado School of Mines, *Mineral and Energy Resources,* 29(2):1-25.
5. Rumble, M.A., and A.J. Bjugstad, 1986. Uranium and radium concentrations in plants growing in uranium mill tailings in South Dakota. *Reclamation and Revegetation Research,* 4:271-277.
6. Garten, Jr., C.T., E.A. Bondietti and R.L. Walker, 1981. Comparative uptake of uranium, thorium, and plutonium by biota inhabiting a contaminated Tennessee floodplain. *J. Environmental Quality*, 10(2):207-210.
7. Dreesen, D.R., M.L. Marple, and E. Kelly, 1978. Contaminant transport, revegetation, and trace element studies at inactive uranium mill tailings piles, pp. 111-140. *In: Uranium Mill Tailings Management,* Proceedings of a Symposium, Volume 1, Civil Engineering Department, Colorado State University, Fort Collins, CO, pp. 172.
8. Roybal, G., and R.W. Eveleth, 1982. Overview of surface mining in New Mexico and ongoing reclamation projects, pp. 2-7. *In:* E.F. Aldon and W.R. Oaks (Eds.), *Reclamation of Mined Lands in the Southwest, A Symposium,* Soil Conservation Society of America, New Mexico Chapter, P.O. Box 2142, Albuquerque, NM, pp. 218.

9. Packer, P.E. and E.F. Aldon, 1978. Revegetation techniques for dry regions, pp. 425-450. *In:* F.W. Schaller and P. Sutton (Eds.), *Reclamation of Drastically Disturbed Lands.* American Society of Agronomy, Madison, WI, pp. 742.
10. Aldon, E.F., 1984. Vegetation parameters for judging the quality of reclamation on coal mine spoils in the southwest. *Great Basin Naturalist,* 44(3):441-446.
11. Aldon, E.F., 1982. Vegetation stability on mine spoils in the southwest, pp. 198-200. *In:* E.F. Aldon and W.R. Oaks (Eds.), *Reclamation of Mined Lands in the Southwest, A Symposium,* Soil Conservation Society of America, New Mexico Chapter, P.O. Box 2142, Albuquerque, NM, pp. 218.
12. Woods, L.E., and G.E. Schuman, 1986. Influence of soil organic matter concentrations on carbon and nitrogen activity. *Soil Science Society of America J.,* 50:1241-1245.
13. Laycock, W.A., 1980. What is successful reclamation? — A look at the concepts of adaptability, productivity, cover, and diversity of seeded species. Northwest Colorado land reclamation seminar II, Proceedings. USDA, ARS, Crops Research Lab., Colorado State University, Fort Collins, CO, pp. 17.
14. Breslin, A.J. and H. Glauberman, 1970. Investigation of radioactive dust dispersed from uranium tailings pile, pp. 249-253. *In:* W.C. Reinig (Ed.), *Environmental Surveillance in the Vicinity of Nuclear Facilities.* Health Physics Society, Charles C. Thomas, Springfield, IL, pp. 465.
15. U.S. Environmental Protection Agency, 1986. Final rule for radon-222 emissions from licensed uranium mill tailings. EPA 520/1-86-009, Office of Radiation Programs, Washington, D.C., pp. 204.
16. Coffey, P.S., W.S. Scott, and K.J. Summers, 1986. The effects of tailing dam profiles on relative wind erosion rates. *J. Environmental Quality,* 15(2):168-172.
17. Oak Ridge National Laboratory, 1979. A radiological assessment of radon-222 released from uranium mills and other natural and technologically enhanced sources. National Technical Information Service, PB 293654, NUREG/CR-0573, Springfield, VA, pp. 216.
18. Nero, A.V., M.B. Schwehr, W.W. Nazaroff, and K.L. Revzan, 1986. Distribution of airborne radon-222 concentrations in U.S. homes. *Science,* 234:992-997.
19. Edling, C., H. Kling, and O. Axelson, 1983. Radon in homes—A possible cause of lung cancer, pp. 123-149. *In: Lung Cancer and Radon Daughter Exposure in Mines and Dwellings.* Linkoping University Medical Dissertations No. 157. Linkoping University, Linkoping, Sweden.
20. Whicker, F.W., 1978. Biological interactions and reclamation of uranium mill tailings, pp. 141-154. *In: Uranium Mill Tailings Management,* Proceedings of a Symposium, Volume 1, Civil Engineering Department, Col-

orado State University, Fort Collins, CO, pp. 172.
21. Rumble, M.A., 1982. Biota of uranium mill tailings near the Black Hills, pp. 278-292. *In: Proceedings, Western Association of Fish and Wildlife Agencies,* Las Vegas, NV, pp. 622.
22. Cataldo, D.A., D. Paine, C.E. Cushing, R.M. Emery, and B.E. Vaughan, 1977. Radionuclide transport, pp. 6.1-6.13. *In:* L.E. Rogers and W.H. Rickard (Eds.), *Ecology of the 200 Area Plateau Waste Management Environs: A Status Report.* National Technical Information Service, Accession PNL-2253, UC-11, Springfield, VA, pp. 118.
23. Dreesen, D.R., and M.L. Marple, 1979. Uptake of trace elements and radionuclides from uranium mill tailings by four-wing saltbush *(Atriplex conescens)* and alkali sacaton (*Sporobolus airoides*), pp. 127-143. *In: Uranium Mill Tailings Management,* Proceedings of a Symposium, Civil Engineering Department, Colorado State University, Fort Collins, CO.
24. Stanley, M.A., G.E. Schuman, F. Rauzi, and L.I. Painter, 1982. Quality and element content of forages grown on three reclaimed mine sites in Wyoming and Montana. *Reclamation and Revegetation Research,* 1:311-326.
25. Doerges, J.E., and G.S. Kennington, 1984. Final report on the radiobiological investigation of uranium mining and processing operations in Wyoming. University of Wyoming, Laramie, WY.
26. Schuman, G.E., D.T. Booth, J.W. Waggoner, and F. Rauzi, 1986. The effects of grazing reclaimed mined lands on forage production and composition, pp. 163-164. *In:* P.J. Koss et. al. (Eds.), *Rangelands: A Resource Under Siege,* Proc., Second International Rangeland Congress, Australian Academy of Science, Canberra, A.C.T. Australia.
27. Yamamoto, T., 1982. A review of uranium spoil and mill tailings revegetation in the western United States. USDA Forest Service, Rocky Mountain Forest and Range Exp. Sta., GTR-RM-92, Fort Collins, CO, pp. 20.
28. Sandoval, F.M., and J.F. Power, 1977. Laboratory methods recommended for chemical analysis of mined-land spoils and overburden in western United States. USDA, Agricultural Handbook No. 525, Washington, D.C., pp. 31.
29. Bauer, A., W.A. Berg, and W.L. Gould, 1978. Correction of nutrient deficiencies and toxicities in strip-mined lands in semiarid and arid regions, pp. 451-466. *In:* F.W. Schaller and P. Sutton (Eds.), *Reclamation of Drastically Disturbed Lands.* American Society of Agronomy, Madison, WI, pp. 742.
30. Redente, E.F., and N.E. Hargis, 1985. An evaluation of soil thickness and manipulation of soil and spoil for reclaiming mined land in northwest Colorado. *Reclamation and Revegetation Research,* 4:17-29.
31. Barth, R.C., 1984. Soil depth requirements to reestablish perennial grasses on surface-mined areas in the northern Great Plains. Colorado School of

Mines, *Mineral and Energy Resources,* 27(1):1-20.
32. USDA Forest Service, 1979. User guide to soils: Mining and reclamation in the west. Intermountain Forest and Range Exp. Sta., GTR-INT-68, Ogden, UT, pp. 80.
33. Sandoval, F.M. and W.L. Gould, 1978. Improvement of saline- and sodium-affected disturbed lands, pp. 485-504. *In:* F.W. Schaller and P. Sutton (Eds.), *Reclamation of Drastically Disturbed Lands.* American Society of Agronomy, Madison, WI, pp. 742.
34. Slick, B.M. and W.R. Curtis, 1985. A guide for the use of organic materials as mulches in reclamation of coal minespoils in the eastern United States. USDA, Forest Service, Northeastern Forest Exp. Sta., GTR-NE-98, Broomall, PA, pp. 144.
35. Thornburg, A.A., 1982. Plant materials for use on surface-mined lands in arid and semiarid regions. USDA, Soil Conservation Service, SCS-TP-157, Washington, D.C., pp. 88.
36. Oaks, W.R., 1982. Reclamation and seeding of plant materials for reclamation, pp. 145-150. *In:* E.F. Aldon and W.R. Oaks (Eds.), *Reclamation of Mined Lands in the Southwest, A Symposium,* Soil Conservation Society of America, New Mexico Chapter, P.O. Box 2142, Albuquerque, NM, pp. 218.
37. Wasser, C.H., and J. Shoemaker, 1982. Ecology and culture of selected species useful in revegetating disturbed lands in the west. USDI, Fish and Wildlife Service, FWS/OBS-82/56, Washington, D.C., pp. 347.
38. USDA Forest Service, 1979. User guide to vegetation: Mining and reclamation in the west. Intermountain Forest and Range Exp. Sta., GTR-INT-64, Ogden, UT, pp. 85.
39. Packer, P.E., C.E. Jensen, E.L. Noble, and J.A. Marshall, 1982. Models to estimate revegetation potentials of land surface mined for coal in the west. USDA, Forest Service, Intermountain Forest and Range Exp. Sta., GTR-INT-123, Ogden, UT, pp. 25, 2 maps.
40. Ries, R.E., and A.D. Day, 1978. Use of irrigation in reclamation on dry regions, pp. 505-520. *In:* F.W. Schaller and P. Sutton (Eds.), *Reclamation of Drastically Disturbed Lands.* American Society of Agronomy, Madison, WI, pp. 742.
41. Verma, T.R. and J.L. Thames, 1978. Grading and shaping for erosion control and vegetative establishment in dry regions, pp. 399-409. *In:* F.W. Schaller and P. Sutton (Eds.), *Reclamation of Drastically Disturbed Lands.* American Society of Agronomy, Madison, WI, pp. 742.
42. Richardson, B.Z., 1985. Reclamation in the Intermountain, Rocky Mountain region, pp. 175-192. *In:* M.K. McCarter (Ed.), *Design of Non-Impounding Mine Waste Dumps.* American Institute of Mining, Metallurgical, and Petroleum Engineers, Inc., New York, NY, pp. 216.
43. Ferguson, R.B., and N.C. Frischknecht, 1985. Reclamation on Utah's Emery

and Alton coal fields: techniques and plant materials. USDA Forest Service, Intermountain Forest and Range Exp. Sta., RP-INT-335, Ogden, UT, pp. 78.
44. Gifford, G.F., M.W. Kress, G.A. Van Epps, J. Briede, C.L. Gifford, C.M. McKell, 1984. Rehabilitation of disturbed sites and uranium mine spoil piles in southeastern Utah, completion report. Institute for Land Rehabilitation, Utah State University, Logan, UT, pp. 105.
45. Van Epps, G.A., and C.M. McKell, 1980. Revegetation of disturbed sites in the salt desert range of the Intermountain West. Utah Agricultural Exp. Sta. Land Rehab. Series No. 5, Utah State University, Logan, UT, pp. 35.
46. Nielson, R.F., and H.B. Peterson, 1978. Vegetating mine tailings ponds, pp. 645-652. *In:* F.W. Schaller and P. Sutton (Eds.), *Reclamation of Drastically Disturbed Lands.* American Society of Agronomy, Madison, WI, pp. 742.
47. Dean, K.C., L.J. Froisland, and M.B. Shirts, 1986. Utilization and stabilization of mineral wastes. USDI Bureau of Mines, Bulletin 688, Washington, D.C., pp. 46.
48. Beedlow, P.A. 1984. Designing vegetation covers for long-term stabilization of uranium mill tailings. Rpt. No. NUREG/CR-3674, PNL-4986, U.S. Nuclear Regulatory Commission, Washington, D.C.
49. Hartley, J.N., H.D. Freemen, E.G. Baker, M.R. Elmore, D.A. Nelson, C.J. Voss, and P.C. Koehmstedt, 1981. Field testing of asphalt emulsion radon barrier system, pp. 319-339. *In: Uranium Mill Tailings Management,* Proceedings of a Symposium, Civil Engineering Department, Colorado State University, Fort Collins, CO.
50. Matthews, M.L., 1981. The research effort of the uranium mill tailings remedial actions project, pp. 33-40. *In: Uranium Mill Tailings Management,* Proceedings of a Symposium, Civil Engineering Department, Colorado State University, Fort Collins, CO
51. Gee, G.W., J.T. Zellmer, M. Dodson, R. Kirkham, B. Opitz, D. Sherwood, and J. Tingey, 1981. Radon control by multilayer earth barriers: 2, Field studies, pp. 289-308. *In: Uranium Mill Tailings Management,* Proceedings of a Symposium, Civil Engineering Department, Colorado State University, Fort Collins, CO
52. Brookins, D.G., 1981. Caliche-cover for stabilization of abandoned mill tailings, pp. 309-318. *In: Uranium Mill Tailings Management,* Proceedings of a Symposium, Civil Engineering Department, Colorado State University, Fort Collins, CO.
53. Phillips, W.F. and D.A. Bell, 1982. Diffusion of radon gas from uranium mill tailings. *J. Energy Resources Technology* 104:130-133.
54. Farmer, E.E., 1983. Hydrologic consequences of mined land disturbance in the western United States, pp. 430-433. *In: New Forests For a Changing World,* Proc., 1983 Society of American Foresters National Convention,

Society of American Foresters, Washington, D.C.
55. Houghton, R.L., G.S. Anderson, S.R. Hill, J.L. Burgess, J.D. Wald, D.P. Patrick, R.L. Hall, and J.D. Unseth, 1987. Prevention of ground-water degradation during reclamation of a uraniferous lignite mine, North Dakota, pp. G-4-1 to G-4-19. *In: Proceedings, Fourth Annual Meeting, American Society for Surface Mining and Reclamation.* American Society for Surface Mining and Reclamation, Princeton, WV, Sections A - M.

Environmental Consequences of Energy Production: Problems and Prospects. Edited by S. K. Majumdar, F. J. Brenner and E. W. Miller. © 1987, The Pennsylvania Academy of Science.

Chapter Fourteen

DEVELOPMENT OF PROJECT APPRAISAL METHODOLOGY IN AIR POLLUTING INDUSTRIES IN EASTERN EUROPE

NENAD STARC
Institute of Economics
Kenedijev trg 7
41 000 Zagreb, Yugoslavia

This paper deals with air pollution in Eastern Europe and discusses the possibilities of development of environmental planning in air-polluting industries. Sulphur dioxide is concentrated upon as the most significant polluter. Its emission in European countries and environmental effects of acid rain are presented. In the remainder, cost-benefit analysis and environmental impact assessment are presented and discussed from the point of view of a socialist methodologist. The two methods are viewed as suitable starting points of development of environmentally unsound project appraisal methodologies in Eastern European socialist countries.

THE SCOPE OF THE PROBLEM

Postwar development in Eastern Europe has implied all types of pollution from its very beginning. The patterns of socio-economic development in these countries differ much from the patterns in the west, but air, waterflows, seas and ground have nevertheless been loaded with all kinds of economically significant quantities of waste. The Danube is equally polluted in its upper and lower flow, and wastes are in the Wisla and the Sava as they are in the Rhine and the Po. Herbicides and pesticides are not applied in quantities usual in the west but their use has been increasing and is likely to reach western levels soon. Rapid industrialization and urbanization have caused many development problems including environmental ones so that industrial and urban areas are congested and noisy, frequently exceeding permissible concentrations of various pollutants in the air. Population in these areas has been experiencing these environmental consequences for decades.

Air pollution has been the most pressing environmental problem in Eastern European countries. Gases and particles eminating from stacks and cars contain a complex set of chemical compounds harmful in one way or another. Due to atmospheric movements, they easily overcome local and regional considerations and become a national problem. Since European countries are small in size, the problem easily becomes an international one as well. This type of environmental pollution, therefore, received the most attention both in the west and in the east. Various studies on the subject claim that NO_x and lead are probably the most harmful components of industrial smoke, but the most attention has been given to sulphur dioxide and its impacts. Due to its high proportion in the structure of gases (European coals contain up to 10% sulphur) SO_2 stands as a major pollutant appearing all over Europe in many millions of tons per year. As stated in Table 1, the main emitters of SO_2 in 1978 in Eastern Europe were the European part of the USSR and East Germany. If the size of particular countries is taken into account, and the emission per km^2 calculated, Czechoslovakia and East Germany appear as the most significant polluting areas.

For various reasons most of the data on particular emitters throughout eastern Europe are either unobtainable or incomparable. In most of the references, however, coal-fired power stations, other industries and heating are treated as three main sources of SO_2 pollution; coal-fired power stations contributing up to 50 percent of the overall emission. Since industrial emission other than energy production comes from a large number of relatively small and dispersed production units, coal-fired power stations, which are fewer, are considered the most hazardous polluters. Almost everywhere, flying particles in their smoke are captured by electroprecipitators, but NO_x and other gases besides SO_2 are poured into the atmosphere. In Yugoslavia, for instance, the share of coal-fired power stations in the overall emission of SO_2 is likely to soon reach 50 percent. The emission from coal-fired power stations ranges from 2.3 up to 87.4 kg/MWh, none of them having any desulphurization devices. Such high SO_2 rations cause high concentrations of emitted gases. At the top of the stack the concentration of SO_2 can reach as high as 20,000 mg/m^3 (Ćurković 1986). The stations are distributed fairly evenly across the country so that there are no regions completely saved from SO_2 impacts. The distribution pattern in other Eastern European countries is more or less the same, and their overall emission is even higher. There are numerous references about proposals for various desulphurization technologies, but no such devices have been installed thus far.

Environmental consequences of SO_2 pollution are significant, in places even bordering on catastrophy. Sulfur dioxide affects human health and attacks buildings and other materials. The effect which is most frequently stressed and which even arises furious diplomatic arguments, is acid rain and the disaster

TABLE 1

SO₂ Emission in European Countries in 1978.

Country	Acreage 10³km²	SO₂ Emission 10³ tons	Population 10⁶	SO₂ Emission kg/km²	SO₂ Emission kg per capita
USSR-European Part	3363.4	16200		4820	
Poland	1312.7	3000	34.0	3290	88
East Germany	108.2	4000	16.9	37000	236
Czechoslovakia	127.9	300	14.6	23460	206
Hungary	93.0	1500	10.7	16130	140
Yogoslavia	255.8	1250	22.3	4890	56
Romania	237.5	2000	22.3	8421	90
Bulgaria	110.9	1000	8.9	9017	112
Albania	28.7	100	2.7	3500	37
Finland	337.1	540	4.8	1600	113
Sweden	450.0	550	8.1	1222	68
West Germany	249.6	3600	61.8	14425	58
Austria	83.9	380	7.6	4530	50
Italy	301.2	4400	55.8	14600	78
Greece	132.0	704	9.6	5333	73
Turkey	460.8	1000	44.9	2170	22
Norway	323.9	150	4.1	463	37
Denmark	43.1	456	5.1	10580	89
Holland	41.0	480	13.7	11700	36
Belgium	30.5	760	9.8	24900	78
Luxemburg	5.2	48	0.36	9230	134
France	544.0	3600	52.6	6617	68
Switzerland	41.3	116	6.4	2810	18
Spain	500.0	2000	35.5	4000	56
Portugal	76.7	168	9.9	2190	17
Great Britain	244.0	4980	56.0	20410	88
Ireland	68.9	174	3.4	2525	51
Iceland	60.7	12	0.2	1980	60

Source: Buchner 1981. Knezevic 1983.

it has caused in forests throughout Europe. Estimates of the effects of acid rain on Eastern European forests and crops are almost completely incomparable and even less reliable than those of SO_2 emissions. Various sources point out, however, that all European countries are severely stricken. The figures for Eastern Europe are the following: in the USSR 360,000 ha of forests have been acidified. Since the European Section is much more industrialized than the Asian region, most of the damaged forest occurs there. Besides, 15 percent of the harvest in 41 million ha of land west of Moscow was affected by acid rain and other forms of air pollution (Elseworth, 1984). In Poland 390,000 ha of woodlands have been damaged to various degrees (Muntingh, 1983) and in East Germany, 12 percent of forest is affected (Muntingh, 1983). In Yugoslavia, estimates were made only in the north-western parts of the country where 52,000 ha of forest appeared to be seriously affected (Muntingh, 1983, Kauzlarić, 1986). In Czechoslovakia, pollution seems to be reaching the crisis point. According to various reports, 480,000 ha of forest (37% of all forest) is already either irreparably spoiled or dead (Elseworth 1984). The figures for Rumania, Hungary and Bulgaria are unreliable, and no figures exist for Albania.

Virtually every European country stands as a significant emitter of sulphur dioxide (Table 1). In Western Europe, the largest quantities are emitted in Great Britain and Italy. Having the highest concentration per km^2, Belgium also stands as a major emitting area. It is clear that SO_2 produced at some point in Europe is likely to penetrate the air space of surrounding countries and this indeed has been going on for decades. "Exporting" and "importing" SO_2 has thus become an important subject of monitoring and modeling. It was soon discovered that industrial smoke becomes subject to long distance transport through the atmosphere, and that SO_2 and other gases produced in eastern Europe can harm the western European environment and vice versa. Estimates of the extent of this transport differ due to unreliable input data on particular countries' emissions and different meteorological models, but they all agree on "net importing" and "net exporting" countries (Highton and Chandwick, 1980. Umwelt, 1982). In the beginning of 1980, the belt formed by East Germany, Czechoslovakia, Hungaria, Rumania and Bulgaria appeared as an area emitting more SO_2 than it was receiving. Poland and the European section of the USSR lying north of the belt appear as net importers as do Yugoslavia and Albania in the south. However, SO_2 can travel through air as far as 2000 km so that it is almost impossible to detect whose SO_2 has fallen where.

Slow implementation of desulphurization technology in Eastern Europe countries takes place in parallel with slow development of environmental legislation. In the USSR, emission is not standardized so that investment projects are given environmental permits according to expected ground concentrations of the air pollutant in question and data on existing air pollution in the area. High stacks are prescribed if there is no other solution, which in the case of SO_2 means that expected high ground concentrations will ultimately lead to a high stack (Krstić 1981). Other countries have similar, even less restrictive regulations. In Yugoslavia, no regulations on SO_2 emission have been passed so that certain government recommendations and the World Health Organization criteria for ground concentrations are taken into account during project appraisal. There are no grounds as of yet for proposing a set of standards that would be obeyed in Eastern Europe as a whole. An internationally coordinated and unified monitoring system which such standards would imply also awaits implementation. Information about pollution episodes, environmentally hazardous accidents, and similar events, is not satisfactory. It has provoked furious accusations in the west which culminated in May 1986 after the Chernobyl nuclear accident. Official informing was judged as slow, inaccurate and insufficient.

Several months later, however, a major accident in a chemical factory near Basel, Switzerland was followed by an even slower and less accurate official informing. There are quite a few such examples all over Europe including cover ups of radioactive leakage in nuclear power stations.

Even the most superficial analysis of causes of environmental pollution in Eastern European countries points to the question of environmental aspects of socialist planning after World War II. All countries in Eastern Europe except Greece proclaimed a socialist socio-economic system, adopted central planning and based their further development on long-run and medium-run plans. Yugoslavia soon turned to a self-managed economy and other countries relaxed their systems to some extent, but the idea of social control over overall production and other development factors was never abandoned anywhere. Pollution, however, appeared at the very beginning and did not receive much attention until recently. This discrepancy (some authors [Bahro 1984], refer to it as system contradiction) certainly requires a longer and more detailed explanation, but it is nevertheless clear that central planning, as practiced in Eastern European socialist countries, is mainly macroeconomic while other aspects including even microeconomic ones are more or less neglected. In this way postwar urbanization took place without much planning. Physical planning was introduced with a delay and then for many years implemented separately from economic planning, and imports of foreign environmentally unsound technology fell out of control from the very beginning. As of this date, environmental planning which could have been introduced decades ago does not exist, which rounds out the picture of an incomplete planning system. The problem of air polluting industries in Eastern Europe appears therefore in the context of fast and not successfully planned socio-economic development. Solution to the problem requires changes in technology, improvements of the planning system and much better developed project appraisal methodology. However, the same holds true for other environmentally unsound industries so that particular problems of air polluting projects appraisal appear in the context of a rudimentary overall development methodology which has never had its environmental aspect.

POSSIBILITIES OF DEVELOPMENT OF PROJECT APPRAISAL METHODOLOGY

Reconsideration of planning systems in Eastern European socialist countries from an environmental point of view reveals several possibilities and constraints to further environmentally sound development. In the case of air polluting industries, the possibilities are mainly technological and to some extent organizational. A low level of ecological awareness and undeveloped investment appraisal methodology seems to be the main constraint.

One technological possibility is a more rational usage of produced energy.

Energy/GNP ratios in Eastern European economies can not be directly compared to those in the west due to different GNP accounting but it is nevertheless evident that they are much larger. It is also evident that they could be lowered, not by major technological changes but by a more efficient organization of production. Such improvements do not require much investment and they are feasible within the existing framework of the organization of production. The most promising possibility, however, comes from certain changes in the prevailing general concept of technology. Existing technologies in air polluting industries can be summarized as processes of taking raw materials, extracting certain chemical elements or compounds with a considerable use of energy and throwing the unused residues into the atmosphere. The damage is thus twofold: the ecosystem is damaged and in most cases the potentially usable part of the resource is wasted. In most cases this unnecessary concept can and should be replaced with another one based on decomposition of resources. If the resource is decomposed at the very beginning of the process so that its components can be used as materials for the production of final products, as catalysts, or brought to a chemical reaction that produces nonharmful compounds, the gain is twofold: the resource is used completely or almost completely and the environment is saved. This simple and appealing concept has been recognized in Eastern European countries (Kos 1984), while experience has been gained mainly in the west. It shows that its application most often does not require new technologies, but rather new combinations of existing ones, and that there is no need for fundamental research. Experience also indicates that such projects are frequently acceptable both on macroeconomic and microeconomic levels. This point is of particular importance in Eastern European socialist countries where the strategy of environmentally sound development is still in formation. In these countries, an opinion prevails that a low level of development does not allow for environmental considerations and that costly "clean" development has to be postponed until better times. Although not always feasible, the decomposition concept serves as a strong argument against such opinions. In some of the countries in question some experience has already been gathered. For the time being its wider application is, however, constrained by a low level of coordination between existing experts, internally inconsistent legislation, and above all by the prevailing low level of ecological awareness. Also the concept seemingly contradicts fast industrialization which in most of the socialist countries still stands as the major development task. In this way it is not likely to become an important component of overall development strategy soon.

There is another possibility for improvement, which is clear in concept and demands even less technological adjustment. It refers to an information system necessary for efficient planning of environmentally sound development. The requirement that everyone who has been or is likely to be affected by pollution should be fully and continuously informed, and that he/she should have a constant opportunity to communicate proposals, is surely consistent with basic ideas of socialism. In theory, a political system that allows for this exists in all Eastern European countries. In spite of that, an information system adjusted to processing and communicating information on environmental matters has yet to be established. The problem does not appear to be conceptual or technical because an adequate hardware has been developed and has been fully available at a relatively low cost. The reasons are similar to those that constrain introduction of decomposition technology.

The question of project appraisal designed to support environmentally sound planning in air polluting industries is, however, both conceptual and dependent on existing ecological awareness. Besides, it has to be stated as a particular as well as a general methodological question. Given the state of microeconomic methodology there is no point in discussing an air or any other pollution project without coping with general problems of environmentally unsound project appraisal. It is easily shown that economic planning in Eastern European socialist countries is implemented with a considerable lack of microeconomic methodology. This has been so from the very beginning of the central planning system in the USSR in the 1930's when balance sheets, rudiments of later developed input-output analysis, were introduced. This method was a genuine socialist contribution to macroeconomic methodology and planning in general. However, analysts of the Soviet system already pointed out in the 1930's that balance sheets normed the aggregate growth of industries while the question of allocation of investments between firms was taken care of rather superficially. This had not changed for decades and since the Soviet model was almost completely adopted by other Eastern European socialist countries after World War II, the microeconomic methodology remained rudimentary. Environmental considerations that were introduced some 10-15 yrs ago, only increased the weight of an already heavy problem. This situation has been recognized so that more or less fruitful methodological discussions are here and there raised in all the countries in question. Although not many authors deal with it, they all generally agree that a genuine socialist methodology should be developed. However, when it comes to the question of the application of methodology that has already been developed in the west, opinions diverge. They range from strong claims that western methodology is completely inapplicable in socialism to a priori statements that methods are universal and ideology-free.

The so called western methods that are at the disposal of a socialist methodologist are numerous. There are three methods (or sets of methods) that represent the methodology developed in western capitalist countries: cash flow analysis, cost-benefit analysis and methods and procedures used in environmental impact assessment of investment projects. Cash flow analysis has been in use in some Eastern European socialist countries; combined with methods of market research it proved superior to other methods that have been in use. It is clear, however, that cash-flow analysis is generally unsuitable for the evaluation of projects that could harm the environment. The projects are approached from the investor's point of view only, so that only financial flows of the project are covered. Environmental and other socially relevant effects remain out of scope. Environmental protection can enter the analysis through increased investment and/or operational cost. In such cases the method is, however, justified only under some rigid assumptions (perfect environmental legislation, perfect control) which do not seem to hold even in highly developed countries. This type of analysis could therefore be used only as a part of a much wider project appraisal procedure. On the other hand, cost-benefit analysis and environmental impact assessment seem to encounter social effects of investment projects and therefore require more detailed elaboration.

COST-BENEFIT ANALYSIS

Cost-benefit analysis is generally defined as an investment appraisal method which deals with all possible effects of an investment from the social perspective. In a rudimentary form it had already been in use in the 19th Century in the USA but it was not shaped for broad usage until the New Deal. After the World War II, (much like EIA three decades later) it existed as a more or less defined procedure used mainly in the public sector but lacked a specific theoretical foundation. However, since the early 1950's, it has been receiving serious theoretical treatment. Today, CBA analysts use a well defined and theoretically founded development method with numerous completed studies. However, cost-benefit analysis as described in numerous guidelines and manuals and as usual prerequisite for granting international loans is a method that has been designed in highly developed countries and founded on neoclassical economic theory. When it comes to its application in Eastern European countries whose socio-economic systems are defined in a different theoretical manner, this is an important point.

Although theoretical aspects of cost-benefit analysis usually provoke long and complex academic discussions, its basic idea is simple. Some project (i.e. some proposed socio-economic activity that requires a certain initial investment) is expected to produce various effects. In order to find out whether the project is socially desirable, the effects are divided into social benefits (i.e. those that increase social welfare) and social costs (i.e. those that decrease it). The classical notion of revenue is obviously insufficient to cover such effects so that the term social benefits is introduced, while the calculation of mere financial outlays is replaced with the measurement of social opportunity cost (i.e. a relative measure which refers to benefits lost had the sources in question been used for some other purpose).

Identifying relevant benefits and costs of a project does not create major problems to a CBA analyst. Problems do appear when it comes to measurement. Effects that result from investment projects have some value and this has to be expressed in common units of measurement. The choice of money as supposedly the most convenient unit created, however, the distinction between tangible and intangible effects because benefits such as "more intelligent electorate" or costs such as "reduced quality of life" do not appear on the market, and therefore, do not have a price that would according to neoclassical economic theory reveal their value. This has created a lot of problems on the operational level and has given rise to various measurement techniques. Intangibles have been approached indirectly via experiments or questionnaires. They can be referred to as private substitution that occurs if the effect is not provided; some notion of their importance can be obtained by reference to costs saved by the project. They can also be assumed absolutely desirable, which reduces the analysis to its cost-effectiveness form such as searching for the project that will provide some proposed benefit at the least cost. As the analysis developed, intangible benefits and costs were given more importance so that the problems of their measurement have been accumulating. The most subtle and laborious steps of the method have nevertheless remained the same: detect all relevant benefits and costs, and express them in terms of money. Once this has been done, the last steps are simple. All the benefits are summed up, the same is done with the costs and the difference $\Sigma B - \Sigma C$ and the ratio $\Sigma B / \Sigma C$ are formed. If $(\Sigma B - \Sigma C) > 0$ and $\Sigma B / \Sigma C > 1$, the project is socially desirable and should be implemented. Problems of pricing the effects can not be fully acknowledged without reviewing the theoretical concepts that cost-benefit analysis is based upon. They are all drawn from neoclassical economic theory, mainly from welfare economics. The basic concept is one of general equilibrium. Then follow the concepts of public goods and externalities and their internalization which have been introduced to explain some violations of equilibrium proposals that appear in a given capitalist economy. These concepts also justify government intervention into the economy. The question of distribution is taken care of according to the so called Pareto criterium and the concept of consumers' surplus, while values of whatever is to be evaluated are obtained by so called "shadow pricing". Finally, there is a concept of social rate of discount which describes social evaluation of effects expected in the future. CBA is often described as an operation/application of these concepts.

There is an important point implied in cost-benefit analysis and its final results. The perspective from which the project is evaluated is claimed social, all relevant impacts are grasped, and cost-benefit analysts are defined as unbiased, value-free practical scientists who do not have their own interests in the project. If the project is thus judged socially acceptable the decision maker should not do anything but confirm the CBA result and start taking care of project implementation. Any decision different from that (except of course implementing projects with better B/C ratios) would negate an evident social betterment and push the economy away from equilibrium.

Unlike cash-flow analysis CBA seems to be an appropriate method for evaluation of environmentally unsound projects. It was not primarily designed for this purpose but this in itself does not make it inappropriate. It is claimed that the method deals with all socially relevant tangible and intangible effects, and environmental changes clearly fall into that category. They are also implied in theoretical foundations of CBA. All kinds of pollution and various congestions can be regarded as social goods (or "bads") and/or externalities so that theoretical justification for their methodological treatment could be designed before environmental protection became an issue at all. Indeed, during the last two decades many such projects have been evaluated in this way. The socialist methodologist deals therefore with a tempting option of applying a ready made method rather than starting off a long development of a new one. The theoretical foundations of CBA and foundations on which the socialist socio-economic system have been based are, however, quite different so that a careful analysis is necessary. The general approach is straight-forward: if there is a method that is consistently derived from some theory, the adoption of the theory implies the adoption of the method as well. Rejection of the theory does not, however, imply rejection of the method. There is always a possibility that the method, viewed as an algorithm for achieving something, will be completely or partially accepted. In this way, theoretical concepts that CBA is based on have to be examined first. If they are rejected there is still a need to examine the algorithm of the method.

The socialist methodologist finds the neoclassical economic theory unsuitable as a basis for development of socialist investment appraisal methodology. The concept of general equilibrium with its a posteriori estimation of allocated resources is opposed to the socialist notion of planned allocation. The concepts of public goods and externalities describe certain undesirable phenomena but do not provide a basis for planning. The Pareto criterium presents a bourgeois notion of justice. However, the method, viewed simply as an algorithm for collecting information and arriving at a decision, consists of steps that could be found useful. The socialist methodologist who deals with rudimentary methods and techniques in his/her everyday work will find CBA techniques, questionnaires, and data processing, very informative and appealing. He/she will also have serious remarks. First of all, cost-benefit analysis is not a planning method. In its algorithm there is no step in which other projects in the economy that are being evaluated at the same time are taken into account. Besides, positive difference $\Sigma B - \Sigma C$ does not automatically verify that the project is acceptable. Its contribution to social welfare may throw out of balance the set of ongoing projects; its benefits may undesirably change the structure of the overall production, etc. The attempt to express everything in monetary terms can hardly be harmonized with principles of socialist development because market criteria are extended to non-market spheres of the economy and society. This point becomes important in the appraisal of environmentally unsound projects. The socialist methodologist will not accept that a partial loss of a resource like air or water be rejected or accepted on the basis of a simple sum of individual evaluations expressed in money. The socialist methodologist would rather go for a democratic procedure of confronting attitudes and opinions of those expected to be affected by the project. A socialist methodologist can not regard them as mere technicians that collect and process individual decisions in the name of society but rather as one of many participants in the social decision-making process.

There is an additional argument frequently raised in discussions of this type. Project appraisal in Eastern European socialist countries is generally undeveloped and completely lacking in environmental aspect, so that it seems that any method is better than none. In the case of cost-benefit analysis this can hardly be accepted however. Having in mind principles of socialist development, a socialist methodologist should be able to approach CBA in a pragmatic manner and detect applicable ideas and/or steps of its algorithm.

ENVIRONMENTAL IMPACT ASSESSMENT

It is interesting that in spite of the existence of CBA which looks quite suitable when environmentally unsound projects are in question, another method or rather set of methods and procedures referred to as Environmental Impact Assessment has experienced a rapid development in the last two decades. EIA is generally defined as a systematic examination of effects that projects, policies and programs are expected to have on the environment, (Clark et al. 1980). The practice of EIA started in the 1960's and in the beginning, studies were limited to reviewing effects on various ecosystems and presenting the results to decision makers. After some experience had been gathered the scope was extended to cover social and economic effects so that EIA studies today start off with a very broad definition of environment. The overall impact of an investment includes changes in climate, flora and fauna, health risks, induced urbanization, interregional migrations, changes in employment and the like. Their main objective, however, is to provide decision makers with alternative courses of action before the decision is made, and also with a scheme of follow up after the project has been implemented. Much like an early form of cost-benefit analysis, EIA is not an academic product but rather a practical response to a pressing issue. There have been no significant attempts as yet to codify general principles and supply a particular theoretical foundation of its procedure. The literature is, however, large and still growing so that refinements and new partial methods have been accumulating.

There are a number of varied proposals of EIA procedure. They all tend to cover every possible step of an environmentally unsound project appraisal starting with a first rough proposal of the project down to some kind of post-auditing. Some of the steps are widely adopted, however, an EIA will always imply impact identification, a baseline study, a proposition of mitigation measures and follow up of the implemented project.

There is also a requirement that the results of an EIA be well presented. Environmental Impact Statement (EIS) (i.e. a document which presents the results in a comprehensive way) is supposed to inform individuals of various professions, interests and educational levels. At least, such a document should be understandable to an intelligent layman.

Inclusion of decision making into EIA procedure does not seem to be agreed upon. If EIA is developed by a team of experts, then decision making is not included because they are supposed to submit their environmental impact statement to someone who will finally decide about the investment. However if the procedure is viewed as an activity developed by more than one agent, decision making is clearly a part of it.

Identification, classification and measurement of impacts with which EIA may be concerned require a particular methodology which, due to the variable set of impacts and their character, is not generally defined. Methods differ from project to project so that only a list of mutually overlapping methods, none of which is superior to others, can be made. Those most often used are ad hoc methods, checklists, two dimensional matrices, overlays, networks, quantitative or index methods, and as the most recent method, models.

Problems of measurement become particularly serious when a considered set of impacts has its environmental, social and economic subsets. If it is proposed that a common unit of measurement should be money, the procedure will face pricing, rate of discount and other problems otherwise quite familiar in an economic analysis performed from a social perspective.

The relationship between cost-benefit analysis and environmental impact assessment does not seem difficult to trace. Their procedures have most steps in common while the steps included in CBA and not in EIA and vice versa clearly point out the overall difference. First of all, CBA is an investment method with a strictly defined procedure. EIA is a rather relaxed procedure with a number of steps that can be accomplished by the use of various methods. However, due to overlapping of steps of the two procedures, an EIA team that has decided to use CBA during impact evaluation may easily end up doing solely the cost-benefit analysis.

Both CBA and EIA require certain preliminary activities. EIA usually involves more experts, teams and even government bodies so that it requires more preliminaries. Since this is a question of more work this step can be regarded as the same in both cases.

Impact identification, quantification and evaluation are the most important steps in both procedures. Compared with the early form of EIA that dealt only with environmental impacts, cost-benefit analysis appears more complex, but since EIA also covers social and economic aspects, the analyzed set of impacts is equally complex and numerous in both procedures. A further overlapping of CBA and EIA is in the quantification of impacts. Here CBA relies on medical, ecological and other studies that identify and measure impacts in physical units, as does EIA. At this point the procedures, however, diverge because CBA has a strict requirement for evaluation in monetary terms, while EIA may adopt the same approach but allows for other ranking and weighting criteria as well. If pricing is chosen as an evaluation mode, EIA and CBA are equalized and present a method that explicitly points out the social desirability of the project in question. A further point of difference between the two procedures is caused by the assessment of mitigation measures. These measures are considered during EIA by its definition while CBA considers the proposed project only. If the main project has revealed unacceptable values of indicators B-C and/or B/C, and some mitigation i.e. some alternative technology or course of action is proposed, this will constitute a new project and CBA procedure will be repeated. This can be accomplished for any number of alternatives so that a set of mutually exclusive project appraisals is offered. Since their costs and benefits are expressed in the same units, ranking within such a set is straightforward and the most socially adequate project is identified. Therefore, if CBA is to deal with mitigations, the procedure will take more time but no essential changes in methodology will be needed. The crucial methodological point, however, is decision making. In comparison with EIA the role of decision makers in CBA is severely reduced (theory of CBA in fact annuls it implicitly). On the other hand, EIA seems to be a procedure that tries to adjust to development practices so that it has never adopted such a rigid methodological approach. A set of recommendations rather than a ready made decision is usually submitted to decision makers so that some room is left for discussions and opened confrontation of interests. When it comes to decision making, CBA and EIA diverge to the point of mutual exclusiveness. If it is decided that CBA will be used in the course of a particular EIA and if the usual approach to decision making has remained unchanged, the procedure will tend to be internally inconsistent. Problems of decision making are closely connected with the question of public involvement during project appraisal and particularly in the moment when the decision is made. Both procedures approach the public more or less indirectly, but none of them requires that everyone likely to be affected once the project is implemented take an active part in actual decision making.

The socialist methodologist will find the EIA approach interesting and more acceptable than cost-benefit analysis. The reasons are mainly in the diversity and simplicity of EIA methods and their lack of ambition to take over the whole decision making process. In fact, the two methods have to be considered simultaneously. A methodologist who is to examine their acceptability in the socialist socio-economic system is confronted with an array of ideas, algorithms and procedures that could very well serve as a starting point for development of a socialist methodology for evaluating environmentally unsound investment projects. In fact, the number of references and evaluated projects in socialist countries over the last couple of years shows that this development has already started. Due to rising pollution on one hand and rising environmental concern on the other, it is likely to grow rapidly.

LITERATURE CITED

Bahro, R. 1984. *From Red to Green.* Verso Editions, London.
Buchner, G. 1981. *Möglichkeiten der reduktion von SO_x* emissionene, Vorgetragen anlass lich der Technisch-Weissen-schaftlichen Woche Moskau. 16-25 September.
Clark, B.D., R. Bisset, and P. Wathern, (Ed.) 1980. *Environmental Impact Assessment.* Mansell, London.
Ćurković, J. (Ed.) 1986. *Procjena prizemnih koncentracija SO_2 u okolini TE Plomin 1 i 2 s ciljem utvrdivanja lokacija dinamčkog monitoringa.* Institut za elektroprivredu, Zagreb.
Elseworth, S. 1984. *Acid Rain.* Pluto Press, London.
Grenzüberschreitender und weitrauminger transport von Luft verunreinigungen, *Umwelt* 8/1982.
Highton, N.H. and M.J. Chandwick. 1980. The Effects of Changing Patterns of Energy Use on Sulphur Emissions and Depositions in Europe, *Ambio.* Vol. XI, No. 6. Stockholm.
Kauzlarić, K. 1986. Stetan utjecaj sumpornih oksida na sume *In: XIV znanstveni skup "Susreti na gragom kamenu".* Liber, Zagreb.
Knežević A. 1983. Kolika je emisija SO_2 u Jugoslaviji Zaštita atmosfere 11.
Kos, V. 1984. Izbor procesa pogodnog za odsumporavanje dimnih gasova termoelektrana u ČSSR. Zaštita atmosfere, 3.
Krstić, M. 1981. Normiranje emisije i kvaliteta vazduha u SSSR. Zaštita atmosfere, 10.
Muntingh, H. 1983. *Information note on acid rain.* European Parliament Committee on the Environment, Public Health and Consumer Protection, Brussels.

Environmental Consequences of Energy Production: Problems and Prospects. Edited by S. K. Majumdar, F. J. Brenner and E. W. Miller. © 1987, The Pennsylvania Academy of Science.

Chapter Fifteen
ENVIRONMENTAL ASPECTS OF THERMAL DISCHARGES TO AQUATIC HABITATS

ROBERT DOMERMUTH
Pennsylvania Power & Light Company
Ecological Studies Laboratory
4417 Hamilton Blvd.
Allentown, PA 18103

Of all water quality issues, none has been more closely identified with the electric utility industry than thermal discharges. Electrical generation appears to be one of many factors contributing to other environmental concerns such as acidification of surface waters and releases of toxic compounds to lakes, rivers and streams. However, industrial releases of waste heat to the aquatic environment largely occurs as a result of electrical generation. Laws (1981) estimates that approximately 75 to 80% of the industrial waste heat produced in the United States is derived from electric power plants. Thus regulatory agencies and the public have come to recognize effects of thermal discharges as a problem closely linked with electrical generation.

Scientists and environmentalists played the early, active role in calling public attention to the issue of thermal enrichment of the nation's waters. As summarized in Schubel and Marcy (1978), such individuals recognized the potential effects that thermal discharges from electrical generating stations could have on aquatic flora and fauna, and led a vigorous campaign to develop regulations and laws to address the issue. At various times regulations have taken the form of maximum allowable temperature rises across condensers, maximum temperature limits on thermal discharges, defined mixing zones within receiving water bodies outside of which specified temperatures could not be exceeded, and prescribed heat rejection rates based on defined low flow rates of the receiving water body coupled with a maximum targeted temperature rise above ambient conditions.

Regulatory efforts culminated with the passage of the 1972 Federal Water Pollution Control Act Amendments (PL 92-500). This law proclaimed national goals of eliminating the discharge of pollutants to navigable waters, and achieving water quality sufficient to insure propagation of fish, shellfish and wildlife resources and protection of recreational uses. The act established the National

Pollutant Discharge Elimination System, which required that a permit be obtained on all point source discharges, including thermal discharges, to navigable waters of the United States. In addition, the law contained two sections specifically addressing electrical generating station cooling water issues: Section 316(a) permitting owners/operators of electric generating stations to obtain alternative, less stringent thermal discharge limitations if they could successfully demonstrate that existing thermal discharges insured protection of aquatic biota; Section 316(b) requiring that the location, design, construction and capacity of cooling water intake structures reflect the best available technology to minimize entrainment/impingement of aquatic organisms.

Studies to address sections 316(a) and 316(b), constituted the first large scale industrial response to national environmental laws. As a result, issues surrounding thermal impacts were scrutinized not only by environmental professionals, but by the public as well. Despite such scrutiny, many misconceptions still exist in regards to thermal discharges and associated environmental impacts.

To gain some understanding of the effects of thermal enrichment, it is necessary to focus on three topics:

1) How electricity is generated;
2) How waste heat from electrical generation is dissipated to the environment;
3) How biota respond to the introduction of waste heat.

The principals of electrical generation and dissipation of waste heat are largely generic, and therefore applicable to any locality where the processes occur. However, responses of aquatic biota to thermal enrichment are subject to regional differences. Keeping these points in mind, the following discussion treats electrical generation and dissipation of waste heat in general terms, while limiting examples of aquatic biota responses to those documented for temperate zone climatic conditions such as occur in Pennsylvania and surrounding states.

ELECTRICAL GENERATION

As depicted in Figure 1, components of a steam electrical generation system include an energy source, boiler, turbine and generator. The energy source, typically coal, oil or nuclear fuel, is used to heat water within the boiler to produce steam. Pressurized steam directed to the turbine blades imparts the force needed to turn the turbine shaft. As the turbine shaft spins, it rotates a large magnet within tightly coiled wires housed in the generator. This action produces an electrical current.

Steam from the turbine is cooled to yield water for recirculation to the boiler. This occurs in the condenser (Figure 1) which functions as a huge heat exchanger. Water at ambient temperature is pumped through pipes within the condenser

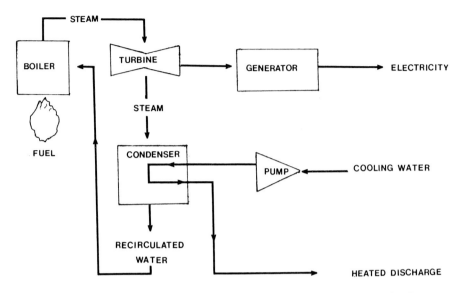

FIGURE 1. Components of a steam electrical generation unit (modified from Schubel and Marcy 1978).

and cools the steam. Heat is transferred to the cooling water and the steam condenses to water. The warmed cooling water is then returned either directly to the receiving water body, or sent to cooling towers or cooling ponds if the generating station is so equipped.

The increase in temperature (termed delta T) of the cooling water within the condenser depends primarily on the pump rate, which effects the contact time between cooling water and steam. Schubel and Marcy (1978) presented a histogram of delta T's for proposed and operating nuclear plants (also applicable to fossil-fueled units) that depicted most units operating at pump rates yielding between 6 and 17°C temperature increases across the condenser. Regardless of how much cooling waters are heated, the process occurs quickly. Schubel et al. (1978) estimated that heat is added to cooling waters for only 2 to 20 seconds. Thus the temperature rise in condenser cooling waters can be viewed as a nearly instantaneous process.

DISSIPATION OF WASTE HEAT

The amount of waste heat produced during electrical generation is largely a function of unit efficiency, fuel type used and plant load. Most electrical generation units operate at 33 to 40% efficiency (Rimberg 1974, Schubel and Marcy 1978). In other words, 60 to 67% of the heat produced is nonproductive and must be dissipated. Approximately half the heat in fossil-fueled units escapes directly to the atmosphere via stacks, with the remainder released to cooling waters. Nuclear units which do not have stacks, dissipate virtually all waste heat to the cooling medium.

The utility industry commonly employs three methods of cooling water disposal: a) direct return to the receiving water body termed a once-through system; b) diversion to cooling ponds and canals prior to release; c) diversion to cooling towers for atmospheric dissipation of heat. A variety of methods are normally utilized by a company. For example, Pennsylvania Power & Light (PP&L) operates 2 nuclear and 14 fossil-fueled major generation units. Both nuclear and 4 fossil-fueled units are equipped with cooling towers, 2 units utilize once-through cooling and discharge through a diffuser and the remainder employ once-through cooling systems discharging to canals.

The Environmental Protection Agency considers cooling towers the "best available technology" in regards to disposal of waste heat, since such systems avoid the discharge of heated effluents to receiving water bodies and generally minimize consumptive water use. However, cooling towers are expensive to construct and operate and not without environmental impacts. In addition, many older plants constructed before the period of environmental awareness have mitigated impacts associated with once-through cooling systems by the incorporation of cooling ponds and canals, and/or the utilization of multi-port diffusers or submerged jets to speed dilution with ambient waters.

Schubel and Marcy (1978) reported 60% of operating fossil-fueled plants in 1973 employed once-through cooling systems, 6% cooling ponds, 21% cooling towers and the remainder combined systems, while nearly 75% of operating nuclear plants utilized once-through cooling and 5% cooling towers. Undoubtedly these ratios have changed in favor of cooling tower equipped units in recent years with the retirement of older plants, and construction of new facilities required to employ best available technology for heat dissipation. However, generating stations utilizing once-through systems are still common. Since such heat dissipation methods clearly present the greatest potential for thermal impacts, concerns surrounding waste heat disposal are still a timely environmental issue.

RESPONSES OF BIOTA TO THERMAL DISCHARGES

Impacts to biota associated with electrical generating station cooling water systems occur at both the intake and discharge points (Figure 1). Intake related effects involve the copious amounts of water required as condenser coolant. This water pumped from rivers, lakes, impoundments, etc., passes into the intake structure and then through a series of screens and strainers designed to minimize debris damage to pumps and condenser components. Of course, the water also contains a variety of phytoplankton, zooplankton, drifting macroinvertebrates and fish eggs, larvae, juveniles and even adults which may be drawn into the intake structure. Larger organisms can be swept against fine mesh screening (termed impingement) eventually succumbing to stress and physical abrasion. Plankton and nekton tend to pass through screens and strainers and are carried to the condenser (termed entrainment), ultimately passing to the discharge point. During this passage organisms encounter nearly instantaneous rises in temperature, physical buffeting and abrasion, considerable pressure changes and even exposure to biocides (i.e. chlorine) utilized to inhibit condenser biofouling.

Valid concerns relating to entrainment/impingement issues have constituted a focal point for many intensive exchanges between the utility industry and regulatory agencies. However, such topics are not detailed in this chapter, since impacts are related to many physical and chemical factors in addition to thermal effects and are often very site-specific. In a classic sense, environmental effects associated solely with waste heat dissipation involve effluents from generating stations with once-through cooling systems. In such situations, discharge of heated waters is continuous subject of course to generating unit availability. This continuous discharge creates a plume of thermally enriched water which covers some portion of the receiving water body. Responses of flora and fauna to such thermal inputs form the basis of the remaining discussion.

Waste heat differs from most other environmental pollutants in two respects: a) effects are nonresidual and disappear quickly following cessation of the thermal input; b) heated discharges may have a stimulatory and even beneficial effect on aquatic flora and fauna up to a level closely approaching the upper lethal limit. As exemplified in Figure 2 (after Sweeny 1984), the stimulatory nature of thermal discharges involves the sharp increase in activity rates (and all physiological processes) which typify poikilothermic organisms exposed to increasing water temperatures. Therefore, effects of thermal discharges may be positive or negative depending on when and how much temperatures are raised above ambient levels.

Negative impacts may involve the elimination from or avoidance of the thermal plume by organisms as lethal temperatures are reached. Lethal temperatures may be encountered as low as 20°C for cold water, small stream mayfly and stonefly species (Widerholm 1984). However biota characteristic of larger rivers, lakes and impoundments, the preferred sites for electrical generation facilities, will generally tolerate temperatures up to 30°C. Above this level reductions in species diversity, overall abundance and biomass can be expected (Widerholm 1984), as the tolerance range of various floral and faunal elements are exceeded. Table 1 contains some examples of upper thermal tolerance values for organisms typically found in temperate region streams, rivers and lakes.

Often negative impacts associated with thermal discharges are more subtle

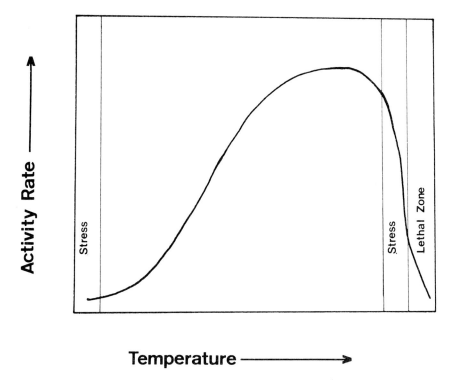

FIGURE 2. Relationship between temperature and activity rates of poikilothermic organisms (modified from Sweeny 1984).

in nature involving sublethal effects on various physiological processes. The scientific literature abounds with such examples. This is not surprising, since temperature is of critical importance to nearly all life functions of aquatic biota (Widerholm 1984; Hutchison 1976). Examples of sublethal effects include:

- alterations in nutritional requirements, reproduction, growth and development, regeneration and encystment and formation of abnormal cellular forms among protozoa (Cairns 1974);
- increased respiration patterns and decreased resistance to toxins among gastropods (Harman 1974);
- changes in metabolism and growth patterns for aquatic insects which may cause elimination or replacement of species with subsequent alterations in community structure (Widerholm 1984);
- early emergence of aquatic insect adult life stages into inhospitable environmental conditions with subsequent death or reproductive failure (Nebeker 1971; Rupprecht 1975);

TABLE 1

Upper thermal tolerance values for selected organisms characteristic of temperature region rivers, lakes and streams.

Organism	Upper Thermal Tolerance Value (°C)	Reference
Leeches	33 to 35	Sawyer (1974)
Crayfishes	35 to 37	Hobbs and Hall (1974)
Zooplankton	31 to 42	Carlson (1974)
Aquatic Insects	30 to 40	Widerholm (1984)
Fishes:		
Carp (egg development)	35 to 42	Frank (1974)
Chain pickeral	33 to 37	Hokanson et. al. (1973)
Bluntnose minnow	35	Robbins and Mathur (1976)
Spotfin shiner	34.4	Robbins and Mathur (1976
Channel catfish	36.6	Peterson and Schuysky (1975)
Bluegill	37.5	Talmage and Opresko (1981)
Smallmouth Bass	36.3	Talmage and Opresko (1981)
Darters	31 to 32	Kowalski et. al. (1978)

- malnutrition of fishes associated with increased metabolic demand accompanied by a reduction in available food supply (Graham 1974);
- inhibition of vertical migratory behavior by zooplankton in lakes (Gehrs 1974);
- shift in dominance from diatoms to green and blue-green algae within thermal plumes (Patrick 1974; Rankin et. al. 1974);
- avoidance of thermally enriched areas by migratory waterfowl (Brisbin 1974);
- blockage of migratory fish movement by the thermal plume.

Obviously many additional examples exist and could be cited. However, all share a common pattern: physiological processes and/or behavior are altered in response to elevated temperature caused by the thermal discharge.

To summarize, negative impacts associated with effluents from once-through cooling systems generally occur when temperatures reach or surpass 30°C, or when the discharge value exceeds ambient temperatures sufficiently to invoke detrimental physiological and/or behavioral changes in the biota. Temperature increases below these thresholds tend to have a beneficial rather than negative effect.

Beneficial responses have been documented for zooplankton (Carlson 1974), fishes (Bennett and Gibbons 1974; Coutant and Cox 1976) and algae (Patrick 1974) as well as other components of the aquatic community. Enhanced growth, diversity and biomass within thermally enriched areas are often mentioned as evidence of the stimulatory and positive effects of thermal discharges. Perhaps the best example of the potential benefit afforded by thermal effluents is the creation of winter recreational fisheries. As ambient water temperatures decrease invertebrates and fishes are attracted to thermal discharges (Dahlberg and Conyers 1974; Benda and Proffitt 1974; Yoder and Gammon 1976; Denoncourt 1983). Anglers often have excellent success fishing in the thermal plume, catching a variety of pan and gamefishes which continue to actively feed throughout the winter. The recreational opportunity created can be substantial. To illustrate, Denoncourt (1984) valued the winter fishery in the thermal discharge created by PP&L's Brunner Island Steam Electric Station at over $650,000.

The attraction of biota to thermally enriched areas in the winter is not without risk to the organisms. Electrical generating units are subject to rapid operational shutdowns when problems develop in critical systems. Such shutdowns are necessary to protect the generating unit. However during winter months, they may lead to a rapid temperature decrease within the thermal plume as condenser cooling water is replaced by water at ambient temperature. Fishes occupying the thermal plume are subjected to this temperature change and usually have insufficient time to either adjust or leave the area. As a result, fishes undergo thermal stress (termed "cold shock") which eventually may lead to death. The exact mechanism causing stress and mortalities is not documented, but research by Block (1974) on channel catfish suggests osmoregulatory collapse. Wintertime "cold shock" fish kills have been and continue to be a highly visible environmental impact associated with thermal effluents.

Beneficial aspects of heated effluents can be heightened by changing excess heat from a waste to a resource through the development and application of innovative technologies. Rimberg (1974) and Lee and Sengupta (1977) provide many examples of how waste heat from electrical generation is or could be utilized. These include spray irrigation, soil heating, environmental control in animal shelters and greenhouses, aquaculture, improved wastewater treatment and centralized space heating for industries and residences. Many problems confront attempts to utilize heated discharges for these and other purposes, but programs have developed beyond the demonstration phase in some cases. For example, one of the largest greenhouse complexes in the world using waste heat now exists near PP&L's Montour Steam Electric Station in Montour County, Pennsylvania. Although the percentage of waste heat utilized as a resource is currently very small, further development of innovative technologies appears promising. Both the public and regulatory agencies should encourage such programs as a means of limiting thermal discharges to aquatic environments.

LITERATURE CITED

Benda, R.S. and M.A. Proffitt. 1974. Effects of thermal effluents on fish and invertebrates. Pages 438-447 *in* J.W. Gibbons and R.R. Sharitz, eds. Thermal Ecology. National Technical Information Service, Springfield, VA.

Bennett, D.H. and J.W. Gibbons. 1974. Growth and condition of juvenile largemouth bass from a reservoir receiving thermal effluent. Pages 246-254 *in* J.W. Gibbons and R.R. Sharitz, eds. Thermal Ecology. National Technical Information Service, Springfield, VA.

Block, R.M. 1974. Effects of acute cold shock on the channel catfish. Pages 109-118 *in* J.W. Gibbons and R.R. Sharitz, eds. Thermal Ecology. National Technical Information Service, Springfield, VA.

Brisbin, Jr., I.L. 1974. Abundance and diversity of waterfowl inhabiting heated and unheated portions of a reactor cooling reservoir. Pages 579-593 *in* J.W. Gibbons and R.R. Sharitz, eds. Thermal Ecology. National Technical Information Service, Springfield, VA.

Cairns, Jr. J. 1974. Protozoans. Pages 1-28 *in* C.W. Hart Jr. and S.L.H. Fuller, eds. Pollution ecology of freshwater invertebrates. Academic Press, New York, NY.

Carlson, D.M. 1974. Responses of planktonic cladocerans to heated waters. Pages 186-206 *in* J.W. Gibbons and R.R. Sharitz, eds. Thermal Ecology. National Technical Information Service, Springfield, VA.

Coutant, C.C. and D.K. Cox. 1976. Growth rates of subadult largemouth bass at 24 to 35.5°C. Pages 118-120 *in* G.W. Esch and R.W. McFarlane, eds. Thermal Ecology II. National Technical Information Service, Springfield, VA.

Dahlberg, M.D. and J.C. Conyers. 1974. Winter fauna in a thermal discharge with observations on a macrobenthos sampler. Pages 414-422 *in* J.W. Gibbons and R.R. Sharitz, eds. Thermal Ecology. National Technical Information Service, Springfield, VA.

Dennoncourt, R.F. 1983. Fish distribution in the vicinity of Brunner Island Steam Electric Station. Proc. PA Acad. Sci. 57: 165-172.

Dennoncourt, R.F. 1984. Recreational/sport fishery benefits associated with a fossil fuel generating station. Pages 170-190 *in* S.K. Majumdar and E.W. Miller, eds. Solid and liquid wastes: management, methods and socioeconomic considerations. The Pennsylvania Academy of Science.

Frank, M.L. 1974. Relative sensitivity of different developmental stages of carp eggs to thermal shock. Pages 171-176 *in* J.W. Gibbons and R.R. Sharitz, eds. Thermal Ecology. National Technical Information Service, Springfield, VA.

Gehrs, C.W. 1974. Vertical movement of zooplankton in response to heated water. Pages 285-290 *in* J.W. Gibbons and R.R. Sharitz, eds. Thermal Ecology. National Technical Information Service, Springfield, VA.

Graham, T.P. 1974. Chronic malnutrition in four species of sunfish in a thermally loaded impoundment. Pages 151-157 *in* J.W. Gibbons and R.R. Sharitz, eds. Thermal Ecology. National Technical Information Service, Springfield, VA.

Harman, W.N. 1974. Snails (Mollusca: Gastropoda). Pages 275-312 *in* C.W. Hart

Jr. and S.L.H. Fuller, eds. Pollution ecology of freshwater invertebrates. Academic Press, New York, NY.

Hobbs, Jr. H.H. and E.T. Hall Jr. 1974. Crayfishes (Decapoda: Astacidae). Pages 195-214 in C.W. Hart Jr. and S.L.H. Fuller, eds. Pollution ecology of freshwater invertebrates. Academic Press, New York, NY.

Hokanson, K.E.F., J.H. McCormick and B.R. Jones. 1973. Temperature requirements for embryos and larvae of the northern pike, Esox lucius (Linnaeus). Trans. Am. Fish. Soc. 102: 89-100.

Hutchinson, V.H. 1976. Factors influencing thermal tolerances of individual organisms. Pages 10-26 in G.W. Esch and R.W. McFarlane, eds. Thermal Ecology II. National Technical Information Service, Springfield, VA.

Kowalski, K.T. et al. 1978. Interspecific and seasonal differences in the temperature tolerance of stream fish. J. Thermal Biol. 3: 105-108.

Laws, E.A. 1981. Aquatic pollution. John Wiley & Sons, New York, NY. 482 pp.

Lee, S.S. and S. Sengupta, eds. 1977. Proceedings of the conference on waste heat management and utilization. Three Volumes. National Aeronautics and Space Administration, et al.

Nebeker, A.V. 1971. Effect of high water temperature on adult emergence of aquatic insects. Water Res. 5: 777-783.

Patrick, R. 1974. Effects of abnormal temperatures on algal communities. Pages 335-349 in J.W. Gibbons and R.R. Sharitz, eds. Thermal Ecology. National Technical Information Service, Springfield, VA.

Peterson, S.E. and R.M. Shutsky. 1975. Temperature tolerance studies on freshwater fishes. Ichthyological Associates, Inc., Drumore, PA. in Proc. NE Fish. Wildl. Conf., New Haven, CT. February 23-26, 1975.

Rankin, J.S., J.D. Buck and J.W. Foerster. 1974. Thermal effects on the microbiology and chemistry of the Connecticut River—A summary. Pages 350-355 in J.W. Gibbons and R.R. Sharitz, eds. Thermal Ecology. National Technical Information Service, Springfield, VA.

Rimberg, D. 1974. Utilization of waste heat from power plants. Noyes Data Corporation, Park Ridge, NJ. 171 pp.

Robbins, T.W. and D. Mathur. 1976. Supplementary materials prepared for the Environmental Protection Agency 316(a) Demonstration for PBAPS units No. 2 and 3 on Conowingo Pond. Report submitted to Philadelphia Electric Co., Philadelphia, PA. 251 pp.

Rupprecht, R. 1975. The dependence of emergence period in insect larvae on water temperature. Verh. Internat. Verein. Limnol. 19: 3057-3063.

Sawyer, R.T. 1974. Leeches (Annelida: Hirudinea). Pages 81-142 in C.W. Hart Jr. and S.L.H. Fuller, eds. Pollution ecology of freshwater invertebrates. Academic Press, New York, NY.

Schubel, J.R. and B.C. Marcy Jr., eds. 1978. Power plant entrainment—A biological assessment. Academic Press, New York, NY. 271 pp.

Schubel, J.R., C.C. Coutant and P.M.J. Woodhead. 1978. Thermal effects of entrainment. Pages 19-93 *in* J.R. Schubel and B.C. Marcy Jr., eds. Power plant entrainment—A biological assessment. Academic Press, New York, NY.

Sweeny, B.W. 1984. Factors influencing life-history patterns of aquatic insects. Pages 56-100 *in* V.H. Resh and D.M. Rosenberg, eds. The ecology of aquatic insects. Praeger Scientific, New York, NY.

Talmage, S.S. and D.M. Opresko. 1981. Literature review: response of fish to thermal discharges. EPRI Rept. No. EA-1840. Electric Power Research Institute, Palo Alto, CA.

Widerholm, T. 1984. Responses of aquatic insects to environmental pollution. Pages 508-557 *in* V.H. Resh and D.M. Rosenberg, eds. The ecology of aquatic insects. Praeger Scientific, New York, NY.

Yoder, C.O. and J.R. Gammon. 1976. Seasonal distribution and abundance of Ohio River fishes at the J.M. Stuart Electric Generating Station. Pages 284-295 *in* G.W. Esch and R.W. McFarlane, eds. Thermal Ecology II. National Technical Information Service, Springfield, VA.

Environmental Consequences of Energy Production: Problems and Prospects. Edited by
S. K. Majumdar, F. J. Brenner and E. W. Miller. © 1987, The Pennsylvania Academy of Science.

Chapter Sixteen
ENVIRONMENTAL IMPACT OF FLY ASH DISPOSAL

JAMES P. MILLER, JR.
Professor Emeritus of Civil Engineering
University of Pittsburgh
Pittsburgh, PA 15219

INTRODUCTION

With the Industrial Revolution came improvements in the standard of living for all persons, but it also created social and environmental problems. Energy and natural resources are requirements of most industrial processes, but the energy production and the use of the natural resources results in waste products. The waste products can be disposed of in the environment by incineration, land filling or cover, and injection or leaching into water resources.

PROBLEM

All of these disposal methods result in pollution of the air, water, or ground resources. Many times land disposal is favored because the waste is buried with an "out of sight out of mind philosophy" only to be found at a later date to be an environmental hazard. The same concept can be used to evaluate government regulations. The long term effects of regulations are not properly considered, or not properly evaluated, or knowledge of long term environmental effects are not known or understood.

To prevent damage to the environment, the best method of disposal of any waste product is to make use of it as a resource for some other product. The abundance of energy and other natural resources has hampered the use of industrial by-products in the past. Industry has favored the more expedient and short-run cost effective methods of air, land, or water disposal.

REVIEW OF ASH PRODUCTION

Fly ash from electric energy generation is one of these waste products. In the 1920's, the advent of suspension fired furnaces caused the problem of ash flying from the power plant chimneys. Mechanical and electrostatic precipitators were installed to collect this ash. Later it was discovered that the gases eminating from the chimneys were harmful to the public. Flue gas desulfurization processes to remove the harmful constituents, created another waste product and its inherent storage handling and sludge disposal problems. At the power plants the primary solid wastes are coal-cleaning wastes, ash, and flue gas desulfurization (FGD) sludge.

A special Report of National Geographic (1981) stated that coal will be the main source of energy for power generation into the distant future. In United States known useable reserves of coal are estimated to be 786 billion tons of coal, one-fourth of the world's coal reserve. Anthracite is 2 percent of the U.S. coal reserve, bituminous 52 percent, subbituminous 38 percent and lignite 8 percent.

Geologists have also estimated that the U.S. has deposits of 1.7 trillion tons at depths under 1000 m (3000 ft), but unfortunately it is either not accessible with present technology or is not economical at present day prices. Of the 680M tons of coal mined in 1979, 77 percent was burned by the electric utility industry to produce 60 percent of all the power generated. This coal comes from approximately 6000 mines, scattered over 26 states that employ over 250,000 persons.

John Faber, (1974) stated that coal burned by the electric utility industry rose from 284 million tons in 1970 to 440M tons in 1975, then increased by 170 percent to 750M tons in 1985. In 1990, 900M tons are expected to be about 13.8 percent of coal burned. In 1970, 39.2M tons of ash were produced, 60M tons in 1975, 90M tons in 1985, and 125M tons is expected by 1990.

ASH GENERATION PROCESSES

The configuration of the furnace determines the nature of the ash produced. Today there are three major types of furnaces used for power generation. Cyclone-fired boilers produce an ash that is 80 to 90 percent a glassy slag due to the coal being combusted at a high temperature causing a fusing of the ash. This ash falls into a water-filled slag hopper. The pulverized coal furnaces burn finer coal which produce mostly fly ash; so called because it becomes entrained in the flue gas. The third type of furnace, the wet-bottom pulverized-coal burner, produces 50 percent of the ash as fly ash. Slags are collected in either dry or water filled hoppers below the boilers, then crushed or ground and transported either wet or dry to a disposal site.

As flue gas leaves the furnace, the larger particles are removed by passing the gas through a mechanical collector—usually a cyclone. The finer particles are then removed by passing the flue gas through a bag house or an electric static precipitator. These devices will remove more than 99 percent of the entrained fly ash. If the flue gas is scrubbed to remove the sulfur oxides, the fly ash may be removed in the FGD system. The fly ash that is collected may be transported to its disposal site in ether a wet or dry state.

Of the total ash produced, 71 percent is fly ash, 21.5 percent bottom ash, and 7.5 percent boiler slag. In 1985, out of the 90 million tons of ash produced, 63.9M tons were fly ash. Utilization of a waste can reduce its environmental impact. Unfortunately, estimates for the utilization of fly are 17.4 percent as compared with 34 percent for bottom ash and 58.7 percent for boiler slag. On the plus side is that the total overall percentages for ash utilization have been increasing from 13 in 1970 to 16.4 in 1975, to 40 in 1985, and estimated over 50 in 1990.

In the mid-1980's, 259 new coal-fired generating facilities have been completed or are expected to come on line. Many of these are cited to take advantage of the sub-bituminous and lignite facilities in the Southwest and Rocky Mountains areas. Utilizations data has been estimated for lignite ash to be 17 percent, 19 percent for sub-bituminous ash, and 21 percent for bituminous ash.

The ratio of 70/30 for fly ash to bottom ash is expected to remain constant. There is emphasis on the control of sulfur oxides by the use of washed coal. The total amount of ash may tend to decline with the burning of lower ash coals.

Scrubber sludge and fluidize bed residues are increasing. The fluidized bed boiler has been developed to burn high sulfur and highly corrosive coals. These residues have exhibited good pozzolanic characteristics because of the limestone processes that are used. They have been used as a sub-base on paving projects, as structural stabilizing agents in land fills, including solid waste applications. Because of the high lime content, these ashes are also attractive for agricultural use. The increased use of sub-bituminous and lignite coals produce a fly ash that is described as more reactive due to the high calcium content.

PHYSICAL PROPERTIES

The major portion of ash consists of thin-walled spheres which can be hollow (cenospheres) or filled with smaller solid spheres (plerospheres). The cenospheres can float and cause problems of settling in ponds. Some other spheres have crystal like needles that adhere to the surfaces giving them a tendency to aglomerate. In the coarse fraction of the fly ash there are irregularly shaped particles which may consist of partially combusted carbon, glass, or magnetic particles. The slag particles produced in the bottom of the furnaces are black or grey and are angular in shape with porous surfaces (Ray and Parker 1977).

While the size of ash particles vary with the configuration of the furnace, coarse bottom ash particles are about 10,000 times as large as fly ash particles. The bottom ash particles are in the size range of 0.1 to 10mm. The greatest portion of fly ash particles from an ESP are between 5 and 100 μ in diameter. Within a single sample of either type of ash the larger particles will be about 200 times the diameter of the smaller particles.

The specific gravities of bottom ash or slag and fly ash are about the same, ranging from 1.6 to 3.1 with the majority falling between 2.2 and 2.6. The dry bulk densities range from 46 to 99 lbs/ft^3 with values between 80 and 90 lbs/ft.^{3} the most common. When unit processes such as ion exchange, adsorption or vapor condensation are significant, specific surface area is an essential consideration. It is expressed as area per unit mass. Smaller particles have higher specific surface areas than larger particles of the same shape. Most values for specific surface area are between 0.2 to 0.8 M^2/g. The specific surface area increases in the order of bottom ash the smallest, then mechanical hopper ash, to ESP ash which is the largest (Kaakinen 1975).

Permeability is important in measuring the migration of leachates and solutes. The higher the permeability, the more water is able to pass through the system to produce a leachate. The boiler slag has the highest permeability equivalent to that of fine gravel. Bottom ash is somewhat lower in the range of 10^{-3} to 10^{-1} cm/sec. The coefficient of permeability for fly ash is in the order of 10^{-6} to 10^{-4} cm/sec., which is similar to that for sandy or silty clay. Fly ash is composed of smaller particles and are usually alkaline, which will develop the pozzolanic or cementing reaction when wetted.

The saturate moisture content of ash is a measure of how much water the material can take up before the production of leachate can begin. Fly ashes from Western coal have reported values of 50 percent based on the weight of the ash. Values of 50 to 110 percent are reported for ponded fly ashes.

Shear strength and angles of internal friction are the two most usual parameters to measure the strength of fly ashes. Shear strength is related to the zeta potential or the attraction between fine particles due to the electrostatic forces. This is in effect a cohesive force. Dry non-alkaline fly ashes and bottom ashes show no cohesion. Dry alkaline fly ashes have some cohesion. 9 to 15 psi, and if mixed with moist calcium sulfate/sulfite salts, they have considerable cohesion when cured, 22 to 173 psi. The compacted fly ash will increase in shear strength with age as the pozzolans become cemented (Grag and Lin 1972). The angle of internal resistance is a measure of the frictional resistance between the particles. It is reported as an angle whose tangent is equal to the coefficient of friction. The values reported range from 25 to 45 degrees. Both fly ash and bottom ash show angles of internal friction about the same which is similar to those of clean graded sand. Many times this internal angle is spoken of as the angle of repose. Both this angle and the shear strength are important parameters to determine the stability and usability of any land fill (GAI Consultants 1979).

CHEMICAL COMPOSITION

Chemical composition of coal ash should not only include the elemental composition but also its compound and mineralogical phases. To accurately estimate the potential for release of an element by leaching the composition of the solid phases in the ash and the composition of the associated surface, deposits must be known.

The National Research Council (1980) in their publication, "Trace Element Geochemistry of Coal Development Related to Environmental Quality and Health" presents an exhaustive bibliography of analytical methods used in trace element analysis of coal and coal ash. The most often used techniques are atomic absorption spectrometry (AAS) and neutron activation analysis (NAA). Wet chemistry techniques and procedures such as sodium fusion are used but require extensive sample preparation prior to analysis. Also used are such analytical techniques as inductively coupled plasma emission spectroscopy (ICPES), spark-source mass spectrometry (SSMS) and optical emission spectroscopy (OES). The National Bureau of Standards (NBS) has issued two reference fly ash standards (Standard #1633 and #1633a). The NBS has also certified the concentrations of most of the major, minor, and trace elements in these reference materials. Boron, silver, and fluorine or fluoride are among the elements not certified in fly ash because they have judged the analytical techniques not to be sufficiently reliable. Thallium, molybdenum and selenium methods are certified but present considerable analytical difficulty due either to very low concentrations or to analytical interferences (Gladney 1980).

It should be noted that the median values of almost all of the elements are one-third or less of the maximum values. The median values are nearer to the low value of the range than to the high value indicating the distributions are skewed to the right and most of the data is clustered at the low end of the range. Many of the elements have large ranges, i.e. mercury had over three orders of magnitude.

Ash composition variance can be attributed primarily to feed coal composition. Firing temperature and boiler configuration also effect ash composition. Some studies have indicated that the arsenic, lead, manganese, mercury, and selenium in the feed coal have been reduced by more than one half by the coal cleaning process (Page 1979). The source of the coal also has an impact on the elements in the ash. Geographic source has a primary effect, but even coal taken from different locations in the same mine have been shown to be quite variable.

The organic matter in coal ash is usually found in only trace amounts. The range of carbon fractions in ash usually run from just under 1 to 4.75 percent of the total ash mass. Most of this carbon is present as organic carbon with none, or very little, inorganic carbon being present.

The organic carbon present is in the form of unbranched alkanes, C15 to C34, and polycyclic aromatic hydrocarbons. The latter come from incomplete combustion of fossil fuels. Many samples that have been tested have shown (PAHs) below detection limits. Benza (a) pyrene (BAP), 3-methylcholanthrene and 7, 12-dimethylbenz (a) anthracene have been detected up to levels of 0.2 mg/kg. Blank entries indicate that the compound was not detected in the sample.

Radionuclides have been detected in fly ash but have shown very low levels of radioactivity. For alpha decay fly ash varied from a high of 36 picocuries per gram (p Ci/g) to under 26 p Ci/g for Eastern coal. For Beta decay values in terms of p Ci/g varied from a high of 54 from Western Lignite to under 36 for both Western and Eastern coal. Radium 226 has been measured in fly ash by gamma counting. Values ranged from 2.6 p Ci/g for Western coal ash to 3.9 p Ci/g for both Eastern and Western lignites (Utley 1982).

The solid phase of fly ash contains and holds the trace elements. It is from this phase the trace elements are released in the leachate. Those elements held in the soluble surface salts are released more quickly than those included in the glass or crystalline phase. The alkali and rare earth elements are held primarily in the glass phase. The glass phase contains the As, Se, Pb, Hg, Ba, and Sr. In the mullite phase the $+3$ cations V, Cr, Fe, Ga, Mo and the $+4$ cations Ti and Zr. The quartz phase trace elements are typically Li^+ and Na^+ (Hulett et al. 1980).

The alumino-silicate phase is the largest and is made up of the glass mullite and quartz phases. This phase will generally be 60 to 75 percent of the total weight of the fly ash; the magnetic phase is 17 to 20 percent and the carbon phase 8 to 15 percent (Hulett et al 1981).

Under the Resource Conservation and Recovery Act of 1976 (RCRA), the Environmental Protection Agency has set up guidelines governing the definitions and control of hazardous wastes. The Federal EPA does not classify fly ash, bottom ash, slag or flue gas emission control as a hazardous waste. However, many of the elements make these wastes a potential hazard if minimum toxicity thresholds are exceeded.

RCRA classifies a waste as toxic if the leachate or supernatant contains 100 times the level of any dissolved constituent specified in a primary drinking standard. The Environmental Protection Agency's Extraction Procedure (EP) is required for regulating purposes in assaying the supernatant from fly ash for toxicity (Ham 1978). The toxicity thresholds of the EPA designated elements based on the geometric means of 17 analyses of Eastern or Midwestern Coal fly ash and 23 analyses of Western coal fly ash, indicate that the fly ash from Western coal was higher in barium, chromium and mercury, and lower in arsenic, cadmium, and lead, while the selenium and silver were about the same as Eastern coal.

LEACHATE TESTING

Some researchers have felt that long term tests, both batch and columnar, would simulate field conditions. The long term batch tests have been useful in estimating whether solubility limits may be reached. The leaching trends observed in the long term tests can give an insight to the kinetics of reaction rates as well as the chemical factors that dominate the leaching process.

When fly ash slurry is stirred continuously for six months, the pH stabilizes at 4.1 for fly ash and 3.8 for slag end, it is assumed chemical equilibrium has been reached (Talbot 1978). These pH values were for air equilibrated atmosphere. The ones equilibrated under an argon atmosphere had slightly higher pH values. None of the slag leachate had a concentration above the limit of detection for any of the RCRA eight elements. Lead and cadmium were detectable in the supernatant of the fly ash. Fly ash showed slightly higher concentrations than slag extracts, although often trace metal concentrations were below detection limits.

The initial effect of the fly ash with water is to raise the pH but then a gradual decrease in the pH was noted. Attempts have been made to extrapolate results of batch tests to a longer term leachate test (Vander Sloot 1982). Major constituents Ca, Mg, Na and Al concentrations were found to differ by a ratio 2 to 1. Also the minor and trace constituents differed from the extrapolated values and showed no consistent pattern.

Long term column and batch tests have not been standardized but they have shown a rapid dissolution of surface phases; the primary constituents released are Ca, Na, and SO^4. The slower dissolution of solids in the matrix tend to release Cd, Mg, and Al. Potassium appears to leach in both processes at a constant rate.

In the flue gas desulfurization (FGD) processes the sludge is primarily made up of calcium carbonate, calcium sulfate, and calcium sulfites. The inert materials include fly ash and grit. A wide range of trace elements have been observed in FGD sludge. The amount of trace elements assayed has been proportional to the amount of fly ash in the sludge.

Median concentrations of As, Cr, Cu, Cd, Hg, Pb, and Se are between 0.3 and 2.7 mg/kg in sludge solids. In the sludge liquor, all of these elements show trace concentrations of less than 0.1 mg/L except for boron and flouride (Tetratech Inc. 1983). All of these are below RCRA standards. Because of the concentrations of total dissolved solids of greater than 1000 mg/L can cause some concern. The reason for this concentration is the soluble salts, such as the sulfates.

SUMMARY

Fly ash, once thought of as a waste product and a disposal problem, is now becoming less of an environmental problem. EPA air pollution requirements of 1.2 lbs. of SO_2 per 10^6 BTU requires a high degree of scrubbing which results in more FGD sludge. Because of the pozzolonic properties of fly ash in FGD sludge, the advantages as an additive to cement and concrete will increase. Also, the same will be true for fly ash used as a soil stabilization agent.

The trace elements in fly ash that would be hazardous are all under the allowed RCRA standards. The economic processes that have been and are being developed for the removal and separation of the trace elements will reduce all of these elements below even the detection level. Much of this decomposes in the leachate process and is below detectable limits in leachate and soil samples of fly ash. The radioactivity of fly ash is many times lower than allowable levels. It would appear that with the need for more concrete to meet construction demands fly ash can help satisfy, thereby reducing the need to import cement. With the continued research in the utilization of fly ash, we may reach a point where coal is burned to generate fly ash.

LITERATURE CITED

Faber, J. 1979. A U.S. Overview of Ash Production and Utilization. Proceedings Fifth International Ash Utilization Symposium. U.S. Dept. of Energy, Morgantown, W. Virginia. METC/SP-79/10 (Pt. 1).

GAI Consultants. 1979. Coal Ash Disposal Manual, EPRI Report No. FP-1257. Electric Power Research Institute, Palo Alto, Ca.

Gladney, E.S. 1980. Compilation of Elemental Concentration Data for NBS Biological and Environmental Standard Reference Materials. Los Alamos Scientific Laboratory LA-8438-M.S. Los Alamos, N.M. pp. 119.

Grag, D.H. and Y.K. Lin, 1972. Engineering Properties of Compacted Fly Ash. Soil Mech. Found. Div. ASCE. SM4., April 1972, pp. 361-380.

Ham, R.K., M.A. Anderson, R. Stanforth, and R. Stegmann. 1978. The Development of a Leaching Test for Industrial Wastes. D.W. Schultz (ed.), Land Disposal of Hazardous Wastes: Proceedings, Fourth Annual Research Symposium. U.S. Environmental Protection Agency, Cincinnati, Ohio. EPA-600/9-78-016.

Hulett, L.D., A.J. Weinberger, K.J. Northcutt, and M. Ferguson, 1980. Chemical Species in Fly Ash from Coal-Burning Power Plants. Science 210:1356-1358.

Hulett, L.D., A.J. Weinberger, K.J. Northcutt, and M. Ferguson, 1981. Trace Element and Phase Relations in Fly Ash. Prepared by ORNL, Analytical Chemistry Division, Oak Ridge, TN, for Electric Power Research Institute, Palo Alto, Ca. EPRI EA-1822, pp. 63.

Kaakinen, J.W., R.M. Jorden, M.H. Lawasani, and R.E. West. 1975. Trace Element Behavior in Coal-Fired Power Plant. Envir. Sci. and Tech. 9:862-869.

National Geographic. Energy—Special Report. February 1981.

National Research Council, 1980. Trace-Element Geochemistry of Coal Resource Development Related to Environmental Quality and Health. National Academy Press, Washington, D.C. 153 pp.

Page, A.L., A.A. Elseewi, and I.R. Straughn, 1979. Physical and Chemical Properties of Fly Ash from Coal Fired Power Plants with References to Environmental Impacts. Residue Reviews 71:83-120.

Ray, .S.S and F.G. Parker. 1977. Characterization of Ash from Coal-Fired Power Plants. USEPA Industrial Environmental Research Laboratory, Research Triangle Park, N.C. EPA. 600/7-77-010, 130 pp.

Talbot, R.W., M.A. Anderson, and A.W. Andrew. 1978. Qualitative Model of Heterogeneous Equilibria in a Fly Ash Pond. Envir. Sci. and Tech. 12:1056-1062.

TetraTech Inc., 1983. Physical-Chemical Characteristics of Utility Solid Wastes. EPRI Report No. EA-3236, Research Project 1487-12. Electric Power Research Institute, Palo Alto, Ca.

Van der Slott, H.A., J. Wijkstra, A. Van Dalen, H.A. Das, J. Slanina, J.J. Dekkers, and G.D. Wals, 1982. Leaching of Trace Elements from Coal Solid Waste. Energie on der Zoek Centrum Nederland. ECN-120. Petten, The Netherlands, pp. 201.

Utley, D. and G. Beall, 1982. Draft Report: A Radiochemical Survey of U.S. Coals and Coal Combustion By-Products. Radian Corp. Austin, TX., pp. 43.

Environmental Consequences of Energy Production: Problems and Prospects. Edited by S. K. Majumdar, F. J. Brenner and E. W. Miller. © 1987, The Pennsylvania Academy of Science.

Chapter Seventeen
ACID RAIN

JOHN J. CAHIR
Professor of Meteorology
and
Associate Dean for Resident Instruction
College of Earth and Mineral Sciences
The Pennsylvania State University
University Park, PA 16802

The term "acid rain" describes an environmental problem that is by now familiar to most people, but many details about it are not known, not even to experts. In the contest of energy production, electric utilities in the United States emit to the atmosphere more than 20 million tons of sulphur dioxide, more than 10 million tons of nitrogen oxides, and substantial amounts of reactive hydrocarbons each year. Other industries contribute as well. Also, even greater amounts of nitrogen oxides are emitted by automobiles, trucks, and busses, and sulphur and nitrogen compounds have been found to be important in precipitation that is more acidic than normal. Those emissions which are not involved in precipitation also return to earth, and may be involved in later chemical transactions. However, the atmospheric chemistry is very complex, involving many reactions and products. Thus, attempts to predict the effects of various control strategies are subject to large uncertainties, and some strategies are expensive. Thus, both energy producers and many others are intensely interested in the outcome of the current debate on what to do about acid rain.

ACID RAIN: WET AND DRY DEPOSITION

Acid rain was formerly thought of in terms of the acidity of precipitation, but that is now recognized as being only part of the issue. Emissions from industrial and transportation sources are mainly gaseous, and they can return to the surface without ever seeing the clouds at all. Also, the gasses may undergo chemical transformation into finely divided liquids or solids, called aerosols, and these, in turn, may be intercepted by features at the earth's surface. Thus, acid rain encompasses wet deposition, which is mostly rain and snow, but can include interception of fog by hills, buildings and trees, and also dry deposition. It often matters little to the receptor whether the chemical was delivered in a hydrated state.

The partitioning between wet deposition and dry is quite uncertain, both globally and at particular locations, but they are believed to have comparable magnitudes over land. Thus, to use sulphur as an example, the atmosphere must deliver tens of millions of tons of sulphur earthward annually, with substantial fractions occurring in precipitation and as dry deposition. Both constitute acid rain.

SOURCES OF CHEMICALS RELATED TO ACID RAIN

Globally, more than half of the sulphur is of anthropogenic (human) origin, but in addition to that, natural sources also put tens of millions of tons of sulphur from sea spray, from decaying organic material, and a small amount from volcanoes into the atmosphere each year. Because organic decay is favored in marine environments, we should expect that over the oceans, the natural sources predominate, and over land, the greater share of sulphur is put in by humankind.

Accordingly, while there is some uncertainty about the amount of exchange of sulphur in the atmosphere above the continental boundaries, it is clear that many of the millions of tons of atmospheric sulphur that result from fossil fuel combustion return to the continents each year. It should not be surprising to find it in the rain. For the eastern half of the United States, about a third of the anthropogenic sulphur is deposited as dry deposition, about a third as wet deposition, and about a third is exported. Thus, by our definition, about two-thirds is involved in the acid rain problem.

Furthermore, what is true for sulphur is also true for nitrogen and carbon. The atmosphere acts as both a reservoir and a conveyor of their compounds, some originating from natural sources and some from human ones. There is no reason to believe that all the material that is put in comes back out in any given period of time, and there is evidence that atmospheric carbon is, in fact, increasing. However, when large quantities are put in each year, we can expect that large quantities come out, even if there is not a balance.

In the case of nitrogen, there are substantial natural sources—lightning fixes nitrogen compounds in its hot channels and biologic activity in soils generates nitrogen compounds—but the atmosphere must still flush out millions of tons from anthropogenic sources, again primarily over land where it is put in. The global and regional nitrogen budgets are less well-known than the sulphur budget, however, so it is difficult to estimate the fractions that are deposited wet and dry and exported beyond saying that the acid rain involvement is substantial.

Although we recognize that acid rain includes both wet and dry deposition, we tend to focus more on the wet side, mainly because observations of acidic properties of precipitation are far more prevalent. Thus, in the discussion that follows, we will focus on those observations and on that chemistry, but the dry deposition should not be forgotten.

WET DEPOSITION

Water solutions are characterized as acid according to the concentration of hydronium ions—hydrated hydrogen ions—which are water molecules with an extra proton. The pH scale is used as a measure of acidity, with more acid solutions having a lower value because pH is the negative logarithm of the concentration of hydronium ions. A neutral solution is said to have a pH value of 7.0; because the scale is logarithmic, ten times as much hydronium concentration as that of pure water corresponds to a pH value of 6.0. A 4.0 pH solution has 1000 times the concentration of a 7.0 sample.

If pure water has a pH of 7.0, what of the rain or snow that falls far from human contaminating influence? The answer is rather surprising—first, that there is no precise figure and second, the range is between about 4.5 and 6.0. That is, natural rain is acidic. Strictly, the wet deposition part of acid rain is precipitation that has been made "more acid" by human activity.

Natural rain is acidic because the atmosphere contains a number of trace gases that are very important chemically, including carbon dioxide, ozone, nitrogen oxides, sulphur oxides, methane, and many more. Of them, carbon dioxide, with a well-mixed concentration of about 340 ppm, assures that natural rain will be a solution of carbonic acid. Furthermore, sulphate and nitrate ions, which are also naturally abundant, act the same way to increase hydronium concentration and drive the pH still lower.

While carbon dioxide is well-mixed, the nitrates and sulphates are more variable, coming as they do from highly unevenly distributed sources. Either one can be present in significant amounts or they may be virtually absent from a sample of air. Thus, we can expect that measured pH of natural precipitation would indeed be acidic, but it could be quite variable, depending on the location and on the recent experiences of air samples that produced the rain. Further, we can be sure that natural acidic rain has been around through the eons, and the degree of its acidity was probably highly variable from time-to time and place-to place, as well.

GLOBAL OBSERVATIONS

But what is observed today? Fig. 1 is a map of the global picture of observed pH. The figure shows a range at some places and a single figure at others, because

FIGURE 1. Global map of observed pH values in precipitation, based on data collected by the U.S. Dept. of Energy. (published in Research/Penn State, June, 1986).

the data are not always strictly comparable, having been collected over different sampling periods. Further, there could have been important differences in sampling techniques. Therefore, we have to approach these data with some caution, but they do show some interesting features.

The observations show that globally, the rain is acidic, even in places like the Indian Ocean and Pacific Islands, but some significant deviations can be noted. Japan, the eastern United States, and western Europe all show low (acidic) pH values. Because these are both heavily populated and industrially active regions, there is good reason to believe that human activity is contributing to the low readings there. We have seen that human activities do generate significant quantities of the precursors of the nitrates and sulphates, and other products which are definitely involved in the chemistry of rainwater, greatly strengthening that belief. But the processes and reactions involved in the chemistry of precipitation are very complex, and that complexity is at the heart of much of the controversy concerning the extent of human effect, the true importance of various sources, and the appropriateness of various mitigating strategies.

Before we turn away from the global picture, it is worthwhile to note an interesting continental effect. At some points on the Eurasian and North American continents, and at some points in Japan, rather high (neutral or basic) pH values are typically observed. At least in part, this complication arises from cations such as calcium, magnesium, potassium, and ammonium—the first three from dry soils, the latter from biological processes—being mixed into the air by the winds to ultimately react in clouds and precipitation to neutralize them. Thus, to a crude first approximation, we might expect continental precipitation to be more alkaline than corresponding oceanic values, and the map seems to confirm that expectation. This makes the very low values noted above all the more striking. However, none of this is simple. Much of the rain in these three areas falls from marine-origin air masses.

Measurements in clouds show the mix of chemicals that can be present. In an August sample of a cumulus cloud at 6000-foot elevation downwind of Buffalo, New York, Mohnen found the constituents shown in Table 1, illustrating the significant contribution of calcium and other cations in ameliorating the acidic effect of a substantial amount of sulphate and nitrate present in the cloud water, resulting in a quite normal pH of 5.4.

In contrast, Table 1 also shows a very different situation on Whiteface Mountain, New York, when such cations were relatively sparse. Here, the pH was very acidic, at about 3.7 despite lower concentration of sulphate and nitrate in both absolute terms, and in terms of acid-forming potential, which is expressed in micro-equivalents per liter. In this case, the basic cations were nearly absent.

Thus, interpretation of pH observations requires consideration of likely sources of natural ions, both positive and negative. For the United States, it is not surprising to find lower pH in eastern precipitation than in the arid West, where alkaline dust is more likely to be involved in precipitation chemistry. The issue is, how much more has humankind increased the natural east-west contrast.

TABLE 1

Comparison of two clouds — August 1980.

	Ca^{2+}	Mg^{2+}	K^+	Na^+	NH_4^+	SO_4^{2-}	NO_3^-	Cl^-	PO_4^{3-}	pH
				ppm (meq/liter)						
Downwind of Buffalo	18.1	2.1	1.6	5.9	3.5	31.1	23.4	8.0	0.1	5.42
	0.90	0.17	0.044	0.26	0.21	0.65	0.38	0.20	0.003	—
Whiteface Mountain	0.5	0.1	0.1	0.1	2.2	1.3	6.3	0.2	0	3.67
	0.025	0.008	0.003	0.004	0.10	0.27	0.10	0.005	0	—

WET DEPOSITION ON THE EASTERN UNITED STATES AND CANADA

Figure 2 is a map depicting the mean annual pH of precipitation observed during 1982. Observations were taken at over 150 stations in the National Atmospheric Deposition Program (NADP) National Trends Network (supported by the Environmental Protection Agency) and at other locations supported by consortia from government and industry. It can be seen that pH values averaged little more than 4.0 in western Pennsylvania, eastern Ohio, western New York, and southeastern Ontario. Further, west-to-east variation in acidity can be seen to involve approximately a ten-fold increase in hydronium concentration.

Taken together with Fig. 3, which shows the emissions of sulphur dioxide and nitrogen oxides (for 1980), it is not difficult to see that the upper Ohio Valley is the anthropogenic locus of the acid rain controversy. NO_x (which can include NO_1, NO_2, N_2O, NO_2, NO_3 and N_2O_5) and SO_x (primarily SO_2) emissions are of comparable magnitude in the greater Ohio Valley, with the sulphur coming

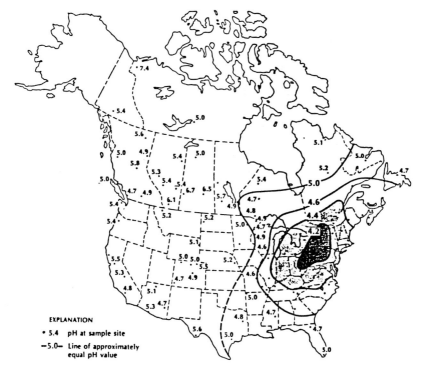

FIGURE 2. Distribution of acidity of precipitation in pH units for 1982. Dots are locations of reporting stations, with reported value plotted except in highlighted region of low pH. (from Weatherwise, October, 1984).

about two-thirds from power plants and one-third from other industries, while half of the nitrogen compounds originate from transportation and about one-third from power plants.

When the low-pH rain is analyzed in these regions, these gasses are implicated, as compounds that derive from them are typically encountered. The reactions are complex, but various oxidants act on them to form sulphate (SO_4^{--}) and nitrate ions (NO_3^-), which form sulphuric and nitric acids (H_2SO_4 and HNO_3) when hydrated.

Figure 4a shows a plot of hydronium ions versus sulphate in Pennsylvania rain samples. It can be seen that the two are correlated and that the presence of hydronium is strongly related to the presence of sulphate.

If we also consider the concentration of nitrate (Fig. 4b), we can see that sulphate and nitrate, taken together, account for a near balance with the hydronium. In fact, they actually over-compensate slightly, and the best balance is achieved by also considering the ammonium cation (NH^{++}). The conclusion is that in the region of low pH, it is sulphate and nitrate that are the responsible ions.

We know further that there is a very large input of SO_2 and various nitrogen

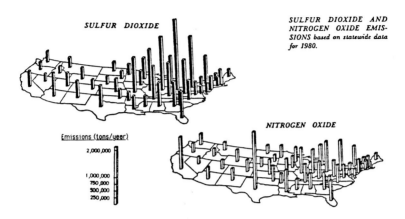

FIGURE 3. Emissions of sulphur dioxide and nitrogen oxides by state for 1980. (from Weatherwise, October, 1984).

compounds in these regions, and that these do come to earth, with substantial amounts doing so in the region. Is there a clear path from NO_x and SO_x to sulphate and nitrate?

LONG RANGE TRANSPORT

It is probably true that local pollutants have been a serious problem throughout the industrial age, but a combination of lower environmental awareness and a perception of locally-balanced benefit probably suppressed complaint. When it became desirable to clear the local air, tall stacks were frequently employed. Because the wind increases quite strongly with height in middle latitudes, both because friction decreases and because the pressure forces usually increase with height, outgassed material can be transported far downwind. A sample of gas rising to some higher level from the top of a 200 m stack can easily encounter 10 ms^{-1} average winds and some of it might travel hundreds of km downwind without settling back to earth, depending on the vertical motions and turbulent eddies encountered. Because most of the gasses and the aerosols that form from them are heavier than air, about half of the mass will slowly settle back to earth within 300 km of the source. That fraction which encounters updrafts and gets involved with cloud formation can be carried to substantial height, which lengthens its travel distance, its residence time and its opportunity for chemical reaction. The clouds may dissipate, but the chemical residue remains, either to settle earthward at some great distance, or to become involved in new cloud and precipitation formation later. Thus, the products can affect regions far from the source.

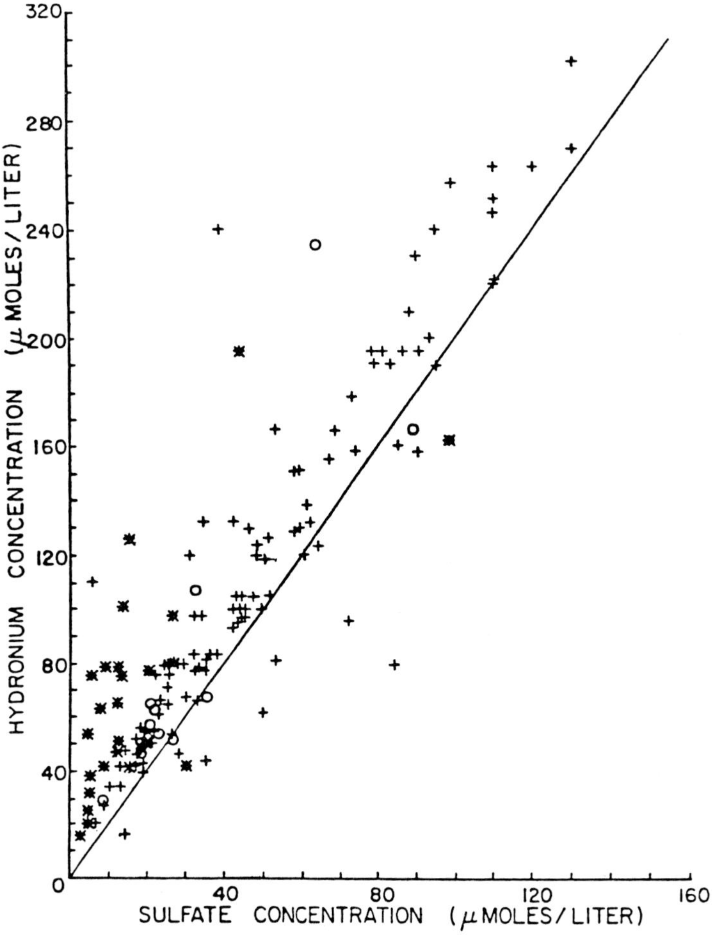

FIGURE 4a. Free hydronium concentration vs. sulfate concentration for precipitation events of 1977-1978 in central Pennsylvania (from Bowersox and de Pena, 1980).

COMPLEX REACTIONS

Gas phase, gas-liquid, gas-solid, and liquid phase reactions can all take place during the atmospheric residence. With respect to wet deposition, most of the nitrate ends up in precipitation after a gas-phase path, in which photochemical oxidation of nitric oxide (a significant combustion product) by ozone (O_3) and the OH radical is a major path to nitric acid. The acid is hydrolyzed in cloud elements or swept out by falling rain or snow, which is called washout. Snow appears to have a somewhat greater affinity than rain for nitrate (relative to sulphate). Intermediate products are NO_2, NO_3 and N_2O_5 in the reactions, some

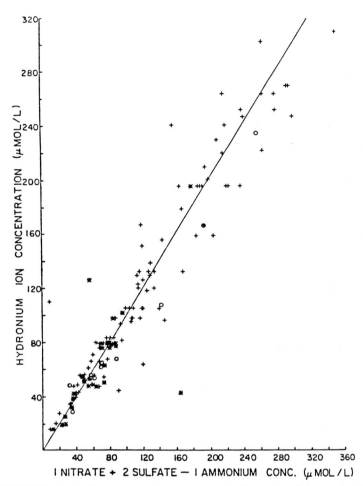

FIGURE 4b. Free hydronium concentration vs. the chemical equivalent balance of sulfate plus nitrate minus ammonium for precipitation events of 1977-1978 in central Pennsylvania. (from Bowersox and de Pena, 1980).

of which are reversible, and several of which require sunlight in the near ultraviolet to proceed.

In contrast, gas phase oxidation of SO_2 does not play the dominant role in the sulphate that is deposited as production from SO_2. Mixed-phase and aqueous solution reactions are also significant. The gas phase part is primarily photo oxidation, principally by hydroxyl, peroxide, and acetyl radicals, all of which require oxone for their production or regeneration. The hydroxyl radical, which plays a significant role in the rate at which many trace gasses are removed from the atmosphere, is formed in many ways, but ozone is a frequent participant.

The dominant liquid phase reaction involves the presence of strong oxidants.

Hydrogen peroxide is often a product of these types of reactions, so three powerful oxidizing agents are commonly present in cloud droplets that have absorbed SO_2, namely O_3, H_2O_2 and OH. Metallic catalysts (iron and Manganese) can speed the reactions, and the pH itself can effect the reaction rates.

It is interesting to note the frequent appearance of H_2O_2, which itself is formed in a chain of reactions that begin with sunlight, ozone, and the OH radical and involve carbon monoxide (CO) and nitric acid (NO). The conclusion is that the conversion rate for sulphur dioxide to sulphate depends heavily on factors such as ozone concentration, sunlight, nitric acid, and hydrocarbon (for hydrogen and catalysts), as well as the ambient pollution level, so that the unstable oxidants can exist in sufficient concentration.

SOME REGULAR VARIATIONS

In these circumstances, it should not be too surprising that measurements show wide variation in the pH of rainfall, in sulphate concentrations, and in the presence of various reactants. Within the daily cycle, ozone and sulphate typically show a peak shortly after noon, related to sunlight and possibly to the downwash (fumigation) of SO_2 from tall stacks which occurs when solar heating mixes low-level air with air at stack levels.

Similarly, sulphate and hydrogen ion concentrations show strong summertime peaks in the rainfall at many eastern U.S. locations, with July/August values 30-50 percent higher than the annual mean. The atmospheric sulphate burden is about 3 times in summer that of winter, despite somewhat higher SO_2 emissions in winter. Because nitrate does not show the seasonal variation and because it has a higher affinity for snow, nitrate takes on a greater relative role in winter precipitation.

LINEARITY

Since nitric oxide, ozone and the hydroxyl radical are competitors in the production of nitrate, it is argued by some that an SO_2 reduction would not produce a proportionate response in hydronium concentration, because nitrate would, in effect replace sulphate. Much progress has been made in recent years in understanding this chemistry, but some uncertainties remain. In brief, we know that the atmosphere makes sulphate and nitrate from the oxides, but we're not sure how efficient it is in doing so. If it doesn't "use" the NO_x and SO_x efficiently, then a fractional reduction in emission might not do much, or any good. In particular, to reduce sulphate and not nitrate could result in nitrate taking up the slack, with little accomplished. Further, nitrates may well be involved in producing ill effects that don't involve sulphate at all.

This so-called linearity problem is at the heart of much of the current debate. Will a one-third reduction in SO produce a proportionate reduction in hydronium-ion concentration in observed precipitation? While there is reason to be doubtful, there is evidence to support the linearity idea, as well. For example, the (molar) ratios of SO_x to NO_x in emissions are about the same as those measured in precipitation. This suggests that available oxidants are sufficient in the atmosphere for the principal reactions involving both groups. Most observers agree that over large regions and over long times, there must be some proportionality between emission and the content of the material returned to earth. But when individual sources and particular receptors are considered, the linearity problem becomes more difficult. Recently, however, quite strong correlations were found between sulphur dioxide emissions from non-ferrous smelters in the western U.S. and sulphate concentration in precipitation at several NADP sites, up to 1000 km away in several western states. Whether such associations can be found in the East, where there are myriad sources and where the chemistry may be more complicated, is problematical. On statistical grounds, it would appear to be extremely difficult to identify individual sources from analyses of rainfall collected at receptor points, but investigators in Rhode Island have reported some success in identifying proportions in terms of broad regional sources. They find substantial contributions from both local and mid-Western sources.

SULPHATE IN THE ATMOSPHERE

Atmospheric sulphate is usually found in particulate form as finely divided salt (mostly ammonium) in the 0.1 to 1.0 micrometer size range, as the end product of oxidations that began with anthropogenic SO_2 and natural sulphur compounds that went through an SO_2 stage. Those sizes don't fall very fast, and would have long lives in the atmosphere if precipitation did not remove them. Most continental clouds actually start by condensation on such particles, but more importantly, in-cloud scavenging is quite efficient in removing sulphate as the droplets (often they are super-cooled) and ice crystals grow to precipitation-sized elements. Below-cloud washout also plays a role, but usually a lesser one.

Thus, the association of more-acidified rain or snow (together with substantial dry deposition) with anthropogenic pollution sources over the eastern United States and very likely in other major industrialized regions is in little question. The problem of identifying individual sources and matching them with individual receptors is much more vexing, in part because so many complex reactions can take place.

SOURCE RECEPTOR RELATIONSHIPS

Tracer studies have shown that long-range transport certainly occurs and is generally consistent with what one might expect. Because dry deposition is presumably a nearly continuous process, the dry deposition patterns could be somewhat different from the wet ones. This is related to precipitation falling preferentially in southwesterly flow aloft over middle latitudes of the Northern Hemisphere while the mean winds are generally westerly. Also, dry deposition may occur closer to the sources, on average. Thus, the total patterns of acid rain, as defined, may be somewhat different from those of Fig. 2, but the gross picture is probably similar.

Details of the source-receptor relationships and of the total acid rain deposition patterns will undoubtedly emerge when atmospheric scientists are able to deploy wind-measuring devices (Doppler-radars and lasers of various types) for continuous remote sensing of atmospheric winds. If one wishes to describe and model a reactive flow through the free atmosphere, it is very undesirable to have winds-aloft observations available every 12 hours, but unfortunately, up to early 1987, that has been case. Many unseen turns and changes can occur during the 12-hour black-out intervals; air can easily move 500 km on a curving, descending path during that time. Because wind directions and speeds vary with height and horizontal location, errors associated with such calculations are simply too large to permit them to be useful, easily exceeding 100 km uncertainties in 12 hour trojectory calculations.

There is reason to be encouraged, however. Experimental continuous wind-measurement equipment has operated for several years in Colorado, Pennsylvania and elsewhere. The National Weather Service and other federal agencies are well along in plans to procure and deploy new radars and wind profilers which can be used in such studies during the next 5-10 years.

TRENDS AND CONTROL STRATEGIES

The major questions that we have not considered are the related ones of trends and control strategies. Although there are moderately large uncertainties, it is quite clear the SO_2 emissions tripled during this century over the United States, but they have probably declined somewhat since about 1970, remaining at high levels. Future tendencies depend strongly on our decisions on control. Nitrogen oxide emissions have probably increased by a factor of ten and because they are more related to transportation, are continuing to increase. They now are approaching SO_2 levels in gross tonnage. Trends in the acidity of precipitation have not been measured for very long, but tend to show either no increase or a slight decrease for the period when SO_2 emission has decreased slightly. Because the chemistry of NO_x and SO_x is coupled there is a good case for trying to control both, if major expenditure is made to control one of them. However, doing either is a multi-billion dollar *per year* proposition, and the uncertainty of the effect associated with the complexities of the processes that we have touched on have been a deterrent to making those expenditures.

The focus in this chapter has been on the eastern United States, where more data have been collected than most of the places. However, acid rain, including dry deposition, is known to be of concern in other regions of the world, notably Europe. However, there is some evidence that the effects of the other trace gasses and reactants, such as ozone, are of great significance in those regions.

In both the United States and elsewhere, there is a problem. But doing something about it is not straightforward. It is not a matter of just getting the sulphur out, but of understanding a very complex chemistry in both the atmosphere and after reception, then deciding what steps are acceptable. There is disagreement about strategies, but there is reason to be hopeful that good science can provide better guidance.

REFERENCES

Bowersox, V.C. and R.G. de Pena, 1980: Analysis of precipitation chemistry at a central Pennsylvania site. *Journ. Geophys. Res.* 85(C10): 5614-5620.

Chameides, W.L. and D.D. Davis, 1982: Chemistry in the troposphere. *Chemical and Engineering News*, Special Report 39-52.

Machta, L.M. 1983: Acid rain: Controllable? *Transactions, Amer. Geophs. Union* 64:953.

Marro, C. and N. Brown, 1986: Acid Rain. *Research/Penn State*, 7:20-32.

Middleton, P. 1984: Acid rain control: Going beyond SO_2. *Weatherwise*, 37, 242-3.

Miller, J.M., 1984: Acid Rain. *Weatherwise*, 37, 233-239.

National Academy of Sciences, 1983: Acid deposition: Atmospheric processes in eastern North America. *National Academy Press*, Washington D.C. 375 pp.

Oppenheimer, M., C.B. Epstein, and R.E. Yuhnke, 1985: Acid deposition, smelter emissions and the linearity issue in the western United States. *Science*, 229:859-861.

Pena, R.G. 1982: Sulphur in the atmosphere and its role in acid rain. Earth and Mineral Sciences Bulletin. 51:61-66.

Wagner, M. 1983: Clarifying the scientific unknowns. *EPRI Journal*, 8:6-15.

Environmental Consequences of Energy Production: Problems and Prospects. Edited by S. K. Majumdar, F. J. Brenner and E. W. Miller. © 1987, The Pennsylvania Academy of Science.

Chapter Eighteen
ACID DEPOSITION ABATEMENT AND REGIONAL COAL PRODUCTION
D.G. AREY, J.A. CRENSHAW, and G.D. PARKER[1]
Southern Illinois University at Carbondale
Carbondale, ILL 62901

INTRODUCTION

Over the next few years legislation will most likely be enacted by the U.S. Congress with the objective of reducing acidic deposition in Northeastern United States and Southeastern Canada. Based on past proposals, the emphasis will be on reducing sulfur dioxide (SO_2) emissions from power plants in the 31 states east of or adjacent to the Mississippi River. When comparing competing legislation, three major issues will be considered. First, the amount of emission reduction and the costs associated with achieving that reduction will be at issue. A second consideration will be the impact of reduced sulfur emissions on the nation's environment, and the third concern will be the effect of the legislation on the regional distribution of coal production.

Mathematical simulation models which forecast the likely course of future events and test the probable outcomes of possible sets of actions are in ubiquitous use in nearly every phase of human decision making. The role of simulation models in forecasting the probable results of proposed changes in the policies or legal requirements surrounding the combustion of coal is perhaps the most notable recent example of model use.

Policies which require widespread reductions of emissions of SO_2, and the choice of legal mechanisms to achieve such reductions, are the subject of the most contentious debate in the history of the nation's efforts to deal with man's

---

[1]Funding for the research on this article was provided by the U.S. Department of Energy, Pittsburgh Energy Technology Center through the Coal Technology Laboratory and the Coal Extraction and Utilization Research Center at Southern Illinois University-Carbondale, DOE Contract No. DE-FCOl-83FEGO339. Dr. Arey is Associate Professor of Geography and Assistant Director of the Coal Research Center, Drs. Crenshaw and Parker are Associate Professors of Mathematics.

The authors gratefully acknowledge the assistance of Barbara Pearson, Elizabeth Rusk and Elizabeth Cornwall in the preparation of the manuscript, and Daniel Irwin for preparation of the maps.

impact on the environment. The intricacy of the acid rain problem presents a situation so complex that simulation modeling is virtually the only way to evaluate the problem and the many potential means to deal with it. For this reason, agencies and legislative bodies at both the state and federal level, industries and public utilities, and special interest groups of all kinds have sought and relied upon the results of such models to make decisions or to shore up arguments for or against a particular action. The fact that models have produced diverse results and require all sorts of simplifying assumptions has not detracted from their use. In fact, just the opposite has happened. Major efforts are underway to refine existing models and develop new ones in nearly every facet of our attempts to deal with the acid rain problem. These models are being used to answer complex questions regarding the behavior of the environment and the methods by which people manipulate the environment or manipulate man-made systems to minimize environmental impacts and economic dislocations.

The purpose of this paper is to introduce a model of the U.S. coal market which can be used to analyze some of the issues associated with acid rain. The model differs from most of the other coal market models currently being used.[1,2,3,4] Its main attractions are its moderate size, its CPU run times of one to two minutes, and its position in the public domain.

MODEL DESCRIPTION

In 1983 the Energy Information Administration (EIA) developed the Coal Supply and Transportation Model (CSTM)[5] to interact with their Intermediate Future Forecasting System (IFFS) model.[6] CSTM was designed to provide a coal production and transportation module for the larger IFFS model which considers all energy sources. IFFS produces a collection of coal demands (referred to as "jobs") which are used as input to CSTM. The model uses piecewise linear supply curves, a transportation network, and flue gas desulfurization (FGD) costs to compute the cost to buy, ship, and scrub coal to meet the given demand as constrained by environmental regulation. It employs a heuristic solution method to reach a coal production and distribution pattern which would minimize the cost to society to produce, transport, and scrub the coal.

The model was modified by EIA in 1984 and a copy was furnished to the authors. Additional modifications were made by the authors at the Coal Technology Laboratory (CTL) at Southern Illinois University to improve FGD cost estimates and alter the way the model handles air emission restrictions. The resulting model, hereafter referred to as CTL/CSTM, is able to forecast sulfur dioxide emissions and the costs associated with their control.

CTL/CSTM contains 32 domestic coal supply regions, each with a single supply centroid. Each region may produce several of 60 different coal types.

Acid Deposition Abatement and Regional Coal Production 247

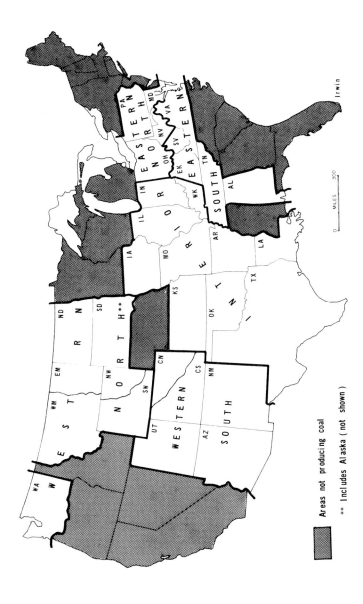

FIGURE 1. CTL/CSTM Supply Regions and Aggregations for Continental U.S.

These types distinguish between deep and surface mines, five BTU levels, and six sulfur levels. For each coal type existing in a supply region, there is a piecewise linear supply curve which was developed from the Resource Allocation and Mine Costing Model (RAMC).[7,5] These curves include the cost of some limited coal cleaning. Average BTU content for each coal type varies from region to region. The 32 coal supply regions, and the aggregations of these regions used later in this chapter are shown in Figure 1.

There are 48 demand regions in the model, each with a single demand centroid. The domestic demand regions are illustrated in Figure 2. As can be seen, some states have been divided into subregions because of their economic geography, while others have been combined. Exports are restricted to two European and two Asian demand centers. Exports to Canada are assumed to move through the Great Lakes.

In the transportation network, besides the supply and destination nodes there are 181 rail nodes, 56 water nodes, 561 rail links, 150 water links and 47 transshipment links between rail and water nodes. Both barge and deep water collier traffic is included. The transportation rate along each link is an increasing function of the tons of coal carried on the link. This transportation network, which is highly detailed, sets the model apart from many other coal market models.

The coal demand forecasts for 1995, derived from IFFS, were put in the form of "jobs." Each job is an aggregation of several coal users within the same demand region into an economic sector. Table 1 provides a description of these sectors. For utility jobs, demands were aggregated if they could use similar coals and if they had similar emissions standards. Each job has associated with it a group of coal types that may be used to satisfy the demand.

The various utility jobs are distinguished by the general coal type they use (bituminous, sub-bituminous, lignite), the presence of scrubbers, and the applicable emission regulation given in terms of pounds of sulfur dioxide per million British thermal units of heat input. (lbs./MBTU) The emission regulations are New Source Performance Standards (NSPS) the Revised New Source Performance Standards, and State Implementation Plan (SIP) standards.

The NSPS apply to all coal fired electric power stations which began construction after August 17, 1971 and mandate that emissions of SO_2 must not exceed 1.2 lbs. of SO_2 per million BTU's of heat input in the coal being burned. This is the reason that sulfur content and heating value are critical variables in the coal supply inputs to CSTM. The adoption of the NSPS led to increased demand for so-called "compliance coal" in the early-70s, that is, coal which could be burned without flue gas clean-up and still comply with the 1.2 lb. standard. A large proportion of the compliance coal in the U.S. is to be found in the western states although some exists in the East as well, especially in southern Appalachia.

The Clean Air Act amendments of 1977 added a new element to the NSPS. The Revised NSPS (RNSPS) are therefore applied to some of the utility jobs

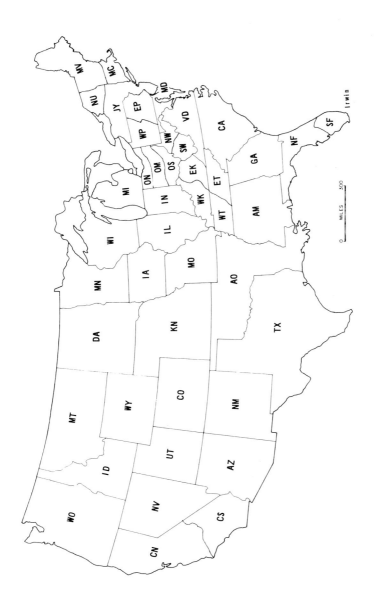

FIGURE 2. CTL/CSTM Demand Regions.

in the model. The RNSPS require those coal-fired electric power stations which began construction after September 19, 1978 not only to comply with the overall emission standard of 1.2 lb. SO_2/MBTU but also require a fixed percentage of the potential emissions of SO_2 to be removed. This percentage can be as high as 90 percent or as low as 70 percent depending on the sulfur content of the coal being used. The RNSPS are often referred to as the "universal scrubbing" standard because all new coal fired power plants are required to install flue gas desulfurization equipment even if they intend to use low sulfur coal. The story of how the RNSPS came into being is well told by Ackerman and Hassler in their book *Clean Coal-Dirty Air*.[8]

For utility jobs which include power stations constructed prior to 1971 the relevant SO_2 regulations are derived from State Implementation Plans (SIP), required by the Clean Air Act of 1970, which allow less stringent emission standards for older power plants. It is the emissions from this population of older boilers, heavily concentrated in the Ohio Valley, which are the target of potential regulation in connection with reducing acidic deposition, especially in the form of sulfates, in the Northeast. Jobs covered by SIPs are distinguished in the model by an implied emission standard determined by the highest sulfur content coal they may burn without scrubbing.

A total of nine different emission levels are used in CTL/CSTM to simulate the NSPS, RNSPS, and SIPs. Each utility sector is assigned an emission level. Levels for RNSPS and NSPS jobs are determined by state and federal regulations.[3] The level for each SIP job was determined by the coal group assigned in CSTM by EIA. For purposes of computing scrubbing costs, it was necessary to distinguish between RNSPS plants, plants that already had scrubbers in place, and those that did not. This led to 23 different indices for emission/scrubber levels. One is assigned to each utility job.

TABLE 1

CTL/CSTM Economic Sectors

Residential/Commercial
Existing Industry
New Industry
Metallurgical Premium
Metallurgical Blend
Export Premium
Export Blend
Export Steam
21 Utility SIP sectors[1]
6 Utility NSPS sectors[2]
3 Utility RNSPS sectors[3]

1. SIP-State Implementation Plan.
2. NSPS-New Source Performance Standards.
3. RNSPS-Revised New Source Performance Standards.

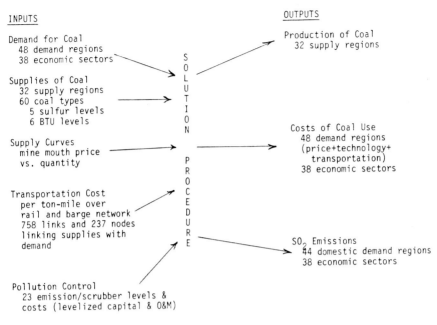

FIGURE 3. Overview of CTL/CSTM.

The total amount of annual emissions of SO_2 from each demand region is computed based upon the sulfur and heat content of the coal used, taking into account the amount of SO_2 removed by scrubbing. It is assumed that non-utility jobs do not scrub. Costs of scrubbing were computed based on Electric Power Research Institute data for new and retrofit plants.[9]

Total costs, SO_2 emissions, and coal production levels are computed after CTL/CSTM has run through its solution procedure to determine which region/types of coal are purchased for each job. The objective of the solution procedure is to purchase, deliver, and scrub the coal at the least cost per BTU for each job. Because the model does not assume cooperation between individual jobs to achieve lowest costs, the solution process gives a local minimum for each demander instead of a global minimum for the nation. This inherently leads to a different solution than that found by mathematical linear and nonlinear programming methods which impose cooperation between demanders. A summary of inputs to and outputs from CTL/CSTM is shown as Figure 3.

MODEL VERIFICATION

The veracity of simulation models is often determined by comparison of model results and actual statistics which measure the "real world" system that

TABLE 2
Model and Actual Coal Production, 1985

Federal Region	Model Production (million tons)	Actual Production (million tons)
New England (ME, VT, NH, MA, CT, RI)	—	—
New York, New Jersey (NY, NJ)	—	—
Middle Atlantic (PA, DE, MD, VA, WV)	258	237
South Atlantic (KY, TN, NC, SC, GA, FL, AL, MS)	168	169
Midwest (MN, WI, IL, MI, IN, OH)	124	122
Central (IA, MO, NB, KS)	7	4
Southwest (NM, TX, OK, AR, LA)	67	69
North Central (MT, ND, SD, WY, VT, CO)	215	225
Northwest (WA, OR, ID)	3	5
West (CA, NV, AZ)	11	10
Total	853	841

TABLE 3
Model and Actual SO_2 Emissions from Coal-Fired Utility Boilers, 1985

Federal Region	Model Emissions (thousand tons)	Actual Emissions (thousand tons)
New England (ME, VT, NH, MA, CT, RI)	155	149
New York, New Jersey (NY, NJ)	355	341
Middle Atlantic (PA, DE, MD, VA, WV)	2381	2531
South Atlantic (KY, TN, NC, SC, GA, FL, AL, MS)	3755	3904
Midwest (MN, WI, IL, MI, IN, OH)	5932	5832
Central (IA, MO, NB, KS)	1270	1496
Southwest (NM, TX, OK, AR, LA)	875	742
North Central (MT, ND, SD, WY, VT, CO)	436	267
Northwest (WA, OR, ID)	42	77
West (CA, NV, AZ)	85	89
Total	15286	15428

the model seeks to replicate. As a check on the functional capability of CSTM the actual 1985 figure for overall demand for coal was used as an input and the model was run in the so-called "base case" mode. The results for coal production and SO_2 emissions in ten standardized federal regions were then compared with actual coal demand and SO_2 emissions. The outcome is quite encouraging as shown in Tables 3 and 4. The model appears to closely reflect the true behavior of the coal market, at least in the case where existing environmental regulations are applied.

SCENARIOS

CTL/CSTM has been used to forecast the reduction in SO_2 emissions, the costs associated with this reduction, and the change in the regional distribution of coal production to be expected in 1995 for several acid rain legislative proposals. For all scenarios examined, the emphasis was placed on the utility jobs. They were the only sources to have their SO_2 emissions constrained. By changing their emission/scrubber level index, various jobs were required to chocse between switching to a low sulfur coal or retro-fitting scrubbers to meet reduced emission levels.

The "Base Case" scenario assumed that no major changes will occur in SO_2 emission standards or in transportation rates by 1995. This scenario set a standard against which emission reductions and costs can be compared.

The second scenario, "Free Choice", was designed to approximate legislative proposals which call for a reduction of SO_2 emissions by utility plants in the 31 states east of or adjacent to the Mississippi River ("Eastern U.S.") by a specific tonnage. The most frequently mentioned target for SO_2 emissions has been 10 million tons per year below 1980 emission levels. The method by which this reduction is to be made is left up to the utilities; hence, the name "Free Choice." To model this scenario, all utility jobs in the 31 eastern states with emission rates greater than 1.2 lbs. SO_2/MBTU were required to reduce emissions to achieve a 1.2 lbs. SO_2/MBTU standard. The model indicates that utilities acting independently would choose to shift to low-sulfur coals to achieve the required reductions. Recent legislative initiatives such as HR 4567, introduced in the 99th Congress, allow free choice at the state level of government by application of a concept known as the "bubble policy." Under this policy each state, as opposed to each emitter, would be required to achieve an overall emission rate of no more than 1.2 lbs. SO_2/MBTU of coal burned. Each state regulatory agency would need to design a plan to reach this goal by whatever means it deemed feasible. For instance, some sources within a state could be allowed to emit more than 1.2 lbs./MBTU so long as there were sufficient sources below that level to compensate for the higher emissions. The free choice scenario discussed here differs in that no state government intervention in utility decision making is assumed.

The third scenario examined is entitled "Forced Scrubbing." This simulates proposed legislation which forces scrubbing on some utilities. As in the second scenario, all eastern utilities were required to meet a 1.2 lbs. emission standard. However, the heavy emitters, those emitting at a level of 3.3 lbs. SO_2/MBTU or higher in the Base Case, were required to install scrubbers to meet the new standard, rather than being allowed to switch to low sulfur coal. Those utilities which would emit less than 3.3 lbs. SO_2/MBTU level had a free choice of methods to meet the 1.2 lbs. standard, but they generally chose to switch to low sulfur coals.

Finally, a "Mixed Case" scenario was constructed. As before, all eastern utilities were required to meet a 1.2 lbs. standard. Jobs were broken into three parts: one part was forced to scrub, a second was forced to switch to low sulfur coal, and the third part of each job had a free choice. This simulated the fact that some utilities cannot switch to low sulfur coal because of boiler and precipitator limitations while others cannot retrofit scrubbers because of space limitations or other reasons. The relative proportions of the three categories were constant within each demand region and were based, in part, on a survey of eastern utilities[10] and a report prepared for the Department of Energy.[11] The proportion of a job that was required to scrub varied from 10 to 54 percent with an average of 38 percent.

This last scenario lies between the second and third scenarios. All three scenarios require the same limitation on emissions rates but with differing levels of required scrubbing. Because each demander tries to minimize his costs, free choice usually leads to a switch to low sulfur coal. However, some utilities will choose to scrub for reasons that cannot be endogenously modeled in CTL/CSTM. This scenario compensates for this exogenously. As a result, the Mixed Case scenario is perhaps more realistic than the Free Choice scenario as a model of free choice legislation which would allow freedom of choice but in which choices would be constrained by technical and other considerations.

TABLE 4

Summary of Model Results for the Four 1985 Scenarios
(All figures in millions)
(See Figure 1 for location of regions)

Name	Total Cost in $	Coal Emissions Million Tons of SO_2			Regional Coal Production in Million of Tons					
		Total Coal U.S.	Total Coal East	Total Utility Coal East	Eastern North	Eastern South	Inter.	Western North	Western South	Total
Base Case	38,637	22.7	20.3	17.2	200.1	321.9	254.0	253.2	92.7	1121.9
Free Choice	45,185	12.0	9.5	6.6	124.5	437.9	136.0	259.2	158.3	1115.9
Forced Scrubbing	47,162	12.5	9.9	6.7	215.0	330.8	236.9	253.3	92.7	1128.7
Mixed Case	44,768	12.4	9.9	6.6	178.1	390.0	184.9	253.4	116.9	1123.2

TABLE 5

Costs of Emission Control Strategies Compared to 1995 Base Case

Case	Increased Total Cost above Base Case ($M)	SO_2 Emission Reduction (M tons)	Average Cost/Ton SO_2 removed ($/ton)
Free Choice	6,548	10.7	612
Forced Scrub	8,525	10.2	836
Mixed Case	6,131	10.3	595

RESULTS OF MODELING SCENARIOS

CTL/CSTM forecasts the cost associated with using coal, the total emissions, and the pattern of regional coal production. By comparing the results from the scenarios one may deduce the cost to reduce emissions, the number of tons of SO_2 emission reduction obtained, and the geographic shifts in coal production. Summary results from the four scenarios are presented in Table 4.

The three emission control scenarios modeled, which were designed to control eastern U.S. utility coal SO_2 emissions, may be compared to each other with respect to their total costs, their average cost per ton of SO_2 removed, and their total SO_2 emissions. As Table 5 indicates, the least expensive strategy is the Mixed Case. The selective retro-fitting of scrubbers, while not always giving the lowest cost for a specific utility (else scrubbing would have been chosen in the Free Choice scenario), yields the lowest overall cost to society.

It should be noted that the emission reductions of more than 10 million tons per year, as shown in Table 5, are from a comparison with the projected 1995 Base Case. In other words, these reductions are not reductions below the SO_2 emissions levels for 1980 which are the most frequently cited legislative target. Nationwide emissions of SO_2 in 1980 were in the neighborhood of 26.5 million tons. These emissions originated not only from coal-fired electric utilities, but also from oil-fired utilities, transportation, and other sources as well. In 1980, the contribution of coal-fired power plants in the eastern 31 states was about 15 million tons of SO_2.

In the context of legislative targets then, the maximum reduction in SO_2 emissions for coal-fired eastern utilities in the CTL/CSTM model would be 8.4 millions tons (15 mt minus 6.6 mt). This means that sources other than coal-fired power plants would have to be controlled if a 10 million ton reduction below 1980 levels remains as a goal of acid rain legislation.

BENEFITS OF CONTROLLING ACIDIC DEPOSITION

CTL/CSTM shows that the cost of partially achieving the target of reducing SO_2 emissions by 10 million tons is between 6 and 9 billion dollars annually, depending on the control strategy selected. A key question, therefore, is what are the annual benefits expected from such an action? The process of estimating the benefits of reducing acid deposition is complex, and is distinct from the research presented here.

In general, estimating the benefits from environmental regulation is a three-stage process. First, a link between sources of pollutants and areas of deposition must be established. Secondly, the effects of pollutants on the environment, both natural and man-made, must be understood. Finally, a method must be derived to translate these physical effects into economic impacts.

CTL/CSTM, and other models of the coal market which estimate emission reductions are of assistance only in the first of these steps and then only partially so. The general nature of emission reductions which would occur in the various demand regions, given different control strategies, can be estimated. For instance, the Mixed Case scenario generated by CTL/CSTM indicates total SO_2 emissions from Ohio, Indiana and Illinois at 1.6 million tons in 1995, a reduction of 3.2 million tons below the 1980 emission levels in these states. But, this says nothing about the relation between emissions in these states and deposition elsewhere, and this source-receptor relationship is one of several focal points in the debate over acid deposition control.

This issue and others are summarized in the annual reports of the National Acid Precipitation Assessment Program (NAPAP). From the most recent NAPAP report and a study by the Congressional Office of Technology Assessment,[12,13] it is evident that research into environmental effects of acid rain is proceeding along several paths including aquatic ecosystems, forests, agriculture, man-made materials, health, and visibility. To date, no widely accepted data on the economic consequences of these effects is available. Perhaps indicative of the complexity of the problem, a range of damages from 1 to 15 billion dollars annually has been cited.[14]

SHIFTS IN REGIONAL COAL PRODUCTION

Models such as CTL/CSTM are particularly useful in estimating the impacts of various pollution control policies on regional coal production. With data on projected production compiled from Table 4, combined with actual production statistics from earlier years, it is possible to ascertain the impact of the legislative scenarios on the long-term changing pattern of regional coal production in the U.S. Table 6 shows the historical pattern of coal production for the five aggregated regions used in this chapter along with projections of possible future patterns.

The Eastern North region, which has generally high sulfur coal supplies, has experienced a steady decrease in market share from 1965 to the present. Some of this change may be attributed to the increased dependence on petroleum as a boiler fuel in the late 1960's. By the mid 1970's, however, the initial impact of the 1970 Clean Air Act is also apparent. Decline in demand for metallurgical coal can also be cited as an important factor in the 1980's. This region's market share is likely to continue its downtrend under all but one of the cases investigated by way of CTL/CSTM. Acid rain legislation, other than regulations which force a technological solution to the problem of reducing SO_2 emissions, is likely to reinforce, if not exaccerbate, previously existing trends.

The same cannot be said for the Eastern South region. Proximity to market and the existence of lower sulfur reserves in southern West Virginia and eastern

TABLE 6
Historical and Projected Coal Production Market Shares
1965-1995

Region	Actual Percentage Market Share by Region with Percentage Change in Market Share from 1965						Projected 1995 Percentage Market Share with Percentage Change from 1985 Production Pattern							
	1965	1975		1985			1995 Base		1995 Scrub		1995 Choice		1995 Mix	
Eastern North	34	27	−7	19	−8		18	−1	19	−0	11	−8	16	−3
Eastern South	39	34	−5	32	−2		29	−3	29	−3	39	+7	35	+3
Interior	23	25	+2	20	−5		23	+3	21	+1	12	−8	16	−4
Western North	1	9	+8	21	+12		23	+2	22	+1	23	+2	23	+2
Western South	2	5	+3	7	+2		8	+1	8	+1	14	+7	10	+3
Total Production (million tons)	512.1	648.4		870.6			1121.9		1128.7		1115.9		1124.2	

Kentucky put this region in a position of increasing its market share under either the Mixed or Choice scenarios. In other cases a continued erosion of market share can be expected.

The Interior region shows a pattern of change somewhat similar to the Eastern North, except that the region's market share held up well between 1965 and 1975. Major down turns in demand for the region's generally high sulfur reserves can be predicted in the Choice and Mix Scenarios. It should be noted that the Interior market share in 1985 is buoyed up by the inclusion of Texas lignite which was not a factor in 1965 or 1975 but which increases in importance in all of the 1995 cases because it is generally low in sulfur content and is captive to the expanding lignite fired electric generating capacity in the state.

The Western North region is a particularly interesting producing area, with a steadily growing market share over the past twenty years. This increase is primarily the result of increased demand for low-sulfur coal, especially in the Midwest, resulting from application of the NSPS and SIP standards. For instance, in Illinois, the application of relatively stringent SIP standards has resulted in the use of western coal to meet fully half of the state's demand for steam coal. It is of interest that the market share captured by the Western North region is virtually unperturbed by the acid rain legislative scenarios investigated using CTL/CSTM. This means that coal production in Wyoming and Montana is likely to continue to rise through the end of the century, independent of any acid rain legislation.

The fifth region considered in this study is the Western South region. Like the rest of the West, it has experienced twenty years of expansion in coal mining. Expansion is likely to continue as well. However, the choice of acid rain control strategy could markedly effect production. The most substantial increases in production could be anticipated with the Choice scenario.

CONCLUSIONS

The model illustrates well the expected trade-offs between energy costs, emissions and control costs, transportation costs, and the dislocation of regional coal markets. Conclusions can be drawn regarding the influence of alternative policies on these elements, and about the model itself.

Changes in the geographic distribution of coal production depicted by the model are not surprising given that one of the strategies chosen for emission control, the Scrub scenario, is designed to minimize disturbance of the existing pattern of coal market while the others do not have that objective. The scenario which required scrubbing involved far less dislocation of the market pattern than would be expected in the absence of emission controls over those already required. The Mixed Case scenario fell between the extremes of forced scrubbing and extensive switching to low sulfur coal.

The largest regional reduction in coal production is predicted in the Interior region under the Free Choice scenario. The largest increase is predicted in the Eastern South region in the same scenario. These extremes indicate why interregional politics have been a major stumbling block to acid rain legislation.

It is to be expected that legislation which would require scrubbing will be the most expensive but will minimize economic impacts on coal producing regions. The least cost solution would be the Mixed Case that allows free choice provided there are a reasonable number of emitters who would choose to scrub. When some emitters scrub, the cost of low-sulfur, high-BTU eastern coal is lower due to decreased demand. Likewise, transportation costs for western low-sulfur coal would be lower since fewer tons are being shipped.

The least cost, Mixed Case scenario takes the middle road between the major disruption of coal markets inherent in the Free Choice option and the higher costs associated with forced scrubbing. Additionally, CTL/CSTM results show that the Free Choice option would require a large increase in mining activity in the Eastern South supply region. This implies that along with coal production dislocations, there would also be environmental disruptions in the East associated with the increased production. This would be over and above that which would occur in the western U.S. with its steadily increasing production even without legislation.

Turning to the efficacy of targeting coal burning utilities as the principal contributor to emission reductions, it should be noted that none of the scenarios modeled meets the desired objective of reducing SO_2 emissions by 10 million tons below 1980 levels. Recently proposed legislation such as HR 4567, mentioned earlier, indicated recognition that industrial emissions must also be controlled.

Finally, observations about the model should be made. First, as indicated above, the costs resulting from various scenarios run by the CTL/CSTM will be higher than those projected by linear programming models since CTL/CSTM does not force cooperation between coal users. Second, the emission constraints are handled differently. Linear programming models place a binding constraint on total SO_2 emissions and the solution dictates what emission rates must be set. On the other hand, the CTL/CSTM uses fixed emissions rates and the solution indicates how many total tons of SO_2 are emitted. Thus, with legislation which is written along the lines of setting SO_2 emissions rates, the CTL/CSTM is very useful for market impact analysis especially considering the low CPU times for its runs. It is also useful, but somewhat less so, for forecasting the general location and amount of SO_2 reduction to be anticipated under differing legislative approaches.

REFERENCES

1. ICF, Inc. 1981. *Capabilities and Experience in the Coal and Electric Utility Industries.* Washington, D.C.
2. Universities Research Group. 1984. *The State-Level Advanced Utility Simulation Model: Analytical Documentation.* Universities Research Group on Energy Project Office, University of Illinois at Urbana-Champaign: Urbana, IL.
3. Federal Energy Administration. 1976. *The National Coal Model: Description and Documentation.* FEA/B-771047. Prepared by ICF, Inc., for the Federal Energy Administration: Washington, D.C.
4. Morrison, Michael B. and Edward S. Rubin. 1985. "A Linear Programming Model for Acid Rain Policy Analysis," *Journal of the Air Pollution Control Association,* Vol. 35, No. 11, November 1985: pp. 1137-1148.
5. U.S. Department of Energy, Energy Information Administration. 1983. *Coal Supply and Transportation Model: Model Description and Data Documentation.* DOE/EIA-0401. Washington, D.C.
6. Gass, Saul I., Frederick H. Murphy, and Susan H. Shaw, editors. 1983. *Intermediate Futures Forecasting System.* NBS 670. U.S. Department of Commerce, National Bureau of Standards: Washington, D.C.
7. Energy Information Administration. 1982. *Documentation of the Resource Allocation and Mine Costing (RAMC) Model.* DOE/NBB-0020. Washington, D.C.
8. Ackerman, B., and W. Hassler. 1981. *Clean Coal: Dirty Air,* Yale University Press, New Haven.
9. Electric Power Research Institute. 1984. *Retro-fitting FGD Cost-estimating Guidelines.* CS 36-96. EPRI: Palo Alto, CA.
10. Arey, D., B. Dziegielewski, J. Crenshaw, G. Parker, and D. Primont. 1985. *Coal Supply and Transportation Model Analysis of the Future U.S. Coal*

Market. Southern Illinois University at Carbondale, IL: Coal Extraction and Utilization Research Center.
11. PED Co. Environmental Inc. 1983. *Acid Rain Boiler Population Retrofit Analysis.* Contract No. DE-ACO1-82 EP12067. Arlington, TX.
12. National Acid Precipitation Assessment Program. 1984.*Annual Report*, 1984 to the President and Congress, Washington, D.C.
13. U.S. Congress, Office of Technology Assessment. 1984.*Acid Rain and Transported Air Pollutants,* Implications for Public Policy, Washington, D.C.
14. Lareau, T. 1984. "Benefits of Reducing Acid Deposition," paper presented at University of Minnesota Conference on Acid Rain, St. Paul, MN.

Environmental Consequences of Energy Production: Problems and Prospects. Edited by S. K. Majumdar, F. J. Brenner and E. W. Miller. © 1987, The Pennsylvania Academy of Science.

Chapter Nineteen
IMPACTS OF ATMOSPHERIC DEPOSITION ON FOREST-STREAM ECOSYSTEMS IN PENNSYLVANIA

JAMES A. LYNCH[1], DAVID R. DEWALLE[2], and WILLIAM E. SHARPE[3]

[1]Associate Professor of Forest Hydrology
[2]Professor of Forest Hydrology
[3]Associate Professor of Forest Hydrology

School of Forest Resources
College of Agriculture
and
Institute for Research on Land and Water Resources
The Pennsylvania State University
University Park, PA 16802

A consequence of energy production through the combustion of fossil fuels has been an increase in the amount of sulfur dioxide and nitrogen oxides emitted to the atmosphere. Precipitation, by cleansing the air of these and other pollutants, plays an important role in biogeochemical cycling delivering these dissolved substances to aquatic and terrestrial surfaces. Atmospheric pollutants are also returned to earth surfaces via dry deposition (solid particulate fallout, aerosol impaction, gaseous adsorption). Through these combined processes, atmospheric pollutants can affect water quality and impact aquatic and terrestrial ecosystems. Forest ecosystems are particularly significant in Pennsylvania because they cover almost 60 percent of the land area and are highly valued for their timber, wildlife, recreation, and water resources.

ATMOSPHERIC DEPOSITION

The geographic relationship of Pennsylvania to the Ohio River Valley and its concentration of fossil fuel power generation plants and the relatively high sulfur dioxide and nitrogen oxides emitted by utilities and industries in Pennsylvania have resulted in Pennsylvania being the recipient of significant quantities of atmospheric sulfate and nitrate deposition (Figures 1 and 2). Although prior to 1980 wet atmospheric deposition (deposition via precipitation) data are generally lacking in Pennsylvania, statewide wet sulfate and nitrate deposition during the 1980's have averaged approximately 31 kg/ha/yr and 21 kg/ha/yr, respectively. Some areas in western Pennsylvania receive as much as 45 kg/ha of wet sulfate and 25 kg/ha of wet nitrate deposition, annually. The least wet sulfate and nitrate deposition generally occurs in the northcentral portion of the Commonwealth (Lynch and Corbett, 1983; Lynch, et al. 1984; Lynch, et al. 1985; Lynch, et al. 1986b). Hydrogen ion deposition in the state averages approximately 0.9 kg/ha/yr. Hydrogen ion concentrations in precipitation are strongly correlated with sulfate and nitrate concentrations. Since the establishment of the National Atmospheric Deposition Program (NADP) in 1978, Pennsylvania has consistently ranked as one of the top three recipients of atmospheric sulfate, nitrate, and hydrogen ion deposition in the United States (NADP, 1985; 1986; 1986a).

Statewide atmospheric deposition patterns can also be discerned through the study of snowpack chemistry. Surveys of snowpack chemistry in 1979-81 (DeWalle et al., 1983) have revealed a zone of high snowpack calcium content in the Ridge and Valley Province of Pennsylvania (Figure 3). High calcium was related to occurrence of limestone quarry operations and was associated with higher snowpack pH in this region of the state. Median snowpack pH in Pennsylvania was 4.3. Snowpack sulfate concentrations generally were greater in the southwest corner of the state (Figure 4) where sulfur dioxide emission densities were greatest. However, snowpack pH was more closely related to snowpack nitrate concentrations which showed no clear regional trends in the state. Thus, it appears nitric acid, rather than sulfuric acid, may be responsible for acid snow in Pennsylvania.

The geographic distribution of sulfate deposition in the state appears to be strongly associated with local sulfur dioxide emissions. Areas 50 to 75 miles downwind of major coal-fired power generation plants in western Pennsylvania generally receive the highest amounts of wet sulfate deposition; amounts decrease as distance from the plants increases beyond this 50 to 75 mile zone. This apparent relationship between local sulfur dioxide emissions and sulfate deposition lends credence to the theory that local, short-distance transport of atmospheric emissions plays an important role in determining atmospheric deposition loading rates in certain portions of the state. It is these local emissions that, when combined with long-distance transported pollutants from points to our west, have resulted in a significant sulfate deposition gradient in western and northcentral Pennsylvania (Figure 1). Wet sulfate deposition across this 75 to 100 mile gradient has consistently differed by a factor of two.

In addition to wet deposition, terrestrial ecosystems in Pennsylvania also

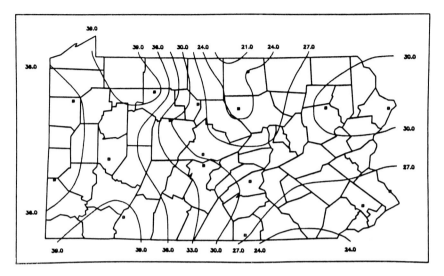

FIGURE 1. Spatial distribution of annual wet sulfate deposition (kg/ha) in Pennsylvania in 1985.

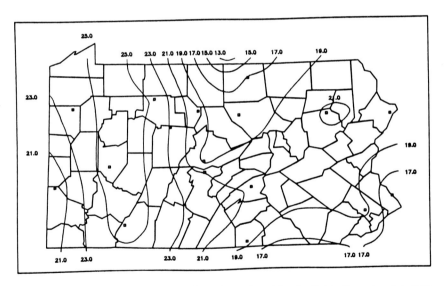

FIGURE 2. Spatial distribution of annual wet nitrate deposition (kg/ha) in Pennsylvania in 1985.

receive significant amounts of sulfate and nitrate as a result of dry deposition processes. Although difficult to quantify, dry deposition has been estimated to contribute up to an additional 50% of the wet deposition depending upon the physical and chemical properties of the atmosphere and the presence of local emission sources. Kostelnik (1986) estimated that dry sulfate deposition

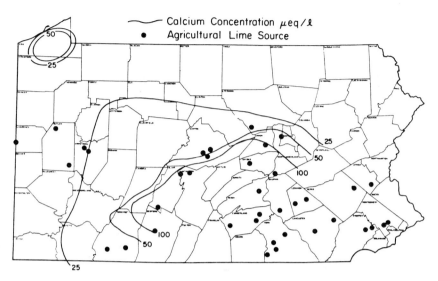

FIGURE 3. Distribution of mean snowpack calcium concentrations (μeq/L) across Pennsylvania in 1979-81.

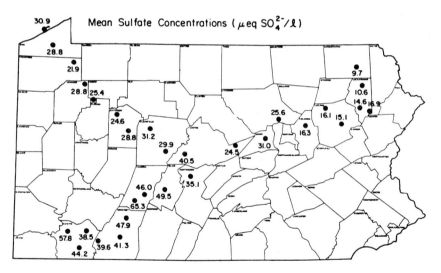

FIGURE 4. Mean snowpack sulfate concentrations (μeq/L) at sampling sites in Pennsylvania in 1979-81.

to a forest canopy in central Pennsylvania may represent up to an additional 32% of the total annual wet sulfate deposition and 11% of the total annual wet nitrate deposition. Data from DeWalle and Sharpe (1985) also show that dry deposition was contributing a maximum of about 30% of the total annual sulfate deposition on the Laurel Ridge in southwestern Pennsylvania based upon comparison of wet deposition and throughfall measurements. These estimates approximate the annual excess sulfate exported from the Leading Ridge Experimental Watersheds in central Pennsylvania, where annual wet sulfate deposition was compared to annual sulfate export in stream water (Dann, 1984). Dry deposition also influences snowpack chemistry. Dry deposition rates on snow for sulfate of 12 mg m^{-2} d^{-1} to essentially zero as a daily average have been observed near State College, Pennsylvania (DeWalle et al., 1986).

VEGETATIVE INTERACTIONS

Very little is currently known of the impacts of atmospheric deposition on forest vegetation in Pennsylvania. Both direct and indirect mechanisms for deleterious impacts of atmospheric deposition on forest vegetation have been proposed. Direct foliar damage by acidic wet and dry fallout leading to tissue lesions and excessive leaching of tissue could occur. Indirect growth declines could also occur due to loss of soil fertility caused by cation leaching in the soil or other interferences with nutrient cycling in the soil or canopy. Although evidence for such impacts exists elsewhere (McLaughlin, 1984), mainly high elevation spruce-fir forests in eastern United States, no studies demonstrating impacts on the deciduous mixed hardwood forests of Pennsylvania have been published.

The first interaction between atmospheric deposition and forest vegetation occurs in the canopy as wet fallout is converted to stemflow and throughfall. Generally, throughfall represents about 85 to 90% of total precipitation in the open, while stemflow amounts to less than 5% of open precipitation. Several studies of throughfall chemistry (DeWalle and Sharpe, 1985; Kostelnik, 1986; Edwards, 1983; Halverson et al., 1982) and stemflow chemistry (Halverson et al., 1982) have been conducted in Pennsylvania. Typical results for throughfall

TABLE 1

Ratio of mean annual ion concentrations in bulk precipitation and throughfall (from DeWalle and Sharpe, 1985).

Ion	Throughfall/Bulk Precipitation		
	Fork Mtn., WV	Peavine Hill, PA	Sand Mtn., PA
H$^+$	0.64	0.89	0.93
NH$_4^+$	0.83	0.69	0.66
Ca^{++}	1.97	2.00	1.64
Mg^{++}	2.23	1.89	2.14
K$^+$	10.1	14.0	5.57
Na$^+$	2.0	1.60	1.00
SO$_4^=$	1.33	1.25	1.29
NO$_3^-$	1.22	1.08	1.19
Cl$^-$	1.70	1.57	1.08

chemistry for three Appalachian forest sites, are given in Table 1. In general, hydrogen and ammonium ions are retained in the canopy, while concentrations of other major cations and anions found in precipitation increase after passage through the canopy. Species composition, canopy density and closure, surface roughness, age, physiology, phenological status, atmospheric loading rates, and the frequency, intensity, and chemistry of precipitation events influence the degree of alteration that occurs. Mobile cations such as calcium, magnesium, and potassium often exhibit the greatest increases beneath the canopy (Figure 5). These increases appear to occur as a result of the concentrating effect of canopy evaporation (maximum effect is roughly 10-15% increase in concentration), wash-off of dry deposition, micro-organism activity in the canopy, and leaching processes where hydrogen ions in precipitation exchange with foliar bound cations (Laikhani and Miller, 1980; Lovett et al., 1985; Lindberg and Lovett, 1986). The significant reduction of hydrogen and ammonium ions in precipitation below the canopy emphasize their role in the canopy exchange process. Tissue leaching is best illustrated for potassium, where throughfall concentrations average up to 14 times greater than in bulk precipitation. Stemflow concentrations of K$^+$ and other ions are even greater than in throughfall, but the small volume of stemflow water makes this hydrologic component of lesser significance when evaluating impacts on aquatic ecosystems (Halverson et al., 1982). Although nutrient cycling in throughfall and stemflow is a natural process, acidic atmospheric deposition may accelerate such leaching of canopy materials thereby causing stress in the plants. However, no data showing that this is occurring are currently available.

Wash-off of dry deposition is the physical process involving the rinsing off by precipitation of superficially deposited particles and adsorbed gases from

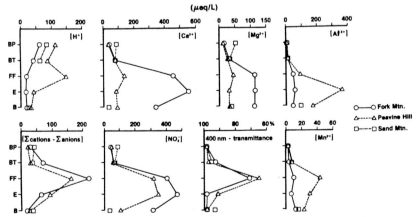

FIGURE 5. Comparison of mean annual ion concentrations and related parameters (11/30/83 to 11/14/84) in bulk precipitation (BP), bulk throughfall (BT), forest floor leachate (FF), E-horizon soil leachate (E), and B-horizon soil leachate (B) at three Appalachian forest sites (from DeWalle and Sharpe, 1985).

forest canopies that contributes to the increase in sulfate and nitrate concentrations in precipitation beneath a canopy. Such increases add to the total atmospheric deposition received by a watershed. Although some of the sulfate and nitrate deposition falls directly into the stream, the vast majority of it must first infiltrate and pass through the soil profile before reaching a stream. Within the soil profile, additional alteration may take place. Chemical alterations of the percolate ultimately determine what impact atmospheric deposition has on a stream ecosystem including its biological component.

SOIL INTERACTIONS

Two points of view can be used when discussing the interactions between atmospheric deposition and forest soils in Pennsylvania. One is the short-term impact of the soil on the chemistry of the acidic water as it moves through the forest floor (litter and humus collectively) and mineral soil horizons or layers. Over the long-term such chemical interactions can also affect the ability of the soil to modify the acidic water. Thus, another topic to consider is the long-term impact of atmospheric deposition on the soil.

In general, forest soils in Pennsylvania are infertile, rocky and acidic. Forest areas in the state are generally relegated to steep terrain, rocky areas where soils are derived from sandstone and shale rocks which only contain small amounts of basic cations such as calcium and magnesium. Soil pH in three Appalachian forest soils studied ranged from pH 3.7 to 4.6 in surface horizons to pH 3.8 to 5.1 in deeper horizons (Table 2). Base saturation in these types of soil is less than 10%, with less than 1.5 meq/100gm of exchangeable Ca^{++} plus Mg^{++} (Leibfried, 1982). Hydrogen ions in infiltrating rain or snowmelt can be exchanged for Ca^{++} or Mg^{++} in the soil which tends to neutralize atmospheric inputs of acid. However, forest soil reserves of exchangeable Ca^{++} and Mg^{++} are often too limited for adequate neutralization. The rate at which Ca^{++} and Mg^{++} can be released from decay of organic materials at the soil surface or weathering of rock fragments from within them becomes critical in determining the impact of the soil on the acidity of the percolating water.

In studies of nutrient cycling at the three Appalachian forest sites in Table 2, changes in pH or hydrogen ion concentration in water as it moved through the soil were related to soil exchangeable Ca^{++} and Mg^{++} (Figure 5). At Fork Mountain large increases in dissolved Ca^{++} and Mg^{++} and reductions in H^+ occurred in soil leachate as water moved through the forest floor and mineral soils. This implied exchange of H^+ for Ca^{++} and Mg^{++} in the soil at Fork Mountain. In contrast, at Peavine Hill, H^+ concentrations actually increased in the forest floor, in part due to organic acids, and no major increases in Ca^{++} or Mg^{++} were indicated. Both the sum of cations minus anions and 400-nm transmittance, which can be taken as an index to organic acid concentration (DeWalle et al., 1985),

TABLE 2

Comparison of selected chemical properties of soils at three Appalachian forest sites (from DeWalle and Sharpe, 1985).

	Horizons/Properties	Fork Mtn., WV	Peavine Hill, PA	Sand Mtn., PA
	Series	Dekalb	Hazleton	Hazleton
A, E	pH	4.6	3.7	3.9
	Mg (meq/100gm)	0.25	0.025	0.1
	Ca (meq/100gm)	0.75	0.78	0.55
B	pH	5.1	4.3	3.8
	Mg (meq/100gm)	0.38	0.00	0.10
	Ca (meq/100gm)	0.65	0.60	0.72
C	pH	5.1	4.3	4.4
	Mg (meq/100gm)	0.60	0.00	0.10
	Ca (meq/100gm)	1.30	0.50	0.05
	Series	Berks	Leck Hill	
A, E	pH	4.7	3.8	—
	Mg (meq/100gm)	0.25	0.10	—
	Ca (meq/100gm)	1.30	0.50	—
B	ph	5.0	4.3	—
	Mg (meq/100gm)	0.40	0.10	—
	Ca (meq/100gm)	0.60	0.53	—
C	pH	4.9	4.4	—
	Mg (meq/100gm)	0.35	0.10	—
	Ca (meq/100gm)	0.90	0.55	—

peaked in the forest floor confirming the influence of organic acidity at this level.

The major role of Ca^{++} and to a lesser extent Mg^{++} at Fork Mountain was substantiated by greater exchangeable soil levels of these nutrients at this site than either Peavine Hill or Sand Mountain (Table 2). The consequence of less available Ca^{++} or Mg^{++} is greater mobilization of Al^{++} and Mn^{++} in soil leachate at Peavine Hill and Sand Mountain (Figure 5). Aluminium concentrations in soil leachate at these sites were nearly 200 $\mu eq/L$ beneath the B-horizon. Recent research near Peavine Hill has indicated dissolved aluminium greater than about 20 $\mu eq/L$ is toxic to trout (Gagen, 1986).

Another major mechanism by which the impact of acidic atmospheric deposition on soil and percolating water can be minimized is sulfate adsorption. If anions such as sulfate are mobile in the soil leachate and are not adsorbed, significant leaching of Ca^{++} and Mg^{++} from the soil can occur in order to maintain charge balance (Ca^{++} plus $Mg^{++} \cong SO_4^{=}$) in the percolating water. If sulfate is adsorbed within the soil, Ca^{++} and Mg^{++} will not be excessively leached. Since atmospheric deposition is often acidified by H_2SO_4 which dissociates to H^+ and $SO_4^=$, the ability of the soil to adsorb the added sulfate is also important in determining the impact of acid deposition on soil fertility.

Appalachian forest soils currently have very limited capacity to adsorb additional sulfate, which indicates there is potential for leaching of Ca^{++} and Mg^{++} from the soil by mobile sulfate anions (Table 3). Data for soils at three Appalachian forest sites (Kleckner-Polk, Diane; personal communication) show amounts of sulfate currently adsorbed (see AS data in Table 3) are typical of that found in similar soils elsewhere in the Appalachians (Johnson and Todd, 1983). Adsorption capacity was positively correlated with the amount of iron and aluminium oxides and clay in the soil and negatively correlated with the sand and organic matter content. Thus, Pennsylvania soils which tend toward a sandy loam texture had only low to moderate adsorption capacity. Surface horizons (A and E) with higher organic matter content also generally show less adsorption than deeper B and C horizons (Table 3). Potential adsorption data for these same soils, which is a measure of total capacity to adsorb sulfate, (see PS data in Table 3) showed no trend for significantly more adsorption even though there are differences between individual pairs of AS and PS data. Fork Mountain did show some tendency for additional adsorption, but overall these data showed that even though soil sulfate adsorption occurred, the adsorption capacity of these soils is currently satisfied.

Field studies also show that soil sulfate inflow and outflow are roughly in balance, which implies no additional sulfate adsorption in taking place in these Appalachian forest soils (DeWalle and Sharpe, 1985). At Fork Mountain, Peavine Hill and Sand Mountain, inputs of sulfate to the soil surface measured in bulk

TABLE 3

Actual (AS) and potential (PS) adsorbed sulfate content of soil at three Appalachian forest sites (Klechner-Polk, Diane: personal communication)

Horizons	Adsorbed Sulfate (mg SO_4-S/kg)					
	Fork Mtn., WV		Peavine Hill, PA		Sand Mtn., PA	
	AS	PS	AS	PS	AS	PS
A, E	27.7	4.8	40.8	22.6	5.9	2.5
B	8.9	13.5	106.9	94.8	18.9	2.9
C	7.4	7.8	45.3	32.3	0.4	0.8
Series	Dekalb		Hazleton		Hazleton	
A, E	6.1	2.2	0.4	0.8	—	
B	51.6	32.4	100.2	63.9	—	
C	69.0	93.2	34.9	28.1	—	
Series	Berks				—Leck Hill	

throughfall for the period November 30, 1983 to November 14, 1984 were 74, 88, and 51 kg/ha, while estimated outflows beneath the B-horizon were 68, 111 and 79 kg/ha, respectively. Adsorption was indicated at the Fork Mountain site, which is consistent with laboratory data, but net loss of sulfate from the soil was actually indicated at the Peavine Hill and Sand Mountain sites. Since soil sulfate adsorption is reversible when ambient sulfate concentrations are reduced, a net loss of sulfate implies a reduction in input sulfate concentration for this year at the latter two sites. However, several years of similar data are needed to infer trends.

No data exist to indicate changes in soil fertility and acidity due to acid rain in Pennsylvania, but Ciolkosz and Levine (1983) evaluated the sensitivity of Pennsylvania soils to acid rain using a simulation model. They found a group of non-sensitive soils, chiefly in glaciated or limestone regions of the State, which were sufficiently buffered to resist change in pH and/or dissolved aluminium concentration for up to 90 years of simulated rainfall. Another group of soils, chiefly sandstone and shale-derived ridge-top and Appalachian Plateau soils, were currently rated as being very-sensitive. More importantly, results of simulations showed that base saturation, exchangeable cation concentrations, and soil pH were much better correlated with sensitivity to acid rain than sulfate adsorption capacity. This corroborates the previously discussed research results showing that sulfate adsorption capacity of Appalachian soils was essentially fulfilled and not protecting the soils against cation leaching by mobile anions.

Long-term trends in soil properties due to acid rain in Pennsylvania remain unresolved. However, available evidence shows that leaching of Ca^{++} and Mg^{++} probably is directly related to levels of sulfate deposition. In soils where gains of Ca^{++} and Mg^{++}, by either weathering of minerals or decomposition of organic materials, can not keep pace with leaching losses, declines in soil pH and fertility will occur over the long term.

GROUNDWATER

The opportunity for periodic recharge of acidic water from the soil to subsurface groundwater aquifers could have detrimental effects on stream chemistry and drinking water supplies. In the Appalachians, groundwater often supplies stream baseflow with relatively high quality water and provides buffering against inputs of more acidic storm water during rain or snowmelt events. Dinicola (1982) showed that in the Laurel Hill area of Pennsylvania, groundwater quality was largely controlled by rock type, as represented by first-order tributaries originating from the rock units (Figure 6). Consequently, streams receiving groundwater from alkaline Loyalhanna Limestone were protected against large pH declines during events and trout mortality was not observed. Streams receiving groundwater from acidic sandstones and shales of the Pottsville and

Impacts of Atmospheric Deposition on Forest-Stream Ecosystems in Pennsylvania 271

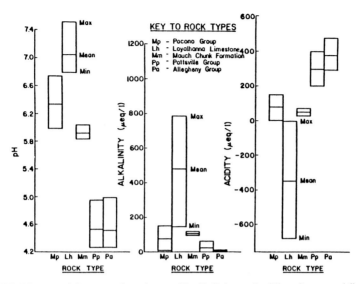

FIGURE 6. Mean, minimum, and maximum pH, alkalinity, and acidity of seeps and first-order tributaries originating from five predominant rock units of the Laurel Highlands.

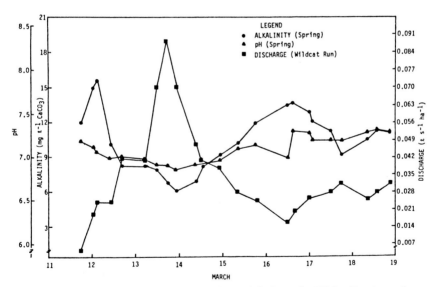

FIGURE 7. Alkalinity and pH for unnamed spring and discharge for Wildcat Run in southwest Pennsylvania.

Allegheny Groups were quite acidic and fishless.

Acid rain can either cause temporary or long-term impacts on groundwater quality. Evidence for both exists for Pennsylvania. Sharpe and Clarkson (1984)

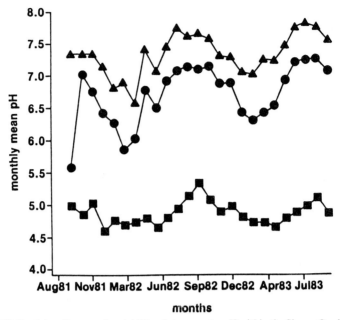

FIGURE 8. Spatial and temporal variability of streamwater pH within the Shaver Creek watershed. pH data collected at the mouth of the watershed (diamonds), at its mid-point (circles), and at the headwaters (squares).

showed that the pH and alkalinity of a near surface spring discharge did temporarily decline during and immediately after acidic runoff episodes in southwest Pennsylvania (Figure 7). Cole and Taylor (1986) found that of ten Pennsylvania groundwater data bases studied, 80% showed a decline in pH, four of seven showed a decrease in alkalinity, and four of five showed an increase in sulfate concentrations over a 20-year period. The causes of all these changes cannot be attributed solely to acid rain, but the implications are clear.

SURFACE WATERS

A major potential environmental consequence of atmospheric deposition of sulfate and nitrate is a decline in pH and acid neutralizing capacity (alkalinity) of surface waters and a corresponding alteration of the biological communities inhabiting these streams. Prolonged and continued atmospheric deposition, particularly of the quantities received by portions of Pennsylvania, can result in long-term changes in stream quality and increased sensitivity to episodic inputs. Unfortunately, long-term declines in stream pH and alkalinity are difficult to detect and quantify because of the lack of reliable historical records. Where long-term data exist, changes in analytical procedures make it difficult to include these data in trend analysis. Adding to these problems is the fact that stream pH (Figure 8) and alkalinity (Figure 9) may exhibit significant temporal and spatial variability (Wagner, 1985) making it difficult to detect changes without repetitive sampling over long periods of time. Compounding these problems are changes in stream quality due to other anthropogenic inputs as a result of changes in land-use.

Very few studies have been able to conclusively demonstrate the long-term acidification of headwater streams; nor have they been able to address its possible relation to atmospheric loading. Long-term potentiometric pH records from the Leading Ridge Experimental Watersheds in central Pennsylvania were analyzed by Chevallier (1985). He reported a statistically significant decline in stream pH of 0.24 unit during the 10 years (1974-1984) of record analyzed. Colorimetric measurements (1962-1973) from this watershed were also analyzed. Although the results indicated a decline of 0.28 pH unit over the 11 years of record, the imprecision of colorimetric measurements and the lack of comparability of colorimetry and potentiometry for poorly buffered waters precluded combining these data sets (Chevallier, 1985). Although atmospheric deposition data are available on the Leading Ridge Watersheds since 1973, no relationship was found between atmospheric deposition and the decline in stream

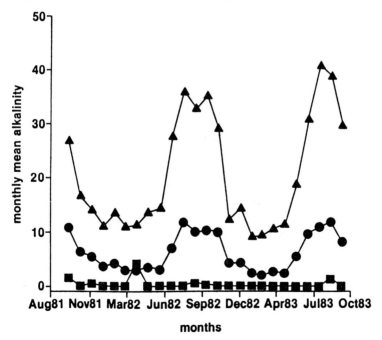

FIGURE 9. Spatial and temporal variability of streamwater alkalinity (as mg/L $CaCO_3$) within the Shaver Creek watershed. Alkalinity data collected at the mouth of the watershed (diamonds), at its mid-point (circles), and at the headwaters (squares).

pH. This fact is not surprising considering that stream acidification would occur only after depletion of available buffering capacity on the watershed (see discussion on soil interaction), a process that would involve considerably longer duration than the 10 years of this study.

Other researchers (reviewed by Haines, 1981) have noted the acidification of surface waters in Scandinavia, Ontario, and other areas of the northeastern United States. Still others have reported no significant change in stream pH (Ritter and Brown, 1981; Helvey et al., 1982). Geological, pedological, and hydrological characteristics in the various regions as well as atmospheric deposition loading rates contribute to these apparent conflicting reports.

Qualitative methods have also been used to assess long-term changes in water chemistry. Fish disappearance and changes in plankton community structure have been reported (Sharpe et al., 1984; Haines, 1981). These studies, as well as studies involving lake sediment records, indicate recent changes in water chemistry. Changes in water chemistry have also been reported based upon comparison of a few measurements separated by ten or more years (Beamish and Harvey, 1972; Likens and Butler, 1981). The reliability of trends in stream pH derived from such comparisons is highly questionable because of the high degree of spatial and temporal variability associated with stream pH (Figure 8). This variability also precludes the use of a single year of data for assessing trends in stream chemistry.

The inability of researchers to demonstrate (quantify) widespread declines in stream pH and alkalinity or show a relationship between observed decreases and atmospheric deposition loading should not be viewed or interpreted as an indication that elevated sulfate and nitrate emissions have not or are not causing acidification of streams in Pennsylvania. The abundance of low alkalinity streams in the State as determined from Fish Commission records (Arnold et al., 1980; Johnson, 1983) in those regions of Pennsylvania that receive the greatest amount of sulfate deposition, the reported loss of fish populations in the Laurel Ridge (Sharpe et al., 1984), the continued loss of waters acceptable to trout (Kimmel, 1984), the abundance of relatively low acid neutralizing capacity of many watersheds, and the observation by the authors of numerous acidified (pH < 5.0) and low alkaline (< 200 μeq/L of $CaCO_3$) streams throughout the state are strong indications that the chemistry of Pennsylvania streams has and is changing. Of particular interest is the Fish Commission study (Johnson, 1983) that estimated that approximately 8,000 km of stocked trout streams in the state are considered vulnerable to the impacts of acidic deposition (based on alkalinity values of 200 microequivalents per liter or less) as are 5,800 km of unstocked trout streams.

Although long-term changes in stream pH and alkalinity have been observed, but not quantitatively linked to atmospheric deposition, short-term, episodic fluctuations have. From a biological point of view, episodic changes in stream chemistry are important because they produce temporarily lethal effects to aquatic fauna and flora. When such fluctuations coincide with vulnerable life stages of aquatic organisms in the affected water, they can be particularly devastating. Such fluctuations may be the cause of significant reductions in fish populations in some portions of Pennsylvania (Sharpe et al., 1984; Sharpe et al, 1987a).

Episodic depressions in pH and alkalinity occur when inputs of precipitation/snowmelt to a stream are sufficient to overcome the ability of the stream or watershed to neutralize the acids in storm runoff (Figures 10 and 11). Two hydrologic processes are involved, with the relative importance of each being dependent upon the antecedent moisture content and the hydrologic properties of the watershed (Lynch et al., 1986a). When high antecedent moisture storage capacity exists, fluctuations in stream pH and alkalinity for small, first order streams may primarily be the result of direct channel interception of precipitation. If the storm is sufficiently large to satisfy available soil moisture storage capacity or if the storm occurs during low moisture storage capacity, rapid subsurface movement of infiltrated water to the stream channel becomes the dominant hydrologic factor. It is this rapid subsurface movement of water through macropores in the soil profile that accounts for the greatest depressions in stream pH and alkalinity. Watersheds in Pennsylvania that are characterized as having thin, porous soils derived from noncarbonate sedimentary rocks and lie in steep, rocky terrain are particularly sensitive to episodic depressions (Sharpe et al., 1987a; Lynch, et al., 1986a; Wagner, 1985). Such watersheds not only respond rapidly to inputs of acidic precipitation/snowmelt, but

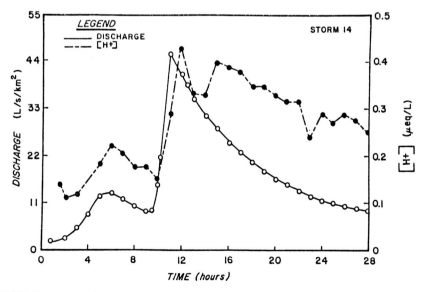

FIGURE 10. Comparison of stream pH and discharge during rainfall event on the Leading Ridge Experimental Watersheds in central Pennsylvania (from Lynch et al., 1986a).

also exhibit very low ability to neutralize these acids (as discussed in soil interactions). Many of Pennsylvania's sandstone, ridge watersheds are particularly sensitive.

The variability of episodic responses between and within watersheds is strongly influenced by the soil, geology, and land use (Wagner, 1985; Dinicola, 1982; Sharpe et al., 1987a). This is particularly true of headwater streams in Pennsylvania that originate in areas of resistant bedrock and infertile, low base status soils and flow into areas with more fertile and/or calcareous soils underlain by less resistant geologic formations. Stream pH and alkalinity of such streams generally increases from the headwaters to the mouth of the watershed (Wagner, 1985). From an episodic response point of view, the largest short-term fluctuations in pH occur in areas of medium buffering capacity where large storms can deplete existing buffering capacity and cause larger pH depressions than in the headwaters (where naturally low baseflow pH and alkalinity exists) or at the more highly buffered lower portions of the watershed (Wagner, 1985). On some watersheds, changes in pH as large as one pH unit have been measured over distances of less than 200 meters, where sharp changes in soils and geology occur.

Short-term fluctuations in stream pH and alkalinity (Wagner, 1985; Hanna, 1983; Lynch et al., 1986a; Lynch and Corbett, 1980; Sharpe et al., 1984) and aluminum concentration (Sharpe et al., 1984) have been observed in many streams in Pennsylvania. The largest reported depression, 2.37 pH units (7.32 to 4.95), occurred on the Leading Ridge Experimental Watersheds in central

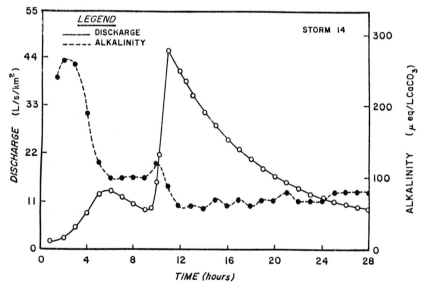

FIGURE 11. Comparison of stream alkalinity and discharge during rainfall event on the Leading Ridge Experimental Watersheds in central Pennsylvania (from Lynch et al., 1986a).

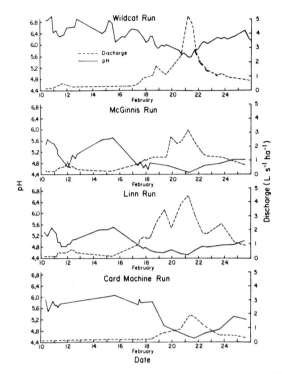

FIGURE 12. Change in pH with discharge in four headwater streams in Laurel Hill in southwestern Pennsylvania during a rain on snow event in February, 1981 (from Sharpe et al., 1984).

Pennsylvania (Lynch and Corbett, 1980). Stream H^+ concentration increased 234 times, while alkalinity decreased from 480 to 20 µeq/L as a result of this 4.38-inch storm. pH depressions to as low as 4.2 have been observed in other Pennsylvania streams (Wagner, 1985). Although these pH depressions occur rapidly, ranging from 1 or 2 to 36 hours, depressed levels of alkalinity and pH may persist for more than a week (Lynch and Corbett, 1980).

Streams in southwest Pennsylvania have also exhibited large measured changes in H^+ and aluminum concentrations during storm runoff events (Sharpe et al., 1984). Three streams studied have exhibited changes in H^+ concentration from about 3 µeq/L (pH 5.5) prior to storm runoff to 32 µeq/L (pH 4.5) at peak flow for a major rainfall-snowmelt event (Figure 12). Concurrent increases in aluminum in these three streams (Figure 13) produced conditions toxic to trout.

Although the occurrence and the most severe pH depressions are often thought to occur during spring snowmelt, the results of a 4-year study on the Leading Ridge Watersheds indicate the maximum pH depressions occur in late summer and early fall following extended dry periods (Corbett and Lynch, 1985). The

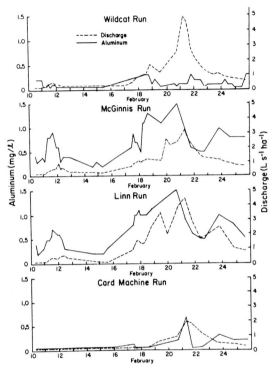

FIGURE 13. Change in total aluminum concentrations with discharge in four headwater streams on Laurel Hill in southwestern Pennsylvania during a rain on snow event in February, 1981 (from Sharpe et al., 1984).

previously mentioned pH depression of 2.37 units occurred in October, while another similar depression (2.10 pH units, pH 7.30 to 5.20) was observed in August. The study also revealed that over the four-year study, an average of 33 episodic stream pH depressions of 0.3 pH unit or greater occurred each year. The frequency of occurrence of stream pH depressions increased from January (lowest) to June (highest, 5.6/year) and then decreased through December. During the study period, storm caused pH depressions of 1.26 pH units or greater occurred at least once for all months except January and September.

Given the frequency and magnitude of pH depressions reported in many Pennsylvania streams and the large number of infertile, headwater streams in the state (Arnold et al., 1980; Johnson, 1983), there exists a substantial possibility for continued and severe loss of fisheries from many of these streams.

AQUATIC BIOTA

A considerable amount of attention has been focused on the impacts of atmospheric deposition on aquatic biota. Of the biota studied, salmonid species such as brook trout (*Salvelinus fontinalis*), brown trout (*Salmo trutta*), and rainbow trout (*Salmo gairdneri*) have received the most attention. These species have been widely introduced and are highly prized for sport fishing. Other aquatic biota have also been studied including bottom dwelling insects and some amphibians.

Acid precipitation has been implicated in fish loss in eastern North America and several Scandinavian countries. In Pennsylvania, persistent reports of fish losses from the Laurel Hill area continue and many streams are no longer capable of supporting trout fisheries because of severe acidification (Sharpe et al., 1984). Detailed analysis of water quality conditions coupled with in-situ bioassays of brook, brown and rainbow trout have indicated lethal physiologic stress due to high concentrations of aluminum ions and the low pH of acidified streams (Sharpe et al., 1983). These conditions prevail during acid runoff episodes accompanying periods of heavy rain and/or melting snow. The toxicity of acid runoff episodes to mottled sculpin (*Cottus bairdi*) has also been demonstrated.

In one stream, trout survival was enhanced by acidification reversal with the addition of alkaline groundwater from three wells that had been developed for this purpose (Gagen, 1986). The resulting pH adjustment caused reduced dissolved aluminum concentrations due to precipitation of soluble aluminum and greatly increased trout survival. This experiment confirmed the role of natural groundwater in the mitigation of headwater stream acidification on the Laurel Hill and demonstrated the role of aluminum and pH in the death of trout. The relationship between pH, aluminum and the presence or absence of brook trout is illustrated by the data of Table 4 (Sharpe et al., 1987a).

A survey of the bottom dwelling insect populations of 11 streams on the Laurel Hill demonstrated the effects of acidification (Perlic, 1985). These data are summarized in Figure 14 (from Sharpe et al., in press). The data of Figure 14 indicate a trend toward declining species diversity with increasing acidification. In addition, the order Ephemeroptera (mayflies) becomes nonexistent in the most severely acidified streams while the order Diptera (true flies) becomes

TABLE 4

Summary of Laurel Hill trout survey data (Sharpe et al., 1987a)

Category	No. of Streams	Mean pH	Mean Alk (mg $CaCO_{3/L}$	Mean Al (mg/L)	Mean SPC (umhos)	Mean No. of Trout
Trout Absent	12	4.97	0.37	0.45	36.0	0
Remnant Trout Populations	4	6.36	4.39	0.05	38.5	9.5
Trout Present	33	6.73	6.97	0.02	44.6	38.6
Culturally impacted (highways, impoundments, etc.)	12	6.73	12.90	0.03	99.0	10.0

relatively more important in the insect community. An intensive side-by-side comparison of the bottom dwelling insects in one severely and one slightly acidified stream by Kimmel et al. (1985) showed some striking differences in community structure (Table 5). The more acid-tolerant species tended to do better in the severely acidified stream. Strong seasonal differences in bottom dwelling insect abundance appeared to be correlated with the frequency and severity of acid runoff episodes occurring in the two streams. Decomposition of white ash (*Fraxinus americana*) leaves was significantly slower on the more severely acidified stream.

Freda (1986) summarized acid rain impacts to amphibian populations. He did not find a relationship between inputs of acidity from atmospheric deposition and the acidity of water in shallow ponds. Freda commented on the possible interference of acidification to amphibian life cycles and stressed the sensitivity of these organisms to small changes in pH and/or reactive aluminum.

Historically, aquatic biota have been the sentinels for acidification change to the landscape. The impacts of atmospheric deposition to these animals has been quite profound, is continuing and is likely to become more extensive in the future.

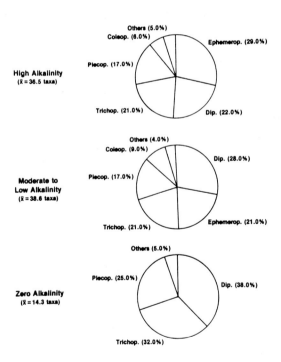

FIGURE 14. Mean percentage distribution of bottom dwelling insect taxa (usually genus) by order for streams in all alkalinity categories (from Sharpe et al., in press).

DRINKING WATER

Although a considerable amount of research into acid rain impacts has been accomplished, very little attention has been focused on specific impacts to drinking water supplies; consequently, there is little quantitative information available. However, it can be generally agreed that the impacts will be negative in that they will act to shift drinking water quality in the direction of lower pH, and higher trace metal concentrations. The consequence of greater corrosivity is

TABLE 5
Taxonomic composition of the bottom dwelling insect communities of Wildcat and McGinnis runs by season (Kimmel et al., 1985).

Taxa	Wildcat Run (less acidic)				McGinnis Run (more acidic)			
	Spr.	Sum.	Fall	Win.	Spr.	Sum.	Fall	Win.
Oligochaeta *spp.*		X		X				
Asellus sp	X	X		X	X	X	X	X
Cambarus sp	X	X	X					
Allonarcys sp		X	X	X				
Peltoperla sp			X					X
Taeniopteryx sp	X	X		X				
Leuctra sp	X	X	X	X	X	X	X	X
Capniidae *sp*				X	X			
Isoperla sp	X	X	X			X		
Chloroperlidae *sp*	X	X	X	X		X	X	X
Epeorus sp	X	X	X	X				
Heptagenia sp	X	X	X					
Stenonema sp	X	X	X				X	X
Baetis spp	X	X	X	X			X	
Paraleptophlebia sp	X	X	X	X				
Ephemerella sp	X	X	X	X				
Gomphidae *sp*		X						
Chimarra sp		X	X	X			X	
Potamyia sp				X				X
Parapsyche sp								X
Hydropsyche sp	X	X	X	X		X	X	X
Lepidostoma sp								X
Rhyacophila sp	X	X	X	X			X	X
Glossosoma sp	X	X	X					
Limnephilidae *sp*	X							
Neophylax sp	X							
Stenelmis sp	X	X	X	X				
Hexatoma sp	X	X	X	X				X
Tipula sp						X		
Antocha sp		X						
Simulium sp	X	X	X	X				
Chironomidae *spp*	X	X	X	X	X	X		X
Ceratopogonidae *sp*		X						
Empididae *sp*		X	X	X			X	

greater concentrations of dissolved metals in drinking water due to mobilization of metals from plumbing systems.

It is well to note that even unpolluted precipitation is corrosive. Precipitation in its purest form contains very little calcium carbonate; consequently, it is corrosive. However, acid precipitation is more corrosive than unpolluted precipitation. The problem is not that sulfuric and nitric acids in acid precipitation have made precipitation corrosive, but that they have made it more corrosive than it otherwise would be. The data of Table 6 (Young and Sharpe, 1984; Sharpe and Young, 1984) help to illustrate this point.

At this writing, there is still a considerable amount of uncertainty about the effects of atmospheric deposition on the quality of Pennsylvania's drinking water. Leibfried et al. (1984) reported lower pH, higher corrosivity and increased aluminum content in source waters of public water supplies following acid runoff episodes. By comparing acid runoff episode response for two areas, one receiving high amounts of acidic deposition, the other somewhat less, Leibfried et al. showed that the area with the highest current deposition had the most corrosive water.

Sharpe and DeWalle (1985) and Sharpe et al. (1987b) have attempted to relate acid precipitation to the drinking water quality of rainwater cisterns and small surface water reservoirs. Sharpe and DeWalle present a model that shows cistern tapwater copper concentrations in excess of drinking water limits when precipitation pH dips below that normally associated with the unpolluted case (5.0-5.6). Sharpe et al. (1987b) have shown that acid runoff episodes are capable of increasing the corrosivity of stored drinking water in small surface water supplies. They also present data (Figure 15) that appear to show increased tapwater corrosivity subsequent to acid runoff episodes. Presumably, the acid runoff episodes were the result of acid precipitation.

Only deep groundwater and areas dominated by carbonate geology and residual soils weathered from carbonates produce drinking waters that appear

TABLE 6

Comparison of best and worst case and mean LI[1] conditions of bulk precipitation collected weekly during 1979, 1980, and 1981 in Clarion and Indiana Counties, Pa. (from Sharpe and Young, 1984; Young and Sharpe, 1984).

Parameter	Best Case	Mean	Worst Case
pH	5.29	3.87	3.40
Alkalinity (mg/L CaCO$_3$)	2.22	0	0
Specific Conductance (umhos/cm)	127	70	220
Calcium (mg/L)	12.8	1.14	0.23
LI @ 20°C	-4.7	-7.4	-8.1

[1] LI - Langelier Saturation Index negative values indicate a tendency to dissolve calcium carbonate (corrode).

to be immune to the impacts of acid precipitation. Ground and surface waters in these areas have enough dissolved calcium carbonate to resist pH change subsequent to acid runoff episodes and groundwater recharge events, although Cole and Taylor (1986) have published data suggesting trends toward increasing acidification of some Pennsylvania groundwater supplies. Some glacial tills also produce well buffered water that is likely to be less affected by acid rain.

The combustion of fossil fuels (primarily coal) has resulted in the gradual acidification of many forested landscapes. The acidification process has resulted in measurable adverse impacts to the quality of drinking water obtained from forested watersheds. Drinking water obtained from rainwater cistern systems has also been negatively impacted by atmospheric deposition (Young and Sharpe, 1984), and there is some evidence to indicate that groundwaters in some regions of Pennsylvania and parts of Sweden may also be affected.

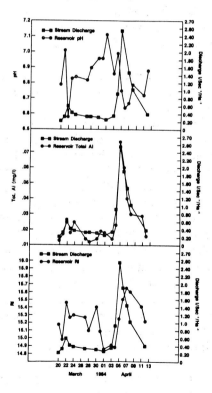

FIGURE 15. Change in reservoir pH, RI (a corrosion index in which water with values greater than 7.5 are increasingly corrosive) and aluminum concentrations with stream discharge.

LITERATURE CITED

Arnold, D., R. Light and V. Dymond. 1980. Probable effects of acid precipitation on Pennsylvania Waters. U.S. E.P.A., EPA-600/3-80-012.

Beamish, R.J. and H.H. Harvey. 1972. Acidification of the La Cloche Mountain lakes, Ontario and resulting fish mortalities. J. Fish. Res. Board of Canada. 29(8): 1131-1143.

Chevallier, E. 1985. Long-term pH trend analysis of a headwater stream in central Pennsylvania. The Penna. State Univ., Graduate School. M.S. Thesis in Forest Resources. 72 p.

Ciolkosz, E.J. and E.R. Levine. 1983. Evaluation of acid rain sensitivity of Pennsylvania soils. The Penna. State Univ., Inst. Res. on Land and Water Resour. Tech. Completion Report A-058-Pa. 106 p.

Cole, C.A. and F.B. Taylor. 1986. Possible acidification of some public groundwater supplies in Pennsylvania. Water Qual. Bull. 11(3): 123-130, 171.

Corbett, E.S. and J.A. Lynch. 1985. Frequency and magnitude of episodic stream pH depressions on a forested watershed. *In:* Proc. Muskoka Conference 85 an Int'l Symp. on Acid Precipitation. Toronto, Canada Sept. 1985.

Dann, M.S. 1984. Assessing methods of stream nutrient export determination. The Penna. State Univ., Graduate School, M.S. Thesis in Forest Resources. 59 p.

DeWalle, D.R., H.G. Halverson, and W.E. Sharpe. 1986. Snowpack dry deposition of sulfur: a four-day chronicle. Proc. Eastern Snow Conf., June 5-6, 1986. Hanover, NH. 5 p.

DeWalle, D.R., G.C. Ribblett, J.D. Helvey, and J. Kochenderfer. 1985. Laboratory investigation of leachate chemistry from six Appalachian forest floor types subjected to simulated acid rain. J. Environ. Qual. 14(2): 234-240.

DeWalle, D.R. and W.E. Sharpe. 1985. Biogeochemistry of three Appalachian forest sites in relation to stream acidification. The Penna. State Univ., Environ. Resour. Res. Inst., Final Report Coop. Agreement 23-829. 36 p.

DeWalle, D.R., W.E. Sharpe, J.A. Izbicki, and D.L. Wirries. 1983. Acid snowpack chemistry in Pennsylvania, 1979-81. Water Resour. Bull. 19(6): 993-1001.

Dinicola, R.S. 1982. Geologic controls on the sensitivity of headwater streams to acid precipitation in the Laurel Highlands of Pennsylvania. The Penna. State Univ., Graduate School. M.S. Thesis in Forest Resources. 69 p.

Edwards, P.J. 1983. Spatial distribution of nutrients in throughfall beneath the crowns of three urban tree species. The Penna. State Univ., Graduate School. M.S. Thesis in Forest Resources. 141 p.

Freda, J. 1986. The influence of acidic pond water on amphibians: a review. Water, Air and Soil Pollution: 30(½): 439-450.

Gagen, C.J. 1986. Aluminum toxicity and sodium loss in three Salmonid species along a pH gradient in a mountain stream. The Penna. State Univ., Graduate School, M.S. Thesis in Ecology. 87 p.

Haines, T.A. 1981. Acidic precipitation and its consequences for aquatic ecosystems: a review. Trans. Amer. Fish. Soc. 110: 669-707.

Halverson, H.G., D.R. DeWalle, W.E. Sharpe, and D.L. Wirries. 1982. Runoff contaminants from natural and man-made surfaces in a non-industrial urban area. Proc. Internat. Symp. on Urban Hydrology, Hydraulics, and Sediment Control. Univ. Kentucky, Lexington, KY, July 27-29, p. 233-238.

Hanna, C.M. 1983. Watershed responses and stormflow chemistry changes following precipitation in central Pennsylvania. The Penna. State Univ. Graduate School. M.S. Thesis in Forest Resources. 100 p.

Helvey, J.D., J. Hubbard, and D.R. DeWalle. 1982. Time trend in pH and specific conductance of streamflow from an undisturbed watershed in central Appalachians. Proc. of the Canadian Hydrology Symp. 1982. Fredericton NB. pp. 637-651.

Johnson, F. 1983. Acid precipitation. The Pennsylvania Fish Commission, July. 1983.

Johnson, D.W. and D.E. Todd. 1983. Relationships among iron, aluminum, carbon, and sulfate in a variety of forest soils. Soil Sci. Sco. Am. J. 47: 792-800.

Kimmel, W.G. 1984. An assessment of realized and potential impacts of acid deposition on salmonid fishery resources of Pennsylvania. Unpublished report to EPA/NCSU Acid Deposition Program (Contract No. ADP-A002-1984). 25 p.

Kimmel, W.G., D.J. Murphy, W.E. Sharpe and D.R. DeWalle. 1985. Macroinvertebrate community structure and detritus processing rates in two southwestern Pennsylvania streams acidified by atmospheric deposition. Hydrobiologia 124: 97-102.

Kostelnik, K.M. 1986. The cycling of acid precipitation through a mixed hardwood forest canopy in central Pennsylvania. The Penna. State Univ., Graduate School. M.S. Thesis in Forest Resources. 107 p.

Laikhani, K.H. and H.G. Miller. 1980. Assessing the contribution of crown leaching to the element content of rainwater beneath trees. In: Effects of Acid Precipitation on Terrestrial Ecosystems. Hutchison, T.C. and M. Havas (eds.) Plenum Press, N.Y.

Leibfried, R.T. 1982. Chemical interactions between forest soils and acidic precipitation during a season on Wildcat Run Watershed in Southwestern Pennsylvania. The Penna. State Univ., Graduate School, M.S. Thesis in Forest Resources. 86 p.

Leibfried, R.T., W.E. Sharpe, and D.R. DeWalle. 1984. The effects of acid precipitation runoff on source water quality. J. Amer. Water Works Assoc. 76(3): 50-53.

Likens, G.E. and T.J. Butler. 1981. Recent acidification of precipitation in North America. Atmos. Environ., 15(7): 1103-1109.

Lindberg, S.E. and G.M. Lovett. 1986. Atmospheric deposition and canopy interaction of major ions. Science. 231: 141-145.

Lovett, G.M., S.E. Lindberg, D.D. Richter, and D.W. Johnson. 1985. The effects of acid deposition on cation leaching from three deciduous forest canopies. Can. J. For. Res. 15: 1055-1061.

Lynch, J.A. and E.S. Corbett. 1980. Acid precipitation—a real threat to aquatic ecosystems. Fisheries 5(3): 8-13.

Lynch, J.A. and E.S. Corbett. 1983. Atmospheric deposition: spatial and temporal variations in Pennsylvania—1982. Res. Proj. Annual Report to Penna. Dept. of Environ. Resour. Penna. State Univ. Inst. for Res. on Land and Water Resources, Univ. Park, Pa. 204 p.

Lynch, J.A., C.M. Hanna, and E.S. Corbett. 1986a. Predicting pH, alkalinity, and total acidity in stream water during episodic events. Water Resour. Res. 22(6): 905-912.

Lynch, J.A., E.S. Corbett, and K.M. Kostelnik. 1986b. Atmospheric Deposition: spatial and temporal variations in Pennsylvania—1985. Res. Proj. Annual Report to Penna. Dept. of Environ. Resour. Penna. State Univ., Inst. for Res. on Land and Water Resour., Univ. Park, Pa. 239 p.

Lynch, J.A., E.S. Corbett, and G.B. Rishel. 1984. Atmospheric deposition: spatial and temporal variations in Pennsylvania — 1983. Res. Proj. Annual Report to Penna. Dept. of Environ. Resour. Penna. State Univ., Inst. for Res. of Land and Water Resour., Univ. Park, Pa. 228 p.

Lynch, J.A., E.S. Corbett, and G.B. Rishel. 1985. Atmospheric deposition: spatial and temporal variations in Pennsylvania — 1984. Res. Proj. Annual Report to Penna. Dept. of Environ. Resour. Penna. State Univ., Inst. for Res. of Land and Water Resour., Univ. Park, Pa. 236 p.

McLaughlin, S.B. 1985. Effects of Air Pollution on Forests, A Critical Review. Journal Air Pollution Control Association 35(5): 512-534.

National Atmospheric Deposition Program. 1985. NADP/NTN Annual data summary of precipitation chemistry in the United States—1982. NADP/NTN Coordinator's Office, NREL, Co. State Univ., Fort Collins, CO. 134 p.

National Atmospheric Deposition Program. 1986. NADP/NTN Annual data summary of precipitation chemistry in the United States—1983. NADP/NTN Coordinator's Office, NREL, Co. State Univ., Fort Collins, CO. 209.

National Atmospheric Deposition Program. 1986a. NADP/NTN Annual data summary of precipitation chemistry in the United States—1984. NADP/NTN Coordinator's Office, NREL, Co. State Univ., Fort Collins, CO. 240 p.

Perlic, T.E. 1985. A comparison of benthic insect communities in streams of varying baseflow alkalinity in the Laurel Highlands of Pennsylvania. The Penna. State Univ., Graduate School. M.S. Thesis in Environmental Pollution Control. 84 p.

Ritter, J.R. and A.E. Brown. 1981. An evaluation of the effects of acid rain on low conductivity headwater streams in Pennsylvania. U.S. Geol. Sur. Report 81-1025. 33 p.

Sharpe, W.E. and T.W. Clarkson. 1984. Ground, surface, and cistern waters as affected by acidic deposition p. 6-31 to 6-56. *In:* The Acidic Deposition Phenomenon and Its Effects, Critical Assessment Review Papers. Vol. II, Effects Science. U.S. EPA, Office of Res. and Devel., Washington, D.C. EPA-600/8-83-016BF.

Sharpe, W.E. and D.R. DeWalle. 1985. Potential health implications for acid precipitation, corrosion, and metals contamination of drinking water. Environ. Health Perspectives 63: 71-78.

Sharpe, W.E., D.R. DeWalle, R.T. Leibfried, R.S. Dinicola, W.G. Kimmel and L.S. Sherwin. 1984. Causes of Acidification of Four Streams on Laurel Hill in Southwestern Pennsylvania, Journal of Environmental Quality, 13(4): 619-631.

Sharpe, W.E. and E.S. Young, Jr. 1984. The Corrosivity of Cistern Water in an Area Impacted by Acid Precipitation and its Relationship to Tapwater Copper Concentration. Proceedings of the Second International Conference on Rain Water Cistern Systems. St. Thomas. VI. E7-1 -E7-10.

Sharpe, W.E., E.S. Young, and D.R. DeWalle. 1983. In-situ bioassays of fish mortality in two Pennsylvania streams acidified by atmospheric deposition. Northeastern Environ. Sci. 2(¾): 171-178.

Sharpe, W.E., V.G. Leibfried, W.G. Kimmel, and D.R. DeWalle. 1987a. The relationship of water quality and fish occurrence to soils and geology in an area of high hydrogen and sulfate ion deposition. Water Resources Bulletin 23(1): 37-46.

Sharpe, W.E., C.L. Spangenburg, D.R. DeWalle and H.G. Halverson. 1987b. The effects of acid precipitation runoff on reservoir and tapwater quality in a small Appalachian Mountain water supply. (Unpublished report) Environmental Resources Research Institute. The Penna. State Univ. 17 p.

Sharpe, W.E., T.E. Perlic, W.M. Tzilkowski and W.G. Kimmel. Status of headwater benthic insect populations in an area of high hydrogen ion and sulfate deposition. Northeastern Environ. Sci. (in press).

Wagner, T.M. 1985. Spatial and seasonal variations in the pH and alkalinity of several central Pennsylvania headwater streams. The Penna. State Univ., Graduate School. M.S. Thesis in Forest Resources. 127 p.

Young, E.S. and W.E. Sharpe. 1984. Atmospheric deposition and roof-catchment cistern water quality. J. Environ. Qual. 13:38-43.

Environmental Consequences of Energy Production: Problems and Prospects. Edited by S. K. Majumdar, F. J. Brenner and E. W. Miller. © 1987, The Pennsylvania Academy of Science.

Chapter Twenty

MANAGEMENT AND ENVIRONMENTAL IMPACTS OF ELECTRIC POWER TRANSMISSION RIGHTS-OF-WAY

W.C. BRAMBLE,[1] W.R. BYRNES[1] and R.J. HUTNIK[2]

[1]Department of Forestry and Natural Resources
Purdue University
West Lafayette IN 47907
and
[2]School of Forest Resources
Pennsylvania State University
University Park, PA 16802

As a basis for discussion of impacts, a brief description is given of commonly used Right-of-Way (ROW) management techniques. The impact of these techniques on vegetation varies greatly with the type used; aerial and other broadcast methods producing the most severe impacts. ROW have had a favorable effect on deer habitat and ROW songbird populations. Impacts on soil conditions may be expected to be negligible; as will the impacts on streams where a buffer zone is established to protect streambank vegetation. Impact on visual quality can be reduced by special attention to the type of ROW clearance used and to the method of vegetation control and will be greatest after aerial or other broadcast spray methods.

INTRODUCTION

Rights-of-way (ROW) constitute a large and growing category of land use which affects the local environment in many areas. High voltage transmission lines of 115 kV and above stretched for 164,022 km (273,370 miles) in 1978 and it was estimated that 67,788 km (112,980 miles) would be added by the year 2000 when they would occupy 2.1 million ha (5.2 million acres) of land (Young and Fisher 1981). Public concern has been voiced regarding destruction of vegetation, wildlife habitat alterations, and visual impacts among other problems. Use of herbicides to control unwanted vegetation has been a special cause for concern.

ROW have been used for a variety of purposes in addition to the primary one of electric power transmission (Randall 1973). Typical of these adjunctive uses are: snowmobile and motor bike riding, horseback riding, berry picking, hiking, skiing, birding, pasturing, and growing various crops compatible with electric power transmission such as Christmas trees.

Although ROW may traverse diverse land areas including agricultural fields, urban and suburban areas, and forests, only the environmental impacts of ROW that pass through forested areas will be discussed in this paper. Such ROW have the potential for important impacts as they must initially be cleared and constantly maintained to control tall-growing trees and shrubs that may interfere with electric transmission.

MANAGEMENT OF ROW

ROW must be managed intensively to insure safe and reliable transmission of electric power. This begins with initial clearance of the forest and continues with ROW maintenance to control undesirable trees and shrubs. As management is also charged with production of minimum adverse impacts on the environment, a sizeable group of trained personnel that may include foresters, arborists and ecologists must be employed. For the important program of ROW vegetation control, the work of this group must be highly organized and computerized for effective operations (Rossman 1972). Furthermore, to guide field operations, utilities have developed carefully designed specifications for guidance of utility personnel and contractors (Penelec Forestry Committee, 1986).

INITIAL ROW CLEARANCE

A power line through a forested area begins with clearance of a ROW corridor of sufficient width to insure reliable transmission. The width of ROW may vary from 21-30 m (70-100 ft) for 115 kV lines to 90 m (300 ft) for higher voltage lines. However, a width of between 45 to 60 m (150-200 ft) was typical of the major transmission lines on which this paper is based. Most ROW are established by clear-cutting all woody vegetation to a height of 7.5-10 cm (3-4 inches). The resultant slash may be piled in various configurations, or simply lopped to reduce height to less than 1 m (3 ft). However, in recent years ROW have also been cleared by selective cutting of only tall-growing tree species (Ulrich 1976). Trees have also been retained in deep ravines, for aesthetic purposes as road and river screens, and to reduce general visual impacts.

Resprouting of cut trees may be reduced by herbicide pretreatment of trees before cutting, or by treatment of freshly cut stumps. Sometimes trees are allowed to sprout and grow for 1-2 years before maintenance treatments are applied.

ROW MAINTENANCE

After a ROW has been established, it must be maintained to control tall-growing tree species and undesirable shrubs not compatible with electric power transmission. Failure to do proper maintenance may result in contact of trees with wires with a resultant "line outage" and costly loss of transmission. Maintenance may be carried out through use of a number of standard methods which may be briefly described as follows: (Penelec Forestry Committee, 1986, Environmental Consultants, Inc. 1984)

Handcutting — A ROW may be clearcut as a maintenance operation to remove all trees and tall shrubs to leave 7.5-10 cm (3-4 inches) stumps. Or ROW may be cut selectively to remove only tall-growing trees. Slash is lopped to reduce its height to less than 1 m (3 ft); or it may be piled on the ROW. In some cases the slash is chipped and may be removed from the ROW in areas such as road screens.

Mowing — All woody brush (trees and shrubs) on the ROW is mowed with a Hydro-axe, or other brush cutter. Woody vegetation is reduced to the lowest stump height possible to leave a shattered stubble one foot or less in height. To reduce resprouting, the cut stubs may be sprayed with a herbicide such as picloram immediately after mowing.

Cut and stump treatment — Tall-growing tree species are cut early in the growing season (May-June) to a low stump height, usually 10 cm (4 inches) or less. A herbicide such as picloram + 2,4-D is then applied to the sapwood and inner bark of freshly cut stumps with a squirt bottle; or the entire stump and root collar along with exposed roots may be sprayed with a mixture of picloram + 2,4-D + triclopyr in oil.

Dormant season and summer basal treatments — These selective treatments may be applied as dormant or growing season sprays with spray wands that produce a coarse spray of a herbicide such as triclopyr in oil to the lower 45 cm (18 inches) of stems of undesirable trees. The herbicide must be applied in sufficient volume to cover the root collar and the stem must be completely encircled. As an alternative, a concentrated solution of herbicide may be carefully applied at a low volume to the bases of tree stems.

Stem-foliage spray — Trees are selectively sprayed during the growing season with a herbicide mixture such as triclopyr + picloram + 2,4-D in water which is applied to a tree in two steps. First, the stem is thoroughly wetted using a narrow spray stream, then the foliage is wetted using a broader stream.

Aerial spray — A helicopter is used that is equipped with a microfoil boom that produces large droplets with minimum drift of a herbicide such as triclopyr + picloram + 2,4-D. The entire ROW is thoroughly covered during the growing season with spray during periods of low wind movement.

Pellet or other broadcast applications — Tall-growing trees are treated during the growing season with a pelleted herbicide such as picloram. Applicators walk the ROW and cast the pellets by hand into the crowns of trees to be eliminated. One to two handfuls (approximately 3 oz. per tree) is applied with oak and ash being given special attention. Pellets may also be broadcast over the ROW at a rate of 11 to 97 kg/h (10-85 lb/A).

As picloram pellets may no longer be available in the future, a broadcast spray similar to that described above for stem-foliage application may be used to produce a herb-grass cover.

Although the methods described above are usually applied as single treatments on a specific ROW segment, a different concept has been developed which calls for differentiation of a ROW into a wire zone, ROW area under the transmission wires, and two border zones, ROW areas on each side of the wire zone (Bramble *et al.* 1985). This permits a special treatment that removes all trees and tall shrubs from under the wires and leaves two shrubby borders. Several vegetation types may be developed such as a herb-grass cover in the wire zone with shrubby borders. This approach helps insure reliable transmission of electric power and produces type diversity and interspersion favorable to wildlife.

To accomplish differential treatment of ROW wire and border zones, a number of combinations have been used (Bramble *et al.* 1985). The simplest of these was to clearcut the wire zone with a selective cut on the borders. The same approach has been used with herbicides by broadcast treatment of the wire zone with a selective spray on the borders. A combination of mechanical with herbicide treatments could use mowing in the wire zone and selective basal spray on the borders.

ENVIRONMENTAL IMPACTS OF ROW

The following sections on environmental impacts of ROW will concentrate on 5 important facets of the natural environment that may be affected in forested areas: vegetation, wildlife habitat, soil, streams, and visual quality. The impacts may be positive or negative, that is they may reflect an improvement or a degradation of the environment. The aim of current ROW management is to minimize the negative impacts through use of maintenance methods designed to protect the environment (Olenik and Rossman 1977).

IMPACTS OF ROW ON VEGETATION

The initial clearance of a ROW has the general effect of producing a continuous narrow corridor through the forest from which all tall-growing tree species have been reduced to stumps by cutting. It has the characteristics of a linear forest opening. The negative impacts of this initial clearance may be ameliorated by selective cutting of undesirable trees whereby desirable small tress such as dogwood (*Cornus florida*) and shrubs such as witch-hazel (*Hammamelis virginiana*) are left intact (Ulrich 1976).

ROW vegetation in the years immediately after clearance will typically be made up of the forest species plus species of open areas such as sweetfern (*Comptonia peregrina*) that were present in forest openings (Bramble and Byrnes 1982). In the following years, a proclimax vegetation more typical of ROW will develop on xeric, mesic, and hydric sites (Johnston and Bramble 1981).

IMPACT ON TARGET TREES

Periodic maintenance treatments to control tall-growing tree species (target trees) are begun shortly after initial ROW clearance using one or more of the methods described in the preceeding section on management.

A major difference between impacts of common ROW treatments on target trees occurs between mechanical and herbicide methods (Bramble and Byrnes 1982). Handcutting and mowing remove trees and leaving stumps that produce sprouts which require recutting at 5 to 8 yr intervals, or less. And when such treatments are repeated over time, a dense tree thicket may develop.

Herbicide treatments, on the other hand, will cause a significant and important decrease in the number of trees on the ROW. This reduction in number of trees per acre should range from 70 to 90 percent after a single herbicide treatment. When herbicide treatments are repeated in 5 to 10 yr cycles, the number of trees per acre over 1 m (3 ft) in height should be reduced to 500 or less.

IMPACT ON NONTARGET VEGETATION

The impacts of ROW maintenance on nontarget vegetation (desirable shrub and herbaceous species that do not need to be controlled) varies considerably with the treatment method used. From results of research in the northeastern United States, (Bramble and Byrnes 1982, Carvell and Johnston 1978), the following predictions may be made for common types of ROW maintenance methods:

Handcutting — On a ROW with medium to high tree density (over 3750 trees/ha (1500/acre), handcutting will lead to development of a dense thicket of resurging trees that will tend to suppress sparse shrub and herb layers. Blackberry (*Rubus allegheniensis*), raspberry (*Rubus idaeus*), and meadowsweet (*Spiraea latifolia*) will occur in openings. Dewberry (*Rubus hispidus*) may form a dense ground cover. Maple-leaf viburnum (*Viburnum acerifolium*) and wild grape (*Vitis spp.*) will also be prominent species. Grasses (*Graminae*), sedges (*Carex spp.*), goldenrods (*Solidago spp.*), hayscented fern (*Dennstaedtia punctilobula*), and bracken (*Pteris acquilina*) will be common herbaceous species, accompanied by wild sarsaparilla *(Aralia nudicaulis)*, wild strawberry (*Fragaria spp.*) dogbane (*Acpocynum spp.*) twisted stalk (*Streptopus spp.*), and wild lily-of-the-valley (*Maianthemum canadense*).

On a ROW with low tree density (less than 3750 stems/ha (1500/acre), handcutting will lead to development of a scattered tree cover with many large openings. Shrubs will occupy about one half of the ROW with herbaceous species occurring in small to large patches intermingled with trees and shrubs.

Selective basal and stump sprays — Where tree density is high (over 3500 stems per acre), the impacts will be similar to those produced by broadcast sprays. However, where tree density is light to medium (less than 8750 stems/ha (3500/acre), only non-target vegetation at the bases of sprayed stems will be affected.

On basal-sprayed areas an existing plant community will remain undisturbed over most of the ROW. Shrubs will occupy about one-third of the ROW area and herbaceous plants about two-thirds. Blackberry, raspberry, meadowsweet, and witch-hazel will be the most common shrubs. Grasses and sedges will be prominent species along with hayscented fern and goldenrod. Cinquefoil (*Potentilla* spp.) wild strawberry, and aster (*Aster* spp.) will also be common.

Stem-foliage spray — Where tree density ranges from medium to high (over 3750 stems/ha (1500/acre), stem-foliage sprays will have impacts similar to broadcast sprays. Grasses, sedges, hayscented fern, goldenrod, and cinquefoil will develop a complete ROW cover. Total shrub cover will be reduced slightly and will slowly increase over the years between treatments. Species such as blackberry, raspberry, and meadowsweet will increase in importance as they spread vegetatively underground.

Where tree density is low (less than 3750 stems/ha (1500 /acre), a stem-foliage spray will produce impacts similar to a selective spray, that is, vegetation will be killed only in the vicinity of sprayed trees.

Aerial or other broadcast sprays — These sprays will be followed by a marked increase in grass-herb cover and a decrease in shrub cover. Shrub species such as blackberry, raspberry, and meadowsweet that spread vegetatively underground will spread slowly. Some herbaceous species such as hayscented fern will spread dramatically after spraying, while goldenrod will be reduced in abundance to scattered clumps and small patches. Blueberry (*Vaccinium* spp.) will be severely damaged but will recover slowly if selective sprays are used in subsequent maintenance.

Repeated aerial sprays will result in development of a grass-sedge-herb community with scattered clumps, or patches, of shrubs. The most common herbaceous species will be grasses along with hayscented fern, wild buckwheat, (*Polygonum scandens*), skullcap (*Scutellaria* spp.), goldenrod, and bracken.

IMPACT ON RARE, ENDANGERED AND PROTECTED PLANTS

Possible impacts of ROW on rare, endangered, and protected plant species were considered in a study of 18 ROW in New York State (Environmental Consultants, Inc. 1984). Although rare and endangered species were not recorded on the mesic ROW areas studied, it was considered possible that some microhabitats where such plants grow could possibly occur at ROW locations such as limestone areas, swamps or bogs, and shaded rock outcrops. "Protected" plant species are those attractive plants that cannot be collected without the owner's consent. Thirty-seven such species listed in New York's state regulations were recorded on the 18 ROW where their presence before and after treatments by herbicidal and mechanical methods was found to be highly variable and not related to treatment. For example, of the 24 protected species found on hand-cut areas, 13 were present before and after cutting, 5 were not found after cutting, and 6 new species appeared. Of the 25 protected species on summer basal sprayed areas, 14 were present both before and after spraying, 7 were not found after spraying, and 4 new species appeared. Of the 20 protected species on aerial sprayed areas, 10 were present before and after spraying, 8 were not recorded after spraying and 2 new species appeared. Of special interest is that purple trillium (*Trillium erectum*) was observed on ROW after herbicide treatments in New York where shrubs had been left, and pink lady's slipper (*Cypripedium acaule*) on a selectively sprayed ROW in central Pennsylvania.

IMPACT OF ROW ON WILDLIFE

ROW may be expected to have three important desirable impacts on wildlife in forested areas (Bramble *et al.* 1985). First, there will be an increase in cover type diversity brought about by creation of a linear opening in the forest when the ROW is cleared. This opening will be covered with a diverse shrub-herb-grass ROW plant community. Second, the two different plant communities, forest and ROW, will be in close juxtaposition so that wildlife can move readily from one community to another. And finally, shrubby ROW-forest edges may develop to add a highly desirable wildlife habitat component. On the negative side, initial clearance for a ROW will remove a strip of trees that could furnish food and cover for tree-inhabiting wildlife species. This impact would be serious only in very small woodlands.

IMPACT ON WHITE-TAILED DEER

Excellent deer habitat has been produced on ROW through careful vegetation maintenance. In central Pennsylvania, for example, deer use of a ROW treated with herbicides and handcutting increased for the first 5 years after initial maintenance treatments, and then remained at a high level through the following years (Bramble and Byrnes 1972). Common herbaceous species such as bracken, goldenrod, and loosestrife were browsed by deer during the growing season; woody plants such as sweetfern, witch hazel, blackberry, and bear oak furnished browse throughout the year.

Deer presence on a central Pennsylvania ROW remained at a high level following use of a special type of ROW maintenance which permitted the development of shrubby borders (Bramble *et al.* 1985). Deer habitat values remained high after both handcutting and herbicide treatments.

IMPACT ON SONGBIRDS

During the 1970's the effects of ROW on songbirds reported that initial clearance had the general effect of causing a decrease in breeding forest birds on the immediate ROW area accompanied by an increase in shrubland species (Bramble *et al.* 1984). The net result was an increase in total bird species.

Although it appears that, in some cases, maintenance by selective spraying of herbicides may be more favorable for songbirds than aerial spraying or mowing, (Malefyte 1982); repeated mowing on a 4-year cycle has produced in one case, at least, an increase in bird species, numbers, diversity and density as compared to a forest area (Kroodsma 1982).

In general, it appears that the total number of birds on ROW is highly consistent regardless of the type of maintenance (Carvell and Johnston 1982, Bramble *et al.* 1986). This consistency in total numbers of birds on ROW before and after various maintenance treatments was further documented by recent research in central Pennsylvania (Bramble *et al.* 1986).

A shift occurred, however, in the number of individuals among dominant bird species after both handcutting and herbicide treatments. Thus, while common species still remained numerous on the ROW, species typical of shrub-grass habitats became the most abundant after treatments owing presumably to an increase in grassy openings.

The Pennsylvania research indicated that the total number of bird species on the ROW treatment areas remained high before and after treatment. And that the diversity index was not significantly different between treatments or between ROW and adjoining forest (Bramble et al. 1986). This similarity was due in large part to use of the ROW by forest-inhabiting species for various activities, and use of the forest by brushland-inhabiting species for perching and escape cover.

IMPACT ON SOIL

The impact of ROW on soils will be considered under two important categories: soil erosion and soil compaction (Byrnes et al. 1982). The impact of ROW management activities on erosion was found to be negligible on the general ROW in forested regions due to protective plant cover and organic mulch. Serious soil erosion problems on ROW were associated only with areas disturbed by tower construction and access roads. Such construction impacts are usually mitigated by topsoil replacement and seeding of protective plant cover, and by proper road construction.

It is important to note in this connection that the humus types found on ROW were observed to be essentially similar to those in the adjoining forest, although the source of organic matter was from different types of plants (Holewinski 1979). Also, little difference was found in the depth of organic layers on ROW and in the forest.

Soil compaction on the general ROW area not influenced by construction was found to be of small magnitude, and only 3 percent of the average ROW was actually transversed by the spray truck during maintenance (Environmental Consultants, Inc. 1984). A hydro-axe used for mowing affected 22 percent of the ROW area. Some erosion occurred in deeply rutted wheel tracks caused by vehicle use on the ROW when the soil was very wet. Although erosion did occur in such wheel tracks, it decreased the following year and was not a serious condition as it was usually localized and restricted to the ROW. However, when access roads occur on steep slopes and are not properly constructed to lessen runoff, serious erosion may occur and the resultant sediment will cause siltation of small streams that cross the ROW. After a ROW has been established, nearly all of the necessary vehicle travel occurs on an access road. However, excess use of such roads by unauthorized off-road vehicles and horse riders may cause serious soil erosion problems in hilly terrain.

IMPACT ON STREAMS

The impact of ROW on streams is dependent upon many variables such as streambank vegetation, channel and bank characteristics, rate of flow, stream depth, and adjacent topography (Carvell and Johnston 1978). In most cases, only a small stream segment is affected, unless excess herbicide was allowed to directly enter the stream. To prevent the latter, a stream buffer zone, 15 to 30 m (50-100 ft) wide, is handcut to remove tall-growing tree species that may grow into transmission wires (Penelec Forestry Committee 1986).

As most streams cross ROW, rather than flow along them, only a small segment is affected by temperature changes, and any thermal modification is transient (Carvell and Johnston 1978). Maintenance of a streamside vegetation that will shade small streams keep the normal temperature gradient unaltered; large, deep streams usually will remain "summer cool." On nine small to large streams that crossed ROW studied by Carvell and Johnston (1978) mean temperature changes because of exposure ranged from -0.3 C to almost 3 C on clear, warm summer days. Broadcast application of herbicides appeared to cause more deterioration of stream habitat than selective spraying, or selective cutting, of trees. Repeated broadcast spraying resulted in a herbaceous riparian vegetation that was inadequate for stream protection.

A study of nine diverse streams in New York state indicated that the general effect of ROW on water temperature of free-flowing streams was negligible, (Holewinski 1979) although some streams were only partially shaded by shrubs and herbs rather than trees. Water temperatures downstream of the ROW ranged from 1.8 C less to 2.5 C greater than those upstream. The maximum stream temperature recorded below the ROW was 18.5 C.

Most stream banks on ROW in New York were well protected and hence did not contribute materially to stream sedimentation or cause deterioration of stream quality. Where sedimentation did occur as a consequence of a ROW, it was caused by flow from poorly maintained access roads, or by erosion from construction areas.

IMPACT ON VISUAL QUALITY

A great deal of attention has been paid to possible impacts of ROW on visual quality (scenic impacts) at the time in the management process when the route is being selected for a new transmission line. For example, 24 papers on ROW routing considerations which included visual impacts were presented at the Second Symposium on Environmental Concerns in Rights-of-Way Management held in Ann Arbor, Michigan, in 1979. Papers given at that meeting included reports on new tools that have been developed to assess visual impact (Boundy 1979) and on selection of new designs and colors for transmission towers

(Howlett 1979). Tower colors have been found to be important in helping to conceal transmission lines within the natural landscape. A tan to brown color typical of wooden poles and dark green that blends with vegetation ranked high as desirable colors.

In New York state, two factors proved to be most important in production of visual impacts: lighting and condition of the vegetation (Environmental Consultants, Inc. 1984). Vegetation conditions that most affected visual impacts were height and density of vegetation. ROW vegetation conditions produced by maintenance treatments that caused visual impacts were brownout, dead stems, and slash accumulation. Hand-cutting produced a negative impact owing to slash accumulation; mowing owing to dead stubble; cut and stump herbicide application to slash accumulation; and other herbicide applications owing to brownout and dead stems.

Aerial spraying, which by its very nature, calls for a blanket coverage of the ROW with herbicide spray, is a method that invariably causes a complete brownout and dead stems. Stem-foliage spraying of dense tree cover causes similar impacts. Although the browning of the vegetation is temporary and usually of only one year's duration, it causes a severe visual impact that must be repeated with each spray application. However, aerial sprays are excellent to begin the reduction of a tall, dense tree cover to lower densities. As the tree density is reduced other herbicide techniques should be employed that are more selective. Where a grassy ROW is desirable, aerial spraying has proven to be the best way to get vegetation converted to a grass-herb community.

As may be expected, the height to which trees are permitted to grow before treatment and their density on the ROW greatly affects the extent and severity of visual impacts. Therefore, management should aim to reduce the tree density on the ROW to less than 1000 per acre, and ROW should be treated when the dominant canopy of trees is not over 3-5 m (10-15 ft) in height.

LITERATURE CITED

Boundy, J.G. 1979. A visual approach to utility planning. *Proc. 2nd Symp. on Environ. Concerns in Rights-of-Way Manage.* Mississippi State Univ., Mississippi State, MS p. 18-1-18-6.

Bramble, W.C. and W.R. Byrnes. 1972. A long-term ecological study of game food and cover on a sprayed utility right-of-way. *Purdue Univ. Agric. Exp. Sta. Research Bull. 885.* 20p.

Bramble, W.C. and W.R. Byrnes. 1982. Development of wildlife food and cover on an electric transmission right-of-way maintained by herbicides: A 30-year report. *Purdue Agric. Expt. Sta. Research Bull.* 974. 24 p.

Bramble, W.C. and W.R. Byrnes. 1983. *Annual Report to Cooperators on the Gamelands 33 Research Project.* 39 p.

Bramble, W.C., W.R. Byrnes and M.D. Schuler. 1984. The bird population of a transmission right-of-way maintained by herbicides. *Jour. Arboriculture* 10(1): 13-20.

Bramble, W.C., W.R. Byrnes, and R.J. Hutnik. 1985. Effects of a special technique for right-of-way maintenance on deer habitat. *Jour. Arboriculture* 11(9): 278-284.

Bramble, W.C., W.R. Byrnes and M.D. Schuler. 1986. Effects of special right-of-way maintenance on an avian population. *Jour. Arboriculture* 12(9): 219-226.

Byrnes, W.R., W.W. McFee and C.C. Steinhardt. 1982. Soil compaction related to agricultural and construction operations. *Purdue Agric. Exp. Sta. Bull. 397.* 164 p.

Carvell, K.L. and P.A. Johnston. 1978. Environmental effects of right-of-way management on forested ecosystems. *Report EA-491. Elect. Power Res. Inst.* Palo Alto, CA 269 p.

Environmental Consultants, Inc. 1984. Cost comparison of right-of-way treatment methods. *Research Report EP 80-5.* 134 p.

Holewinski, D.E. 1979. Environmental and economic aspects of contemporaneous electric transmission line right-of-way management techniques. *Proc. 2nd Symp. on Environ. Concerns in ROW Manage.* Elect. Power Res. Inst. Palo Alto, CA p. 35-1 to 35-14.

Howlett, B.E. 1979. Selecting designs, materials and colors for transmission structures in different environments. *Proc. 2nd Symp. on Environ. Concerns in Rights-of-Way Manage.* Ann Arbor, MI 19-1 to 19-12.

Johnston, P.A. and W.C. Bramble. 1981. Vegetation distribution associated with right-of-way habitats in New York. *Proc. 2nd Symp. on Environ. Concerns in Rights-of-Way Manage.* Mississippi State Univ., Mississippi State, MS 44-1 to 44-15.

Kroodsma, R.L. 1982. Effect of power-line corridor on the density and diversity of bird communities in forested areas. *Proc. 3rd Symp. on Environ. Concerns in Rights-of-Way Manage.* San Diego, CA 689 p.

Malefyte, J.J. de Waal. 1982. Effect of vegetation management on bird population along electric transmission rights-of-way. *Proc. 3rd Symp. on Environ. Concerns in Rights-of-Way Manage.* San Diego, CA 689 p.

Olenik, C.J. and W.R. Rossman. 1977. Foresters keep the vegetation in vegetation management. *Ind. Veg. Manage.* 9(2): 3-6.

Penelec Forestry Committee. 1986. Specifications for line clearance. *Pennsylvania Elec. Co. Manual.* 40 p.

Randall, W.E. 1973. Multiple use potential along power transmission rights of way. *Proc. AIBS Colloquium on Biotic Management Along Power Transmission Rights of Way.* Carey Arboretum, Millbrook, NY 89-113.

Rossman, W.R. 1972. Power line right-of-way management. *Ind. Veg. Manage.* 4(3): 2-6.

Ulrich, E.S. 1976. Selective clearing and maintenance of rights-of-way. *Proc. 1st Nat. Symp. on Environ. Concerns in Rights-of-Way Manage.* Mississippi State Univ. Mississippi State, MS 206-219.

Young, F.S. and J.S. Fisher. 1981. Land use planning issues and future rights-of-way from a researcher's prospective. Proc. *2nd National Symp. on Environ. Concerns in Rights-of-Way Manage.* Miss. State Univ., Mississippi State, MS. 6-1 to 6-9.

Environmental Consequences of Energy Production: Problems and Prospects. Edited by S. K. Majumdar, F. J. Brenner and E. W. Miller. © 1987, The Pennsylvania Academy of Science.

Chapter Twenty-One
IMPACTS OF HYDROPOWER DEVELOPMENT ON RIVER RESPONSE

DARYL B. SIMONS and ROBERT K. SIMONS
Simons and Associates, Inc.
375 East Horsetooth, Suite 103, Shores 4,
Fort Collins, Colorado 80525

PHYSICAL PROCESSES GOVERNING CHANNEL RESPONSE

The major physical processes in an alluvial channel are those related to water discharge, channel slope and shape, geology and soils, sediment transport, vegetative effects, and man's influence. As in the watershed, each of these is intertwined in a set of complex relationships. Bedrock channels have a restricted set of conditions while alluvial channel conditions are more dynamic.

Water discharge is a key process, resulting from other processes, that affects channel shape and sediment transport. Hydraulic variables, such as velocity, depth, and flow area, are important in analysis of channel response. Coinciding with these are channel shape, channel slope, and flow resistance from grains of bed and bank material, bed forms and bars. Geology and soils of the channel bottom and banks help determine the relative erodability of the system, and therefore its response to other changes or alterations. Local variations in geology, soils, vegetation, and flow rate play important roles in determining bank stability in channels. Sediment transport in channels differs between watersheds because there is often more material available for transport than can be carried. Vegetation is an important variable in channel systems. Often it is bank vegetation that resists erosion during periods of high flow. Similarly, vegetation tends to stabilize interchannel bars and islands, thus producing an added resistance in the channel. These banks and bars can substantially alter flow lines and channel movement.

A more complete understanding of the physical processes governing channel response can be attained through consideration of important river mechanics concepts and related variables. A brief discussion of several important concepts and variables follows.

GOVERNING VARIABLES AND MECHANICS OF CHANNEL RESPONSE

Because of the number of interrelated variables that can react simultaneously to natural or imposed changes in a river system, river response to both natural and man-imposed forces is complex and varied in nature, but it is predictable. Simons and Richardson (1963) detailed the variables affecting alluvial channel geometry and bed roughness and concluded that the nature of these variables is such that, unlike rigid boundary hydraulics problems, it is not possible to isolate and study the role of an individual variable. For example, evaluation of the effects of increasing channel depth on average velocity is hampered because related variables such as flow resistance also respond to the changing depth. Not only will velocity respond to a change in depth, but also form of bed roughness, channel cross-section shape and sediment discharge. Position and shape of alternate, middle, and point bars can also be expected to change.

Major variables that influence alluvial channel flow include velocity, depth, slope of energy grade line, density of water-sediment mixture, representative fall diameter of bed material, gradation of bed material, density of sediment, shape factor of particles, shape factor of the reach of the stream, shape factor of the cross-section of the stream, seepage force in the bed of the stream, concentration of bed-material discharge, fine material concentration, and particle terminal fall velocity. Dimensional analysis of these variables verifies the importance of the Froude number, Reynolds number and a relative roughness parameter.

ALLUVIAL CHANNEL BED CONFIGURATIONS

It is known that for flow in channels composed of erodible granular material, a strong physical interrelationship exists between the friction factor, the sediment transport rate, and the geometric configuration assumed by the bed surface. The changes in bed forms result from the interaction of the flow, fluid, and bed material. Thus, the resistance to flow and sediment transport are functions of the slope and depth of the stream, the viscosity of the fluid, and the size distribution of the bed material. The interaction between the flow and bed material and interdependency among the variables make the analysis of flow in alluvial sand-bed streams extremely complex. However, with a general understanding of the different types of bed forms that may occur and a knowledge of the resistance and sediment transport associated with each bed form, the engineer can begin to analyze and understand alluvial channel flow.

HYDRAULIC GEOMETRY

Hydraulic geometry is a general term used to denote relationships between bankfull discharge, channel morphology, hydraulics and sediment transport. In natural alluvial channels, the morphologic, hydraulic, and sedimentation characteristics of the channel result from a variety of factors. Generalized hydraulic geometry relations apply to channels within a physiographic region and can be computed from data available for gaged rivers. Hydraulic geometry relations express the integral effect of all hydrologic, meterologic, vegetative, and geologic processes in a drainage basin.

Hydraulic geometry relations of alluvial streams are necessary for river engineering computations and river modeling. Forerunners of these relations were the regime theory equations of stable alluvial channels. A generalized version of hydraulic geometry relations was developed by Leopold and Maddock (1953) for different regions in the United States and for different types of rivers. Hydraulic geometry relations can be stated as $W = aQ^b$; $y_o = cQ^f$; $V = kQ^m$; $Q_T = PQ^j$; $S = tQ^z$; $n = rQ^{y_o}$; where W is the channel width, y_o is the channel depth, V is the average velocity of flow, Q_T is the total bed-material load, S is the energy gradient, n is Manning's roughness coefficient, and Q is the discharge. Leopold and Maddock (1953) have shown that in a drainage basin, two types of hydraulic geometry relations can be defined: (1) those relating W, y_o, V and Q_T to the variation of discharge at a station, and (2) those relating these same variables to the discharges of a given frequency of occurrence at various stations on a channel. Because Q_T was generally not available, Leopold and Maddock used Q_s, the suspended load transport rate. The former relations are called *at-station* relationships and the latter, *downstream* relationships.

Hydraulic geometry relations were theoretically developed at Colorado State University (Li et al., 1976). These relations are almost identical to those proposed by Leopold and Maddock (1953). The at-station relations derived at Colorado State University are:

$$W \propto Q^{0.26} \quad (1.1)$$
$$y_o \propto Q^{0.46} \quad (1.2)$$
$$S \propto Q^{0.00} \quad (1.3)$$
$$V \propto Q^{0.30} \quad (1.4)$$

Equation 1.3 implies that slope is constant at a cross-section. This is not precisely true. At low flow the effective channel slope is determined by the thalweg which flows from pool through crossing to pool. At higher stages the thalweg straightens somewhat, thus shortening the path of travel and increasing the local slope. In extreme cases river slope approaches the valley slope at flood stage. During high floods, flow often cuts across the point bars, thus developing chute channels. This direction of travel means the water takes a shorter path and a steeper channel prevails under this condition.

Similarly, the derived downstream relations for bankfull discharge are:

$$y_b \propto Q^{0.46} \tag{1.5}$$
$$w_b \propto Q^{0.46} \tag{1.6}$$
$$S \propto Q^{-0.46} \tag{1.7}$$
$$V_b \propto Q^{0.08} \tag{1.8}$$

where the subscript b indicates the bankfull condition.

LONGITUDIAL STREAM PROFILE

The longitudinal profile of a stream shows its slope, or gradient, and is one of the more important variables influencing channel response. It is a representation of the ratio of the fall of a stream to its length over a given reach. Because a river channel is often steepest in its upper regions, most river profiles are concave upward. As with other channel characteristics, shape of the profile is a result of a number of interdependent factors, representing a balance between the transport capacity of the stream and the size and quantity of the sediment load supplied.

Shulits (1941), among others, provided an equation describing the concave horizontal profile of a channel in terms of distance along the stream as:

$$S_x = S_o e^{-\alpha x} \tag{1.9}$$

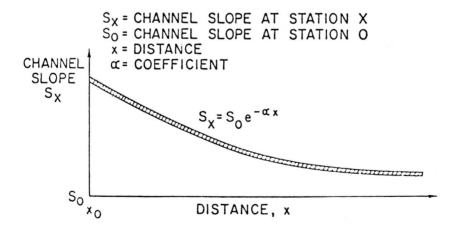

FIGURE 1. Generalized variation of channel slope with distance.

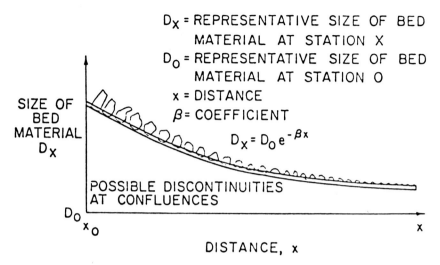

FIGURE 2. Generalized variation of size of bed material with distance.

where S_x is the slope at any station a distance x downstream of a reference station, S_o is the slope at the reference station and $\propto$ is a coefficient of slope reduction (Fig. 1).

Similarly, the grain size of the bed material decreases in a downstream direction. Transport processes reduce the general size of sediment particles by abrasion and hydraulic sorting. Abrasion is size reduction by mechanical actions such as grinding, impact, and rubbing, while hydraulic sorting results in differential transport of particles of varying sizes. For sedimentary particles of similar shape, roughness and specific gravity, the result of these processes is the observed reduction of bed-material size along the direction of transport. The change in particle size with distance downstream can be expressed as:

$$D_{50x} = D_{50o}e^{-\beta x} \qquad (1.10)$$

where D_{50x} is median size of bed material at distance x downstream of a reference station, D_{50o} is median size of bed material at the reference station, and β is a wear or sorting coefficient (Fig. 2). This trend is found in large and small channels.

SEDIMENT

Sediment Transport

Sediment transport is an important process with in a river system. The sediment in a river channel almost always originates in the drainage basin itself.

Eroded material is carried into the river and along the river's course by flowing water. Ultimately, this material is deposited in the lower reaches of the river, on the river delta, or for the finer material, in the sea. This constant displacement of material implies a slow but continuous change in the longitudinal profile of the river. As a result, it must be anticipated that large quantities of sediment will pass through a river system each year.

Sediment Yield

The quantity of sediment delivered to the channel depends on watershed processes previously discussed. The capacity of a stream to transport sediment depends on the hydraulic properties of a stream channel. Variables such as slope, roughness, channel geometry, discharge, velocity, turbulence, fluid properties, and size, and gradation of the sediment are closely related to the hydraulic variables controlling the capacity of the steam to carry water. The total sediment load of a stream is the sum of the bed-material load and the wash load, the sum of bed load and suspended load, or the sum of measured and unmeasured load.

Aggradation, Degradation and Lateral Migration

Alluvial channel systems generally exhibit significant changes in depth, width, alignment, and stability particularly during floods of long duration and as a consequence of water resources development. These changes may be defined as local scour, general scour, degradation or aggradation, and lateral migration. Local scour is caused by disturbances in the flow, such as vortices and eddies as observed around piers, dikes, and other obstructions to flow. General scour is due to contractions causing increased velocities across the entire contracted width. General scour can be caused by spur dikes, embankments, piers, and the accumulation of debris at bridge openings. Degradation and aggradation are the long-term vertical downcutting or rising, respectively, of the river bed due to changes in controls, such as dams, changes in sediment transport or river geomorphology. These effects are generally additive; local scour can occur while scour due to contraction and degradation or aggradation of the stream is taking place. Likewise, lateral migration results from bank line shifting and bank sloughing, while bank sloughing can be accelerated by excessive degradation. Therefore, degradation and lateral migration are interdependent.

VARIABLES RELATED TO BED AND BANK PROCESSES

Vegetation

Vegetation on the banks of a river can have a significant effect on stabilizing the system. The root structure of plants, bushes, and trees helps to maintain and develop a stable soil structure and serves as an erosion-retarding force.

Whenever water flows by or over noncohesive banks with only the forces of gravity to resist motion (no vegetation), erosion can be severe. Trends in vegetative growth can be indicative of future bank stability. Using aerial photographs, the changes in vegetation over a given time period can be classified into groups such as (1) no change, (2) vegetation increasing, (3) vegetation damaged, and (4) vegetation destroyed. These classifications could also include bank stability.

Bed Material

Bed material is the sediment mixture which composes the streambed. Bed material ranges in size from huge boulders many feet in diameter to fine clay particles. The erodibility or stability of a channel largely depends on the size of particles in the bed. It is often not sufficient to know the median bed-material size (D_{50}) in determining the potential for degradation; knowledge of the bed-material size distribution is important. As water flows over the bed, smaller particles are more easily transported, while larger particles remain, armoring the bed. This armor layer can serve as a control until a larger flow event occurs.

Bank Material

Bank material is generally made up of smaller or the same sized particles as the bed. Thus, banks are usually more easily eroded than the bed because of particle size availability and channel migration tendencies unless protected by vegetation, inherent cohesiveness, or some type of man-made protection. River banks or reaches of river banks can be classified according to stability by looking at vegetation, cohesiveness, frequency of protection, lateral migration tendencies of the stream, etc.

RIVER CLASSIFICATION

Geomorphic Systems

In contrast to the lack of a widely used watershed classification system are the numerous schemes applied to rivers. Davis (1899) first suggested that rivers could be divided into three stages: youth, maturity, and old age. These stages were later subdivided based on the presence or absence of rapids, falls, meanders, oxbow lakes, flood plains, canyons, and other factors. In general with this system, a stream can range from youth to old age as it flows from its watershed to the ocean.

Although this system does not account for spatial variability, it provides an initial common classification language. Thornbury (1969) presented common valley classifications based on their development on the surface of the land. Although not precisely analagous, these patterns are often used for stream classification. These patterns are recognized as antecedent, superposed, consequent, and subsequent. Antecedent streams or valleys antidate structures that they cut across, such as a valley through an updomed area or a fault zone. A superposed stream extends across structures that are older than the stream but were covered by the original stream bed material. Erosion eventually removed this material and superposed the river on the structure. Consequent streams are a result of the initial land slope while subsequent streams have been shifted from their original consequent courses to ones following weaker belts of rocks.

All rivers can be separated into two major groups depending on their freedom to adjust their channel. Bedrock channels are confined between rock outcrops such that the material forming their bed and banks determines the channel morphology. Alluvial channels are free to adjust their pattern and gradient in response to hydraulic changes, and they flow through a channel with bed and banks of material transported under present flow conditions. Alluvial channels are of primary interest here because of their dynamic behavior in response to spatial and temporal changes of natural and man-induced processes.

PLANFORM CLASSIFICATION

Rivers, or segments of rivers, can also be generally classified as straight, meandering, braided, or some combination of these (Fig. 3). Reaches of a river

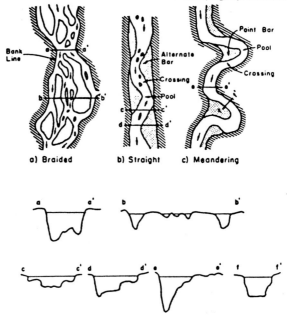

FIGURE 3. Typical river channel patterns (after Simons et al. 1975).

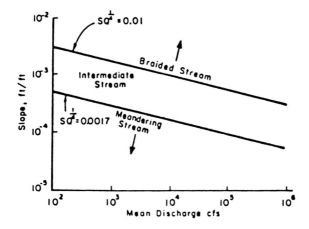

FIGURE 4. Slope-discharge relation for raiding or meandering in sand-bed streams (Lane 1957).

that are relatively straight over a long distance are generally unstable, as are divided flow reaches and those in which bends are rapidly migrating. Long straight reaches can be created by natural or man-made cutoffs of meander loops when long reaches of sinuous meandering channels with relatively flat slopes are converted to shorter reaches with much steeper slopes. Straight reaches can also be man-induced by contraction works such as dikes and revetment used to reduce or control sinuosity (Simons et al., 1975).

A braided channel is relatively wide with poorly defined unstable banks. It is characterized by a steep, shallow water course with multiple channel divisions around alluvial islands. Braiding is one pattern which can maintain quasi-equilibrium among the variables of discharge, sediment load, and transport capacity. Lane (1957) concluded the two primary causes that may be responsible for the braided condition generally are: (1) the stream may be supplied with more sediment than it can now carry, resulting in deposition of part of the load, and (2) steep slopes which produce a wide shallow channel where bars and islands readily form.

A meandering channel is one that consists of alternating bends, giving an S-shaped appearance to the plan view of the river. Lane (1957) concluded that a meandering stream has a channel alignment consisting principally of pronounced bends, the shapes of which have been determined predominantly by the varying nature of the terrain through which the channel passes. The meandering river consists of a series of deep pools in the bends and shallow crossings in the short straight reach connecting the bends. The thalweg flows from a pool through a crossing to the next pool, forming the typical S-curve of a single meander loop.

Because of the physical characteristics of straight, braided, and meandering streams, all natural channel patterns intergrade. Although braiding and meandering patterns are strikingly different, they actually represent extremes in a continuum of channel patterns. On the assumption that the pattern of a stream is determined by the interaction of numerous variables whose range in nature is continuous, it is not surprising that a river may exhibit braiding, straight, and meandering forms. Alteration of the controlling parameters in a reach can change the character of a given stream from meandering to braided or vice versa. Studies have quantified this concept of a continuum of channel patterns.

Lane (1957) suggested relationships among slope, discharge, and channel patterns in meandering and braided streams, and observed that an equation of the form

$$SQ^{1/4} = K \tag{1.11}$$

fits a large amount of data from meandering sand streams. Here S is the channel slope, Q is the water discharge, and K is a constant (Fig. 4). If a river is meandering, but with a discharge and slope that borders on transitional, a relatively small increase in channel slope could initiate a tendency toward a transitional or braided character.

DESCRIPTIVE CLASSIFICATION

There are other classification schemes that uses vegetation pattern, sinuosity, and bank characteristics as has been proposed by Culbertson, et al. (1967) to delineate subclassifications within the major types of meandering, straight, and braided channels if a more detailed classification scheme is required.

RESPONSE OF RIVER SYSTEMS TO HYDROPOWER DEVELOPMENT

There are many types of hydropower developments that are utilized to generate energy. Major classifications would include large dams with hydropower, small dams with hydropower, diversion dams with hydropower, run of the river hydropower systems, and thermal and nuclear power systems that require cooling water. The major question is what impacts does development of hydropower have on the river system and the environment. In order to consider this serious question, it is first of all necessary to consider the size of the river system, the climatology experienced by it, the geology, etc. With adequate background information on the physiography and the climatology of the system, one can then determine existing hydrologic conditions in the system and initiate the study to determine impacts of the proposed hydropower on the hydrology. To begin with, the climate is a given but is subject to side variation. The hydrology is basically also a given, however, we must consider the natural hydrologic system as it will be modified by a proposed hydropower system and the hydrologic system as it may be affected by proposed water resources developments that could have a side effect on existing systems.

The impacts on hydrology imposed by hydropower systems depends very significantly on, not only the characteristics of the system, but the size of the system. For example, if one constructs a relatively large dam and storage facility to facilitate hydropower generation then the hydrology may be impacted by the magnitude of the storage and the magnitude and time of the diversions. These types of impositions on the system can alter flood flows, can change base flows and, depending upon the type of hydrosystem, can basically generate surge-type flows. Another important impact that hydrology may have on the hydropower system pertains to the sediment. Basically hydropower development requires utilization of clean water in order to minimize damage to hydraulic machinery. With large reservoirs, literally all of the sediment being delivered to the reservoir may be trapped. This means that there will be a clear water release downstream. The release of clear water can induce erosion in the system and other related responses. If we are dealing with the diversion from a river to the penstock for power delivery, then it is essential to take as clean of water as possible from the river; leaving the remaining system possibly overloaded with sediment which can result in aggradation and other related problems. Still another type of system may be operated to generate peak power. For this type of power system, there can be significant surging of flows. This can induce bank erosion, loss of riparian vegetation, and other related adverse impacts.

As we consider the release of essentially clear water, for example downstream of a major dam, it has been recognized that degradation does occur to some degree depending upon the characteristics of the bed and bank material. Often, however, degradation is offset by the steepening of the energy gradient of tributaries entering the mainstream. It is possible that as degradation occurs in the main river that steepening of the tributaries may result. This change can result in the discharge of additional sediment into the system from the tributaries. In many instances the supply of sediment may be large enough to cause aggradation downstream of these tributaries—a reversal of response.

Many rivers are so fully developed, taking water for a variety of uses, that we end up with a river system tremendously overloaded with sediments subjecting it to severe aggradation and possible change in river geometry subject to severe aggradation. For example, a river system such as the Rio Grande in the western United States is developed to the point where waters are almost fully utilized except during periods of major floods. Simultaneously, all uses generally require clear water, hence we have a system with inadequate flows to convey the ever increasing accumulations of sediment in the mainstream. Hence, in this type of system when major floods during unusually wet years occur, the channel may not be capable of adequately handling such flows and severe flooding may result.

Most river systems involve multipurpose development, that is developments for flood control; hydropower; irrigation; recreation—including fishing, boating, etc.; sand and gravel operations; navigation; wildlife; hunting; water supplies from municipalities and industry and even disposal of wastes. In these types of systems, all are subjected to wet, normal, and dry periods. When large reservoirs are first built, if the storage volume significantly exceeds the annual runoff, the downstream river can be subjected to basically a man-made drought while the reservoir is filling for normal operation. Conversely, if the reservoir is not operated correctly, it is possible that the reservoir may be basically full of water to provide optimum head for power generation during normal periods. Then, if flooding conditions develop on the watershed, it is possible to generate floods in excess of those that would have occurred normally by excessive releases from the reservoir to safeguard the dam.

In summary, consider large reservoirs. These systems during the construction period may have significant local effects. They may disturb the local terrain and may increase the sediment load downstream:

- During periods of time that the reservoir is filling to achieve its anticipated operational level, imposed storage conditions cause low flows downstream and clear water releases from the reservoir. These changes may encourage the encroachment of vegetation downstream and, in general, the construction of a dam and reservoir within the river system can encourage encroachment of vegetation, loss of channel capacity, degradation which will result in steepening of the gradients of tributaries downstream of the dam, and introduction of instabilities in the tributaries. There may be significant downstream degradation. Also as the reservoir fills, there may be significant upstream aggradation as a result of the delta forming in the mouth of the reservoir and, with changes in water levels, there may be local increases in water table levels causing other adverse conditions such as possible salinity and alkalinity problems. Also, such development can create new wetlands and cause the loss of others.

The response of a river system to each hydropower system will be different depending upon such variables as:

- Climate. Characteristics of the climatological period, that is whether it is wet, normal or dry.
- The type of hydropower system being installed.
- The geology and geomorphology of the watershed.
- The location of the system on the watershed.
- The relative stability of the watershed.
- The supply of sediment versus transport capacity considering sediments.
- Environmental factors.
- Government, state, county, and city regulations.

- Environmental activism.
- National, state, and local politics.
- The financial attributes of the project.

There are no set standards for evaluating positive and negative impacts related to hydropower development. Each system must be evaluated on its own merits. Systems and system response is so complex and varied that each project and each system requires its own mathematical models, physical models, environmental evaluation, engineering design, maintenance requirements, etc.

GENERAL SOLUTION APPROACH

General

To more precisely determine the impacts of climatic changes, water resources development and, in this case, possible impacts of hydropower development on channel morphology three basic solution procedures can be applied: (1) qualitative, involving geomorphic concepts, (2) quantitative, involving geomorphic concepts and basic engineering relationships, and (3) quantitative, involving sophisticated mathematical modeling concepts. A mathematical model is simply a quantitative expression of a physical process or phenomenon.

As described previously, knowledge of governing physical processes and the sensitivity of the system play the most important part in deciding an appropriate level of analysis. A qualitative analysis of these processes can provide insight into complicated river response problems. The general knowledge obtained from this analysis provides understanding and direction to a quantitative analysis.

A brief description of each level of the general solution approach is included. To illustrate the application of the approach to a typical alluvial system problem, the multiple level approach is introduced in relation to the processes of aggradation and degradation, or channel "gradation."

QUALITATIVE GEOMORPHIC TECHNIQUES

Geomorphology is the study of surficial features of the earth and the physical and chemical processes changing landforms, while fluvial geomorphology is applied to morphology (and mechanics) of rivers and river systems. Engineers and others concerned with analysis of fluvial systems are mainly interested in hydraulic geometry and other fluvial geomorphologic parameters related to watersheds and channels. In addition, investigation of river response to climatic changes and diastrophic events of the earth's evolution provides information and insight into long-term adjustment to hydrologic conditions (Schumm 1971).

Evidence of bank cutting, shifting of thalweg, lateral migration meander tendencies, vegetation changes, and sediment deposition can be documented

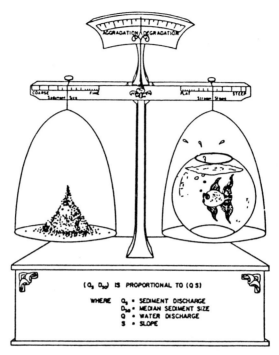

FIGURE 5. Schematic of the Lane relationship for qualitative analysis.

by studying photographs for different years. Changes in stream widths also can be documented by measuring the minimum, maximum, and average stream width in each reach. These measurements utilize monoscopic or stereoscopic techniques, depending on the availability of adequate aerial photographs and the level of detail required in the analysis.

Lateral migration tendencies can also be evaluated qualitatively based on a knowledge of geomorphic concepts. Alluvial channels of all types deviate from straight alignment. The thalweg oscillates transversely and initiates the formation of bends. In general, the river engineer concerned with channel stabilization should not attempt to develop straight channels. In a straight channel, the alternate bars and thalweg (the line of greatest depth along the channel) change continuously, thus the current is not uniformly distributed through the cross-section, but is deflected toward first one bank and then the other. Sloughing of the banks, nonuniform deposition of bed load caused by debris such as trees, and the Coriolis force have been cited as causes for meandering of streams. When the current is directed toward a bank, the bank is eroded in the area of impingement, and the current is deflected away and may impinge upon the opposite bank further downstream. The angle of deflection of the thalweg is affected by the curvature formed in the eroding bank and the lateral depth of erosion.

Qualitative geomorphic analysis of actual gradation changes (aggradation/degradation) is based on the concept of equilibrium. The qualitative approach assumes that rivers strive, in the long run, to achieve a balance between the product of water flow and channel slope and the products of sediment discharge and size. The most widely known geomorphic relationship embodying the equilibrium concept is known as Lane's principle, which is illustrated schematically in Fig. 5.

QUANTITATIVE APPLICATION OF GEOMORPHIC PRINCIPLES

Geomorphic principles applied qualitatively to predict river response provide only a general understanding of the direction of change. Geomorphic principles can also be applied to quantitatively evaluate such problems as aggradation, degradation, and lateral migration. This generally requires collection and analysis of data for at least several years.

It is necessary to establish, by some means, the temporal changes in the water surface elevation for a given discharge, the elevation of the channel bottom, and changes in the slope of the channel bottom and the bed material size at a given location. Information on elevation changes can be gained from a series of maps which were prepared at different times, although these data are not always available. Data from railroad and pipeline crossing surveys upstream or downstream of a highway bridge may also be helpful in determining river bed elevation changes as a function of time.

Analysis of gaging station stage trends is generally easily done and yields useful information on long-term trends. The U.S. Geological Survey and Corps of Engineers have already performed the analysis over large areas, providing excellent records, many of which extend for 30 years or longer.

Another item of useful information usually available from gaging stations is suspended sediment load. Although few stations have continuous sediment data, these data, when available, can provide clues to the presence of gradation problems. By definition, aggradation takes place when sediment inflow to a river reach exceeds sediment outflow. Any change in the long-term sediment load signals an imbalance in the stream system. Such imbalances tend to produce lateral movement, bank sloughing, and gradation problems.

An assessment of relative stability can also be made by evaluating the incipient motion parameters. The definition of incipient motion is based on the critical or threshold condition where the hydrodynamic forces acting on the grain of sediment particle have reached a value that, if increased even slightly, will move the grain. Under critical conditions, or at incipient motion, the hydrodynamic forces acting on the grain are balanced by the resisting forces of the particle. The Shields relation defines the beginning of motion of a sediment particle.

Another method for verifying the presence of gradation changes is stream profile evaluation. This method requires considerable surveying effort. The idea is similar to measuring the change in bed elevation from a bridge deck, but in this case, a longitudinal profile of the thalweg is surveyed and compared to a historic profile. Rough profile analysis can be performed by plotting the elevations of cross-sections from sites such as pipeline crossings or railroad bridges as a function of time. Another source of information is from hydrographic surveys, or U.S. Corps of Engineers' potamology studies on many major streams.

It is expected that the general properties, such as geometry and alignment of a river, will be changed by the passage of floods during the design lifetime of most projects. It is therefore necessary to establish information on the river morphology to estimate these changes. The representative cross-section in the vicinity of a proposed project can be developed by averaging the nearby cross-sections where the channel geometries have been surveyed. From this representative cross-section, the channel geometry relations can be established.

BASIC ENGINEERING RELATIONS

Basic engineering relationships together with qualitative and quantitative geomorphic concepts provide solutions to specific problems such as those related to surface profiles and sediment transport rates. Geomorphic principles are useful for establishing a basic understanding of channel response, prior to an analysis using engineering relationships.

Calculation of water surface profiles is an integral part of a sediment transport analysis for gradation changes. After qualitatively classifying the type of profile, computer programs for the computation of the elevation or depth of flow for water surface profiles can be used. However, most computer programs require a qualitative analysis of general characteristics of back-water curves in order to determine whether the analysis proceeds upstream or downstream.

For large and complex situations involving dams, reservoirs, hydropower, bridges, culverts, and long reaches of river, it is often necessary to use a relatively complex computer program.

Knowledge of sediment transport conditions is essential to channel response analyses. Evaluation of general and local scour resulting from a dam and other structures requires accurate estimates of sediment transport rates. Many formulas and procedures have been developed to predict sediment transport.

Degradation, aggradation, and movement of pollutants are closely related to water and sediment movement. Understanding the physical processes related to water and sediment routing is of fundamental importance for effective analysis of watershed and river response.

QUANTITATIVE ANALYSIS WITH MATHEMATICAL MODELING

A mathematical model is simply a quantitative expression of a process or phenomenon that is being studied. In a conventional method of analysis, a series of manual calculations may be required. With the advancement of numerical techniques and computer technology, a series of tedious computations can be conducted efficiently, repeatedly, and accurately through the formulation and construction of a mathematical model.

With such a model, a whole array of "what-if" questions can be answered with minimum requirement of time and effort. Since no process can be completely understood and observed, any mathematical expression of a process will involve some level of uncertainty. This uncertainty can be minimized if the governing physical processes are considered in the analysis and the model is properly designed, calibrated, and verified. Model development, verification, and application to analysis and design problems require the consideration of the nature of the problem, physical environment, objective of the study, time, manpower, and money. Since time, manpower, and money are always limited, decisions must be made relative to the degree of complexity the model is to have, and the extent of verification to be performed.

The multiple level solution approach stresses that knowledge of governing physical processes plays the most important part in deciding an appropriate level of mathematical analysis. If the governing physical processes are emphasized in the analysis, the degree of complexity required to represent the physical system can be defined.

Physical processes governing watershed and river responses are themselves very complicated. Many past studies have utilized a statistical interpretation of observed response data, e.g., unit hydrograph method for water routing, Universal Soil Loss Equation for soil erosion, and hydraulic geometry equations for stream morphology. However, it is often difficult to predict the response of a watershed to various watershed developments or treatments using such methods, because they are based on the assumption of homogeneity in time and space.

Mathematical simulation of governing physical processes provides a direct estimate of the time-dependent response of fluvial systems. Physical process modeling can be used to analyze basic ecosystem processes and the impact of land use and water resource development management activities on specific processes. It is then possible to predict the cause-effect relationships between management activities and ecosystem response. With the aid of systems analysis techniques, a desirable mix of management activities can then be selected taking into consideration both the environmental and resource goals.

CONCLUSION

In conclusion, one of the most important variables, and yet the one most underrated in significance and the one most poorly understood, is the role of sediment and the dynamic nature of rivers. In future evaluations of hydropower and related systems, more emphasis will be placed on quantifying impacts of changes in both the water and sediment environment on hydropower, irrigation, hydraulic structures, the environment, the performance of hydraulic machines, and, of course, local and national policies and politics. The guidelines presented herein provide a means of quantifying response of river systems to the development and operation of both simple and complex hydropower systems.

LITERATURE CITED

Culbertson, D.M., L.E. Young, and J.C. Bricé, 1967. Scour and Fill in Alluvial Channels, U.S. Geological Survey, Open File Report.

Davis, W.M., 1899, The Geographical Cycle, Geographical Journal, Vol. 14, pp. 481-504.

Leopold, L.B., and T. Maddock, 1953, The Hydraulic Geometry of Stream Channels and Some Physiographic Implications, *U.S. Geological Survey Professional Paper 252*, U.S. Geological Survey, Washington, D.C.

Lane, E.W., 1957, A study of the Shape of Channels Formed by Natural Streams Flowing in Erodible Material, Missouri River Division Sediment Series No. 9, U.S. Army Engineer Division, Missouri River, Corps of Engineers, Omaha, Nebraska.

Li, R. M., D.B. Simons and M.A. Stevens, 1976, Morphology of Cobble Streams in Small Watersheds, *Journal of the Hydraulics Division*, A.S.C.E. Vol. 102, No. HY8, Proc. Paper 12304, pp. 1101-1117, Aug.

Schumm, S.A., 1971, Fluvial Geomorphology: The Historical Perspective, in Chaper 4 of *River Mechanics*, (I. H. W. Shen, Ed.) Fort Collins, Colorado.

Shulits. S., 1941, Rational Equation of River-Bed Profile, *Transactions*, Amer. Geophysical Union, 22: 622-630.

Simons, D.B., P.F. Lagasse, Y.H. Chen and S.A. Schumm, 1975. The River Environment—A Reference Document, prepared for U.S. Department of the Interior, Fish and Wildlife Service, Twin Cities, Minnesota.

Simons, D.B. and E.V. Richardson, 1963, A Study of Variables Affecting Flow Characteristics and Sediment Transport in Alluvial Channels, *Proceed.*, Federal Inter-Agency Sedimentation Conference, Jackson, Mississippi, pp. 193-206.

Thornbury, W.D., 1969, *Principles of Geomorphology*, 2nd Edition, John Wiley & Sons.

Environmental Consequences of Energy Production: Problems and Prospects. Edited by S. K. Majumdar, F. J. Brenner and E. W. Miller. © 1987, The Pennsylvania Academy of Science.

Chapter Twenty-Two
IMPACTS OF HYDROPOWER DEVELOPMENT ON ANADROMOUS FISH IN THE NORTHEAST UNITED STATES

GORDON W. RUSSELL[1] and RICHARD A. ST. PIERRE[2]

[1]U.S. Fish and Wildlife Service
P.O. Box 1518
Concord, NH 03301
and
[2]U.S. Fish and Wildlife Service
P.O. Box 1673
Harrisburg, PA 17105

Dam development in the Northeast United States contributed to reduction or elimination of spawning runs of Atlantic salmon (*Salmo salar*) and American shad (*Alosa sapidissima*) in numerous coastal rivers. Early efforts to restore these valuable fish through hatchery supplementation and fish passage construction largely failed. Hydroelectric development in the early 20th century at historic and new dam sites further restricted access and altered spawning and nursery habitat. In addition to these impacts hydropower operations affect migratory fish restoration programs by killing or stressing fish during downstream passage through turbines, altering natural river flows, degrading water quality and enhancing predation.

Recent legislation encouraging development of small-scale hydroelectric projects, viewed as an opportunity to improve the country's energy self-sufficiency, further exacerbates fish restoration attempts. Difficult decisions are forced on federal regulators trying to resolve the conflicting public demands for both power production and recovery of lost fisheries.

INTRODUCTION

Development of alternate energy sources in the Northeast has resulted in renewed interest in hundreds of former hydropower sites. Many of these are located on rivers containing valuable aquatic resources. Anadromous species of fish, those migrating from the sea to reproduce in fresh water, can be significantly impacted by construction and operation of hydroelectric projects. Although adverse effects can be lessened with measures such as fish passage facilities, multiple damming of rivers results in cumulative losses in riverine habitat and reduces the numbers of migrating fish. In some river drainages, anadromous fish management is not possible or is jeopardized due to increased hydropower development.

In this chapter, we review the impacts of hydropower development on the management and restoration of anadromous fish in the Northeast. Emphasis is given to Atlantic salmon and American shad. In addition to presenting historical and current perspectives on hydropower activity and anadromous fish programs in the Northeast, we describe the impacts of hydroelectric projects on salmon and shad populations. We also discuss the conflict of public policies that has resulted from encouraging hydropower development at the expense of anadromous fish restoration.

STATUS OF THE FISHERIES

Historical Perspective

Atlantic salmon and American shad have a long history of importance in eastern North America. Western Atlantic distribution of Atlantic salmon is from northern Newfoundland to coastal Connecticut (Fig. 1). American shad range from northern Florida to the Gulf of St. Lawrence, being most abundant in mid-Atlantic states. Although spawning habitat and freshwater residency differs for these two anadromous species, factors affecting reproductive success and maintenance of the populations are similar.

Through the 18th century, at least 34 coastal rivers in Maine, the Merrimack and Connecticut rivers and their headwater tributaries supported self-sustaining stocks of Atlantic salmon estimated to number 250,000-500,000 fish (Stolte 1981, 1982, Maine Atlantic Sea-Run Salmon Commission 1984). Commercial harvest of salmon in rivers, particularly at the base of natural falls, is documented as far back as 1628. As colonization and industrial development expanded through the 17th and 18th centuries, spawning habitat was progressively degraded, destroyed and made inaccessible to salmon. The species was extirpated by dam construction in the Connecticut River around 1800, and the Merrimack around 1850. Maine rivers suffered serious declines throughout the 1800's prompting construction of fish passage facilities at dams, institution of restrictive fishing regulations and hatchery supplementation of natural stocks. Nevertheless, commercial landings fell dramatically from an average of over 13,000 salmon per year in the 1880's to fewer than 1,000 fish annually after 1940 (Maine Atlantic

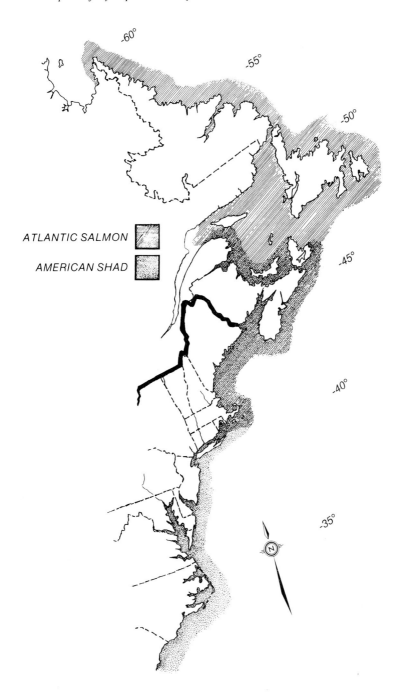

FIGURE 1. Distribution of Atlantic salmon and American shad along the Atlantic Coast of North America. Shad range continues southward to the St. Johns River, Florida (lat. 30°N).

Sea-Run Salmon Commission 1984). The major commercial fishery for salmon in Maine, the Penobscot River, closed in 1948 following a catch of 40 fish.

Dip nets and fykes were used to take American shad for local consumption, but gave way to seines and gill nets as market demand expanded through the 19th century. Construction of dams without suitable fish passage facilities, overfishing near river mouths and at points of concentration, and degradation of water quality from industrial, domestic and agricultural sources brought about serious reduction of shad stocks by the 1860's (Mansueti and Kolb 1953). Numerous states and the federal government embarked on shad culture to supplement the dwindling natural populations, and billions of shad fry were produced and stocked from these efforts during the period 1870-1940 (Robbins and Watts 1982). However, with loss of habitat, commercial landings of shad in New England and mid-Atlantic states mirrored a continued decline in abundance. Harvest of shad decreased from an average annual 14.5 million kg at the turn of this century to less than 1.8 million kg/year in the 1970's (Boreman 1981).

Records of recreational angling for shad and Atlantic salmon in the Northeast United States are sparse prior to 1880. Salmon fishing in Maine expanded in the late 19th century gaining national attention for this "king of game fishes." Major seasonal sport fisheries for shad exist where this species occurs in abundance, notably the Connecticut and Delaware rivers. It is a credit to the fighting abilities and food value of the American shad that it has been dubbed the "poor man's salmon."

Cooperative Restoration Programs

Coordinated efforts to restore shad and salmon to historically important breeding and nursery waters are not new. Fish Commissioners from several New England states met in the late 1860's to initiate shad and salmon restoration programs on the Merrimack and Connecticut rivers. At about the same time, the State of Maine instituted laws to control fishing for salmon and requiring fish passage at dams that blocked migration. One of the principal reasons for establishing the Pennsylvania Fish Commission in 1866 was to improve migratory runs of shad and herring to the Susquehanna River.

Fish culture was in its infancy in North America in the 1860's, frequently referred to as more art than science. Nevertheless, successful hatching and rearing techniques were developed for both salmon and shad with much of the early technology being carried forward to programs of today. The salmon hatchery effort in Maine has continued (and grown) almost without interruption since the 1870's. However, restoration efforts on the Merrimack and Connecticut rivers were terminated in the late 19th century due to ineffectiveness of fish passage facilities, lack of secure sources of salmon eggs, continued water quality degradation, overfishing near mouths of rivers, and emphasis on other fish and wildlife resources. Shad culture efforts grew to peak production levels during the late 1890's and gradually tapered off until the last federal shad station was closed in 1949. By this time it was evident that stocking artificially cultured larval shad in proven nursery waters below dams would not substantially improve the populations.

Many hydroelectric dams were built on major New England rivers and on the Susquehanna River in the first several decades of the 1900's. Other existing (sometimes breached) dams were reconstructed and outfitted for hydroelectric production. In most cases, fish passage facilities incorporated into these structures were not effective for shad, and salmon stocks south of Maine had been lost years before. As fisheries continued to decline and with improvements in water quality and fish passage technology, states and the federal government took a new look at shad and salmon restoration in the 1960's.

Considerable success was realized in fish ladder effectiveness in the northwestern United States during the 1940's and 1950's. American shad, transplanted to the west coast in the late 19th century, and native salmon and steelhead, (*Salmo gairdneri*) made extensive migrations past hydroelectric dams on numerous major rivers, most notably the Columbia. With this improved technology, and passage of the Anadromous Fish Conservation Act of 1965 (16 U.S.C. 757a-757f), state and the federal government renewed their cooperative efforts to restore shad and salmon to select rivers in the Northeast.

In Maine, the Atlantic Sea-Run Salmon Commission (established in 1947), the Departments of Marine Resources and Inland Fisheries and Wildlife, joined with the U.S. Fish and Wildlife Service and the National Marine Fisheries Service to survey salmon habitat, rear and stock salmon smolts in historically important nursery waters, and develop fish passage wherever appropriate. Cooperative efforts to restore salmon and shad on the Connecticut and Merrimack rivers were reinitiated in 1967 and 1969, respectively, with the formation of policy and technical committees for fisheries management. In the past decade, fish passage facilities were built or improved on 6 mainstem and tributary dams on the Connecticut and the lower two dams on the Merrimack. A 1986 agreement reached with the private utility owning the remaining 5 mainstem dams on the Merrimack provides for fishway development on a phased schedule at those projects beginning in 1988. With these improvements, salmon and shad will once again have access to most of their historic spawning waters in both rivers.

In 1984, the Connecticut River Atlantic Salmon Commission replaced the Policy Committee for Fishery Management of the Connecticut River. Restoration goals for that river are 38,000 Atlantic salmon and 2 million shad (Stolte 1981). Combined state and federal annual salmon stocking in the Connecticut system now amounts to about 2 million fry and 400,000 smolts. Restoration targets on the Merrimack River are 5,000-7,000 salmon and 1 million shad at the river mouth (Technical Committee for Anadromous Fishery Management of the Merrimack River 1985). Federal and state hatcheries currently stock about 500,000 to 1 million salmon fry and 150,000 smolts for this effort. Fish and Wildlife Service hatcheries in Maine contribute about 800,000 smolts annually to that state's salmon restoration effort. Additionally, Maine has recently initiated shad restoration on the Androscoggin, Kennebec, and Penobscot rivers with combined goals of over 3 million fish (Atlantic States Marine Fisheries Commission 1985).

An initial agreement regarding shad restoration on the Susquehanna River was reached in 1970 between the Fish and Wildlife Service, fishery agencies from Maryland, Pennsylvania, and New York, and private utility companies operating hydroelectric dams on the river. This led to construction of an interim trap and lift at Conowingo Dam, (the first dam on the river), construction of a shad hatchery on a major tributary, and stocking of prespawned adults and cultured fry and fingerlings above all dams. Financial assistance from the utilities is guaranteed through 1994 for the demonstration program and current annual stocking levels exceed 5,000 adults and 15 million shad fry. The goal on the Susquehanna is to restore self-sustaining runs of 2,000,000 shad above all dams once permanent fish passage is developed (Susquehanna River Anadromous Fish Restoration Committee 1979).

The Pawcatuck River in Rhode Island is receiving 50,000 juvenile salmon annually along with transplanted, pre-spawned adult shad. Additionally, 10 other river systems from New Hampshire to Virginia are undergoing shad restoration attempts. Fish passage is still required at 86 dams in the northeast and mid-Atlantic to provide access to all suitable historically utilized shad spawning habitat (St. Pierre and Howey, in press).

ROLE OF HYDROPOWER IN THE NORTHEAST

The Northeast has a long history of hydropower, beginning with early settlement of the region. Villages and towns were often located along rivers to take advantage of waterpower to run mills and factories. As a result, hundreds of dams were built throughout the region.

The earliest facilities generated mechanical power that was used directly by various local industries. In the early 1900's hydroelectric generating facilities were added, not only making it possible to use electricity in local manufacturing, but also allowing the transmission of energy elsewhere. While hydropower was once the principal energy source in the region, it had become more economical by the mid-20th century to produce power from oil and coal fired steam generation. The use of fossil fuels together with the emergence of nuclear power led to the retirement of many hydroelectric facilities.

In the 1970's the price of imported oil abruptly increased, and what had been previously considered as an almost unlimited supply could no longer be guaranteed. It also became increasingly more expensive and time consuming to build new large-scale fossil fuel or nuclear power plants. Consequently, public awareness and the energy industry turned to alternate sources, including hydropower.

Congress enacted legislation in the late 1970's to encourage the development of small scale hydropower. The passage of the Public Utility Regulatory Policies Act in 1978 (PURPA), 16 U.S.C. 796(17)-(22), 824a-3, 2601-2645, 2701-2708, resulted in high rates being paid by utilities for power generated at newly developed sites. In addition, regulations governing licensing of small hydroprojects were simplified. Finally, a series of financial incentives were established that resulted in a rush of applications for federal approval of small-scale hydropower projects.

In the early 1980's there was a tremendous increase in the number of proposed hydropower projects in the Northeast, because of PURPA and other incentives. However, even if all feasible sites were to be developed, hydroelectric power would still make only a minor contribution to the region's energy needs. In New England roughly 5 percent of electricity is generated by hydropower (New England River Basins Commission 1981). Even with extensive new hydropower development, the contribution of this energy source would be expected to double at most.

The flurry of hydropower activity has diminished recently, due to a number of factors. Earlier tax incentives have been reduced. High interest rates and the temporarily low cost of oil have also caused a drop in development activity. However, a change in any of these conditions could result in a new surge. Furthermore, many of the older projects will need to be relicensed in the next 10 to 15 yrs. (Most hydroelectric projects are under the jurisdiction of the Federal Energy Regulatory Commission). Therefore, hydropower will continue to generate interest by developers and by regulatory and reviewing agencies.

IMPACTS OF HYDROPOWER ON ANADROMOUS FISH

Hydropower development can adversely impact anadromous fish in a variety of ways, including:

1. Blockage or delay of upstream and downstream migrations;
2. Stress, injury, or mortality in turbines or at spillways;
3. Loss of downstream habitat or delays in upstream migration due to reduced or fluctuating flow discharges at the powerhouse;
4. Loss of upstream habitat due to the creation or extension of an impoundment;
5. Predation on migrating fish in impoundments and below dams; and
6. Degraded water quality in impoundments and below dams (Rochester *et al.* 1984).

Anadromous fish restoration programs in the Northeast must deal repeatedly with these impacts, for the river systems there typically contain numerous hydropower dams. In many rivers, anadromous fish restoration will likely be infeasible due to cumulative losses of habitat and the unlikely passage of adequate numbers of fish at multiple dams (U.S. Fish and Wildlife Service 1984).

Blockage of Migrations

Construction of dams eliminated anadromous fish runs in many rivers in the Northeast. Dams usually prevent upstream migrating fish from reaching desired spawning areas, although species such as Atlantic salmon are known for their jumping ability, and may be able to leap over low barriers (Reiser and Peacock 1985). Addition of fish passage facilities at dams can partially mitigate the blockage of fish movement. However, use of these facilities is less than 100 percent, and typically there is a delay in migration at each site. For species such as American shad that are faced with a short, temperature-limited migratory period, delays in passage may result in a shortened upstream run (Knight and Greenwood 1981).

Impoundments created by dams can also delay downstream migrations of anadromous fish, leading to increased losses due to predation. For Atlantic salmon, delays in downstream movements much beyond normal spring runoff can result in higher mortality at the turbines, especially if there are no downstream fish passage facilities. Increasing water temperatures and decreasing streamflow following spring runoff can also cause salmon to undergo behavioral and physiological changes and "desmoltify," thus reverting to an earlier life history stage (parr) and remaining in fresh water (Saunders 1960).

Design of fish passage facilities is intended to maximize passage efficiency and to minimize delay, stress and injury to migrants. Several types of conventional fishways are currently used in the Northeast, including the pool and weir/orifice, vertical slot, Denil and Alaska steeppass. Fish lifts (elevators) have been constructed on the Connecticut, Merrimack, and Susquehanna rivers, and are proposed in other drainages. Selection of a particular type and dimensions of the fish passage facilities depends on the characteristics of the hydroproject and on the species and numbers of fish to be passed. Typically, the Denil and steeppass fishways are used in smaller streams for river herring and salmon where river levels are relatively constant during the migration period. Vertical slot, pool and weir/orifice, and fish lifts are better suited for larger rivers where impoundment levels fluctuate, and where many fish must be quickly passed upstream.

Fish trap-and-haul programs are an alternative means of passing upstream migrating fish around multiple dams (Rochester *et al.* 1984). This is being done in the Northeast using specialized transport systems on major rivers such as the Susquehanna, Connecticut, Merrimack, Androscoggin and Penobscot.

Typical operations include trapping fish at the lowest dam and trucking them to upstream spawning areas. The objective is to increase natural production and out-migration, so that upstream runs will build more rapidly. However, trap-and-haul operations are labor intensive and can subject fish to excessive stress through handling. Consequently, these programs are usually carried out on an interim basis until actual upstream passage facilities can be built at all intermediate barriers.

Loss at Turbines and Spillways

Immature anadromous fish must migrate from freshwater to the sea where they remain before returning to spawn. In addition, shad and Atlantic salmon in the northeast can spawn in successive years, so that downstream migrants typically include post-spawning adults along with the juveniles. Depending on the species and life stage, downstream migrations can occur from the spring through fall. Where such migrations involve passing one or more hydroprojects, turbine-related mortalities may severely reduce the numbers of downstream migrants (Turbak *et al.* 1981).

Fish losses at turbines are a result of acute mortality during passage and delayed mortality due to stress, injuries and predation. The mortality attributable to passage through turbines depends on a variety of factors. Studies have shown losses ranging from 0 to 100 percent (Rochester et al. 1984). Turbine design and efficiency have a major influence on the amount of mortality, as does the species and size of the downstream migrants (Rochester et al. 1984, Ruggles 1985). Although studies are being conducted in the Northeast to assess turbine-related losses, additional information is needed to gain a better understanding of the exact causes of mortality and to assess how great an impact such losses will have on restoration programs in the region.

Downstream migrating fish face a greater chance of suffering turbine-related damage if there is little or no spillage at the dam. This is especially true for juvenile shad and river herring, as they migrate during the summer and fall when river flows are low. In such cases they are faced with delaying their migration or going through the turbines.

Losses of downstream migrants can be reduced by installing screens or louvers in front of the intake structure along with a sluiceway or conduit for bypassing fish over or around the dam. Adequate screening can be achieved at many projects in the region by simply designing the intake trashrack so that the bar spacing is narrow enough to prevent fish from passing through, and by placing it at a proper angle to the flow so that fish are guided to the bypass structures. Screening is not practical at some of the older hydroprojects located on large rivers in the Northeast due to design constraints and cost. In such cases the use of operational measures such as controlled spills is an alternative means of passing large numbers of downstream migrants. Injuries and mortalities can also be inflicted on upstream migrating fish that are attracted into the draft tubes at the powerhouse and come in contact with the turbine blades. Installation of screens or racks in the tailrace can effectively prevent turbine-related impacts to upstream migrating fish.

Reduced or Fluctuating Flow Releases

Hydroelectric projects are usually designed to maximize operating head and water flow (Rochester, *et al.* 1984). To achieve this the project often includes a diversion of flow through a penstock or canal to a powerhouse located downstream. The bypassed natural river channel below the dam can be from less than 33 m (100 ft) to several km in length. Unless adequate instream flow releases are maintained habitat quantity or quality is reduced. This can have an adverse impact on upstream migrating fish that are attracted to these areas below dams and to juvenile anadromous species utilizing the bypassed reaches as rearing habitat.

Natural river flows can also be manipulated by hydroprojects, thus having an effect on anadromous fish populations. Most projects operate in a run-of-river mode where inflows are continuously released from the powerhouse. Some store and release river flows seasonally on a regular basis (cycling mode) or in response to peak electrical demand (peaking project). Where natural flows are interrupted at hydroprojects, downstream habitat may be degraded and fish migrations may be interrupted.

Protection of resident and migrating fish can be achieved through maintenance of adequate instream flow releases below dams and powerhouses. A variety of methods are available for determining instream flow needs (Rochester et al. 1984, Loar and Sale 1981). In New England, U.S. Geological Survey stream flow records are heavily utilized in formulating recommendations for instream flows (U.S. Fish and Wildlife Service 1981), based on the assumption that aquatic organisms have adapted to natural flow conditions, and that they will be maintained as long as such flows are provided.

Establishing an adequate minimum flow does not entirely mitigate the effects of fluctuating discharges below cycling or peaking projects. Sudden changes from the storing to generating mode can result in scouring and dislocation of organisms (New England River Basins Commission 1981). When flows are abruptly reduced, fish and other aquatic life may be stranded or disoriented in the case of upstream migrants. Mitigation of these impacts can be achieved by gradual transitions between minimum and maximum project discharges, often referred to as stepping or ramping (Rochester *et al.* 1984).

Habitat Losses Due to Impoundments

Most of the proposed hydroelectric projects in the Northeast involve dams with existing impoundments. Consequently, addition of new hydroelectric generating facilities at these sites often does not result in the loss of free-flowing riverine habitat. It is only when the project includes building a new dam, re-establishing a breached dam or adding flashboards to enlarge an impoundment that there may be a loss of habitat for anadromous fish. Occasionally the loss of habitat for one species results in increased spawning and nursery areas for another anadromous species. American shad, for example, require free-flowing conditions, whereas the alewife (*Alosa pseudoharengus*) spawns in lakes and ponds. Additional impounded water may favor alewives at the expense of shad.

It is usually difficult or impractical to replace habitat losses that result from impoundments. Occasionally it is possible to offset losses by increasing productivity in another stream segment. This involves channel modifications through use of structures or the addition of spawning materials. However, river systems undergoing anadromous fish restoration in the Northeast are already heavily impounded, and opportunities for improving remaining riverine segments are limited. The only means to replace lost habitat value may be to remove an existing dam or open a formerly inaccessible river segment or tributary. Both options may not be feasible, however, given the high demand for potential hydropower sites.

Predation

Hydropower impoundments and tailwaters often contain populations of piscivorous fish such as pike, perch, and bass. Fish eating birds, including mergansers, bald eagles, ospreys, and cormorants, can also prey heavily on anadromous fish. Juvenile anadromous fish are susceptible to predation as they migrate downstream or temporarily reside in impoundments. Physical damage or disorientation of outmigrant juveniles as they pass through turbines at hydroelectric dams may also induce high losses to predation in project tailwaters.

Pollution

Water pollution was a major factor, along with the construction of dams, for the demise of many anadromous fish runs in the Northeast (U.S. Fish and Wildlife Service 1984). Federal legislation has resulted in the regulation and treatment of much of the region's municipal and industrial discharges. Impoundments, however, continue to act as nutrient sinks for treated organic materials that are discharged into the rivers. In some instances, dissolved oxygen (DO) deficiencies develop behind dams due to high biochemical oxygen demand (BOD). Unless the deoxygenated releases from the dam are reaerated, fish populations downstream can be adversely impacted.

Spillage over dams and natural flow through rapids can restore oxygen levels to saturation levels. Where spillage is eliminated by diverting all flows through turbines and where multiple dams eliminate free-flowing river segments, dissolved oxygen levels remain depressed. Oxygenation can be restored by intentionally introducing air into the water flowing through the turbines, or by simply allowing some of the flow to spill over the dam.

Cumulative Impacts

Restoration of anadromous fish to large river systems in the Northeast requires successful passage at multiple dams. Even with properly designed upstream and downstream fish passage facilities, there will be cumulative reductions in migrating fish. In some instances successive losses at multiple barriers will lower the utilization of available spawning and rearing habitat to the point where restoration may not succeed.

Atlantic salmon restoration is currently proposed by the Fish and Wildlife Service for 8 rivers in New England (U.S. Fish and Wildlife Service 1984). However, cumulative impacts of existing and proposed hydropower projects may ultimately determine where restoration will be feasible. For example, there are 7 operating projects that salmon must pass in the Merrimack River Basin to reach headwater spawning areas. Although fish passage facilities are to be installed at each project, computer simulated population projections suggest that the existing dams would significantly reduce the numbers of returning adults, and that further hydropower development could jeopardize the chances for successful salmon restoration in the Merrimack (Russell 1986).

Lack of adequate fish passage facilities increases the cumulative impact on migrating fish. Restoration of anadromous fish usually involves retrofitting fish passage facilities at existing hydropower developments. Emphasis has typically been placed on building upstream fish passage facilities at these dams without acting concurrently to protect downstream migrants. This is the case on the Connecticut River where downstream losses may be limiting the stability of the shad population despite the construction of new upstream fish passage facilities at upriver dams (Crecco 1983).

HYDROPOWER AND FISH RESTORATION—CONFLICTING MANDATES

Expansion of hydropower has been legislated by Congress, and it enjoys popular support, along with other alternate energy sources. If viewed simply as retrofitting numerous idle mill dams with hydroelectric generating facilities, increased hydropower development does not lead to a perception of significant environmental harm.

Restoration of anadromous fish is also publicly mandated, and is often perceived as the culmination of efforts to rectify past environmental damage such as water pollution. Many hydroelectric projects in the Northeast cause little impact to anadromous fish, due to their location and the fact that they typically involve existing dams and do not call for modifying present impoundment levels or river flows. However, in some river basins, increased hydropower development is in conflict with the goal of restoring self-sustaining runs of anadromous fish (Cavan 1985).

Conflicts between hydropower development and anadromous fish restoration could be avoided or minimized with adequate river basin planning. Comprehensive plans are a requirement of the Federal Power Act (16 U.S.C. 791a-825r), the legislation authorizing federal regulation of private hydroelectric development. Before the Federal Energy Regulatory Commission (FERC) can license a project, they must determine that the proposed development is best adapted to a comprehensive plan for improving the waterway where it is located for various beneficial public uses. Although the FERC has not yet developed comprehensive river basin plans for rivers in the Northeast, there is valuable information in a variety of fish restoration documents (Maine Atlantic Sea-Run Salmon Commission 1984, Stolte 1982, Technical Committee for Anadromous Fishery Management of the Merrimack River 1985, Susquehanna River Anadromous Fish Restoration Committee 1979), and regional (New England River Basins Commission 1981) and state (Maine Office of Energy Resources 1982) hydropower development reports. What is required is the integration of available data so that new hydroelectric projects can be directed to those stream segments having minimal competing uses.

Hydropower will continue to be an important use of rivers in the Northeast. However, unless development is adequately planned and coordinated to allow for competing uses such as anadromous fish restoration, conflicts will be difficult to avoid. Because the public demands that all its interests in waterways be protected, hydropower regulators will be faced with increasingly tough decisions.

LITERATURE CITED

Atlantic States Marine Fisheries Commission. 1985. Fishery Management Plan for the anadromous alosid stocks of the eastern United States: American shad, hickory shad, alewife, and blueback herring. Phase II in Interstate Management Planning for Migratory Alosids of the Atlantic coast. Washington, D.C. XVIII + 347 pp.

Boreman, J. 1981. American shad stocks along the Atlantic Coast. Lab. Ref. Doc. No. 81-40. Nat. Mar. Fish. Serv., Woods Hole, MA. 14 pp.

Cavan, A. 1985. Mandates in conflict: Is reconciliation possible? Hydro Review 4(2): 55-62.

Crecco, V. 1983. Potential problems associated with turbine-induced mortality. Unpubl. report prepared for Technical Committee for Anadromous Fishery Management of the Connecticut River. 3 pp.

Knight, A.E., and J.C. Greenwood. 1981. Special report—habitat criteria for Atlantic salmon and American shad. Compiled for Policy and Technical Committees for Anadromous Fishery Management of the Merrimack River. U.S. Fish Wildl. Serv., Laconia, NH.

Loar, J.M., and M.J. Sale. 1981. Analysis of environmental issues related to small-scale hydroelectric development. V. Instream flow needs for fishery resources. ORNL/TM-7861. Oak Ridge National Laboratory, Environmental Science Div., Publ. 1829. Oak Ridge, TN. 123 pp.

Maine Atlantic Sea-Run Salmon Commission. 1984. Management of Atlantic salmon in the State of Maine—A strategic plan. Augusta, ME. III-3-10.

Maine Office of Energy Resources. 1982. Comprehensive hydropower plan. Submitted as a proposed comprehensive plan under section 10(a) of the Federal Power Act. Vol. I. Augusta, ME. 68 pp.

Mansueti, R.J., and H. Kolb. 1953. A historical review of the shad fisheries of North America. Ches. Biol. Lab. Publ. 97, Solomons, MD. 293 pp.

New England River Basins Commission. 1981. Water, Watts and Wilds: Hydropower and Competing uses in New England. Final Report of the New England River Basins Commission's Hydropower Expansion Study, Boston, MA. 136 pp.

Reiser, D.W., and R.T. Peacock. 1985. A technique for assessing upstream fish passage problems at small-scale hydropower developments. Pages 423-432 in F.W. Olson, R.G. White and R.H. Hamre (Eds.) *Proceedings of the Symposium on Small Hydropower and Fisheries.* The American Fisheries Society, Bethesda, MD. 497 pp.

Rochester, H., Jr., T. Lloyd, and M. Farr. 1984. Physical impacts of small-scale hydroelectric facilities and their effects on fish and wildlife. U.S. Fish Wildl. Serv. FWS/OBS-84/19. 191 pp.

Ruggles, C.P. 1985. Can injury be minimized through turbine design? Hydro Review. 4(4): 70-76.

Robbins, T.W., and D.C. Watts. 1982. A history of shad culture with emphasis on federal hatchery systems. Pages 1-7 in R.G. Howey (Ed.) *Proceedings of 1981 American shad Workshop—Culture, Transportation and Marking.* U.S. Fish and Wildl. Serv. Info. Leaflet 82-01. Lamar, Pa. 24 pp.

Russell, G.W. 1986. Conflict between hydropower development and Atlantic salmon restoration. Paper given at 11th Ann. National Conference on Rivers, April 4-6, 1986, Washington, D.C.

Saunders, J.S. 1960. The effect of impoundment on the population and movement of Atlantic salmon in Elerslie Brook, Prince Edward Island. J. Fish. Res. Bd. Canada 17(4): 453-473.

Stolte, L.W. 1981. *The Forgotten Salmon of the Merrimack.* Dept. Int., NE Region. U.S. Gov. Printing Office, Wash, D.C. 214 pp.

Stolte, L.W. 1982. A strategic plan for the restoration of Atlantic salmon to the Connecticut River Basin. Prepared for Policy and Technical Committees for Fishery Management of the Connecticut River. U.S. Fish and Wildl. Serv., Concord, NH. 56 pp. & App.

St. Pierre, R.A., and R.G. Howey. (in press). Restoration of American shad in Eastern North America. J. Washington Acad. Sci., Wash., D.C.

Susquehanna River Anadromous Fish Restoration Committee. 1979. A strategic plan for the restoration of diadromous fishes to the Susquehanna River Basin. U.S. Fish and Wildl. Serv., Harrisburg, PA.

Technical Committee for Anadromous Fishery Management of the Merrimack River. 1985. Restoration of Atlantic salmon to the Merrimack River 1985 through 1999. U.S. Fish and Wildl. Serv., Concord, NH. 24 pp. & Apps.

Turbak, S.C., D.R. Reichle, and C.R. Shriner. 1981. Analysis of environmental issues related to small-scale hydroelectric development. IV. Fish mortality resulting from turbine passage. ORNL/TM-7521. Oak Ridge National Laboratory, Environmental Science Div., Publ. 1597. Oak Ridge, TN. 112 pp.

U.S. Fish and Wildlife Service. 1981. Interim regional policy for New England stream flow recommendations. Fish Wildl. Serv., Newton Corner, MA. 3 pp.

U.S. Fish and Wildlife Service, 1984. Draft Environmental Impact Statement for the Restoration of Atlantic salmon to New England rivers. U.S. Fish Wildl. Serv., Newton Corner, MA. 105 pp. & App.

Environmental Consequences of Energy Production: Problems and Prospects. Edited by S. K. Majumdar, F. J. Brenner and E. W. Miller. © 1987, The Pennsylvania Academy of Science.

Chapter Twenty-Three

POTENTIAL IMPACTS OF HYDRO AND TIDAL POWER DEVELOPMENTS ON THE ECOLOGY OF BAYS AND ESTUARIES

GRAHAM R. DABORN

Acadia Centre for Estuarine Research

and

Department of Biology

Acadia University

Wolfville, Nova Scotia B0P 1X0

INTRODUCTION

As shown in the previous chapter, the downstream impact of inland hydro developments has been the subject of numerous investigations in recent decades. Rarely, however, has consideration been given to the potential effects such developments have on estuaries and coastal bays near the mouth of the river, perhaps because it has usually been assumed that these receiving waters were too remote from the site of the development. The events following closure of the Aswan Dam in 1964 clearly demonstrated that downstream effects may not only be extensive in space, but have profound effects on the complex and productive ecosystems of the coastal zone (Ben-Tuvia 1983).

The characteristic features of estuaries depend to a large extent upon the interaction between river runoff patterns and tidal or wind-induced movements of the sea (Daborn 1986). Consequently, major hydro developments that influence the seasonality and/or volume of freshwater input to the coastal zone may have important effects that ramify throughout the coastal ecosystem. Furthermore, even small, localized hydro developments on tributary rivers may have observable impacts upon estuaries because their effects are often cumulative (Daborn and Dadswell (in press)): several small projects may have a total effect similar to a single large development.

Failure to recognize the potential long distance effects of hydro developments is only part of the problem. Another important impediment to adequate prediction of the impacts of man-made changes on estuaries and coastal waters is the very complexity of these systems. Each system is unique. Developing sufficient understanding of the properties of the system to permit adequate modeling or prediction of impacts requires a broad range of basic information on hydrology, biogeochemical cycles, biophysical and interspecific interactions, etc. Often these have changed considerably over time for natural or anthropogenic reasons. Nonetheless, concerted, multidisciplinary research efforts may yield sufficient knowledge of a system within a few years to allow such predictions to be made with some confidence. This has been demonstrated, for example, by major projects in the Dutch Delta (Saeijs 1982), and in relation to tidal power proposals in the Severn Estuary (Anon 1981) and the Bay of Fundy (Gordon and Dadswell 1984).

In many respects, tidal power proposals differ from those of more traditional hydro power developments. Located in the zone of tidal influence, the power to be captured comes largely from the ebb and flow of coastal waters, rather than from retention of large volumes of river water in a reservoir. Turbines must be able to operate efficiently under relatively low heads (< 5 m). The time of generation is usually set by the tidal flux, and thus changes from day to day by about 50 min, in contrast to river hydro, where generation may be matched to regional peak demand. In other respects, however, the two types of development are similar. Both modify the timing of natural flow patterns, with wide-ranging effects on sedimentation and erosion processes, on productivity, and on the fate and distribution of benthic animals. Both may intersect the migratory routes of fish, although the estuarine location of a tidal power dam makes this feature more prevalent.

In this chapter I shall use recent experiences gained in relation to large scale tidal power proposals to exemplify some of the major influences that tidal and hydro power developments may have on coastal and estuarine waters.

TIDAL POWER PROPOSALS

Tidal power is an old technology: tidal mills were in operation in 11th century Europe and probably much earlier, and some of these continued to operate until the mid-19th century (Lawton 1972, Charlier 1983). Modern proposals, however, are typically of much greater scale, and involve production of electrical rather than mechanical energy. At the present time, modern tidal power plants of varying size operate in four countries, but there are numerous other sites around the world where developments have been or are being considered (Table 1).

All existing plants and those that have been seriously considered, are of the same conventional design (Fig. 1). This involves a concrete or rockfill dam, containing turbines and sluice gates, that extends completely across an estuary with relatively large tidal range (> 5 M). The dam is used to retain water in a large

TABLE 1

Existing and Proposed Tidal Power Plants

a) *Existing plants*

Country	Site	Completion Date	Output (MW)
France	La Rance	1966	240
U.S.S.R.	Kislaya Bay	1969	0.4
People's Republic of China	> 120	Since 1959	7900 (total)
Canada	Annapolis Royal	1984	20

b) *Proposed developments*

Country	Site	Country	Site
Argentina	San Jose Gulf	Korea	Seoul R.
	San Julian Gulf		Inchon Bay
	Deseedo Estuary		Garolim Bay
	Gallegos Estuary		
	Santo Cruz Estuary	United Kingdom	Severn R.
			Solway Firth
Australia	Secure Bay		Strangford Lough
	Walcott Inlet		Mersey Estuary
	St. Georges Basin		
	George Water	U.S.A.	Cobscook Bay, Maine
			Friar Roads, Maine
Canada	Minas Basin		Half Moon Cove, Maine
	Cumberland Basin		
	Shepody Bay		Cook Inlet, Alaska
	Ungava Bay		
		U.S.S.R.	White Sea
France	Chansey Is.		Penzuina
	Mount St. Michael		Gizhiga
India	Gulf of Kutch		
	Ganges R.		
	Gulf of Cambay		

upstream headpond. This is filled through sluice gates on the rising tide, until sufficient head develops between the headpond and falling sea levels to operate the turbine efficiently. At this point the turbine gates are opened and generation proceeds for about 6½ hrs before rising sea levels once again eliminate the head of water. In most designs less than half of the water in the headpond can be passed during a generation cycle. Consequently, the design leads to retention of a large volume of sea water above the dam (Fig. 1), and a consequent conversion of previously intertidal habitats to subtidal ones. Alternative designs involving vertical axis turbines have been suggested, but these at present are only in early stages of development.

A) SECTION

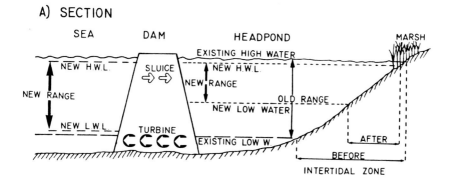

B) PLAN

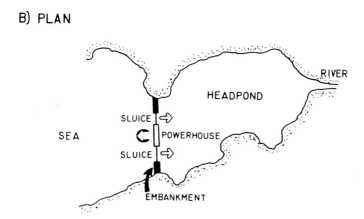

FIGURE 1. Conventional tidal power development.
HWL—high water level   LWL—low water level

In sites with very large tidal range, such as the Bay of Fundy (< 16 m), Severn Estuary (< 12 m), and La Rance Estuary (< 13 m), it is possible to generate power both on rising and falling tides. At La Rance, variable-pitch bulb turbines are used in this "double effect" mode, but economic considerations generally favour "single effect" ebb tide generation. This is partly because the cost of unidirectional turbines is much less than bidirectional ones, and because the total power output may be almost the same since the headpond can be raised to higher levels by sluicing if only ebb generation is considered. Construction costs vary considerably from site to site, and power yielded depends both upon the tidal range and the volume of water passing the point of construction. Consequently cost-benefit analyses produce extremely varied results for

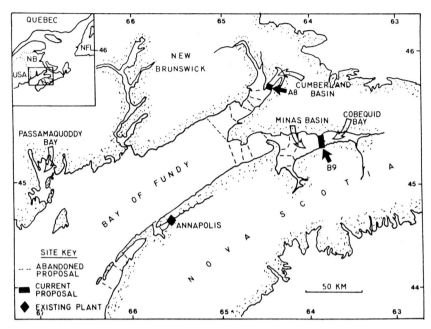

FIGURE 2. Proposed tidal power developments in the Bay of Fundy. (N.B. Some smaller dam sites omitted for the sake of clarity.)

different sites. With the conventional design, the complete dam must be constructed before power can be generated, resulting in a heavy 'front-end' financial loading such as occurs with other hydro developments.

In both the Bay of Fundy (Fig. 2) and the Severn Estuary (Fig. 3), several sites and plans have been considered for their potential suitability for tidal power generation, but the number has generally been reduced to a few which exhibit the best cost-benefit relationships. Since each location is unique in the specific combination of oceanographic and biological properties, environmental effects vary considerably from site to site.

In the Bay of Fundy, proposals for tidal power development were first made in 1910, again in 1919, and in the 1950s with reference to Passamaquoddy and Cobscook Bays in the outer Bay of Fundy region (Gordon 1985). Physical oceanographic and fisheries investigations were stimulated by these proposals, leading to the first systematic ecological studies of these macrotidal estuaries. Eventually the projects were always abandoned for economic reasons (once after some initial construction had begun), although environmental concerns were raised in each case. Interest in new proposals for tidal power developments in the upper Bay of Fundy (notably Minas Basin, Cumberland Basin and Shepody Bay—Fig. 2) was reinforced by the rapid escalation of oil prices in the 1970s. In order to assess the probable environmental consequences, a multidisciplinary

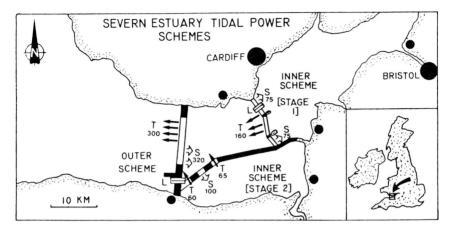

FIGURE 3. Proposed tidal power developments in the Severn Estuary. T—turbines, S—sluices, L—ship locks. Solid bars represent rockfill/concrete embankments.

research program was established involving government and university scientists with a broad mandate to examine all aspects of the Bay of Fundy. This program was coordinated by the Fundy Environmental Studies Committee (Atlantic Provinces Council on the Sciences). A similar multidisciplinary team was created for the Pre-feasibility Study of the Severn Barrage (Anon 1981).

ENVIRONMENTAL EFFECTS OF TIDAL POWER

Physical Effects

Construction of large tidal barriers involves major changes to patterns of water flow throughout the estuary and the coastal waters with which it is connected. Macrotidal estuaries such as Fundy and the Severn are tidally dominated systems: the water column tends to be completely mixed from top to bottom, and to carry high levels of suspended sediment that are maintained in dynamic equilibrium with intertidal and subtidal deposits. Construction of a barrier with sluice and turbine units will necessarily change patterns of water flow, and produce significant decreases in turbulent mixing over much of the new headpond. Where freshwater inputs are large, stratification of the water column should occur in peripheral regions of the headpond, although strong currents generated during sluicing may maintain the mixing and turbidity in central portions. Where stratification occurs, sediments will tend to settle and accumulate, allowing greater light penetration, but converting many firm or sandy substrates to muddy ones. Experience with causeways built across small estuaries in the Fundy system has shown that very large accumulations of fine silts can occur at rates too rapid to permit dewatering. The resulting wet sediments are treacherous and biologically unproductive for many years after the accumulation has begun (Daborn and Dadswell, in press, Amos 1984).

Seaward of the barrage, there will be local changes in the distribution and size of mudflats and sandflats as the pattern and strength of tidal currents will be modified. These are quantitatively difficult to predict without knowledge of the siting of turbine and sluice units.

The most far-reaching physical effects of barrage construction, however, are caused by changes in the resonant frequency of the estuary. Macrotidal estuaries derive their characteristic high tides in part from the near-coincidence of their natural oscillating period (a function of morphometry) with the forcing period of the ocean tides. The natural period of the Bay of Fundy—Gulf of Maine system is about 13 h: shortening the Bay by building a tidal barrage in the headwater region will bring this natural period even closer to the 12.4 h forcing period of the Atlantic tide. Present mathematical models predict that the Minas Basin barrage (B9) will cause a 10-15 percent increase (20-30 cm) in tidal range in the Gulf of Maine because of enhancement of resonance (Greenberg 1984). In effect, the Bay—Gulf system will be "tuned" more closely to the natural tidal frequency. Increases in tidal range will cause greater vertical mixing in shallow areas of the Gulf. This, in turn, will decrease sea surface temperature because of the increased upwelling of cool deeper water, and possibly increase the incidence and extent of coastal fog. It should also increase the recycling of important nutrients, particularly nitrogen and phosphorus, with obvious implications for biological productivity.

Interestingly, precisely the opposite effect is predicted for the Severn barrage (Heaps 1982). The natural period of the Severn Estuary—Bristol Channel system is about 11 h—thus, shortening the system with a tidal barrier should, if anything, decrease the tidal range throughout because the system will be moved further away from resonance.

Biological Effects

Because macrotidal estuaries are physically dominated systems, all changes in physical properties—currents, sedimentation patterns, light penetration, etc.—are expected to have profound and far-reaching biological effects. As with most natural ecosystems, knowledge of biological processes is generally much less satisfactory than that of physical ones, and predictions of biological effects are consequently more generalized and less precise.

Despite the apparently harsh environmental conditions of strong tidal currents, high turbidity, and extreme ice action in winter, the inner Bay of Fundy harbours a biological community of surprising variety and productivity. The major source of primary production fueling the system appears to be the peripheral saltmarshes, particularly the lower *Spartina alterniflora* zone (Daborn

1984). This is supplemented by benthic diatom production on intertidal mudflats. Phytoplankton production is low except in outer regions where turbidity is less. The saltmarshes yield large quantities of leaf material that decompose on the mudflats or in the water column, and are broken up by epibenthic and benthic macroinvertebrates such as mysids, crangonid shrimps and amphipods. The fine fragments are consumed by small particulate feeders such as copepods and larvae of benthic molluscs and polychaetes (Daborn 1984), which in turn provide primary food resources for some resident and migratory fish. Diatom production on the mudflats is consumed by benthic deposit feeders, particularly the amphipod *Corophium volutator*, the clam *Macoma balthica* and some polychaetes. *Corophium* constitutes the major prey of some migratory shorebirds and many fish. Both of these food chains (Fig. 4) are relatively short, so that a great deal of the intertidal biological production is quickly conveyed to the birds and fish foraging in the estuary.

With construction of a tidal power barrage, both sources of primary production will be reduced in the headpond region: the area of intertidal zone will be almost halved, with a corresponding reduction in benthic diatom production, and, unless remedial measures are taken, existing saltmarshes will be overtaken by terrestrial vegetation because the new high water level behind the dam will be lower. Although the latter effect will be transient, saltmarsh development down the shoreline to a new equilibrium position will be a slow process, and therefore the major contribution of the marsh will be reduced for at least a decade after completion of the dam.

Higher production by phytoplankton in the clearer, nutrient-rich waters of the headpond will compensate to some extent for the decline in saltmarsh and benthic diatom production. The shift toward greater pelagic primary production should favour food chains involving planktivorous organisms (e.g., herring, menhaden, cnidarians) and benthic suspension feeders such as mussels (Daborn et al. 1984). Similar predictions have been made on the basis of an ecosystem simulation model (GEMBASE) for the Severn Estuary tidal power proposal (Radford 1981).

Very different concerns apply to the seaward side of the barrage. Because of changed current patterns, the mud- and sandflats will become redistributed, requiring some time (e.g., 10-20 yrs) before they have stabilized and harbour the benthic communities typical elsewhere in the system. There is no anticipation yet that the relative proportion of muddy and sandy environments will change: hence no long-term change in benthic productivity due to sediment type is expected.

This conclusion is of great importance. During the last few years it has become evident that the headwaters of the Fundy system are of immense importance for migratory birds and fish. From mid-July through September the mudflats of the upper Bay are visited by vast numbers of small shorebirds, particularly sandpipers and plovers, that are on their fall migration from arctic

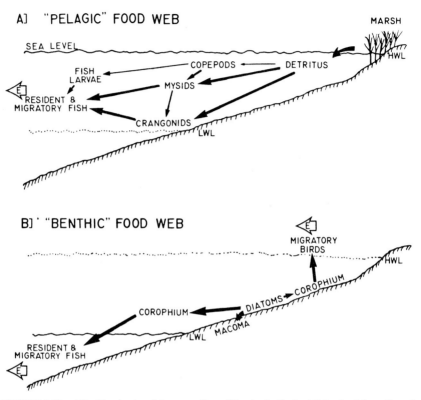

FIGURE 4. Simplified food webs of the upper Bay of Fundy. A—"pelagic" food web based largely on saltmarsh detritus and functioning in the water column; B—"benthic" food web based largely on benthic diatom production when intertidal zone is exposed; E—export of production from the upper Bay; LWL, HWL—low water level, high water level.

breeding grounds to wintering grounds in the Caribbean or South America. It has been estimated that more than half of all the birds nesting in the eastern and central Canadian arctic visit the Bay of Fundy on their way south. While there they forage extensively on the mudflats, each bird often consuming thousands of benthic animals per day. Studies have shown that during an average stay of 2-3 weeks, a semipalmated sandpiper may double its weight (from 16 to 32 g), by consuming hundreds of thousands of *Corophium* (Hicklin and Smith 1984). The fat accumulation is essential for a subsequent non-stop oversea migration of more than 4,000 km to the wintering grounds. Upon arrival, the shorebirds weigh once more what they weighed on first arrival at the upper end of the Bay. Clearly, the sojourn in the upper Bay is of critical importance.

Construction of a tidal power barrage, and the consequent reduction of intertidal zone and redistribution of mudflats, may well have significant effects on these transient consumers. Although estimates of abundance and feeding rate of shorebirds suggest that only a small fraction ($\mp$ 5%) of the available benthic secondary production is currently consumed, the long time for adjustment of sediments, and deleterious effects of overcrowding on remaining feeding grounds (Goss-Custard 1980), may well be damaging to migratory bird populations.

Furthermore, the currently abundant benthic resources are also utilized extensively by fish. During summer months, the waters of the upper Bay system are the feeding grounds for more than 50 species of fish. Some are year-round residents, but most species are part of stocks that move in and out of the Bay from the continental shelf. Included in this group are such commercially important species as herring, cod, halibut, haddock, hake and pollock. Many anadromous fish that spawn in rivers tributary to the upper Bay also feed there—including salmon, striped bass and river herrings. Finally, there is an important group of fish that migrate over large stretches of the Atlantic coast into the upper Bay, where, like the birds, they find abundant and important food resources (Dadswell et al. 1984). The best known example is the American shad, a commercially important species in the U.S., which spawns in east coast rivers from southern Florida to the Gulf of St. Lawrence. Recent research suggests that most, if not all, shad visit the upper Bay of Fundy at least once during the 3-4 years spent at sea before returning to home rivers to spawn (Dadswell et al. 1988). Other examples include striped bass, alewife, and blueback herring.

Direct impact on larger fish is to be expected where natural movements will bring them through the location of a tidal power barrage. During spawning runs, fish may be able to reach home rivers by moving through the sluice gates on the rising tide, but when travelling to sea, the only or most obvious open passage will probably be the turbines during generation. For feeding migrants such as the shad, repeated movements into and out of the headpond may entail successive passages through the turbines (Dadswell et al. 1984). Experiments with the tidal power station at Annapolis Royal have shown that mortalities of large fish such as the 50-cm-long shad may be high: c. 20 percent on each pass (Dadswell et al. 1986). Even juvenile fish suffer unexpectedly high mortalities (Stokesbury 1986).

Although the problems posed by migratory animals have generated much of the interest and controversy that surrounds Fundy tidal power proposals, the most far-reaching consequences may well be associated with changes in the tidal amplitude in the Gulf of Maine, 300-800 km away. As indicated before, the Minas Basin barrage is expected to increase tidal range in the Gulf by about 10-15 percent of the present range, although the magnitude varies somewhat according to the design and manner of operation of the barrage (DeWolfe 1986). Such a change could result in an increase of 15-33 percent in tidal mixing energy (Garrett et al. 1978). Usually, mixing increases pelagic productivity because it recirculates essential nutrients from deeper water to the surface where they may stimulate phytoplankton growth. This enhancement is expected to result in increased fish production in certain portions of the Gulf (Campbell 1986). An independent set of evidence for this has been found in correlations between fish catches in the Gulf and the natural 18.6 year nodal cycle of the tides: landings of some important commercial stocks, such as cod, halibut and haddock, peak following the years of maximum tidal range (DeWolfe and Daborn 1985). The effect is detectable even though the modulation of the tide is only about 3-4 percent—about the increase that would be caused by the proposed Cumberland Basin tidal power development, but considerably less than that for the Minas Basin one. Enhanced fish production in one of North America's most important fishing grounds might be an environmental benefit that partly compensates for other increases in fog, flooding, drainage problems or fish mortality associated with passage through turbines.

Large scale tidal power development thus represents a very mixed bag of environmental effects. What such proposals have shown, however, is the extent to which the river and estuary may be connected with a much larger coastal system. It clearly necessitates a very large scale holistic approach to evaluation of the potential for tidal power.

COMMON FEATURES OF TIDAL AND RIVER HYDRO DEVELOPMENTS

In spite of the highly site-specific nature of tidal and hydro developments, there are some features in common. Both change important characteristics of outflow from the river. Both may influence erosional processes above and below the point of construction, with related influences on sediment accumulation. As far as receiving systems are concerned, changed temporal patterns of water flow may have significant impacts on mixing characteristics of estuaries and coastal waters, with far-reaching biological consequences.

Large scale hydro dams often modify the annual pattern of outflow, particularly where patterns of precipitation are strongly seasonal. The downstream effects of such change on riverine fauna are well known; the impact of such change on estuaries and enclosed seas, however, has often not been considered (Ketchum 1983). A dramatic illustration of this impact has been provided by construction of the Aswan High Dam. Prior to closure in 1964, the majority of annual outflow (c 3×10^9 m^3) from the Nile occurred in September and October, delivering large quantities of phosphates, silicates and silt to the eastern Mediterranean, and stimulating spectacular blooms of phytoplankton (Halim 1960). High plankton densities supported major fisheries for sardines (*Sardinella* spp.) that totalled 18,000 t. in 1962. Following closure, the seasonal flood effect was largely eliminated: phytoplankton concentrations in October dropped to 10 percent of previous values, and the sardine fishery collapsed. It has not recovered (Dickie and Trites 1983).

Large scale modification of seasonal flow patterns from hydro developments that influence bays and coastal waters have occurred in many other river systems. In North America, obvious examples include the St. Lawrence (Dickie and Trites 1983), San Francisco Bay—San Joaquin—Sacramento system (Orlob and Davoren 1984) the Santee—Cooper (Kjerfve 1976), and the Colorado—Gulf of California system (Alvarez-Borrego 1983). Man-made reservoirs act as sediment traps, reducing the important outflow of silt and nutrients into the estuary. While this may reduce the problems of eutrophication experienced in other estuaries receiving excessive nutrient input, and diminish to some extent the deleterious input of heavy metals and pesticides, it also may reduce important nutrient supplies that fuel more nutrient—limited coastal ecosystems.

Changing the seasonal flow pattern from the river can have far reaching effects, not just in the river and estuary. It has been suggested that progressive development of several small reservoirs and one giant one (Manic 5 with a storage capacity of 140 km^3) in the St. Lawrence watershed is having profound effects on the entire ecosystem of North America from the Gulf of St. Lawrence to Cape Cod (Neu 1975). Outflow from the river influences the degree of vertical mixing in the estuary. This has been correlated with changes in the production of haddock, halibut and lobster stocks in the Gulf of St. Lawrence which show increased catches 8, 10 and 9 years (respectively) after years of peak outflow (Sutcliffe 1973). These lag times correspond approximately to the time for development to catchable size in each species.

As with tidal power developments, fluctuations in the amount of river outflow relative to the tidal prism may well produce profound biological changes within the system—influencing the extent of benthic-pelagic coupling, for example (Daborn 1986), or the relative importance of pelagic and benthic food chains. Such modifications, effective over large areas of the coastal zone, might well have unacceptably high costs in terms of decreased fishery productivity. When these are added to turbine-induced mortalities of anadromous or estuarine fish populations, the economic value of large scale hydro or tidal power developments may be viewed very differently.

In the final analysis, however, the justification for interference with natural estuarine and coastal ecosystems by power developments must be determined by recognition of all consequences—environmental, social and economic. In this context it is also essential to recognize the costs of not proceeding with the development. If, for example, the alternative to tidal power development in Eastern Canada is increased use of Nova Scotia coal by thermal generation, the 1-5 percent sulphur content of that coal will create much more negative environmental effects in the atmosphere without the compensatory advantages that might be seen in the tidal power effect on the Gulf of Maine fishery (Conley and Daborn 1983). Prudent development decisions demand a perspective that is both comprehensive and global.

LITERATURE CITED

Alvarez-Borrego, S. 1983. Gulf of California. pp. 427-449. *In* Ketchum, B.H. (Ed.). *Ecosystems of the World 26. Estuaries and Enclosed Seas.* Elsevier Sci. Publ. Co., N.Y. 500 p.

Amos, C.L. 1984. The sedimentation effect of tidal power development in the Minas Basin, Bay of Fundy. Can. Tech. Rept. Fish. Aquat. Sci. 1256: 385-402.

Anon. 1981. *Tidal Power from the Severn Estuary.* Energy Paper No. 46, Department of Energy, HMSO, London. 2 Vols.

Ben-Tuvia, A. 1983. The Mediterranean Sea B. Biological Aspects. pp. 219-238. *In* Ketchum, B.H. (Ed.). Ecosystems of the World 26. Estuaries and Enclosed Seas. Elsevier Sc. Publ., N.Y. 500 p.

Campbell, D.E. 1986. Possible effects of Fundy tidal power development on pelagic productivity of well-mixed waters on Georges Bank and in the Gulf of Maine. pp. 81-107 *In* G.R. Daborn (Ed.). *Effects of Changes in Sea Level and Tidal Range on the Gulf of Maine — Bay of Fundy System.* Acadia Centre for Estuarine Research Publ. 1, Wolfville, N.S. 133 p.

Charlier, R.H. 1982. *Tidal Energy.* Van Nostrand Reinhold Co., N.Y. 351 p.

Conley, M.W. and G.R. Daborn (Eds.). 1983. *Energy Options for Atlantic Canada.* Formac Publishing Co. and Acadia University Institute, N.S. 157 p.

Daborn, G.R. 1984. Zooplankton studies in the Bay of Fundy since 1976. Can. Tech. Rept. Fish. Aquat. Sci. 1256: 135-162.

Daborn, G.R., G.S. Brown and B. Scully. 1984. Impact of a tidal power station on zooplankton-fish interactions in Minas Basin. Can. Tech. Rept. Fish. Aquat. Sci. 1256: 527-533.

Daborn, G.R. 1985. Environmental implications of the Fundy Bay tidal power development. Water, Power & Dam Constr. 37: 15-19.

Daborn, G.R. 1986. Effects of tidal mixing on the plankton and benthos of estuarine regions of the Bay of Fundy. pp. 390-413. *In* Bowman, M.J., C.M. Yentsch and W.T. Pearson (Eds.). *Tidal Mixing and Plankton Dynamics.* Lecture Notes on Coastal and Estuarine Studies No. 17, Springer-Verlag, Berlin, pp. 390-413.

Daborn, G.R., and M.J. Dadswell. (In press) Natural and anthropogenic changes in the Bay of Fundy—Gulf of Maine—the International Symposium on Natural and Man-Made Hazards, Rimouski, Quebec.

Dadswell, M.J., R. Bradford, A.H. Leim, D.J. Scarratt, G.D. Melvin and R.G. Appy. 1984. A review of research on fishes and fisheries in the Bay of Fundy between 1976 and 1983 with particular reference to its upper reaches. Can. Tech. Rept. Fish. Aquat. Sci. 1256: 163-294.

Dadswell, M.J., R.A. Rulifson and G.R. Daborn. 1986. Potential impact of large-scale tidal power developments in the upper Bay of Fundy on fisheries resources of the Northwest Atlantic. Fisheries 11: 26-35.

Davoren, W.T. and J.E. Ayres. 1984. Past and pending decisions controlling San Francisco Bay and Delta. Wat. Sci. Tech. 16: 667-676.

DeWolfe, D.L. 1986. An update on the effects of tidal power development on the tidal regime of the Bay of Fundy and the Gulf of Maine. pp. 35-54. *In* G.R. Daborn (Ed.). *Effects of Changes in Sea Level and Tidal Range on the Gulf of Maine — Bay of Fundy System.* Acadia Centre for Estuarine Research Publ. 1, Wolfville, N.S. 133 p.

DeWolfe, D.L. and G.R. Daborn. 1985. Correlations of fish catches in the Gulf of Maine and northwest Atlantic with long term tidal cycles. Acadia University Institute, Wolfville. 28 p.

Dickie, L.M. and R.W. Trites. 1983. The Gulf of St. Lawrence. pp. 403-425. *In* Ketchum, B.H. (Ed.). *Ecosystems of the World 26. Estuaries and Enclosed Seas.* Elsevier Sci. Publ. Co. N.Y. 500 p.

Garrett, C.J.R., J.R. Keeley and D.A. Greenberg. 1978. Tidal mixing versus thermal stratification in the Bay of Fundy and Gulf of Maine. Atmos—Ocean 16: 403-423.

Gordon, D.C. Jr. and M.J. Dadswell (Eds.). 1984. *Update on the Marine Environmental Consequences of Tidal Power Development in the Upper Reaches of the Bay of Fundy.* Can. Tech. Rept. Fish. Aquat. Sci. 1256: 686 p.

Gordon, D.C. Jr. 1983. Integration of ecological and engineering aspects in planning large scale tidal power development in the Bay of Fundy, Canada. Wat. Sci. Tech. 16:281-295.

Goss-Custard, J.D. 1980. Competition for food and interference among waders. Ardea 68: 31-52.

Greenberg, D.A. 1984. The effects of tidal power development on the physical oceanography of the Bay of Fundy and Gulf of Maine. Can. Tech. Rept. Fish. Aquat. Sci. 1256: 349-369.

Halim, Y. 1960. Observations on the Nile bloom of phytoplankton in the Mediterranean. Rapp. Comm. Int. Mer. Medit. 26: 57-67.

Heaps, N.S. 1982. Prediction of tidal elevations: model studies for the Severn Barrage. pp. 51-58. *In Severn Barrage,* Institution of Civil Engineers, Thomas Telford Ltd., London, U.K. 240 pp.

Hicklin, P.W. and P.C. Smith. 1984. Studies of birds in the Bay of Fundy—a review. Can. Tech. Rept. Fish. Aquat. Sci. 1256: 295-319.

Ketchum, B.H. (Ed.) 1983. *Ecosystems of the World 26. Estuaries and Enclosed Seas.* Elsevier Sci. Publ. Co., N.Y. 500 pp.

Kjerfve, B. 1976. The Santee—Cooper: a study of estuarine manipulations. pp. 44-56. *In* Wiley, M. (Ed.). *Estuarine Processes* Vol. I. Academic Press, N.Y. 541 pp.

Lawton, F.L. 1972. Tidal power in the Bay of Fundy. pp. 1-104 *In* Gray, T.J. and O.K. Gashus. (Eds.) *Tidal Power.* Plenum Press, N.Y. 630 pp.

Neu, H.J.A. 1975. Runoff regulation for hydropower and its effects on the ocean environment. Can. J. Civ. Eng. 2: 583-591.

Orlob, G.T. 1976. Impact of upstream storage and diversions on salinity balance in estuaries. pp. 3-17. *In* Wiley, M. (Ed.). *Estuarine Processes* Vol. II. Academic Press, N.Y. 428 p.

Prouse, N.J., D.C. Gordon Jr., B.T. Hargrave, C.J. Bird, J. McLachlan, J.S.S. Lakshminarayana, J. Sita Devi and M.L.H. Thomas. 1984. Primary production: organic matter supply to ecosystems in the Bay of Fundy. Can. Tech. Rept. Fish. Aquat. Sci. 1256: 65-96.

Radford, P.J. 1981. Modelling the impact of a tidal power scheme on the Severn estuary ecosystem. pp. 235-247. *In* W.J. Mitsch, R.W. Bosserman and J.M. Klopateck (Eds.). *Energy and Ecological Modelling.* Elsevier Scientific Publishing Co., Amsterdam.

Saeijs, H.L.F. 1982. *Changing Estuaries.* Rijkswaterstaat Communications No. 32. Government Publishing Office, The Hague. 413 p.

Stokesbury, K. 1986. Downstream movements of juvenile alosids and preliminary studies of juvenile fish mortality associated with the Annapolis tidal power turbine. Acadia Centre for Estuarine Research Publ. 2, Wolfville, N.S. 28 p.

Sutcliffe, W.H. Jr. 1973. Correlations between seasonal discharge and local landings of American lobster (*Homarus americanus*) and Atlantic halibut (*Hippoglossus hippoglossus*) in the Gulf of St. Lawrence. J. Fish. Res. Bd. Can. 30: 856-859.

Environmental Consequences of Energy Production: Problems and Prospects. Edited by S. K. Majumdar, F. J. Brenner and E. W. Miller. © 1987, The Pennsylvania Academy of Science.

Chapter Twenty-Four
A "BATHTUB" PRIMER—LESSONS FROM WEST VALLEY, NEW YORK

ROBERT F. SCHMALZ
Department of Geosciences
536 Deike Bldg.
The Pennsylvania State University
University Park, PA 16802

Power consumption in the United States is on the order of one billion kilowatts (10-12 watts) today, and is likely to increase in response to the needs of a growing population desiring ever-higher living standards. There appears to be little prospect of meeting the demand with alternative energy sources before the turn of the century, and the economic and environmental costs of burning fossil fuels may soon become prohibitive. Nuclear energy affords an almost limitless source of power at a reasonable cost, but this energy resource cannot be fully utilized until we develop economically reasonable, technically feasible and socially acceptable means of disposing of the radioactive wastes produced by commercial power reactors. Future waste disposal management can benefit from a through understanding of the problems encountered in the "unsuccessful' disposal facilities of the past. The commercial low level radioactive waste disposal operation at West Valley, New York, is one of three such "failed' facilities in the United States.

THE WESTERN NEW YORK NUCLEAR SERVICE CENTER

History
The Western New York Nuclear Service Center was opened in 1963 on a 3,345 acre site near the village of West Valley, in Cattaraugus County, New York (Figure 1). It was operated by Nuclear Fuel Services, Inc. and comprised three distinct functional components: (i) a reactor fuel reprocessing plant, (ii) a high-level radioactive waste disposal facility licensed by the U.S. Atomic Energy Commission (now the Nuclear Regulatory Commission) and (iii) a low-level radioactive waste disposal facility licensed by the State of New York.*

The Center was intended to provide comprehensive spent fuel and waste management services for commercial power reactors, but the operator's license required that the low-level waste disposal facility be made available to all waste generators at reasonable cost. The West Valley site thus became the third public LLRW disposal facility in the United States (Table 1). The first commercial shipment of low-level radioactive waste was accepted for disposal at West Valley in November, 1963; the facility was closed voluntarily by the operator in March, 1975. During the intervening period, 2,360,000 cubic feet of waste were buried there, including "special nuclear material" (reactor fuel), industrial trash, animal carcasses, filter sludge, exchange resins, liquid scintillation "cocktails", absorbent and packing materials, protective clothing, plastics and glassware. The most radioactive waste was embedded in concrete, less hazardous material was packed in steel drums, or in wooden, plastic or cardboard crates, bags and boxes. Some of the problems which developed subsequently at West Valley are the direct result of the heterogeneity of the waste and waste packages.[1]

West Valley was a "shallow land burial" (SLB) site. Waste was buried in a series of trenches, nominally 34 feet wide, 20 feet deep and ranging from 560 to 800 feet in length (Figure 1, insert). Although the early records are incomplete, there seems to have been little effort to prepare the site (grading, etc.) before the excavation of the northern trenches was begun.[1] The excavation was carried out in piece-meal, the length of each trench segment being determined by the

TABLE 1
Commercial LLRW Disposal Sites in the United States

Site	Area (acres)	Volume of Waste through 1980	Year Opened	Current Status
Beatty, Nevada	79	3,182,000 cu.ft.	1962	Open
Maxey Flats, Kentucky	254	4,770,000 cu.ft.	1963	Closed, 1978
West Valley, New York	25	2,360,000 cu.ft.	1963	Closed, 1975
Richland, Washington	100	2,180,000 cu.ft.	1965	Open
Sheffield, Illinois	20	3,196,000 cu.ft.	1967	Filled to Capacity, 1978
Barnwell, South Carolina	259	11,425,000 cu.ft.	1971	Open

From: Clancy, et al., 1981[1]

---

*High-level reactor waste is composed mainly of spent reactor fuel, cladding hulls, and irradiated reactor components. It usually contains substantial quantities of long-lived alpha emitting isotopes (uranium, plutonium, etc.) but ordinarily makes up a small part of the waste volume from a commercial reactor. All high-level radioactive waste is under the jurisdiction of the federal government, and permanent repositories for high-level commercial waste are currently being developed. Low-level radioactive waste (LLRW) makes up the bulk of the waste generated by commercial power reactors. The principal contaminants are isotopes with half-lives of thirty years or less, primarily tritium, strontium-90, and cesium-137. The waste material is in the form of liquid-free "housekeeping" trash (protective clothing, absorbent materials, tools, etc.), filter sludge and exhausted exchange resins. Low level radioactive waste is also generated by a wide range of medical, research and industrial activities, but together these produce about one-tenth the volume of LLRW produced by commercial reactors in Pennsylvania (Table 2).

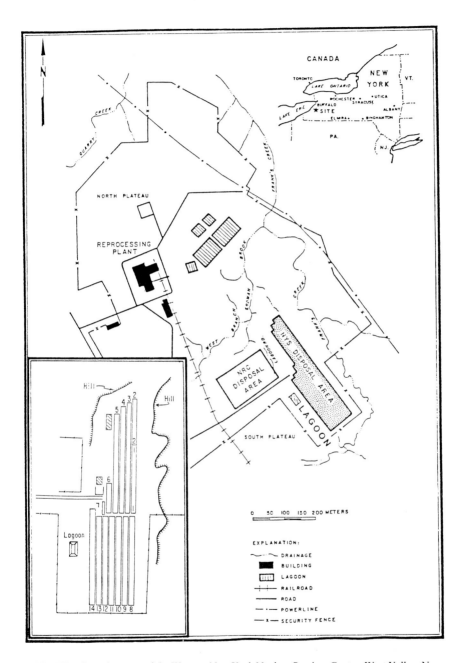

FIGURE 1. Location map of the Western New York Nuclear Services Center, West Valley, New York, showing layout of burial trenches in the state-licensed LLRW disposal area (inset). Modified from Clancy, et al., 1981[1]

TABLE 2

*Low-Level Radioactive Waste
Shipped for Disposal from Pennsylvania in 1984
(Pennsylvania Department of Environmental Resources)*

Generator	Volume (cubic feet)		Activity (curies)	
Nuclear power plants	156,008	(71.3%)	94,370	(99.46%)
Industrial fuel cycle	37,794	(17.3%)	455	(0.48%)
General industrial	5,887	(2.7%)	25.5	(0.03%)
Medical research	1,192	(0.5%)	2.1	
Medical treatment and diagnosis	14,130	(6.4%)	6.2	
Academic	3,461	(1.6%)	20.8	(0.02%)
Government	225	(0.1%)	1.7	(<0.01%)
TOTAL Pennsylvania	218,699		94,881.3	
TOTAL: United States	2,553,730		600,880	

NOTE: These figures change substantially from year to year, depending on the nature of waste shipments from generators. The values shown include several shipments of irradiated components from the decommissioning of the damaged reactor at TMI #2. In 1982, when no such shipments were made, nuclear power plant waste comprised 71.6% of the LLRW volume, but only 32.8% of the curies.

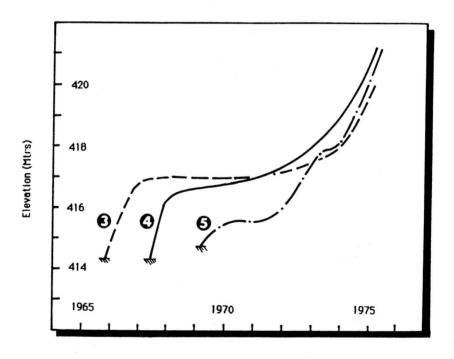

FIGURE 2. Recorded water levels in trenches #3, #4 and #5, 1965-1975. Redrawn from Prudic, 1986[10]

quantity of waste awaiting burial. The waste was stacked or simply dumped in the trench and covered with loosely packed soil. A permanent cap was constructed after the entire trench had been excavated and filled. In the northern area, a single "umbrella" cap 4 feet thick was constructed over trenches #1 through #4; all the subsequent trenches were capped individually. Sumps and standpipes were installed to allow water levels in the trenches to be monitored and to permit the collection of water samples. The operating license required, however, that state authorization be obtained before pumping water from the closed trenches for purposes other than analysis or testing.[1,2] Water removed from the partially filled trenches during the burial operations and leachate extracted (under permit) after closure was collected in lagoons and later treated to precipitate dissolved isotopes. The purified water was then diluted and discharged to a nearby stream.

Operating practices in the southern trenches (#8-14) were modified as a result of experience in the northern portion of the burial area. The southern portion of the site was graded, and most of the soil and weathered surficial materials were removed before the excavation of the burial trenches began! Probably as a result of these changes, at least in part, the southern trenches (exclusive of trench #14) have presented relatively few problems. For this reason, the remainder of this paper will focus on the difficulties encountered in the northern trenches, and in particular on trenches #3, #4 and #5.

During the six to nine months after the northern trenches were filled, covered and capped, water accumulated in them to depths of 5 to 10 feet. Water levels in the trenches generally remained stable until late 1971, when they began to rise again at a steady or accelerating rate (Figure 2).[3] The site operator repeatedly sought permission to pump the affected trenches and treat the leachate, but no action was taken on these requests, and in March, 1975, the accumulating water broke through the cap of trench #4. The necessary permit was issued immediately thereafter, and all the trenches have been pumped at intervals, as needed, since that time.

The rate of seepage from trench #4 during the March, 1975, incident has been estimated at 4 liters/day, and the total volume of contaminated leachate which escaped through the trench cap is though to be less than 300 liters (between 70-80 gallons)! All burial operations at West Valley were stopped as soon as the seepage was detected, and the site has not been operated as a commercial disposal facility since that time.

ENVIRONMENTAL CHARACTERISTICS AND SITE GEOLOGY

The West Valley site is located on a partially wooded terrace abutting the western slope of Buttermilk Creek valley, at an elevation of 1,390 feet. It is surrounded by a glaciated, rolling terrain of farmland and hardwood forest, approximately thirty-one miles south-southeast of the city of Buffalo. The area

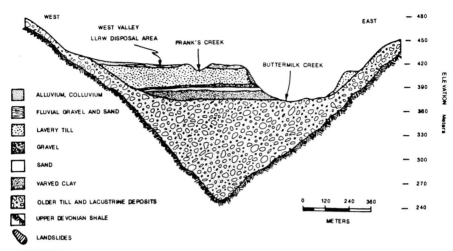

FIGURE 3. Geologic section across Buttermilk Creek Valley at the West Valley LLRW disposal site. Redrawn from WSDP-027, by permission

is characterized by a humid temperate climate modified slightly by the stabilizing effects of Lake Erie, 25 miles to the west. Mean monthly temperatures range from 21 to 67 degrees F, with an annual mean of 45 degrees F. Humidity is moderate (50-80%) and the average annual precipitation of 45 inches is evenly distributed throughout the year.[4,5]

The valley of Buttermilk Creek is filled by 100 to 300 feet of Pleistocene glacial deposits resting upon shale beds of Paleozoic age (Figure 3). These deposits comprise a complex series of pro-glacial lake beds, channel sands and clayey glacial till.[5,6] The state-licensed disposal facility is located on a plateau composed of 80 to 100 feet of till (the Lavery till); the lake beds and channel deposits lie at greater depth and are mainly found in the deeper parts of the valley to the east.[6]

The Lavery till, in which the waste materials were buried, is a poorly sorted, unstratified, dark grey, clay-rich sediment containing scattered pebbles and cobbles. It appears to be ideal for the confinement of waste because of its exceedingly low hydraulic conductivity (on the order of 2×10^{-7} cm/sec or less) and high adsorptive capacity for dissolved ions.[1,2] Although the till is water-saturated, it is not an aquifer; groundwater movement is essentially restricted to molecular diffusion by the low conductivity. The thin silt beds and occasional sand lenses present in the till appear to be too discontinuous and too limited in extent to permit significant water movement either vertically or horizontally.[2,7,8] The exposed upper portion of the Lavery till has been leached and oxidized to a depth of 3 to 10 feet. Locally, the weathered till may be overlain by silty, yellow-brown colluvium or alluvial gravel, but throughout most of the area, it is the oxidized, dark brownish-grey weathered till which forms the surface deposit.[6]

The weathered and unweathered till are cohesive, and during periods of prolonged drought, desiccation cracks may be formed in the till surface. The upper ten feet of the till may also exhibit random fractures which are believed to have resulted from differential settling.[6,9] Both the fractures and the desiccation cracks often exhibit discolored margins which have been interpreted as evidence of the movement of oxygen-rich meteoric water through them.[4,10] Such cracks and fractures have been observed in excavations at depths of 15 feet, and they may extend as far as 50 feet below the surface.[5,10] Because of the cracks and fractures, and the leaching noted earlier, the hydraulic conductivity of the weathered till is several orders of magnitude greater than that of the unweathered till beneath.[4,5,10] The rate of groundwater movement through the weathered till may exceed 100 feet/year, in contrast to flow rates of 0.5 inch/year (or less) in the unweathered till. Following a heavy rainfall or during the spring thaw, the weathered till becomes saturated with water which is, in effect "perched" on the impermeable till below.[11,12] Consequently, small springs and mud flows appear where the perched water is discharged along the contact between weathered and unweathered till where it intersects the surface of the ground, as in the slopes adjoining Frank's Creek and Erdman Brook. The hydraulic characteristics of the weathered till have probably contributed significantly to the water management problems at West Valley.

THE PROBLEMS ENCOUNTERED AT WEST VALLEY

Critics of shallow land burial (and apponents of the nuclear industry generally) describe the West Valley disposal facility as "insecure" and cite the experience there as evidence that shallow land burial cannot succeed in a temperate, humid climate.[13,14] Some more outspoken critics describe the facility as "an accident waiting to happen".[15] Although the principal route of radionuclide release from the West Valley trenches may be as gases produced by the decay of organic waste,[16,17,18,19] the facility has been most severely criticized as an example of the so-called "bathtub" effect. Water seeping into the trenches is trapped by the impermeable till and may leach radioactive materials or other pollutants from the waste with which it is in contact. If not pumped out, the accumulating leachate may fill the trench and escape, "like a bathtub overflowing", to contaminate the surrounding area. Trenches in which water accumulates must therefore be monitored constantly, and pumped often enough to prevent overflow. The leachate must be processed to remove hazardous contaminants before it can be either evaporated or discharged to the environment. The critics argue that long-term monitoring and pumping of the trenches will be extremely costly, and may require an institutional control period far in excess of the one-hundred years specified by the federal regulation.[13,14] Even some proponents of shallow land burial argue that because of the "bathtub" effect, disposal facilities should not be located above impervious soils in a humid climate.[20] The overflow of water from trench #4 in March, 1975, and the 1983 seepage of kerosene-tributyl phosphate from the NRC high-level disposal facility at West Valley have been publicized as evidence of the "bathtub" hazard. Unfortunately, critics have sometimes overstated the resulting environmental contamination and thereby aggravated public concern.[21]

Finally, there is evidence that some radioactive isotopes, especially tritium, may have migrated away from the West Valley trenches in the subsurface.[1,10] Such subsurface migration is offered as definitive "proof" that the West Valley trenches are insecure[13,14] despite the fact that the extent of suspected tritium migration (approximately 10 feet in 11 years) lies well within the range predicted by permeability measurements carried out prior to site licensing.[10,22,23,24,25]

CAUSES OF THE WATER MANAGEMENT PROBLEM AT WEST VALLEY

The immediate cause of the seepage that led to the closing of the LLRW disposal site at West Valley is clearly the delay in granting permission to pump and treat water from the trenches early in 1975. The impending problem was recognized long before it became critical, and the solution was at hand; only regulatory indecision prevented timely corrective action. Today, a similar accident should not occur: it is only necessary that the site operator's license mandate regular monitoring of water levels in burial cells, and stipulate that they be pumped as necessary.

This does not explain how the water entered the trenches, however, and the answer to this question may be the key to successful long-term shallow land burial of waste. Studies have identified several factors which probably contributed to the water management problems at West Valley.[3,4,7,8,10,24,26,27,28,29] Although many of these factors may apply in detail only at West Valley, it is quite certain that the principles involved have broader applicability.

First, some of the waste buried at West Valley was packaged in cardboard boxes, paper drums and even plastic bags—containers which exhibit little or no structural stability. Because disposal costs were low (less than $1 per cubic foot initially) there was little incentive to compact the waste prior to disposal. The waste containers, often only partially filled, were simply dumped into the trenches, with no attempt to fill interstices or to minimize void space. It has been estimated that only 20-30% of the excavated trench volume was actually occupied by waste, and that void space was never less than 35% of the trench volume. In some instances the void space may have been nearly double that figure.[2] After the trenches were covered, settling and compaction of the waste, and decay of organic matter, left the trench caps unsupported. As the caps collapsed, open cracks formed, allowing surface water direct access to the waste beneath.[27] Today, regular inspection of the caps and a program of active cap maintenance have corrected this problem almost entirely at West Valley.[1,10,30] Improved operating procedures coupled with more restrictive standards for waste packaging have largely prevented the development of similar problems elsewhere.[31]

Second, desiccation cracks and fractures in trench caps and in the weathered till at West Valley are believed to form a complex network of seepage channels which may conduct surface water into the waste.[4,10,28] The cracks and fractures may also intersect permeable sand or silt strata below the surface, permitting lateral movement of water into the burial trenches.[4,11,30] Once again, the program of cap maintenance and surface water management at West Valley has effectively controlled and largely eliminated this problem.

Finally, some investigators believe that lenses of sandy or silty sediment within the Lavery till provide permeable aquifers by which surface and subsurface water may gain access to the waste directly.[4,13,30] Although sand lenses may play a critical role in channeling water into some of the West Valley trenches, notably trench #14,[30] two observations suggest that the importance of the sand and silt lenses may have been overestimated: (i) In general, the permeable strata appear to be discontinuous lenses, unconnected to water sources or outlets,[6,7] and (ii) when encountered in the trench walls or in borings, the sand lenses are usually found to be dry, rather than water-saturated as one would expect an active aquifer to be.[9]

Any of the foregoing factors (cap subsidence, desiccation cracks or inhomogeneity of the burial medium) might be encountered at a shallow land burial site anywhere; thus the West Valley experience affords a warning and guidelines for preventive action. There remain, however, some hydrologic anomalies at West Valley which may be unique and which deserve more careful examination.

HYDROLOGIC ANOMALIES AT WEST VALLEY

Perhaps the most interesting feature of the hydrologic history of the northern trenches at West Valley is illustrated by the water-level curves for trenches #3, #4, and #5 (Figure 2). In all three trenches, water accumulated to a depth of five to ten feet within eighteen months after closure. The levels remained nearly stable until mid-1971 or early 1972, when they began to rise rapidly, culminating in the break-through incident at trench #4 in March, 1975.[3,10] Obviously, rising water levels indicate that water is entering the trenches more rapidly than it can escape, whereas stable levels can occur only when the rate of seepage loss is sufficient to balance the influx. Thus, when compared to the initial period of rapid water level rise, the stable interval during the late 1960's must reflect either a sharply reduced influx of water to the trenches, or a very substantial increase in the rate of water loss. A few simple (and necessarily approximate) water budget calculations, using trench #3 as an example, illustrate this point, and lead to important inferences concerning the hydrology of the West Valley site.

Firstly, consider the assertion that cracks in the trench cap provided the primary pathway by which water entered the trenches. For simplicity, assume the dimensions of trench #3 are 600' by 34', and that the void space in the buried waste was 50% at the time the trench was capped. In order to provide for a six foot annual rise in water level during each of the two intervals of rapid change, the net annual accumulation of water in the trench must have exceeded 61,200, cubic feet (600 x 34 x 6 x 0.5). Assuming that groundwater can escape through the till at a rate of approximately 1 foot/year (conductivity = 10^{-7}cm/sec), annual seepage losses through the walls and floor of the trench, based on an assumed wetted surface of 28,000 square feet and porosity of 37%, will be approximately 10,400 cubic feet. Thus it appears that no less than 71,600 cubic feet of water entered trench #3 annually during those periods when the water level was rising. Annual precipitation at West Valley (45") would provide just 76,500 cubic feet of water over the entire surface area of the trench. The similarity of these two water volumes suggests that for practical purposes, the entire trench cap behaved as though it were infinitely permeable or "transparent". To channel this volume of water into the trench through a system of cracks, those cracks would have to receive the entire precipitation burden from an area of more than 19,000 square feet. If allowance is made for ordinary evapo-transpirational losses at West Valley, the necessary drainage area would exceed an acre. There is no record of the development of such a surface drainage system in the vicinity of the trenches at West Valley, however, and a drainage net of this size which discharged into a system of cracks in the trench cap would be certain to attract attention. As important a role as cracks in the trench caps may have played, one must assume that substantial amounts of water may have entered the trenches by some other path.

When rainfall records are compared with the changing water levels in trench #5 (Figure 4), it is evident that surface water has almost immediate access to the trenches. Rarely is there a delay of more than a few hours between the onset of a rainstorm and the first water level response in the trench. The close correspondence between rainfall amounts and changes in water level (0.05 meter of precipitation leads to 0.05 meter rise in water level) is further evidence of a very fundamental process at work. A slightly more detailed water budget for trench #3 affords some insights into the nature of this process.

Annual precipitation of 45" on an area the size of the trench cap will supply 61,200 cubic feet of water, or approximately 1,700 gallons per day, averaged throughout the year. If the Lavery till has a porosity of 37%,[5] then surely the void space in the trench must have been at least 40% at the time it was capped. To raise the water level in the trench 72" per year, allowing for 40% void space, will require a net daily influx of 1,070 gallons, to which must be added an

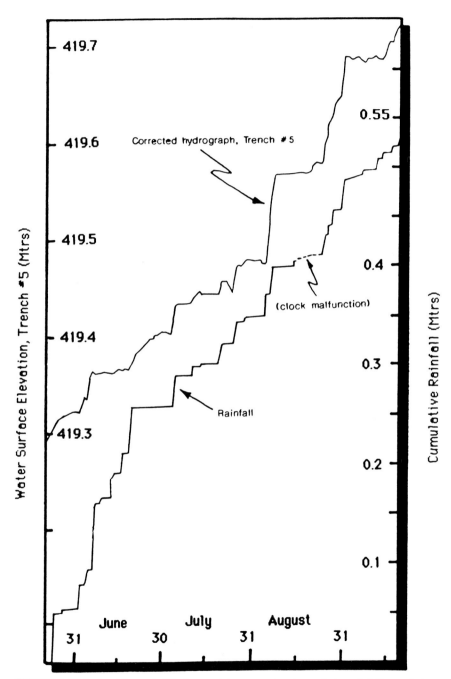

FIGURE 4. Water level (corrected for barometric pressure) and cumulative rainfall records for trench #5, from May 21 through September 20, 1975.
Redrawn from Prudic D.E. and A.D. Randall, 1977[8]

estimated 230 gallons per day to balance the seepage losses estimated earlier. The gross volume of water added to the trench throughout the year must be on the order of 1,300 gallons each day. The remaining 400 gallons/day, the difference between the 1,700 gallon/day supply and the 1,300 gallon/day gross discharge into the trench, is a minimal estimate of daily losses due to runoff and evapo-transpiration (Table 3, A). The resulting estimated water budget for the periods of rapidly rising water levels (1965-67 and 1971-75) is in strong contrast to a similar estimate for the intervening period (1967-70).

During the stable interval, as a first approximation, we may assume that annual precipitation and evapo-transpiration are unchanged (although there was a period of drought in 1971). Since the water level in the trench remains steady, there can be no net influx, although sufficient water must enter the trench to balance seepage losses which are assumed to be approximately 230 gallons/day, as before. The resulting estimate (Table 3, B) leaves 1,070 gallons/day, 63% of the supply, unaccounted for. It also requires that the gross flow into the trench be reduced to less than one-fifth (17.8%) of its previous daily volume, although there is no obvious explanation for the reduction. Not only does this apparent reduction in the gross influx to the trench imply a remarkable improvement in the performance of the trench cap, it also requires a three-fold increase, albeit unnoticed at the time, in runoff from the site (Table 3, B-1).

Alternatively, we might assume that the flow of water into the trench has not changed significantly during the stable period, but that some new factor has caused a substantial increase in the rate of water loss from it; that the "bathtub" is either leaking badly or actually overflowing in the subsurface (Table 3, B-2). Although it is quite probable that increased runoff and evapo-transpiration played a role in changing the water regimen, there are indications that greater subsurface losses also played a part.

Between 1967 and 1970, the stable water levels recorded in trenches #4 and #5 correspond very closely with the boundary between weathered and unweathered till exposed in the trench walls.[10] The significance of this fact is

TABLE 3

Simplified water mass-balance analysis, Trench #3.

Case	Pptn	Trench Net	Leakage Loss	Trench Gross	Evapotrans. + Runoff	Unaccounted For
A	1,700	1,070	230	1,300	400	0
B	1,700	0	230	230	400	1,070
B-1	1,700	0	230	230	1,470	0
B-2	1,700	0	230 + 1,070	1,300	400	0

In example 'A', the water level is assumed to rise at 72"/year, while in case 'B' the water level in the trench is assumed to be made stable. Void space in the capped trench assumed to be 40%. See text for explanation of calculations. (All units are gallons/day.)

evident when one considers that although the hydraulic conductivity of the unweathered till is extremely low (on the order of 10^{-7}cm/sec), that of the weathered till is estimated to be ten- to one-hundred times greater.[9,10,24,25] As noted earlier, this boundary supports a "perched" water table during periods of abundant surface runoff.[5,11] Ground water "perched" on the fresh till is free to move downslope within the weathered zone at rates estimated to be on the order of 100 feet/year.[9,25] Thus, the weathered till constitutes an aquifer through which leachate can escape the trenches in the subsurface. The path becomes effective when the level of the water accumulating in the trench reaches the lowest point at which the boundary between weathered and unweathered till is intersected by the trench excavation (the "spill point"). Thereafter, the water level will remain at the level of the spill point until some external factor substantially alters the influx of water. A crude estimate of the potential discharge from the trench by this path suggests that it may be too important to overlook. Assuming that the water level in the trench stands six inches above the weathered-unweathered till boundary, and that the weathered till has porosity of 50% and hydraulic conductivity of 10^{-5}cm/sec, the daily seepage loss through the weathered till will be on the order of several hundred gallons. The actual discharge will depend upon the slope of the piezometric surface, the cross-sectional area through which the flow is occurring and other factors, but the discharge is certainly not negligible, and may be of the same order of magnitude as the discrepancy in the water budget of Table 3, B.

Subsurface flow along the boundary between weathered and unweathered till is thought to have played a role in the release of plutonium-contaminated kerosene-tributyl phosphate in 1983.[9] It is reasonable to attribute some of the imbalance in the water budget presented in Table 3, (B-2) to the same mechanism. If valid, this model affords a simple explanation of the water level data presented in Figure 2: waterlevels rose rapidly at first as water entered the trenches at a rate of nearly 1,300 gallons per day. When the accumulating water reached the spill point, discharge balanced the influx, and the water level was stabilized at the level of this spill point in each trench. Enlarged cracks in the trench caps observed after the summer drought of 1971 allowed so great an increase in the flow of water into the trenches that the relatively slow seepage through the weathered till could not accommodate the volume, and levels again began to rise until the trench caps were breached in 1975. Regrading of the trench caps after 1975[10] has effectively corrected the problem today.

The foregoing model affords an equally interesting explanation of the influx of water to the trenches. Groundwater perched on the upper surface of the unweathered till would flow down the piezometric gradient along the boundary. Excavations cut into the impermeable till would capture the slowly moving water until filled, but there would be no surface evidence of the flow, even if the accumulating water were derived from a very extensive drainage area. Maps of the upper surface of the unweathered till reveal an undulating surface

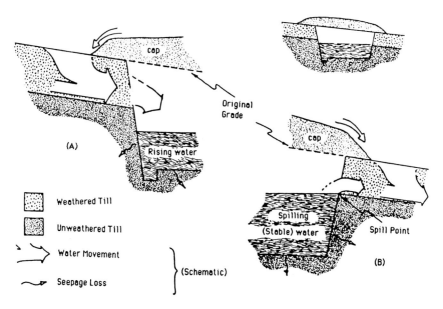

FIGURE 5. Diagram illustrating inferred groundwater movement into and out of West Valley trenches through the weathered till aquifer beneath the completed trench cap.

sloping generally northeastward from the high-level burial ground toward the northern trenches.[9] Although incomplete, these studies suggest that groundwater perched on the unweathered till would generally flow from the extensive "Southern Plateau" area (Figure 1) northward toward Erdman Brook and and the northern trenches of the low-level burial area.

If valid, this perched water hypothesis is made possible only by the peculiar construction techniques used in excavating the early trenches at West Valley. According to the best available information, trenches #1 through #5 were excavated without any preliminary site preparation! Sufficient excavation was carried out to accommodate the waste awaiting disposal, the waste was emplaced, and the temporary cover built. The waste was stacked to within four feet of the original surface grade (level with original grade after 1965), and a significant thickness of weathered till was thus exposed in the trench walls below the level of the final cap. As effective as the cap might be, it served only as an umbrella to divert surface runoff to the margins of the trench; there the water could enter the weathered till aquifer to flow into the trench beneath the cap (Figure 5 (A)). As soon as the accumulating water rose to the level of the spill point, it could escape from the trench through the same aquifer (Figure 5 (B)). In the course of grading the area where the southern trenches are located, most of the soil, alluvium, colluvium and weathered till were removed before excavation began. As a result, relatively little of the permeable material was exposed in the trench walls, and water management problems in the southern trenches are less common.

Obviously, the perched water table at West Valley is not unique. At any location where a modestly permeable soil or regolith rests on a much less permeable deposit, a perched water table will develop at least during periods of abundant surface runoff.[31] Because impermeable deposits are generally considered to offer the most secure sites for hazardous and radioactive waste repositories,[32] the presence of a perched water table may actually indicate a favorable site location rather than a hazard to be avoided.

A situation very similar to that at West Valley was encountered at the low level radioactive waste burial facility at Barnwell, South Carolina. The burial medium at Barnwell is a silty clay which is somewhat more permeable than the Lavery till. It is overlain by a highly permeable sand. To prevent lateral migration of perched water through the sand into the trenches, the site of each burial cell is prepared by grading and by the construction of a clay barrier through which the trench is finally excavated (Figure 6).[33] The incorporation of such "Barnwell Barriers" in the trench design appears to have been highly successful in South Carolina where annual precipitation is comparable to western New York. Had such barriers been built during excavation of the West Valley trenches, many of the water management problems at the facility might have been avoided. Similar barriers should be included in the design of all shallow land burial facilities in humid climates in the future.

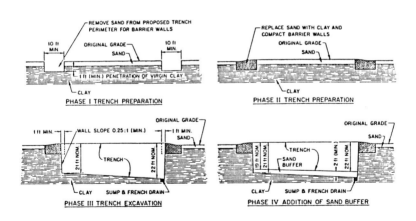

FIGURE 6. Barnwell (S.C.) disposal trench construction technique, showing the preliminary site preparation and construction of the "Barnwell Barrier", and other details of the Barnwell trench design.

From Clancy, J.J., et al., 1981[1]

CONCLUSIONS

The most significant conclusion to be drawn from the West Valley experience is that a low level radioactive waste disposal facility can be established and successfully operated in impervious strata and under a wet and often adverse climate.

Although the West Valley facility is widely regarded as unsuccessful, its performance has generally met or exceeded the design standards and objectives. More than 2.25 million cubic feet of low level radioactive waste disposed of at West Valley have been confined for nearly one-quarter of a century, without contamination of the environment and without endangering the health and safety of the public. This success has been achieved, however, only at the cost of an unanticipated program of continuing water management and trench maintenance. It may have been naive to believe, when the facility was opened, that little more than routine monitoring of the site would be required once the final closure had been completed. The experience at West Valley, like that at Sheffield, Illinois, and Maxey Flats, Kentucky, makes it quite clear that an active maintenance program will probably be necessary at any LLRW disposal facility throughout the institutional control period. Such a program will be extremely expensive, and may involve substantial unanticipated costs. Ample provision must be made to meet these costs through a generous (rather than "adequate") perpetual care surcharge imposed on the wastes received.

It is impossible to foresee all of the problems which may be encountered at a specific site, but the experience at West Valley clearly identifies several guidelines which may eliminate problems and help to reduce operating and maintenance costs at future facilities.

- Gases generated by decaying organic matter (and other biodegradable materials) may provide the principal route by which radionuclides are released to the environment.

  Dehydration and incineration before disposal offer partial solutions, but these processes may release the radionuclides to the atmosphere prior to burial. Embedding or sealing the material in packages may offer a more secure solution.

- Leaching of radionuclides from the waste by ground- and surface water is a second, and perhaps equally important release pathway.

  Operating plans must include means of diverting surface water from the disposal area and preventing the movement of groundwater, including "perched" water, into the disposal cells. Due attention must be given the escape of leachate from the disposal cells as well.

- Collapse of buried waste due to compaction, decay or consolidation of the waste is the primary cause of trench cap failure.

  Exclusion of liquids and biodegradable materials from the waste, pre-compaction of all waste, required use of structurally stable containers, and thorough filling of interstitial spaces with stable packing (sand or grout)

will substantially reduce problems arising from cap failure.
- Facilities located in (on) impermeable materials will characteristically exhibit "perched" groundwater which must be diverted around subsurface disposal units whether they be trenches, vaults or other types.

    Grading the site to the level of the perching aquaclude before construction, or surrounding the disposal units with water barriers which extend into the impermeable strata ("Barnwell barriers") can prevent incursion of perched water into the disposal cells.
- Jurisdictional disputes and regulatory inadequacies may delay effective action to prevent the release of radionuclides to the environment by unanticipated events.

    Licensing requirements and operating regulations must be sufficiently flexible to allow prompt and effective emergency action by the operator or the regulatory agency or both.

The only known environmental contamination from the West Valley LLRW operation resulted from problems of an institutional nature, and the administrative and regulatory changes necessary to avoid recurrences are either already in place or simple to enact. Geologically, the site exemplifies a shallow land burial facility located on impermeable strata in a moist, temperate climate. Despite continuing problems with water management, both above and below ground, radionuclides disposed of at the site have been effectively confined for almost twenty-five years. Limited migration of isotopes out of the burial trenches has been detected, but the extent of migration has never exceeded that deemed acceptable at the time the site was licensed, nor has it violated the limits set by substantially more restrictive current federal regulation (10 CFR 61). Based on experience gained at West Valley, we may conclude that effective confinement of low-level radioactive waste can be achieved in a shallow land burial facility under climatic conditions like those of the humid northeastern United States provided the following requirements are met:

All non-solid waste forms must be excluded.

All waste must be pre-compacted to the maximum practicable extent.

All bio-degradable materials, including organics, must be incinerated, embedded or placed in sealed containers prior to disposal.

Waste packaging standards must ensure long-term structural stability.

Operating standards must ensure minimum void space and long-term structural stability.

Trench design must provide sub-drainage, with monitoring wells and sumps.

Operator/licensee must be authorized to pump trenches into prepared holding/processing facilities as required, without discharge to the environment.

Engineering design must prevent access of perched water as well as ordinary ground-water to disposal trenches.

Surface surface water management plans must divert all surface water away from trenches and groundwater recharge areas.

Provided the foregoing conditions are observed, there is every reason to infer, from the West Valley experience, that low level radioactive waste can be disposed of in shallow land burial facilities in appropriately selected locations in the northeastern states, with full confidence and at reasonable cost.

REFERENCES CITED

1. Clancy, J.J., D.F. Gray & O.I. Oztunali (1981). Data Base for Radioactive Waste Management: Review of Low-Level Radioactive Waste Disposal History. Division of Waste Management, Office of Nuclear Material Safety and Safeguards, U.S. Nuclear Regulatory Commission, Washington, D.C. (NUREG/CR-1759,Vol. 1).
2. United States Department of Energy (1978). Western New York Nuclear Service Center Study, Companion Report (Volume 2) U.S. Department of Energy, Washington, D.C. (TID-28905-2).
3. Kelleher, M.J. (undated). Water Problems at the West Valley Burial Site. New York State Department of Environmental Conservation, Albany, New York.
4. Kelleher, M, J. Lyons, K. Tong, W. Greenman & T. Waldman (1983). "Report: Investigation of a Low Level Waste Burial Site at the Western New York Nuclear Services Center." Power Authority of the State of New York, Albany, New York.
5. West Valley Demonstration Project (1984). "Environmental Evaluation for Disposal of Project Low-Level Waste. (WVDP-027)"
6. LaFleur, R. G. (1979). "Glacial Geology and Stratigraphy of Western New York Nuclear Service Center and Vicinity, Cattaraugus and Erie Counties, New York." U.S. Geological Survey Open File Report 79-989, Washington, D.C.
7. Bergeron, M.P. (1985). Record of Wells, Test Borings and some Measured Geologic Sections near the Western New York Nuclear Service Center, Cattaraugus County, New York. U.S. Geological Survey, Open File Report 83-682.
8. Prudic, D.E. & A.D. Randall (1977). Groundwater Hydrology and Subsurface Migration of Radioisotopes at a Low-Level Solid Radioactive Waste Disposal Site, West Valley, New York. U.S. Geological Survey, Open File Report 77-566.
9. West Valley Nuclear Services, Inc. (1984). "Report: Investigation of Kerosene-Tributyl Phosphate Migration, NRC Licensed Disposal Area, Western New York Nuclear Service Center, West Valley, New York.

10. Prudic, D.E. (1986). Ground-Water Hydrology and Subsurface Migration of Radionuclides at a Commercial Radioactive-Waste Burial Site, West Valley, Cattaraugus County, New York. U.S. Geological Survey, Professional Paper 1325, 83 pp.
11. Schmalz, R.F. (1985). Geological Note on the West Valley LLRW Site. Agenda, PIER Program Coordinating Committee Meeting, 28 September, 1985, West Valley, New York. The Pennsylvania State University, University Park, PA., 5 pp.
12. Hoffman, V.C., R.H. Fickies, R.H. Dana Jr. & V.S. Rayan (1980). Geotechnical Analysis of Soil Samples and Study of Research Trench at the Western New York Nuclear Service Center, West Valley, New York. U.S. Nuclear Regulatory Commission, Washington, D.C. (NUREG/CR-1566).
13. Sierra CLub, Radioactive Waste Campaign Fact Sheet (undated). A "Low-Level" Nuclear Waste Primer. Sierra Club Radioactive Waste Campaign, 78 Elmwood Street, Buffalo, NY, 14201.
14. Sierra Club, Radioactive Waste Campaign Fact Sheet (undated). Insecure Landfills: The West Valley Experience. sierra Club Radioactive Waste Campaign, Box 64, Station G, Buffalo, NY, 14213.
15. Sierra Club, Radioactive Waste Campaign Fact Sheet (undated). WEST Valley: A Challenge for the 80's. Sierra Club Radioactive Waste Campaign, Box 64, Station G, Buffalo, NY, 14213.
16. Matuszek, J.M., L. Hussain, A.H. Lu, J.F. Davis, R.H. Fakundiny & J.W. Pferd (1979). Application of Radionuclide Pathways Studies to Management of Shallow Low-Level Radioactive Waste Burial Facilities, *in* Carter, M.W., A.A. Moghissi & B. Kahn, Eds., Management of Low-Level Radioactive Waste, vol. 2, Pergamon Press, New York.
17. Matuszek, J.M. (1980). Biochemical and Chemical Processes Leading to Radionuclide Transport from Low-Level Waste Burial Sites. Transactions, American Nuclear Society, vol. 34, pp. 155-156.
18. Matuszek, J.M. (1982). Radiochemical Measurements for Evaluating Air Quality in the Vicinity of Low-Level Waste Burial Sites—The West Valley Experience. Symposium on Low-Level Waste Disposal: Site Characterization and Monitoring, Arlington, VA, June, 1982 (NUREG/CP-0028), pp. 423-442.
19. Matuszek, J.M. & L.W. Robinson (1983). Respiration of Gases from Near-Surface Radioactive Waste Burial Trenches. *in* Waste Management '83, Proceedings of the Symposium on Waste Management, Tuscon, Arizona, February, 1983.
20. Mallory, C.W. (1982).Decommissioning of Low-Level Radioactive Waste Disposal Sites—Implications for Future Sites, *in* Waste Management '82, Proceedings of the Symposium on Waste Management, Tuscon, Arizona, March, 1982.

21. Anon. (1984). Don't Reopen W. Valley For Nuclear Wastes (Editorial). The Buffalo News, Monday February 13, 1984, p. C-2.
22. Nuclear Fuels Services, Inc. (1962). Safety analysis, spent fuel processing plant: Part B of license application to U.S. Atomic Energy Commission, volume 1.
23. Dames & Moore (1963). Report. Site Investigation, Proposed Spent Nuclear Fuels Reprocessing Plant, Near Springville, N.Y., Rockville, MD. (Nuclear Fuel Services, Inc., unpublished report) 28 pp.
24. Prudic, D.E. (1982). Hydraulic Conductivity of a Fine-Grained Till, Cattaraugus County, New York. Ground Water, Vol. 20, p. 194.
25. Lu, A.H. (1978). Modeling of Radionuclide Migration from a Low-Level Radioactive Waste Burial Site. Health Physics, vol. 34, pp. 39-44.
26. Spath, J.P. & T.K. DeBoer (1984). "West Valley Low-Level Radioactive Disposal Area." New York State Energy Research and Development Authority, Albany, New York.
27. Prudic, D.E. (1979). Recharge to Low-Level Radioactive Waste Burial Trenches 11 Through 14, West Valley, New York. U.S. Geological Survey, Open File Report 79-990.
28. Prudic, D.E. (1980). Permeability of Covers Over Low-Level Radioactive Waste Burial Trenches, West Valley, Cattaraugus County, New York. U.S. Geological Survey, Water Resources Investigations 80-55.
29. Giardina, P.A., et al. (1977). Summary Report on the Low-Level Radioactive Waste Burial Site, West Valley, New York, (1963-1975). U.S. Environmental Protection Agency, Washington, D.C., February, 1977.
30. Mestepey, J.H., personal communication to the author, December 4, 1986.
31. Freeze, R.A. and J.A. Cherry. Groundwater. Prentice-Hall, Inc. Englewood Cliffs, N.J., 604 pp., 1979.
32. Schmalz, R.F. (1983). Geology—Site Specific. Chapter 6 *in* Witzig, W.F., W.P. Dornsife and F.A. Clemente, Eds., Low Level Radioactive Waste Disposal Siting: A Social and Technical Plan for Pennsylvania, Volume 3. Institute for Research on Land and Water Resources, The Pennsylvania State University, University Park, Pennsylvania (4 vols.)
33. Smith, R.L. (1984). Low Level Radioactive Waste Disposal at the Barnwell Waste Management Facility. *in* Low Level Waste Disposal Handbook, Electric Power Research Institute (NP-2488-LD).

Environmental Consequences of Energy Production: Problems and Prospects. Edited by S. K. Majumdar, F. J. Brenner and E. W. Miller. © 1987, The Pennsylvania Academy of Science.

Chapter Twenty-Five
ENVIRONMENTAL IMPACTS OF NUCLEAR POWER PRODUCTION

WARREN F. WITZIG[1], KANAGA SAHADEWAN[2] and JAMES K. SHILLENN[3]

[1]Professor Emeritus
Nuclear Engineering Department
[2]Project Assistant
[3]Project Coordinator
Energy Technology Projects
The Pennsylvania State University
University Park, PA 16802

This paper analyzes the environmental impacts of nuclear power generation on air, water and terrain. Although the process of nuclear power generation involves many different aspects, this paper will focus on the four main aspects of that process. They are: Construction, Normal Plant Operation, Accidents and Decommissioning. Another aspect is the Mining & Fabrication of the uranium fuel, which is discussed in another chapter titled "Mining and Processing of Radioactive Materials". Analysis of each aspect includes a definition, an examination of the applicable regulatory guidelines, an overview of the present and future technology and a brief comparison to other energy sources.

CONSTRUCTION

The amount of land needed for electrical power generation is often overlooked but as preservation of specific environmental habitats and archeological sites become more important, the subject of land use requirement becomes a greater concern. For nuclear-fueled plants, the suitability of a proposed site is established by considering such factors as the reactor design, safety systems, and expected radioactive material inventory. The site is then evaluated with respect to population density, meteorology, seismological stability, hydrology, and geology. Typical land requirements have been 400-700 acres for 2200 MWe plants comprised of two units. For comparison, the total land use of all categories in the fuel cycle is 22,400 acres[1] during the 30 year life-time of a coal fired power plant as to about 1000 acres for a nuclear power plant. It should be pointed out that some types of stack gas scrubber systems that remove sulfur oxides for coal power plants will require even larger amounts of land use for disposal of sludge. For example, The Bruce Mansfield Plant at Shippingport, PA requires a sludge disposal site of 1,300 acres that will be filled with 200,000 tons of sludge over the next 20 to 25 years.

The type of cooling system must also be considered when evaluating land use. The thermal efficiency of light-water reactors at present is approximately 33%, this means about two-thirds of the energy generated by a power plant is released into the environment. This waste heat is transferred from the condensing steam to the cooling water in the condensers. Until recently, the cool water in the condenser was taken from a nearby river and after it was heated in the condenser it was discharged back into the river. (This is referred to as once through cooling). The result is an increase in water temperature near the discharge point and some ecological effects. To minimize thermal discharge into aquatic habitats, cooling systems have been designed to release most of the waste heat into the atmosphere and a small fraction of the heat into the aquatic environment.

Two major types of cooling systems that are being employed are cooling ponds and cooling towers. Cooling ponds are often used when the land is available at a reasonable price. Cooling towers are most frequently used when cooling ponds are not practical. Three environmental effects of cooling towers that have been intensively investigated are shadowing, drift and fogging. Shadowing is the restriction of sunlight caused by the visible plume, drift is the disposition of detrimental chemicals onto surrounding areas, and fogging which is the restriction of visibility by the visible plume reaching the ground. Experience indicates that as long as waste heat is limited to that from a 2000 MWe installation, shadowing and drift are restricted to the immediate vicinity of the plant and rarely does the plume from large cooling towers reach the ground to cause any restrictions in visibility.

The probability of occurrence of an earthquake or other geological hazard and its potential consequences on a nuclear facility must be taken into account in the selection of a site. However, all areas of potential geological hazard cannot be avoided. Microseismic events occur continually at all points on the earth's surface, with no visible effect on any well-built structure. For example, no active fault is known or suspected in the vicinity of the Peach Bottom Atomic Power Station near Delta, PA, and known faults in the area have been inactive for 180 million years. 10 CFR 100—Appendix A (Code of Federal Regulations) details the seismic and geological siting criteria for nuclear power plants.

The chain of regulatory approval for the construction of a nuclear power plant begins with the obtaining of a construction permit from the Nuclear Regulatory Commission (NRC). 10 CFR 50.10 states that no person shall begin the construction of a production or utilization facility until a construction permit is issued. 10 CFR 50.34 requires that an application for a permit shall include a preliminary safety analysis report and an environmental impact statement.[2] The construction permit stage involves an informal site review, application for license, review by the regulatory staff, review by the Advisory Committee on Reactor Safeguards (ACRS), public hearings, and appeals by the Atomic Safety and Licensing Appeals Board (ASLAB).

After the construction phase has been completed, an application for an operating license is submitted to the NRC. This application is again reviewed by regulatory staff and the ACRS. Public hearings and appeals are held to ensure that the plant can be operated in such a manner that it would not pose an undue risk to the health and safety of the public and that it will create only a minimal impact on the environment.

NORMAL PLANT OPERATION

The most dominant feature in the nuclear fuel cycle is the operating power reactor. It contains a large amount of radioactivity. As of the end of 1986, thirty-five countries had some 540 reactors in operation, under construction or on order. The United States had about one-fifth of the world total in number of operating plants (105) and a power generating capacity of 117000 MWe[3]. In the case of power reactors, the power level is commonly given as the electrical output, or MWe (megawatts electric), which in modern reactors is about one-third the total power produced.

A light water reactor consists of UO_2 (fuel pins), enriched to about 3% in ^{235}U, submerged in water. In the proper geometry, this arrangement permits a chain reaction in which a neutron striking a ^{235}U nucleus induces a fission reaction, which releases neutrons, some of which induce other fission reactions. Each fission reaction releases energy (about 200 MeV), which is rapidly converted into heat, warming the surrounding water. As water is pumped through the reactor at a rate of thousands of gallons per second, it is heated to 600°F. The hot water may be then converted to steam either in the reactor (boiling water reactor, BWR) or in an external heat exchanger (pressurized water reactor, PWR); the steam then drives a turbine which turns a generator to produce electric power.

Three streams of wastes originate from the operation of a power plant. They are Gaseous, Liquid, and Solid waste streams.

Gaseous

Gaseous waste stream consists of fission products, activation products, radioiodines, and carbon-14 particulates. An operating reactor will generate about 80 different primary fissions products, and even more are produced as a result of decay chains.[4] Radioactive noble gases such as Xenon-135 and

TABLE 1

Summary of Gaseous Releases from U.S. Nuclear Plants in 1983[5]

	# of Plants	Curies	Becquerel	Curie/Plant
Fission &	BWR (22)	1.2×10^6	44 PBq	54.5×10^3
Activation Gases	PWR (37)	0.18×10^6	7 PBq	4.9×10^3
I-131 &	BWR (22)	9.31	0.35 TBq	0.43
Particulates	PWR (37)	1.51	0.06 TBq	0.05

$P = Peta - 10^{15}$; $T = Tera - 10^{12}$

Krypton-85 provide the major source of fission product activity. The greatest gaseous releases are from BWRs and, in particular, the earlier models. BWRs release from 100 to 10,000 Curies (Ci) (3.7—370 Tera Becquerels) per Megawatt of electricity.[4] Noble gas releases from PWRs are in general lower than for BWRs, and the total activity released may be two orders of magnitude lower. These numbers have been decreasing yearly due to increased holdup time in gas treatment systems in the newer plants. In both cases, a majority of the noble gas nuclides are short-lived with half-lives ranging from a few minutes to several days.

Activation products are produced by interactions of neutrons with hydrogen and oxygen in the coolant water, oxygen, nitrogen and argon in air dissolved in the coolant, and any materials used in construction. Tritium is an important gaseous activation product, particularly in PWRs and in heavy water reactors (HWR).[2] Gas cooled reactors produce ^{41}Ar from activation and they account for a relatively large amount of activation gases to the environment. Another gaseous nuclide released from the operation of a power plant is radioiodine. The major iodine isotope of concern is Iodine-131. Carbon-14 is produced in power reactors by neutrons reacting with Oxygen-17 and Carbon-13. Production of these nuclides is relatively low, and releases lower still.

Movement of radionuclides through the biosphere to man is very important in determining the biological impacts of radioactivity in the environment. Three routes of importance regarding gaseous waste streams are: 1) direct radiation, 2) deposition on plants & soil, and 3) inhalation by humans and animals.[6] Radioactive nuclides can reach humans directly, by people ingesting the plants where radionuclides were deposited, and by ingesting the milk and meat from animals who either inhaled radionuclides or ate contaminated vegetation. The following diagram illustrates these concepts.

Liquid

The major radioactive constituent in the *liquid waste stream* released to the environment from reactors is tritium, which can amount to 1000 C (10 Tbq) per year. PWRs, largely because of their use of boron in the primary coolant for reactivity control, and Heavy Water Reactors (HWR) because of deuterium activation, are the major sources of reactor produced tritium. The amount and

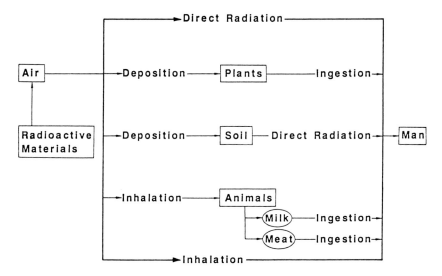

Figure 1. GASEOUS PATHWAYS TO MAN

composition of liquid releases of mixed fission products mainly depend on the design and the operating practice of the reactor. But they have been relatively small and decreasing in their value to about 1 Ci (37 GBq) per year. Liquid effluents also contain low concentrations of fission products that have not been removed fully by ion exchangers and evaporators. In many cases they are diluted by mixing with cooling tower blowdown water. Isotopes of Cesium-134 and Cesium-137 represent about 10% and 35% of the radioactivity concentration in liquid discharges from U.S. PWRs and BWRs respectively. Cobalt isotopes Cobalt-58 and Cobalt-60 contribute about 65% of the activity in PWRs and about 10% in BWRs. Sodium-24 and Strontium-89 are nuclides which consistently contribute to the liquid discharge from light water reactors.

For liquid effluents, the important pathway to man is through drinking water. Tritium, radioiodines, noble gases, cesium, and transition metals (Fe,Co,Ni,Zn,Mn) enter man through drinking water. Man's consumption of fish and shellfish also make him a recipient of some elements in the liquid waste stream.

In the U.S., discharge of radioactive effluents is federally regulated through 10 CFR 20[7]. Utilities operate under a philosophy referred to as "as low as reasonably achievable" or ALARA. ALARA has been defined for design purposes, so that persons in unrestricted areas will not receive more than 10 mrad (100 micro Gy) from gamma radiation and 20 mrad (200 micro Gy) from beta radiation. This ALARA principle has been broadened to include environmental releases. Liquid releases are limited to 5 Ci (185 GBq) per year, except for tritium, or to that quantity that will deliver an annual dose equivalent of no more than 5 mrem (50 micro Sv).

TABLE 2

Summary of Liquid Releases from U.S. Nuclear Plants in 1983[5]

	# of Plants	Curies	Becquerel	Curie/Plant
Mixed Fission & Activation Products	BWR (22)	28.1	1.1 TBq	1.28
	PWR (37)	76.9	2.8 TBq	2.07
Tritium	BWR (19)	8.5×10^2	4 TBq	45
	PWR (37)	180.5×10^2	667 TBq	487

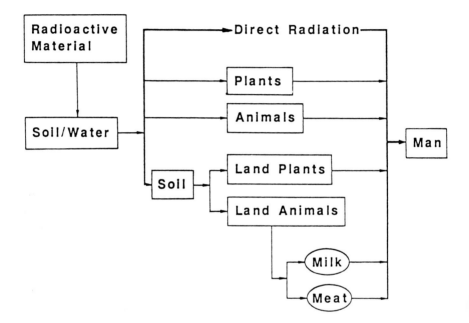

Figure 2. LIQUID PATHWAYS TO MAN

Solid

Out of all the three waste streams, *solid waste stream* has the largest volume and the highest concentration of radioactivity. Solid waste is usually classified as High Level (HLW) or Low Level (LLW). In general the activity of LLW is less than 10 Ci/ft^3 and HLW is more than 1000 Ci/ft^3. Spent fuel and some transuranic wastes (^{239}Pu, ^{240}Pu, ^{241}Pu) are included in HLW.

The term *low level waste* is quite broad, and generally includes wastes from various radionuclides, protective clothing, tools, air filters, and other materials from various operations with radioactivity. LLW accounts for upwards of 90% of the total volume of solid wastes and only 1% or less of the radioactivity.

LLW from all sources is ranked as class A, B, or C depending on its concentration of radionuclides, especially those with long half-lives. For Class A—Safe handling and packaging is required, for Class B—Must be stabilized to keep its size and shape for at least 300 years, and for Class C—with the highest quantity of long-lived radionuclides—also must be stabilized, as well as specially situated, at the disposal site. In 1985, 95% of all LLW shipped[8] was Class A; 2% was Class B; and < 1% was Class C. In the U.S., disposal of low-level waste started in 1962, and to date at six disposal sites [Barnwell, SC; Beatty, NV; Maxey Flats, KY*; Richland, WA; Sheffield, IL*; West Valley, NY* (*—Closed temporarily)] 38.8×10^6 ft^3 of LLW containing 9.5×10^6 Ci (350 PBq) of activity has been buried.[9]

On Jan. 15, 1986, the President signed into law the Low-Level Radioactive Waste Policy Amendment Act of 1985 (Public Law 99-240). This gave additional time to the states to respond to the Low-Level Radioactive Waste Policy Act of 1980, which would have allowed present disposal sites to prohibit the disposal of LLW from other states after January 1, 1986. The Appalachian compact, which was adopted by the Pennsylvania legislature and signed into law in December 1985, provides for the operation of facilities in Pennsylvania for regional disposal of LLW. Four states: Pennsylvania, West Virginia, Maryland, and Delaware are designated as the only eligible states for this compact.

The main component of high level waste from the normal operation of a nuclear power plant is spent fuel. By regulation (10 CFR 60.2)[10], high level waste includes 1) irradiated reactor fuel, 2) liquid waste resulting from the operation of the first cycle solvent extraction system in chemical reprocessing plants, or equivalent, and 3) solids into which such liquids have been converted. So far, spent fuel has been stored in pools within the reactor site with limited space, but a permanent solution to this problem is within view. By the passage of the Nuclear Waste Policy Act of 1982 (NWPA), the Congress has declared that the disposal of high-level waste is a responsibility of the federal government, and it has established a policy and schedule for the siting, construction, and operation of underground repositories for the permanent disposal of high level waste.

TABLE 3

Solid Waste Summary from U.S. Nuclear Plants in 1983[5]

# of Plants	Volume	Activity	
	Cubic Meters	Curies	Becquerel
BWR (22)	2.30×10^4	2.28×10^5	8.4 PBq
PWR (37)	2.24×10^4	7.20×10^5	26.0 PBq

# of Plants		Activity	
	Volume/Plant	Curies/Plant	PBq/Plant
BWR (22)	1.06×10^3	10.36×10^3	0.38
PWR (37)	0.61×10^3	19.60×10^3	0.73

The three candidate sites have been nominated and recommended by DOE and approved by the President for a HLW repository. These sites are, Yucca Mountain, NV; Deaf Smith County, TX; and Hanford, WA. These sites were selected from a group of nine potential sites based upon geological, environmental, economic, and social factors. According to the NWPA, after the three sites are studied in more detail, the President will recommend one site to Congress for construction authorization by March 31, 1987. That repository is scheduled to begin receiving waste shipments by Jan 31, 1998. However, the political aspects of siting will probably delay the schedule. Such a delay will cause difficulty to accommodate local storage for the life of a plant. Local storage may well incur a larger public risk compared with disposal. Overall, High Level Waste disposal is a political problem as the technology for safe disposal has been demonstrated at West Valley and Hanford.

Solid wastes do not enter the biological pathways of man in a major way. Some wastes, which were deposited in the sea may reach man through phytoplanktons and fish and waste which leaked into ground water from repositories may reach man through drinking water and food. The resulting dose levels would be insignificant.

In concluding this section, it is important to note that the NRC annual report summarized the dose received from 71 nuclear power plants, to the population

TABLE 4
Comparison of Radiation Doses from Energy Sources

Energy Source	Critical Nuclide	Maximum* Individual	Average* Individual
Coal-fired electric station, [1,000 MW(e)] Appalacian Coal[a]	230_{Th} (228_{Rn}, 228_{Th}, 232_{Th})	65 (bone)	.5-7 (bone)
Coal-fired electric station [1,000 MW(e)] Western coal[a]	230_{Th} (228_{Rn}, 228_{Th}, 232_{Th})	300 (bone)	5-70 (bone)
Boiling Water Reactor[b]	Noble Gases Tritium Halogens	11 (thyroid) 2 (whole body)	<1 <1
Pressurized Water Reactor[b]	Noble Gases Tritium Halogens	1 (thyroid) 0.8 (whole body)	<1 <1

[a] EPA 520/1-77-009 Radiological Quality of The Environment in the United States, 1977.
[b] EPA 520/7-79-006 Radiological Impact Caused by Emissions of Radionuclides into Air in the United States, 1979.
* Dose to Critical Organ (mrem/yr)

living between 2 and 80 km. These exposures compare favorably with regulations, i.e. a low value of 1×10^{-5} mrem (0.01 micro Sv) to a high of 5×10^{-2} mrem (50 micro Sv). Comparing the impact on the environment, especially on humans, from the operation of a nuclear plant and a coal fired plant gives a clearer idea of the safe and effective operation of a nuclear plant. The following table illustrates the dose in mrem/yr to the critical organs from the various energy sources. It is very evident that operation of a nuclear plant has a lower health impact than the operation of a coal fired plant even when only the radionuclide emissions are considered.

While there is no risk-free method to generate electricity, it is clear that the normal operation of a nuclear power plant carries a very small public risk, one more favorable than a coal plant.

ACCIDENTS

The health effects from the operation and the presence of nuclear reactors has been the subject of much speculation and many myths. Although there have been several accidents involving nuclear reactors and critical assemblies (see Table 4A), four are of major significance from an environmental standpoint. Two of them: TMI 2 and Chernobyl 4 will be examined in detail.

TABLE 4A
List of Nuclear Facility Accidents

Name	Type of Facility	Location	Date	Release Size
Windscale	Reactor/Pu Producer	UK	Oct 1957	Medium
NRU Reactor	Test Reactor	Canada	May 1958	Trace
Y-12 Plant	Chemical Processing	Oakridge	June 1958	Trace
Boriskidrich	Critical Assembly	Yugoslavia	Oct 1958	Trace
Westinghouse	Test Reactor	PA	April 1960	*
SL-1	Military	Idaho	Jan 1961	Small
Plutonium Recycle	Test Reactor	Los Alamos	Sept 1965	Trace
Enrico Fermi	Power Reactor	Michigan	Oct 1966	Trace
TMI2	Power Reactor	PA	March 1979	Small
Chernobyl Unit 4	Power Reactor	Kiev, Russia	April 1986	Large

* Incomplete Data

The first significant reactor accident occurred in October 1957 at Windscale No. 1, a graphite moderated plutonium production reactor located in northwest England. An estimated 20,000 Ci (740 TBq) of Iodine-131 was released along with 600 Ci (22 TBq) of Cesium-137, 80 Ci (3 TBq) of Strontium-89, and 9 Ci (3.3 GBq) of Strontium-90.[14] The release of radioiodine to the environment resulted in the contamination of milk, as well as the exposure of persons through inhalation. The second accident was at the SL-1 Reactor (an experimental military reactor) at the National Reactor Testing station in Idaho. The accompanying steam explosion killed the three-man crew and destroyed the reactor. About 70 Ci (2.6 TBq) of Iodine-131 was released.[1] Neither of these two accidents had containment other than commercial type of buildings.

The third accident, and the first major commercial nuclear power plant accident, occurred at Three Mile Island Unit 2 (TMI 2) near Harrisburg, Pennsylvania. In the early morning hours of March 28, 1979, while the reactor was operating at 97% of full power, a malfunction occurred to components that maintain the flow of coolant water to the steam generators in the secondary loop. This resulted in a loss of ability to remove heat from the primary system. Thus, most of the heat generated by the reactor remained in the reactor vessel and primary loop. This caused the coolant water pressure and temperature to increase rapidly, which in turn, caused a relief valve on the pressurizer to open. Steam and water were discharged to the reactor coolant drain tank located in the basement of the containment building. A key factor in the accident was that the relief valve failed to close when pressure returned to normal, allowing some 32,000 gallons of water to escape. As the water level fell, the fuel became uncovered resulting in intense heat which produced fuel damage and fission product release.[12]

Although the core sustained serious damage, including a partial meltdown, relatively little radioactivity was released to the environment. The total quantity of activity released was about 2.5×10^6 Ci (92.5 PBq), almost entirely short-lived noble gases (Table 5). Less than 30 Ci ($<$ 11 GBq) of Iodine-131 was released.

There have been many estimates and some measurements of the population doses from the TMI 2 accident. Accepted values were prepared by technical staff members of the NRC, Department of Health, Education and Welfare (HEW) and the Environmental Protection Agency (EPA), who constituted an Ad Hoc Population Dose Assessment Group. There was general agreement that the primary exposure came from noble gases which are chemically and biologically unreactive. The radiation dose received by the general public can be summarized as follows:[13]

Highest *Whole Body Dose* to any one person: <0.1 rem (<0.0001 Sv)

Doses within 50 mile radius of TMI
—Collective Whole Body *Population Dose* : 3330 person-rems
   to General Population   (33.3 person-Sv)
—Total Collective Population Dose to   : 1400 to 2800 person-rems
   Thyroid   (14 to 28 person-Sv)

One way of evaluating the potential health impacts of the above whole-body doses is to compare them with the natural background radiation dose of 0.1 rem/yr (0.0001 Sv/yr). The average individual dose of 0.0015 rem (15 milli Sv) to an individual living within a 50 mile radius, therefore, would be equivalent

TABLE 5
Radionuclides Released from TMI 2 Accident[14]

Nuclide	Curies Released	Becquerels Released	Half Life
Xe-133	8.3×10^6	307 PBq	5.3 days
Xe-135	1.5×10^6	56 PBq	9.1 hrs
Kr-88	6.1×10^4	2 PBq	2.8 hrs
I-131	<30	<1 TBq	8 days
Cs-137	4.0×10^{-5}	1.5 MBq	30 yrs
Sr-90	6.0×10^{-5}	2.2 MBq	27.7 yrs
Tritium	147	5.4 TBq	12.3 yrs
Co-58	4.0×10^{-4}	14.8 MBq	71.3 yrs
Co-60	9.0×10^{-5}	3.3 MBq	5.3 yrs

M = Mega—10^6; T = Tera—10^{12}; P = Peta—10^{15}

to less than an additional 5 days exposure to natural background radiation. Furthermore, the total collective dose from the accident to the 2 million population living within 50 miles would be less than 1% of the total radiation these people receive each year from medical and natural sources of radiation.

Another source of radiation that can be used to compare the health impact of the TMI accident is the radioactive emissions from a coal fired plant. The number of fatal cancers due to radioactive particulate emissions from the *normal operation* of a coal fired plant is estimated to be 1.5 per year, compared to 0.7, which is the expected number of cancer fatalities from the TMI 2 accident.[15] One of the prime reasons for the low amount of fission product release from TMI 2 to the environment and the corresponding small environmental effects was the containment structure.

Within months after the accident, the Pennsylvania Department of Health initiated several epidemiological studies to evaluate the possible short term and long term health effects of the TMI 2 accident. The findings indicated that: 1) there were no measurable increases in prematurity, congenital abnormalities, or neonatal deaths, in the population near TMI, 2) the observed cases of infant hypothyroidism were not related to the TMI 2 incident and these types of anomalies are not expected to result from direct or indirect exposure to radiation, and 3) to date, there has been no evidence that the number of cancer deaths in the TMI area has been different than that which would have been expected

TABLE 6
Potential Radiation Health Effects of the TMI 2 Accident to the 2 Million Population Living Within 50 Miles.[16]

	Naturally Occurring	From TMI 2
Total Cancers	541,000	< 1
Fatal Cancers	325,000	< 1
Non Fatal Cancers	216,000	< 1
Genetic Effects	78,000	< 1

had the accident never occurred. The results of the epidemiologic study do not provide any evidence of increased cancer risks to residents near TMI.

The fourth major accident, and so far the most serious one, occurred at the Unit 4 reactor of the Chernobyl Power Station located in the Ukraine about 60 miles north of Kiev near the town of Pripyat (pop. 45,000), on April 26, 1986 at 1:23 a.m. This Unit is a large, 1000-MWe plant commissioned in 1983.

The Chernobyl reactor is a water-cooled, pressure tube, graphite moderated (a moderator slows down neutrons to speeds which make it easier to produce fission) reactor. The core of this RBMK (Russian acronym) type reactor is a graphite cylinder which has vertical penetrations for fuel channels and control rods. The fuel channels are housed within zirconium alloy pressure tubes which hold within them small zircalloy clad fuel pins. The advantage of this type of design is that the reactor can be refueled while operating at full power, eliminating the need to shut down the reactor to refuel as is the case for U.S. power plants. The Chernobyl reactor was designed to be able to produce plutonium for nuclear weapons in addition to producing electricity. The Soviets retrofitted some features that would give protection from a loss-of-coolant accident to some portions of the plant, the big pipes, the pumps, and the steam separators. The core and core outlet, which is where the accident initiated, did not have any containment.

There are 1661 pressure tubes in the RBMK reactor instead of one large pressure vessel which exists in the U.S. plants. Each of these pressure tubes has a sensitive transition weld from zircalloy to stainless steel. A light cylindrical casing encloses the space of the graphite cylinder. If several of these pressure tubes in the reactor suddenly burst due to overheating or failures in the 3322 sensitive transition welds (temperature transient must be less than 30°F/hr), there is ample force and pressure available to lift the cylindrical cover, thus, opening the reactor compartment and the large reactor hall over the top of the reactor. The latter is an ordinary industrial type structure. It was not designed to hold pressure and consequently it failed to block any radioactive releases.

The chernobyl reactor design causes an operating characteristic known as a positive temperature coefficient of reactivity. This creates a positive feedback situation where when the temperature goes higher, the fission reaction increases which increases the temperature, and so on. This makes the control of the reactor difficult, particularly during an excursion. The RBMK type of reactor design would not be licensable in the U.S.

A routine shut down of the reactor was scheduled for April 25, 1986. Prior to the shutdown, an experiment was set to determine the ability of turbine momentum to produce startup electricity for emergency equipment following an interruption in the electrical power supply. Reactor power was gradually reduced from full power level of 3200 MWt to 700-1000 Mwt. However, due to a need for electricity the Unit 4 reactor continued to operate at 1600 MWt without the emergency core cooling system. During this delay, core reactivity was greatly reduced because of the build up of xenon-135 which acts as a reactor "poison", and at one point the reactor power fell to a low of 30 MWt, later on stabilizing at 200 MWt. The operator, in trying to stabilize the reactor took a number of actions, the net effect of which was to increase the void fraction and cause the power level to increase so rapidly that two explosions occured. The steam released by the failure of one or more pressurized tubes reacted with the zirconium, producing the hydrogen that exploded. Parts of the core were scattered about the building as well as on the roofs of the reactor, turbine, and auxiliary buildings producing localized fires.

On the first 6 days after the accident, out of the 80 MCi (2.96×10^{18} Bq) of the Iodine-131 inventory, over 50% was released, and out of the 6 MCi (2.22×10^{17} Bq) of Cesium-137 about 55% was released to the environment.[17] Some of the Iodine-131 was found in the milk supply of a few Scandinavian countries.

Short term effects of the accident can be summarized as follows.

Evacuated
    Pripyat 49,000
    Chernobyl 12,000
    18-Mile Radius 74,000
Acute Radiation Sickness
    Screened 350
    Hospitalized 203
Deaths
    Explosion 2
    Acute Radiation Injury
        Burns 20
        Graft vs. Host Disease 2
        Acute radiation Syndrome 7
    Total 31

There are three components to the total external exposure which may result from the Chernobyl incident, namely that from the cloud itself, from inhalation, and from the deposited radioactivity. The highest reported initial external radiation levels were about 1 mR/hr (0.01 mSv/hr). The collective dose received by the population was 2×10^6 Person-rem (9 MSv) in 1986 and it is estimated that the population will receive about 29×10^6 Person-rem (29 MSv) for the next 50 years. Such an exposure could cause in excess of 6000 cancer fatalities which is less than 0.5% of the number of cancers occurring naturally.

A vast amount of biomedical information will be learned from the Chernobyl accident, including details on the acute radiation syndrome, effects from utero exposure, effects of ingestion and inhalation exposures, and possible late stochastic effects. It is clear that the medical, technical, and the political consequences of this accident will be felt for a long time. The U.S. can indeed learn from the Chernobyl accident, particularly in the area of epidemiology, emergency evacuation planning and recovery methods. However, the accident is not of a type that U.S. reactors could have.

DECOMMISSIONING

Due to technical obsolescence, accompanied by deterioration of mechanical and electrical equipment, and gradual build-up of radioactive contaminants in the reactor system, a nuclear steam supply system will have a finite lifetime. At the time the reactor system is taken out of service a careful assessment must be made of the disposition of the reactor system and its components.

In its proposed decommissioning regulations (regulatory guide 1.86)[18], the NRC definition of decommissioning is: "to remove (as a system) safely from service and reduce residual radioactivity to a level that permits release of the property for unrestricted use and termination of license". Unrestricted use of certain areas near a decontaminated reactor is possible if the radiation levels are below 5 mR/hr, as specified in regulatory guide 1.86. Three modes of decommissioning recognized by the NRC as suitable for reactors are DECON, ENTOMB and SAFSTOR.

— DECON: Equipment, structures and portions of a facility and site containing radioactive contaminants are removed or decontaminated to a level that permits the property to be released for unrestricted use shortly after cessation of decommissioning operations.

• ENTOMB: Radioactive contaminants are encased in a structurally long-lived material, such as concrete. The entombed structure is appropriately maintained, and continued surveillance is carried out until the radioactivity decays to a level permitting unrestricted release of the property. This alternative would be allowable for nuclear facilities contaminated with relatively short-lived radionuclides such that all contaminants would decay to levels permissible for unrestricted use within a period on the order of 100 years.

• SAFSTOR: The nuclear facility is placed and maintained in such condition that the facility can be safely stored and subsequently decontaminated at some later time (deferred decontamination) to levels that permit release for unrestricted use.

No matter what the decommissioning mode is, the ultimate goal includes removal of all spent fuel from the premises.

The selection of the appropriate mode of decommissioning for a particular plant is based on a balanced evaluation of several factors, including: personal exposure (ALARA), safety, environmental impacts, regulatory requirements, land value & potential reuse of the site, technological feasibility and economics.

With the increased interest in decommissioning, the NRC has proposed revisions to 10 CFR Parts 30, 40, 50, 51, 70 and 72. Although the main thrust of the revisions is financial assurance for decommissioning, the proposed rulemaking also includes standardization of the decommissioning mode terminology. When the new rule is final, all power reactor licenses will have two years in which to submit a proposed decommissioning funding plan or a certification that financial assurance for decommissioning will be provided in an amount at least equal to $100 million. One of the main reasons for this is the estimated decommissioning costs ranging from $14 million for SAFSTOR of a facility under 100 MWe to $170 million for a facility of over 1,000 MWe. It should be pointed out that there is still uncertainty as to the actual cost of decommissioning of a large nuclear facility.

Estimated collective radiation doses available for decommissioning are mostly dependent on the mode of decommissioning and the size of the plant.[1] For a single pressurized water reactor (PWR), SAFSTOR gives an estimated collective dose of 448 person-rem (4.48 person-Sv). This value includes 4.311 person-Sv for decommissioning, 0.139 person-Sv for transportation and 0.030 person-Sv for public. For the same PWR, DECON gives a collective dose of 1350 person-rem (13.5 person-Sv).[19]

The first commercial plant to be decommissioned in the U.S. will be the Shippingport Station located near Pittsburgh. This plant has operated since 1957 and according to the Department of Energy (DOE), will provide an opportunity to demonstrate safe, cost-effective dismantlement of a large scale nuclear plant. The Shippingport decommissioning will also provide data such as collective radiation dose, technology, and costs for future decommissioning projects. Other plants such as Hallam and Daryland have been decommissioned; they were experimental, not commercial plants.

According to a recent survey[20] decommissioning of nuclear power facilities will be a very small part of the nuclear industry's activity over the next five years and virtually non-existent during the following 10 year period. This is contrary to several studies which have concluded that the nuclear industry will be experiencing significant decommissioning activity before the year 2000. Of all the power reactors in the United States, only four are planning some kind of decommissioning operation before the turn of the century and these plants represent only 390 MWe of capacity.

REFERENCES

1. Eichholz, G.G. 1978. Power Plant Siting, pp. 437-470. in: *Environmental Aspects of Nuclear Power.* Ann Arbor Science. Ann Arbor, Michigan.
2. U.S. Code of Federal Regulations, Title 10, Part 50, *Domestic Licensing of Production and Utilization Facilities,* 1986.

3. *World List of Nuclear Power Plants,* Nuclear News, August 1986/Vol 28/No. 10.
4. Kathern, Ronald L. 1983. Radioactivity From Nuclear Reactors, pp. 137-170. In *Radioactivity In The Environment.* Harwood Academic Publishers, Chur.
5. Tichler, J., and K. Norden, 1986. *Radioactive Materials Released from Nuclear Power Plants,* USNRC Report. NUREG/CR-2907. Vol. 4.
6. Kathern, Ronald L. 1983. Pathway Analysis and Dose Assessment, pp. 263-281. In *Radioactivity In The Environment.* Harwood Academic Publishers, Chur.
7. U.S. Code of Federal Regulations, Title 10 Part 20,*Standards for Protection Against Radiation,* 1986.
8. Russ, George D., 1986. Reducing the Volume Improving the Form. *Low Level Radioactive Waste: Building a Perspective.* AIF, Maryland.
9. Russ, George D., 1986. Engineered Safeguards, *Low Level Radioactive Waste: Building a Perspective.* AIF, Maryland.
10. U.S. Code of Federal Regulations, Title 10 Part 60,*Disposal of High-Level Radioactive Wastes in Geologic Repositories,* 1986.
11. Kathern, Ronald L. 1983. Radioactivity from Nuclear Reactors / Reactor Accidents, pp. 162-169. In *Radioactivity In The Environment.* Harwood Academic Publishers, Chur.
12. Behling, Hans. U. 1986. Contributing Events Of The Accident, pp. 19-20. In *Radiation And Health Effects; A Report on the TMI-2 Accident and Related Health Studies.* GPU Nuclear, Middletown, Pa.
13. U.S. Nuclear Regulatory Commission. 1979. Dose Assessment From External Exposure. pp. 12-43. In *Population Dose and Health Impacts of the Accident at the Three Mile Island Nuclear Station.* Washington, D.C.
14. Behling, Hans. U. 1986. Radioactivity Released To The Environment, p. 21. In *Radiation And Health Effects; A Report on the TMI-2 Accident and Related Health Studies.* GPU Nuclear, Middletown, Pa.
15. U.S. Nuclear Regulatory Commission. 1979. Summary of Health Effects. pp. 60-64. In *Population dose and Health Impacts of the Accident at the Three Mile Island Nuclear Station.* Washington, D.C.
16. Behling, Hans. U. 1986. Predicted Delayed or Late Health Effects, p. 32. In *Radiation and Health Effects; A Report on the TMI-2 Accident and Related Health Studies.* GPU Nuclear, Middletown, Pa.
17. Hull, Andrew. P. 1986. *Preliminary Dose Assessment of the Chernobyl Accident.* BNL-38550. Safety and Protection Division, Brookhaven National Laboratory, Upton, NY.
18. U.S. Atomic Energy Commission, *Termination of Operating Licenses for Nuclear Reactors,* Regulatory Guide 1.86, June 1974.
19. Moore, Jr. E.b. *Costs and Radiological Impacts of Decommissioning Nuclear Reactor Stations With Delayed Offsite Waste Disposal,* Nuclear

Safety, Vol. 27, No. 3, July-Sept. 1986.
20. Shillenn et al., *A Survey of Commercial Nuclear Facility Decommissioning Plans,* Proceedings of ANS International Topical Meeting on Waste Management and Decontamination and Decommissioning, Sept. 11-14, 1986, Niagara Falls, NY.

Environmental Consequences of Energy Production: Problems and Prospects. Edited by S. K. Majumdar, F. J. Brenner and E. W. Miller. © 1987, The Pennsylvania Academy of Science.

Chapter Twenty-Six
EFFECTS OF THERMAL EFFLUENTS FROM NUCLEAR REACTORS
WILLIAM D. McCORT
Savannah River Ecology Laboratory
Drawer E
Aiken, South Carolina 29802

INTRODUCTION

One of the major consequences of energy production is the discharge of thermal effluents. Thermal ecology, the study of the structure and function of ecosystems as influenced by temperature, (Gibbons and Sharitz 1974, 1981), has become increasingly an important area of research as energy demands increase throughout the world. Evaluation of environmental effects of thermal pollution as "good or bad" is often difficult since effects range from detrimental to beneficial.

The Savannah River Plant (SRP) in Aiken, South Carolina provides a unique opportunity to study the long-term effects of thermal effluents on the environment. Nuclear production reactors discharging cooling water into streams, a recirculating reservoir, and a once-through cooling lake create a full range of temperatures representative of thermal effluents not only from nuclear reactors, but also from other energy production facilities. The objective of this chapter is to highlight areas of research in thermal ecology on the SRP in order to 1) illustrate the overall effects of thermal effluents, 2) evaluate whether the effects are "good or bad", and 3) present recommendations for mitigation.

SITE DESCRIPTION

The 77,701 ha SRP is operated by the U.S. Department of Energy. Nuclear production reactors, in operation since the 1950's, have created long-term thermal conditions in reservoir, stream, and swamp habitats. Two reactors discharge cooling water heated 70 C directly into small streams (Four Mile and Pen Branch Creeks; Fig. 1). Thermal gradients are produced throughout the streams and in a 3,020 ha cypress-tupelo swamp which receives the discharges. Similar cooling water discharges into Steel Creek (Figure 1) ceased in 1968 when one of the

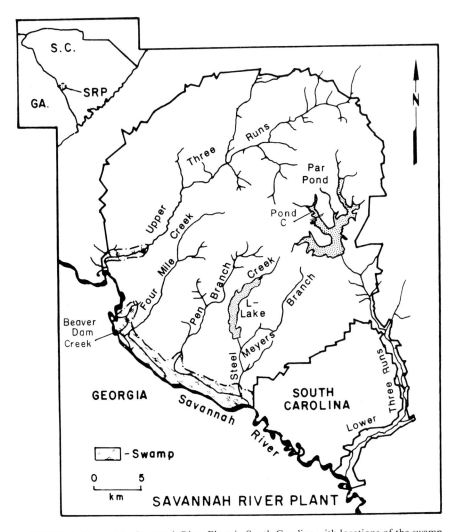

FIGURE 1. Map of the Savannah River Plant in South Carolina with locations of the swamp, streams, and reservoirs.

reactors, L-Reactor, was placed on standby. L-Reactor restarted in 1985 and began discharging cooling water into a newly constructed 405 ha cooling lake, L-Lake (Figure 1). After cooling, water temperatures at midlake are 32.2 C or less.

Par Pond, a 1,100 ha reservoir on the SRP (Figure 1), receives reactor cooling water from a fourth reactor. The thermal effluents are discharged through a series of precooling ponds and a canal system into Pond C (67 ha, Figure 1) for preliminary cooling. Pond C water at 35 to 38 C flows through a dam (Hot

Dam) and enters Par Pond where it cools to near-normal temperatures for reservoirs of the region. The Par Pond water is recycled to the reactor or discharged through a dam (Cold Dam) into Lower Three Runs Creek.

Hence, the SRP offers a wide range of thermal conditions and habitat types for study. Other impacts besides thermal effluents on SRP aquatic ecosystems are limited to on-site SRP activities since the watersheds of the thermally impacted streams and lakes lie within the SRP.

EFFECTS OF THERMAL EFFLUENTS

MORTALITY

The high temperatures (70 C) of the SRP reactor effluents exceed the thermal tolerance of most flora and fauna. Little vegetation survives from the point of reactor discharge to the mouths of the streams. Increased flows from approximately 0.3 to 1.4 m³sß1 natural flow to about 11 m³sß1 during reactor operation has resulted in silt deposition and the formation of large deltas at each stream's entrance into the swamp. Water temperatures can be as high as 25 to 30 C above ambient in the deltas.

The Savannah River Swamp is vegetated primarily with *Taxodium distichum* (bald cypress), *Nyssa aquatica* (tupelo gum), and other bottomland hardwood species. Sharitz et al. (1974a) studied the extent of tree kills from the reactor effluents. They reported that the swamp canopy was closed in 1951. Since the start of the reactors in 1953, 227 ha of the swamp forest were intensely affected (two-thirds to total tree kill). The areas of intense tree kills were predominately in the deltas of the three streams receiving reactor effluents (Four Mile, Pen Branch and Steel Creeks) as well as the delta of a fourth stream receiving heated water from a coal power plant (Beaver Dam Creek, Figure 1). Moderate effects (one-third to two-thirds of the trees killed) were observed in 263 ha of the swamp forest. Another 1,396 ha were slightly affected (< one-third of the trees killed). Tree mortality was found throughout a total of 1,886 ha (60 percent) of the swamp forest.

Although flooding for prolonged periods can kill trees, the cause of the majority of the tree kills in the deltas is probably due to the high water temperatures (Sharitz et al. 1974a). About half of the area of total tree kill in the deltas was also extensively flooded and deposited with silt. In contrast to the deltas, tree kills in the interior of the swamp, where water temperatures are less extreme, were the result of continuous inundation rather than thermal waters.

Mechanisms leading to mortality in the swamp probably resulted from direct injuries (membrane and chemical decomposition) and indirect injuries (biochemical lesion, toxicity, protein loss and starvation) (Levitt 1972, Donovan and McLeod 1985). The loss of the swamp canopy reduced the cooling effect of the canopy and additional areas of the swamp received direct injury to plant tissues. (Donovan and McLeod 1985). Even after 30 years, the extent of the damage to the swamp continues to spread (Scott and Sharitz 1985).

Baldcypress seeds germinated at sites where water temperatures were 42 C. Growth and survival were not reduced. However, only 25 percent of the baldcypress seeds shed into 50 C water survived for 72 hours (Mcleod and Sherrod 1981).

Fish mortality from thermal effluent discharge into SRP streams is low compared to the number of fish invading the streams after cessation of reactor operations (Aho and Anderson 1986a). Records of fish species composition in Steel Creek, the non-thermal portion of Pen Branch and Meyers Branch (a natural non-thermal stream, Figure 1) prior to the operation of the reactors indicate no appreciable change during the past three decades of reactor operation (Aho et al. 1986). The species composition remains high at 11 to 35 species in the streams. However, fish utilization of the heated portions of the streams during reactor operation is minimal (Aho and Anderson 1986a) because the high temperatures of the reactor effluents exceed the thermal tolerance of most fish. When the reactors shut down and the stream waters cool to ambient temperatures, the fish move upstream from the Savannah River Swamp into the streams. Initial reinvasion of the streams occurs quickly, within 12 hrs of reactor shut-down. Reinvasion is not pulsed but a continuous repopulation of the streams throughout the period of reactor outage. Stream utilization is reduced during periods of increased flow and temperature when reactor operations are restarted. It is most likely that fish find refuge from the lethal temperatures by moving downstream into the Savannah River Swamp. Backwater areas along the streams may also provide refuge areas (Aho and Anderson 1986a).

Fish recolonization is hypothesized to be the result of random movement into the streams as temperatures allow rather than a response to increased prey abundance (Aho and Anderson 1986a). Prey species of aquatic insects and other macroinvertebrate populations quickly reinhabit the streams after reactor shutdown (O'Hop et al. 1986), but prey species recolonization takes longer than the usual period of reactor down times (Aho and Anderson 1986a).

In contrast to the low number of fish kills in the streams, large numbers of fish are killed in closed cooling systems on the SRP. Cooling water from a reactor flows through 6.8 km of canals and two small impoundments before entering Pond C where surface temperatures frequently exceed 50 C. The majority of Pond C remains above 45 C during most of the year, and is too hot throughout most of its volume to support aquatic vertebrates when the reactor is operating. During reactor operation, fish are restricted to one of four cooler refuge areas which receive cooler water from streams, underground springs, and/or do not receive direct flow of hot water. When the reactor is not operating, the water temperatures decrease throughout the pond and fish leave refuge areas. Upon restart of the reactor, the fish return to their refuges. However, during reactor outages of normal duration an average of 4,592 (n = 19, range = 28 to 33,425) fish are killed during each reactor restart (Aho and Anderson 1986b). After one unusually long outage, 206,853 fish were killed (not included in average or range above because of its singular occurrence). Water temperatures can increase so rapidly that fish are killed before they can return to the refuge areas. Fish also can become isolated from the refuge areas in pockets of cool water, and eventually be killed by the high temperatures. Small bluegills (*Lepomis macrochirus*) were the species most commonly killed in such conditions. However, one kill included at least 20 largemouth bass (*Micropterus salmoides*) weighing more than 1,000 g each (Siler and Clugston 1975).

Fish kills also may occur after periods of rain (Siler and Clugston 1975). Observations suggest that rain runoff may create a cooler layer of water that flows along the bottom of Pond C to the deepest areas of the pond. Fish may swim from the deep areas into the cool water flows towards the shore. As the rain and runoff stop, the fish can be quickly trapped and killed by the high water temperatures.

PRIMARY AND SECONDARY PRODUCTIVITY

Increased water temperatures can affect primary and secondary productivity. Aboveground biomass and net primary productivity (NPP) of trees varied, inversely with the level of disturbance at three sites along Pen Branch delta (Scott and Sharitz 1985). The most disturbed site had water temperature 5 to 10 C above ambient during periods of low water. Water temperatures at the intermediate and least disturbed sites were at ambient temperatures during low water, but periodically had temperatures 10 C above ambient during periods of flooding. Aboveground biomass was significantly lower at the most disturbed site (1.2 kg m^{-2}) than at the intermediate (16.7 kg m^{-2}) and least disturbed sites (22.9 kg m^{-2}). Aboveground NPP was significantly lower at the most disturbed site (5 g m^{-2} yr^{-1}). The intermediate site (474 g m^{-2} yr^{-1}) was significantly less than the least disturbed site (626 g m^{-2} yr^{-1}).

A similar study as that described above for tree woody biomass was conducted for a herbaceous annual plant, *Ludwigia leptocarpa*. Biomass estimates were made at three sites along a thermal gradient, the highest water temperature site at 40 to 44 C, the intermediate site at 30 to 38 C, and the ambient site at 22 to 26 C. *Ludwigia leptocarpa* shoot height and standing crop, including shoot and shoot mass, were significantly higher in the highest and intermediate sites than in the ambient site plants at the end of each growing season (Christy and Sharitz 1980).

Yellow-bellied turtles *(Pseudemys scripta)* occupying thermally affected areas on the SRP have been reported to have larger body sizes and faster growth rates than turtles in natural areas (Gibbons 1970, Christy et al. 1974). The differences were not directly attributable to elevated temperatures since some of the turtles were from unheated portions of Par Pond and recovering areas that no longer receive thermal effluents. Instead, diet has been suggested as the explanation for the size and growth differences. Data indicate that Par Pond turtles have higher protein diets than turtles from natural areas (Gibbons 1970, Parmenter 1980). Changes in plant community structure in thermal and post-thermal areas (Sharitz et al. 1974b) and opportunistic carnivory in the yellow-bellied turtle (an omnivore) could lead to diet changes with higher protein content (Christy et al. 1974). An additional explanation for the increased growth is that elevated temperatures can increase feeding rates, digestive rates and activity periods (Parmenter 1980). Timing of basking and foraging behavior could be involved in maximizing feeding and digestive rates, and minimizing metabolic costs.

SPECIES COMPOSITION AND DIVERSITY

Elevated water temperatures cause changes in species composition and diversity as heat tolerances of some plants and animals are exceeded. *Typha latifolia* (common cattail) and *T. domingensis* (giant cattail) are the dominate emergent macrophytes in Par Pond. *T. latifolia* grows in < 50 cm water depths, and *T. domingensis* grows in deeper water, 50 to 150 cm. *T. domingensis* cannot tolerate elevated temperatures (34 to 40 C), but *T. latifolia* can grow in areas where temperatures are 34 to 40 C (Sharitz et al. 1984).

Najas guadalupensis (bushy pondweed) and *Potomogeton* spp. (pondweed) compete poorly with *Myriophyllum* and *Eleocharis* at normal temperatures. However, *Najas guadalupensis* and *Potomogeton* spp. are more tolerant of higher temperatures and become the dominant plants at these temperatures (33 to 34 C) (Sharitz and Gibbons 1981).

Diversity of species can be decreased by the elimination of some species unable to tolerate elevated temperatures. An unheated shoreline of Par Pond was found during a summer survey to have 33 species of vascular aquatic plants, 22 fish species, and 9 species of reptiles. In contrast, only 8 species of plants, 5 to 6 fish species (although 12 other non-resident species of fish are found in Pond C; Aho and Anderson 1986b) and 2 reptile species were recorded in the heated waters of Pond C, and no species not found in Par Pond was recorded for Pond C. The relative abundance of shoreline aquatic plants was high for Par Pond but low for Pond C (Parker et al. 1973, Gibbons and Sharitz 1974).

REPRODUCTION

Reproduction is affected by thermal effluents. Research on *Bufo terrestris* (southern toad) along a thermal gradient in Pond C has exemplified thermal effects on reproduction ranging from mortality to growth changes. Tadpoles hatched from eggs laid in water with maximum temperatures of 35 C but died during metamorphosis. Tadpoles surviving at lower temperatures showed differences in rates of development and growth. Warmer temperatures caused larvae to develop significantly faster and to metamorphose at much smaller body sizes than those at lower temperatures (Nelson 1974a). These results were duplicated in laboratory controlled experiments with *Rana utricularia* (Nelson 1974b).

Pseudemys scripta in Par Pond also show reproductive differences from turtles in normal habitats. Female *P. scripta* in Par Pond are significantly larger and have larger clutches than female turtles in other SRP natural habitats (Gibbons 1970). Male *P. scripta* reach sexual maturity at significantly younger ages than males from a natural habitat (Gibbons et al. 1981). These results are representative of either direct or indirect effects of the thermal conditions in Par Pond (Sharitz and Gibbons 1981). Female *P. scripta* increase their individual fitness by increased clutch sizes and males increase their fitness by younger sexual maturity (Gibbons et al. 1981).

Reproductive activity in *Gambusia affinis* is affected by thermal conditions. Female mosquitofish in the heated habitats of Pond C, sampled in February, June, October, and December, had embryos throughout the year. However, females in unheated habitats were in their reproductive periods only during the warmer part of the year. The mean number of embryos in the Pond C females generally increased as water temperatures decreased (Bennett and Goodyear 1978).

EVALUATION OF EFFECTS

Research in thermal ecology has provided a wealth of data on the effects of thermal alterations in the environment. Responses to increases in thermal conditions are varied with enhancement of some biological processes and the decline of others. Lists of effects categorized as beneficial or detrimental illustrate the wide range of responses to thermal effects. Variables such as age, season, and location can change responses of species to particular thermal conditions and complicate evaluation of the effects (Table 1) (Sharitz and Gibbons 1979).

The concept of a subsidy-stress gradient is applicable to many biological responses to increased thermal conditions (Gibbons 1976, Odum et al. 1979). Increased temperatures can be viewed as an energy supplement that could enrich an ecosystem. The response of the ecosystem to elevated temperatures would be enhancement of a biological process at some level of organization. With continued increases in temperatures, an upper treshold would eventually be reached, and any further temperature increases would be deleterious to the process.

TABLE 1

Detrimental and Beneficial Effects of Thermal Effluents in Aquatic Habitats. Classifications are subjective, and although there may be influences of other non-thermal factors, the thermal variable is considered to be the major influence in the table. Modified from Sharitz and Gibbons (1979), Gibbons et al. (1980) and Gibbons and Sharitz (1981).

EFFECT	CONSEQUENCE	SOURCE
Tree kill and elimination of certain other plant species in swamp and reservoir ecosystems	Detrimental	11,27,31,24,28,15
Fish mortality	Detrimental	2,32
Reduced diversity of fish	Detrimental	24,52
Reduced diversity of insects	Detrimental	11,49
Reduced diversity of reptiles	Detrimental	24
Reduced abundance and diversity of waterfowl and other birds	Detrimental	37,54
Increased parasitism and disease in fish	Detrimental	34,39,40,41,42
Emaciation of largemouth bass	Detrimental	40,47
Enhanced growth and reproduction of certain plants	Beneficial	7,38
Enhanced growth rate and increased clutch size in turtles	Beneficial	6,9,13
Enhanced growth in fish	Beneficial	4,35,36,44,50
Increased thermal tolerance of fish	Beneficial	4,43,48,53
Increased thermal tolerance of insects	Beneficial	45,51
Increased abundance and reproduction of mosquitofish	Beneficial	4,44
Enhanced fishing success during winter	Beneficial	46
Increased development of larval amphibians	Beneficial	19

An example of a subsidy-stress gradient is provided by a survey of the distribution and abundance of submerged (0.5 to 1.0 m deep) aquatic macrophytes in Par Pond (Grace 1977). The biomass of the community increased and decreased along a thermal gradient caused by thermal discharges from Pond C into Par Pond. The biomass of the macrophytes in Par Pond increased as thermal discharges from the dam elevated temperatures above ambient (26 C). Macrophyte biomass maximized in water at 31 to 32 C. Biomass decreased as temperatures continued to increase with decreasing distance from the dam. Macrophytes stopped growing about 50-100m from the effluent entry point where temperatures exceeded 33-34 C. No vascular plants were found at the time of the survey along the shoreline at the dam where temperatures were highest (Sharitz and Gibbons 1981).

Caution must be taken not to oversimplify the subsidy-stress model. Each biological process has a particular range of temperatures within which the process can proceed. The determination of a subsidy-stress gradient is complicated by interactions with other environmental gradients such as nutrient gradients, latitudinal gradients, and successional gradients. A further complication is the many hierarchical levels of organization. Beneficial effects at one level of organization could be deleterious at another. Also, effects can be beneficial on a short-term, but detrimental on a long-term time scale (Odum et al. 1979).

Categorical statements about thermal effects are seldom possible. Instead, effects need to be evaluated on a species by species and case by case basis. It is not possible to evaluate a particular thermal condition as beneficial or detrimental without a thorough understanding of the many interacting variables and levels of organization (Gibbons et al. 1980). Nevertheless, evaluations and decisions must be made regarding the costs and benefits of energy production and thermal effluents in the environment. Decisions need to be based on in-depth studies of the environment to be impacted. Gibbons et al. (1980) recommended the choice of species or environmental components be based on 1) the importance and/or "visibility" of the species to the public, 2) attributes well-suited for quantitative measurements and evaluation of stress responses, and 3) ecological value, the importance of the species or environmental component to the functioning of the ecosystem.

MITIGATION

Thermal effluents in aquatic ecosystems affect wetlands. Mitigation of wetland impacts and losses is increasing in frequency as the value of wetlands is becoming more recognized. Wetlands have been described as important for recreation, aesthetics, and as habitats for the survival of wildlife. More recently, wetlands have been recognized as natural diluters, filters and decomposers of industrial and other human societal wastes (Odum and Kroodsma 1976). Hence, state and federal regulatory agencies are more rigorously requiring the mitigation of perturbations, including thermal perturbations, to these important components of our environment. L-Lake, a thermal mitigation project on the SRP illustrates the difficulty of mitigating thermal effects on a large scale.

DOE proposed to restart L-Reactor on the SRP in order to meet increased demands for plutonium. As required by the National Environmental Policy Act, the National Pollution Discharge Elimination System (NPDES) provision of the Clean Water Act and Section 316(a) of the Clean Water Act, DOE prepared an Environmental Impact Statement detailing 33 cooling water alternatives to mitigate the impact of thermal effluent discharges to Steel Creek. The alternatives included a return to the previous operating mode of direct discharge of cooling water into Steel Creek, cooling lakes, cooling towers, and spray cooling systems. In the final EIS, DOE chose a once-through, 405 ha cooling lake, L-Lake, as the preferred alternative (U.S. Department of Energy 1984). Dam construction and the creation of L-Lake in the basin of Steel Creek was completed, and L-Reactor was restarted in October 1985.

The idea behind the construction of L-Lake was that the cooling water from L-Reactor was supposed to cool by spreading over the surface of the lake. Regulatory stipulations require that temperatures at midlake be maintained at 32.2 C or less. Furthermore, Section 316(a) of the Federal Water Pollution Control Act (1972) requires that the lake support "balanced indigenous populations" with the following characteristics:

1. The lake must not be dominated by thermally tolerant species;
2. The lake must support biotic diversity and productivity that are similar to other lakes in the region;
3. The lake biota must include representatives of all trophic levels that are typical of lakes in the region;
4. The biotic communities of the lake must be self-maintaining, not requiring restocking.

Since L-Reactor's restart, L-Lake's performance has been monitored carefully. The compliance temperature of 32.2 C or less at the lake's midline has been attained. However, the reactor discharge sometimes remains in a narrow plume depending on meteorological conditions rather than spreading over the lake's surface. The plume can move from shoreline to shoreline by the prevailing winds. Corrective measures have been made to enhance cooling in the lake. The reactor outfall has been redirected to the northern end of the lake in order to increase the effective cooling surface area of the lake. L-Lake's performance is currently being assessed.

McCort et al. (1987) analyzed the L-Lake project along with other options for mitigating the effects of the L-Reactor thermal effluents. They concluded that the major impact of the L-Lake project was the alteration of the riverine ecosystem, Steel Creek, to a lacustrine ecosystem. This alteration resulted in the manipulation of core factors (as defined by Brinson and Lee 1987) such as hydroperiod, hydrologic energy and nutrient regime. Manipulation of core factors greatly reduces the likelihood of a successful mitigation and should be avoided. Although mitigation alternatives can be conceptualized, implementation of the concept is dependent upon technology. If the technology is experimental or undeveloped, the mitigation is less likely to succeed.

McCort et al. (1987) suggest that whenever mitigation that requires the change of core factors is attempted, the likelihood of mitigation in-kind, on-site is an unrealistic endeavor. During the early planning stages of a project, it is best to resolve that impacts to wetlands will be avoided. In cases where avoidance is not possible, efforts should be focused on careful planning, goal setting, and duplication of wetland functions for successful mitigation.

CONCLUSIONS

There are many environmental effects of thermal effluents. Detrimental effects vary from extensive mortality and alteration of ecosystems to minor perturbations. There are also beneficial effects such as increased primary and secondary productivity. Absolute statements about the effects of thermal effluents are seldom possible. The many levels of biological organization and interactions within and between these levels in particular environments result in a vast array of responses to thermal conditions.

Evaluation of the thermal effects must be based on in-depth studies of the environment to be impacted. In cases where thermal impacts to the environment are deleterious, it is best to avoid the impacts. If the impacts cannot be avoided, then carefully planned mitigation can lessen the impacts and preserve ecosystem functions. Every effort should be made to not accept thermal impacts or mitigations that change or manipulate ecosystem core factors.

Finally, efforts need to go beyond the mitigation of thermal effluents and focus on the amelioration of thermal discharges (Odum and Kroodsma 1976). Thermal discharges are a tremendous loss of energy. We need to discover ways of reducing this loss and of using thermal effluents to our benefit.

ACKNOWLEDGEMENTS

This chapter was supported by the United States Department of Energy, Savannah River Operations, contract DE-AC09-76SROO-819, with the University of Georgia, Institute of Ecology, Savannah River Ecology Laboratory. The author is grateful to Dr. J. Whitfield Gibbons for his critical review of this manuscript.

LITERATURE CITED

1. Aho, J.M. and C.S. Anderson. 1986a. Post-disturbance patterns in fish recolonization of streams receiving thermal effluents from nuclear production reactors, pp. 96-118. *In:* J.M. Aho, C.S. Anderson, K.B. Floyd and M.T. Negus (Eds.) Patterns in Fish Assemblage Structure and Dynamics in Waters of the Savannah River Plant, Comprehensive Cooling Water Study Final Report. SREL-27, pp. 160.
2. Aho, J.M. and C.S. Anderson. 1986b. Dynamics of the fish kills in Pond C, life in a capricious environment, pp. 119-159. *In:* J.M. Aho, C.S. Anderson, K.B. Floyd and M.T. Negus (Eds.) Patterns in Fish Assemblage Structure and Dynamics in Waters of the Savannah River Plant, Comprehensive Cooling Water Study Final Report. SREL-27, pp. 160.

3. Aho, J.M., C.S. Anderson and M.T. Negus. 1986. Structure of fish communities in the SRP streams, pp. 79-95. *In:* J.M. Aho, C.S. Anderson, K.B. Floyd and M.T. Negus (Eds.) Patterns in Fish Assemblage Structure and Dynamics in Waters of the Savannah River Plant, Comprehensive Cooling Water Study, Final Report. SREL-27, pp. 160.
4. Bennett, D.H. and C.P. Goodyear. 1978. Response of mosquitofish to thermal effluent, pp. 498-510. *In:* J.H. Thorp and J.W. Gibbons (Eds.) Energy and Environmental Stress in Aquatic Systems. DOE Symp. Ser. (CONF-771114), National Technical Information Service, Springfield, Va., pp. 854.
5. Brinson, M.M. and L.C. Lee. 1987. In-kind mitigation for wetland loss: Statement of ecological issues and evaluation of examples. *In:* R. R. Sharitz and J.W. Gibbons (Eds.) Proceedings of Freshwater Wetlands and Wildlife: Perspectives on Natural, Managed, and Degraded Ecosystems. In press.
6. Christy, E.J., J.O. Farlow, J.E. Bourque and J.W. Gibbons. 1974. Enhanced growth and increased body size of turtles living in thermal and post-thermal aquatic systems, pp. 277-284. *In:* J.W. Gibbons and R.R. Sharitz (Eds.) Thermal Ecology. AEC Symp. Ser. (CONF-730505), National Technical Information Service, Springfield, Va., pp. 670.
7. Christy, E.J. and R.R. Sharitz. 1980. Characteristics of three populations of a swamp annual under different temperature regimes. Ecology: 61: 454-460.
8. Donovan, L.A. and K.W. McLeod. 1985. Morphological and root carbohydrate responses of bald cypress to water level and water temperature regimes. J. Therm. Biol. 10: 227-232.
9. Gibbons, J.W. 1970. Reproductive dynamics of a turtle (*Pseudemys scripta*) population in a reservoir receiving heated effluent from a nuclear reactor. Can. J. Zool. 48:881-885.
10. Gibbons, J.W. 1976. Thermal alteration and the enhancement of species populations, pp. 27-31. *In:* G.W. Esch and R.W. McFarlane (Eds.) Thermal Ecology II. ERDA Symp. Ser. (CONF-750425), National Technical Information Service, Springfield, Va., 404.
11. Gibbons, J.W. and R.R. Sharitz. 1974. Thermal alteration of aquatic ecosystems. Am. Sci. 62: 660-670.
12. Gibbons, J.W. and R.R. Sharitz. 1981. Thermal ecology: Environmental teachings of a nuclear reactor site. BioScience. 31: 293-298.
13. Gibbons, J.W., R.D. Semlitsch, J.L. Greene and J.P. Schubauer. 1981. Variation in age and size at maturity of the slider turtle (*Pseudemys scripta*). Am. Nat. 117: 841-845.
14. Gibbons, J.W., R.R. Sharitz and I.L. Brisbin. 1980. Thermal ecology research at the Savannah River Plant: A Review. Nuclear Safety. 21: 367-379.
15. Grace, J.B. 1977. The Distribution and Abundance of Submerged Aquatic Macrophytes in a Reactor-cooling Reservoir. M.S. thesis, Clemson Univer-

sity, Clemson, S.C. pp. 143.
16. Levitt, J. 1972. Responses of Plants to Environmental Stresses. Academic Press, New York, NY., pp. 697.
17. McLeod, K.W. and C. Sherrod, Jr. 1981. Baldcypress seedling growth in thermally altered habitats. Amer. J. Bot. 68: 918-923.
18. McCort, W.D., L.C. Lee and G.R. Wein. 1987. Mitigating for large-scale wetland loss: A realistic endeavor? *In:* G. Brooks (Ed.) Proceedings of the National Wetland Symposium: Mitigation of Impacts and Losses. October 8-10, 1986, New Orleans, La., The Association of State Wetlands Managers. In Press.
19. Nelson, D.H. 1974a. Growth and developmental responses of larval toad populations to heated effluent in a South Carolina reservoir, pp. 264-276. *In:* J.W. Gibbons and R.R. Sharitz (Eds.) Thermal Ecology. AEC Symp. Ser. (CONF-730505), National Technical Information Service, Springfield, Va., pp. 670.
20. Nelson, D.H. 1974b. Ecology of Anuran Populations Inhabiting Thermally Stressed Aquatic Ecosystems with Emphasis on Larval *Rana pipiens* and *Bufo terrestris*. Ph.D. dissertation, Michigan State University, East Lansing, Mich., pp. 174.
21. Odum, E.P. and RL. Kroodsma. 1976. The power park concept: Ameliorating man's disorder with nature's order, pp. 1-9. *In:* G.W. Esch and R.W. McFarlane (Eds.) Thermal Ecology II. ERDA Symp. Ser. (CONF-750425), National Technical Information Service, Springfield, Va., pp. 404.
22. Odum, E.P., J.T. Finn and E.H. Franz. 1979. Perturbation theory and the subsidy-stress gradient. BioScience. 29: 349-352.
23. O'Hop, J., B.C. Kondratieff and B.G. Coler. 1986. Recolonization of Four Mile Creek by stream invertebrates following the cessation of heated water discharge. Bulletin of the North American Benthological Society. pp. 82. Abstract.
24. Parker, E.D., M.F. Hirshfield and J.W. Gibbons. 1973. Ecological comparisons of thermally affected aquatic environments. J. Water Pollut. Control Fed. 45: 726-733.
25. Parmenter, R.R. 1980. Effects of food availability and water temperature on the feeding ecology of pond sliders (*Chrysemys s. scripta*). Copeia. 3: 503-514.
26. Scott, M.L. and R.R. Sharitz. 1985. Disturbance in a cypress-tupelo wetland: An interaction between thermal loading and hydrology. Wetlands. 5: 53-68.
27. Sharitz, R.R., J.W. Gibbons and S.C. Gause. 1974a. Impact of production-reactor effluents on vegetation in a southeastern swamp forest, pp. 356-362. *In:* J.W. Gibbons and R.R. Sharitz (Eds.) Thermal Ecology. AEC Symp. Ser. (CONF-730505), National Technical Information Service, Springfield,

Va., pp. 670.
28. Sharitz, R.R., J.E. Irwin and E.J. Christy. 1974b. Vegetation of swamps receiving reactor effluents. Oikos. 25: 7-13.
29. Sharitz, R.R. and J.W. Gibbons. 1979. Impacts of thermal effluents from nuclear reactors on southeastern ecosystems, pp. 609-616. *In:* R.A. Fazzolare and C.B. Smith (Eds.) Changing Energy Use Futures, Vol. II. Second International Conference on Ecology Use Management, Oct. 22-26, 1979, Los Angeles, California. Pergamon Press Inc., New York, pp. 952.
30. Sharitz, R.R. and J.W. Gibbons, 1981. Effects of thermal effluents on a lake: Enrichment and stress, pp. 243-259. *In:* G.W. Barrett and R. Rosenberg (Eds.) Stress Effects on natural Ecosystems. John Wiley & Sons Ltd. New York, pp. 305.
31. Sharitz, R.R., D.C. Adriano, J.E. Pinder, III., J.C. Luvall and T.G. Ciravolo. 1984. Growth and mineral nutrition of cattails inhabiting a thermally-graded South Carolina reservoir: I. Growth and the macronutrients. J. of Plant Nutrition. 7: 1671-1698.
32. Siler, J.R. and J.P. Clugston. 1975. Largemouth bass under conditions of extreme thermal stress, pp. 333-341. *In:* R.H. Stroud (Chairman) and H. Clepper (Ed.) Black Bass Biology and Management. National Symposium on the Biology and Management of Centrarchid Basses, Tulsa, Oklahoma, Feb. 3-6, 1975. Sports Fishing Institute, Washington, D.C., pp. 531.
33. U.S. Department of Energy. 1984. Final Environmental Impact Statement, L-Reactor Operation Savannah River Plant, Aiken, S.C. DOE/EIS-0108. Vols. 1-3.

ADDITIONAL REFERENCES

34. Aho, J.M., J.W. Gibbons and G.W. Esch. 1976. Relationship between thermal loading and parasitism in the mosquito fish, pp. 213-218. *In:* G.W. Esch and R.W. McFarlane (Eds.) Thermal Ecology II. ERDA Symp. Ser. (CONF-750425), National Technical Information Service, Springfield, Va., pp. 404.
35. Bennett, D.H. 1972. Length-weight relationships and condition factors of fishes from a South Carolina reservoir receiving thermal effluent. The Progressive Fish-Culturist. 34: 85-87.
36. Bennett, D.H and J.W. Gibbons. 1974. Growth and condition of juvenile largemouth bass from a reservoir receiving thermal effluent, pp. 246-254. *In:* J.W. Gibbons and R.R. Sharitz (Eds.) Thermal Ecology. AEC Symp. Ser. (CONF-730505), National Technical Information Service, Springfield, Va., pp. 670.
37. Brisbin, I. Lehr, Jr. 1974. Abundance and diversity of waterfowl inhabiting heated and unheated portions of a reactor cooling reservoir, pp. 579-593.

In: J.W. Gibbons and R.R. Sharitz (Eds.) Thermal Ecology. AEC Symp. Ser. (CONF-730505), National Technical Information Service, Springfield, Va., pp. 670.

38. Christy, E.J. 1976. Population Dynamics of Two Semi-aquatic Macrophytes Growing Under Different Temperature Regimes. M.S. thesis, University of Georgia, Athens, Ga., pp. 60.
39. Esch, G.W., T.C. Hazen, R.V. Dimock, Jr., and J.W. Gibbons. 1976. Thermal effluent and the epizootiology of the ciliate *Epistylis* and the bacterium *Aeromonas* in association with centrarchid fish. Trans. Am. Micros. Soc. 95: 687-693.
40. Esch, G.W., T.C. Hazen. 1978. Thermal ecology and stress: A case history for red-sore disease in largemouth bass, pp. 331-363. *in:* J.H. Thorp and J.W. Gibbons (Eds.) Energy and Environmental Stress in Aquatic Systems, DOE Symp. Ser. (CONF-771114), National Technical Information Service, Springfield, Va., pp. 854.
41. Esch, G.W. and T.C. Hazen. 1980. Stress and body condition in a population of largemouth bass: Implications for red-sore disease. Trans. Amer. Fish Soc. 109: 532-536.
42. Eure, H.E. and G.W. Esch. 1974. Effects of thermal effluent on the population dynamics of helminth parasites in largemouth bass, pp. 207-215. *In:* J.W. Gibbons and R.R. Sharitz (Eds.) Thermal Ecology: AEC Symp. Ser. (CONF-730505), National Technical Information Service, Springfield, Va., pp. 670.
43. Falke, J.D. and M.H. Smith. 1974. Effects of thermal effluent on the fat content of the mosquitofish, pp. 100-108. *In:* J.W. Gibbons and R.R. Sharitz (Eds.) Thermal Ecology. AEC Symp. Ser. (CONF-730505), National Technical Information Service, Springfield, Va., pp. 670.
44. Ferens, M.C. and T.M. Murphy. 1974. Effects of thermal effluents on populations of mosquitofish, pp. 237-245. *In:* J.W. Gibbons and R.R. Sharitz (Eds.) Thermal Ecology. AEC Symp Ser. (CONF-730505), National Technical Information Service, Springfield, Va., pp. 670.
45. Garten, C.T. and J.B. Gentry. 1976. Thermal tolerance of dragonfly nymphs. II. Comparison of nymphs from control and thermally altered environments. Physiol. Zool. 49: 206-213.
46. Gibbons, J.W., J.T. Hook and D.L. Forney. 1972. Winter responses of largemouth bass to heated effluent from a nuclear reactor. Prog. Fish Cult. 34: 88-90.
47. Gibbons, J.W., D.H. Bennett, G.W. Esch and T.C. Hazen. 1978. Effects of thermal effluent on body condition of largemouth bass. Nature. 274: 470-471.
48. Holland, W.E., M.H. Smith, J.W. Gibbons and D.H. Brown. 1974. Thermal tolerances of fish from a reservoir receiving heated effluent from a nuclear reactor. Physiol. Zool. 47: 110-118.

49. Howell, F.G. and J.B. Gentry. 1974. Effect of thermal effluents from nuclear reactors on species diversity of aquatic insects, pp. 562-571. *In:* J.W. Gibbons and R.R. Sharitz (Eds.) Thermal Ecology. AEC Symp. Ser. (CONF-730505), National Technical Information Service, Springfield, Va. pp. 670.
50. Lattimore, R.E. and J.W. Gibbons. 1976. Body condition and stomach contents of fish inhabiting thermally altered areas. Am. Midl. Nat. 95: 215-219.
51. Martin, W.J. and J.B. Gentry. 1974. Effect of thermal stress on dragonfly nymphs, pp. 133-145. *In:* J.W. Gibbons and R.R. Sharitz (Eds.) Thermal Ecology. AEC Symp. Ser. (CONF-730505), National Technical Information Service, Springfield, Va., pp. 670.
52. McFarlane, R.W. 1976. Fish diversity in adjacent ambient, thermal, and post-thermal freshwater streams, pp. 268-271. *In:* G.W. Esch and R.W. McFarlane (Eds.) Thermal Ecology II. ERDA Symp. Ser. (CONF-750425), National Technical Information Service, Springfield, Va., pp. 404
53. Murphy, J.C., C.T. Garten, Jr., M.H. Smith and E.A. Standora. 1976. Thermal tolerance and respiratory movement of bluegill from two populations tested at different levels of acclimation temperature and water hardness, pp. 145-147. *In:* G.W. Esch and R.W. McFarlane (Eds.) Thermal Ecology II. ERDA Symp. Ser. (CONF-750425), National Technical Information Service, Springfield, Va., pp. 404.
54. Straney, D.O., L.A. Briese and M.H. Smith. 1974. Bird diversity and thermal stress in a cypress swamp, pp. 572-578. *In:* J.W. Gibbons and R.R. Sharitz (Eds.) Thermal Ecology. AEC Symp. Ser. (CONF-730505), National Technical Information Service, Springfield, Va., pp. 670.

Environmental Consequences of Energy Production: Problems and Prospects. Edited by S. K. Majumdar, F. J. Brenner and E. W. Miller. © 1987, The Pennsylvania Academy of Science.

Chapter Twenty-Seven
GENERATION, PROCESSING AND DISPOSAL OF LOW LEVEL RADIOACTIVE WASTE IN THE UNITED STATES

KEITH G. MATTERN
Power Production Engineer
Susquehanna Steam Electric Station
Pennsylvania Power & Light Company
P.O. Box 467
Berwick, PA 18603

When producing electricity with nuclear power in the United States, the most recognized environmental effect is the shallow land disposal of Low Level Radioactive Waste (LLRW). Although small amounts of radioactive waste in the liquid and gaseous form are released to the local environment surrounding a nuclear facility, the majority of radioactive waste (in radioactive concentration) is disposed of at designated burial sites. These burial sites are usually not in the vicinity of any nuclear power plant. Although catastrophic events at a nuclear facility have a devastating effect on the local environment, as proven by what has happened in the Soviet Union, the local environmental effects from normal operation seem to minimal. Even to this date, the local environmental effects of the Three Mile Island accident have not been substantiated.

To understand why burial sites and alternative disposal options are a critical aspect of commercial nuclear power generation, it is essential to understand how and what wastes are generated and how they are processed before disposal.

WASTE GENERATION

Low Level Radioactive Waste is a result of either operation of the plant or from maintenance of the facility. Waste from plant operation can be in liquid, gaseous or solid form. Since there are strict regulatory restrictions and the concept of As Low As Reasonably Achievable (ALARA) in terms of radiation exposure to workers and environment, the amount and concentration of radioactivity that may be released in liquid or gaseous form to the local environment requires these waste streams to be processed and radioactive concentrations reduced.

The results of this processing, which are usually water purification media such as ion exchange resins and filter media, are then processed again and packaged for disposal in a burial site. This type of waste is commonly referred to as "wet waste" but is not indicative of its final disposed form.

The waste that is generated as a result of maintenance of a facility is in the form of radioactive contaminated trash. The trash usually consists of rags, protective clothing (fabric and plastic), tools and discarded components of the power plant. This type of waste is classified as Dry Active Wastes or "DAW" and is also processed before final disposal.

As with all types of wastes, the quantity and quality becomes a major factor in processing and disposal. In the United States today the current trend still remains to reduce the volume of waste and increase the radioactive concentration so that the number and size of the burial sites will be as small as possible. The amount of waste produced by a facility is usually proportional to the electrical capacity of the station and the radioisotope concentration (quality) is related to fuel (source term) integrity and processing technology employed. The current waste generation rate is approximately 22,000 cubic ft. per year for each reactor and declining steadily.

WASTE PROCESSING

When a utility decides to pursue electric generation with nuclear power, a major aspect of the operation of the plant is devoted to waste processing or "effluent treatment." With the regulatory, political and economic motivations to limit its radioactive effluent, power plants use many advanced technological methods for reducing the liquid and gaseous waste and concentrating their solid waste to be released to the environment.

As described previously, the local environmental impact is reduced by controlling the amount of radioactivity released in the liquid and gaseous form. Liquid effluent releases are usually made to large bodies of water, such as rivers or lakes so that the concentration is reduced substantially. Gaseous effluents are filtered to remove particulate radioactive contamination and stored in delay tanks to allow noble gases to decay before they are released and dispersed into the local atmosphere. The total radioactive amount and concentration limits that are established for releases are based on calculated radiation exposure to a member of the public from the diluted effluents. An elaborate Radiological Environmental Monitoring Program is required by the facility license to ensure that the concentrations of radioactive materials released and levels of radiation the public is exposed to are not higher than expected. The program may consist of periodic radioisotopic measurement of ground and surface water supplies, agricultural samples, livestock samples, and soil samples located near the plant.

To minimize impact on radiation exposure to the local environment, liquid and gaseous wastes are processed, evaluated and monitored at all times.

Methods used to treat gaseous effluents usually consist of filters (fiber or carbons) to remove radioactive particulates and storage volumes to allow radioactive decay before release. When these media effectiveness degrades they are disposed of as Dry Active Waste.

Liquids that will be released to the public are treated to remove the soluble and insoluble radioactive impurities in them. Demineralization, reverse osmosis, evaporation and high efficiency filtration are the most common methods of treatment used today. These processes produce a large volume of waste "sludges" that are then processed as "wet waste" and are processed once again for disposal. It should be understood that no solid radioactive waste is disposed near a power plant and the majority of its radioactive waste is in the solid phase. The utility is forced then into forming a posture signaling the efforts it will make to conform to regulatory limits and minimizing the impact on the local environment in addition to reducing its processed waste.

SOLID WASTE DISPOSAL

After transforming most of its radioactive waste into a solid phase the utility has to prepare the waste for final disposal. With only three active waste burial sites open at the time of this writing and the limited space allocated for each plant, the waste is processed into its final form to reduce the volume and meet regulatory packaging requirements. Listed is the current operational burial sites with its availability for disposal of Low Level Radioactive Waste;

Site	Space Available (area)	Percent Filled
Barnwell, South Carolina	180	33%
Beatty, Nevada	60*	90%
Hanford, Washington	100	20%

*Twenty additional acres are designated for chemical waste disposal.

The present packaging requirements are based on the radioactive concentration of the waste. Wet waste that has a concentration of one micro-curie per cubic centimeter or greater requires that the final waste will be in a "stable" form. This is usually accomplished by mixing a nonradioactive media such as cement, asphalt or polymers with the waste to form a free standing, monolithic, homogeneous solid. Another method is to package the waste in a High Integrity Container (HIC) and remove the free water by vacuum or heat. The purpose of the HIC is to provide the stability to the final waste form without involving the addition of nonradioactive media or creating a chemical reaction with the waste.

Dry Active Waste (D.A.W.) is usually placed in strong type steel containers in preparation for burial. Volume reduction techniques such as shredding and compacting are typically used in the industry.

Once the waste is properly packaged it is then transported to the burial sites and placed in specific trenches depending on the classification of waste. When a trench is filled it is covered with dirt and grass is grown over the top which results in a large area of an open field with sampling pipes protruding from the ground. A marker is placed at the head of the trenches stating contents of the trench which completes the illusion of an eternal graveyard. Recent experiments in growing shallow rooted evergreen trees is an attempt to indicate that there may be a future for the nuclear waste disposal sites.

CURRENT TRENDS IN SOLID WASTE DISPOSAL

With the increasing levels of Low Level Radwaste production in the 1970's and early 1980's it was evident that the host states of burial sites would conclude they were the "dumping ground" for the by-product of nuclear power generation stability and prosperity. Even with increasing burial fees (the economic penalty for utilities) for disposal which is the only way to express the host states displeasures and reward its constituents, the nation was forced into developing a new posture.

It appears the new posture is dilution and not concentration. The method to dilute the environmental, political and economic impact will be by:

1. Creation of regional compacts.
2. Approval of "deminimis" (waste disposal in local landfills).
3. Incineration of waste.

These trends will increase power plants local environmental radiation exposure but will not leave the large scar that the host states must hide. To support this posture will require the present environmental monitoring programs to be expanded or new ones created. If new programs are created, the federal and state governments must decide soon what agencies will implement the programs. These programs will provide the public with some security and the scientific community with data to predict effects of low level radiation.

The biggest challenge in the next decade will be convincing the public that small amounts of radioactive material disposed of in local landfills and small amounts of increase contamination in the air from incineration will have smaller environmental impact and risk on the present average 200 mrem radiation exposure we receive from other sources. Will shallow burial be the final solution?

REFERENCES

1. Brenneman, Faith, N. 1986. A review of low level radioactive waste compacts on a national level, pp. 29-41. In S.K. Majumdar and E.W. Miller (Eds.) *Management of Radioactive Materials and Wastes: Issues and Progress.* The Pennsylvania Academy of Science, pp. 405.
2. Susquehanna Steam Electric Station Facility License NPF-14, Technical Specifications, Appendix A.
3. Electrical World, July, 1986. Haxy Radwaste Rules Hold Up Utility Action.

Environmental Consequences of Energy Production: Problems and Prospects. Edited by S. K. Majumdar, F. J. Brenner and E. W. Miller. © 1987, The Pennsylvania Academy of Science.

Chapter Twenty-Eight
RADIATION PROTECTION ASPECTS OF THE TMI-2 ACCIDENT AND CLEANUP
JAMES E. HILDEBRAND[1] and MICHAEL J. SLOBODIEN[2]

[1]GPU Nuclear Corporation
Parsippany, NJ 07054
and
[2]GPU Nuclear Corporation
Forked River, NJ 08731

INTRODUCTION

On March 28, 1979, the fuel core of the Three Mile Island Unit 2 nuclear reactor was severely damaged. The impact of the accident and the lessons learned have had a profound impact on the nuclear industry. The reader is referred to other reports regarding the causes of the accident and the lessons learned.[1,2,3,4,] The purpose of this paper is to discuss the radiation protection aspects of the accident and the cleanup.

PUBLIC HEALTH IMPACT OF THE TMI-2 ACCIDENT

Although the damage to the TMI-2 nuclear core was severe, only small quantities of radioactivity were released to the environment resulting in small doses to the public. Immediately following the accident, a number of independent scientific groups were established by the Nuclear Regulatory Commission (NRC), State of Pennsylvania, and the nuclear industry to investigate the accident and determine the doses to the off-site population. There was general agreement among these investigative committees that the radiation doses to the general public were small. Among the most noteworthy committees was an Ad Hoc Group consisting of dosimetry experts from the NRC, the United States Department of Health, Education and Welfare, and the Environmental Protection Agency. The Ad Hoc Group report report concluded that the maximum exposure to any one individual was less than 100 mrems (0.1 rem). This group further determined the population dose within a 50-mile radius to be about

3300 person-rems! Based on these findings, it was concluded that there were no immediate health effects and that long-term effects to the public would be minimal. Furthermore, a commission established by President Carter to determine the potential health effects to the population and workers concluded that ". . . since the total amount of radioactivity released during the accident at TMI was so small, and the total population exposed so limited (that), there may be no additional detectable cancers resulting from the radiation. In other words, if there are any additional cancer cases, the number will be so small that it will not be possible to demonstrate this excess or to distinguish these cases among the 541,00 persons (of the 2.2 million population) living within the fifty-mile radius of TMI, who would for other reasons develop cancer in their lifetime."[5]

Following the accident, the Pennsylvania Department of Health initiated several epidemiologic studies to evaluate the possible long-term effects of the TMI-2 accident.[6,7] A TMI population registry was developed to collect information on people living in the immediate area of TMI. In September 1985, the Pennsylvania Department of Health issued a report on one cancer study. This study showed that to date, no increased incidence of cancer had been demonstrated. The Department of Health stated that, "results of our epidemiologic study...do not provide evidence of increased cancer risks to residents near the TMI nuclear plant."[8] These studies will continue since the latency period between exposure to radiation and the development of cancer may be many years.

RADIATION EXPOSURES TO CLEANUP WORKERS

Although causing inconsequential health effects to the public, the TMI-2 accident resulted in hostile radiological environments in the plant, thus requiring effective radiation protection measures to ensure a safe working environment for cleanup workers. The cleanup presented GPU Nuclear Corporation (GPUNC) with radiation protection challenges unprecedented in the nuclear power industry. Even so, throughout the cleanup the commitment of GPUNC has been to perform the difficult and complex activities with the primary objective being the health and safety of the workers and the public. This objective has been achieved in both cases.

External Exposure
Throughout the cleanup, predictions of high worker doses have been the subject of considerable interest. In fact, in the early stages of the cleanup, there were those who were calling for the establishment of a separate worker registry for TMI-2 cleanup workers to allow long-term monitoring of possible worker health effects. It was expected that TMI-2 workers would be a highly exposed population and should be treated differently from other nuclear plant employees. However, through an effective radiation protection program, worker exposures

have been minimized and have been far lower than predicted.

Through 1986, the TMI-2 cleanup collective dose, including the accident recovery dose, has been 3729 person-rems or an average of 466 person-rems per year (Table 1). For perspective, a typical U.S. power reactor would have expended about 5200 person-rems during the eight-year period of the cleanup (1979-1986).

TABLE 1

Collective and Average Dose History for the TMI-2 Cleanup Project

Year	Collective Dose (person-rems)	Maximum Individual Dose(rems)	Average Individual Dose(rem)
1979	488	4.5	.23
1980	193	2.1	.15
1981	138	2.0	.16
1982	384	3.0	.42
1983	373	2.7	.49
1984	514	3.7	.77
1985	722	3.5	.58
1986	917	3.4	.54
Total	3,729		

Also, as shown in Table 1, since 1979, no worker has received an annual whole body dose greater than 4 rems. (The federal limit allows 3 rems per quarter not to exceed 12 rems per year with an average of 5 rems per year). Furthermore, the average individual exposure for the eight years has been about 400 millirems (0.4 rem) per year as compared to an industry average of about 650 millirems (0.65 rem) per year per worker. These data demonstrate that the health risk to TMI-2 workers has been no greater than that to employees of other nuclear plants in the U.S.

Furthermore, the sum of all of the individual occupational doses for the entire cleanup period is less than the sum of non-occupational natural background doses for the same work force and less than half of the total doses from natural and medical radiation sources in the same period. Taken as a whole, the TMI-2 work force has experienced very low risks from occupational radiation exposure. Putting the risk into perspective, the health risk to the average TMI-2 worker is less than the risk associated with being overweight by one ounce or smoking two cigarettes in a year. Such health risks are well within those which we are willing to accept in our everyday lives.

Internal Exposure

Another source of exposure arises from radioactivity taken into the body by ingestion or inhalation. This has been a very minor source of exposure to TMI-2 workers. In the eight-year period of the cleanup, over 36,000 internal radioactivity measurements were made as part of the TMI-2 Radiological Controls program. Fewer than 2% of these measurements indicated internal radioactivity giving rise to measurable doses. The contribution of the total radiation dose due to internal radioactivity has been negligible when compared to those sources outside the body. It is clear that at TMI-2 sources of radioactivity within the body are not significant.

Skin Contamination

Throughout the cleanup, effective measures have been taken to minimize the contamination of workers' skin. Contamination to the skin sometimes occurs when exposed skin comes in contact with surfaces contaminated with radioactive materials. Contamination of the body is minimized by the use of special protective garments during work in areas containing loose surface radioactive contamination. These include cloth coveralls, plastic "wet suits," cloth head covering, and rubber gloves and boots.

However, even with the use of these garments, skin contamination occasionally occurs, When it does, it is usually to a small portion of the skin. In most cases, the contamination is easily removed by washing with soap and water. The actual dose to the skin from the contamination is very small—usually less than 10 millirems (0.01 rem). (The NCR limit for skin contamination is 7500 millirems or 7.5 rems per calendar quarter.)

MANAGEMENT OF THE RADIATION PROTECTION PROGRAM

The Radiation Protection Organization

Prior to the 1979 accident at TMI-2, the radiation protection and plant chemistry functions were combined and reported to the plant administrator. In 1979, a distinct group was formed dedicated to radiation safety and organizationally independent of the plant operations division. The Radiological and Environmental Controls Division (R&EC) was established with a separate reporting chain to the company president through its own division Vice President/Director. R&EC is responsible for radiation protection, radiological and non-radiological environmental monitoring, industrial safety, and corporate medical programs at two operating power reactors in addition to the TMI-2 plant.

The radiation protection charter of the R&EC Division includes the following elements:
- Establish and maintain policies, procedures, standards and practices.
- Provide administrative and technical guidance applicable to radiation protection, use and control of radioactive materials, respiratory protection and radiological work planning.
- Conduct an ongoing program of radiological effluent monitoring.
- Train and qualify technicians in radiation protection procedures and techniques.

- Monitor and control external and internal exposures through a program of workplace and worker monitoring.

A full time professional staff including technical managers, engineers, scientists, first line supervisors and field technicians carries out the division charter. The staff includes persons with bachelor and advanced degrees in biological and physical sciences, engineering and includes professionals certified by the American Board of Health Physics, American Board of Industrial Hygiene, and National Registry of Radiation Protection Technologists.

The radiation protection organization is responsible to develop procedures, plan work in radiologically controlled areas, monitor the workplace and environment to establish radiological conditions, and maintain accountability of radioactive materials. A separate internal assessment function reporting directly to corporate officers is designed to ensure that senior management's attention is drawn to solving problems long before they become serious.

Radiation Protection Training

The safe conduct of work in the presence of ionizing radiation requires special training. Several types of training have been conducted to support TMI-2 cleanup efforts. Most of these programs are standard throughout the nuclear power industry and include General Employee Training (GET), Radiation Worker Training (RWT) and Respiratory Protection Training. In the GET program, all persons employed at the TMI site are give a basic awareness of the nature of radiation and emergency response actions. Persons whose work requires them to enter radiologically controlled areas participate in RWT—a curriculum that addresses the nature of radiation, its effects on the body, comparative health risks, work techniques to minimize exposure to radiation, the nature of contamination and risks. A separate program is devoted to respiratory protection for persons who may require it. The RWT and respiratory protection programs include sessions which involve practical exercises of wearing protective equipment and using monitoring instrumentation.

A special program was established in 1985 for managers, supervisors, and engineers to develop awareness of techniques for design of systems, work planning and conduct of work performance to minimize exposure to radiation. The major thrust of this program is to emphasize that exposure minimization starts with design concepts and is the responsibility of all segments of the project not just the radiation protection professionals.

HEALTH RISK MANAGEMENT

Throughout the cleanup, the focus of the GPUNC radiation protection program has been to minimize the overall health risk to the worker including heat stress, visual and hearing acuity and cardiopulmonary stress, as well as ionizing radiation. This approach, for example, has been considered in specifying protective clothing requirements for work in highly contaminated areas. As protective clothing is prescribed, trade-offs have to be considered between protecting the worker from skin contamination and impacting his ability to perform his intended task at maximum efficiency and with minimal exposure. Sufficient layers of protective clothing could be worn by the worker to preclude any chance of skin contamination. However, by doing so, the worker performs less efficiently, thereby reducing his whole body dose but increasing his heat stress potential.

The same approach has been applied in specifying respiratory protection. Respiratory protection devices have been gradually reduced as the various parts of the plant have been decontaminated with no appreciable internal dose received by workers. Until mid-1984, respirators were required for all work in the Containment Building. However, as airborne levels were reduced, it was no longer necessary for respirators to be required for certain jobs. In fact, as with protective clothing, it was in accordance with the as low as reasonably achievable (ALARA) philosophy to remove respirators since without the respirator the employee works more efficiently, thereby reducing his external dose for the job. In performing the job without a respirator the worker may receive some low level internal dose, but overall the total dose, considering both external gamma radiation and dose from internally deposited radionuclides, will be lower without the respirator.

Because of the unusual nature of many of the cleanup operations, practical solutions to worker comfort problems have had to be developed. Apparent in the prescription of radiological controls measures have been the consideration of the overall risk to the worker.

SIGNIFICANT TECHNICAL ACHIEVEMENTS

Because of the nature and magnitude of radioactivity encountered in the TMI-2 plant, a number of technical protection problems have been overcome through innovative solutions. Some of these are addressed below.

External Dosimetry

One of the most significant unknowns encountered during the cleanup was exposure of personnel to high levels of high energy beta radiation. Because of the fission products released from the TMI-2 core, beta and gamma emitting radionuclides were discharged into the Auxiliary and Containment Buildings. A principal long-lived beta emitting fission product is strontium-90 in equilibrium with its daughter, yttrium-90, which decays with the emission of a high energy beta particle. Prior to the accident, this type of high energy beta emitter had rarely been dealt with by health physicists in the nuclear power industry. Historically, beta exposure to the skin and eyes of workers had never been a significant radiation protection problem in the industry because of the absence of strontium-yttrium-90 contamination in the operating plant. Thus, very little experience existed in the industry in monitoring for and protecting workers against such high energy beta radiation.

Prior to the accident, a thermoluminescent dosimeter (TLD) consisting of two lithium fluoride chips was used to monitor personnel exposures. Primarily because of the existence of high energy beta contaminants in the plant, in 1980 GPUNC began development of a highly automated dosimetry processing system, including computer-controlled hardware and quality control devices. The TLD is able to discern between beta and gamma radiation and is more sensitive than previously available equipment. This TLD system uses a dosimeter with four separate sensing elements to provide data suitable for assessing both the quantity of dose and the type of radiation to which the worker is being exposed. Several independent tests demonstrated that this system performed well and the TLD was put into full operation in 1983. In 1984, the dosimeter was accredited by the Natural Bureau of Standards National Voluntary Laboratory Accreditation Program (NVLAP). The GPUNC dosimetry program was among the first to receive NVLAP Accreditation.

In addition to TLDs which must be analyzed by an automated processor, persons who enter radiologically controlled areas to perform work controlled by a Radiation Work Permit are also provided with a self-indicating pocket ionization chambers. These devices are similar in size to a ball point pen. The device is equipped with an eyepiece which permits examination of a scale so that at any time a worker can determine his own dose. Dosimeters are used to enable the wearer to periodically check and keep track of his own radiation exposure during a job. The official record of radiation dose is, however, obtained from the TLD. Each pocket dosimeter is routinely tested as part of a quality control program.

TLD badges and pocket dosimeters are worn on the portion of the body that is expected to receive the highest dose, normally at the thigh, waist or chest. In some special situations, dosimeters are also worn at other locations, for example, on the hands, fingers or head. In addition, electronic dosimeters which can be set to signal an audible alarm are used for much of the work in the plant and all work in the TMI-2 Containment Building.

Automated Data Management

The data management tasks related to radiation protection are considerable. The utility recognized the need for a comprehensive Radiation Exposure Management system (REM) in 1975. The system which has developed over the past 11 years is now one of the largest of its kind. The REM system operates on a mainframe computer to serve the needs of three nuclear stations and a corporate office. In 1986, the system had accessible storage of about 4 billion bytes and processed over three million transactions.

The features of the REM system include the ability to maintain radiation exposure records for various parts and organs of the body and automatically select the proper dose to assign for a given monitoring period based on the types of radiation to which a person was exposed. The system operates in real time to keep track of entries into and exits from the radiologically controlled areas. Associated with each entry are data elements for radiation dose, total time spent in the area, levels of airborne radioactivity experienced, and the nature of the work performed including system and component. The REM system has supported as many as 100 remote terminals with typical on-line response times of ten seconds or less.

Heat Stress Program

In the first few years of the decontamination work in the Containment Building, a major problem that had to be overcome was the potential for heat stress of workers. Because of the high contamination and airborne radioactivity levels in the building, workers wore layers of protective clothing including respirators and impermeable plastic suits. The result was that the potential for heat stress, rather than radiological conditions, became the limiting stay-time factor. In the summer months when temperatures approached 90 degrees in the building, workers could tolerate less than one hour of work. In response to the problem, GPUNC developed a comprehensive heat stress control program consisting of employee training, administrative controls, and personal cooling devices.

The heat stress index utilized at TMI is a computerized program based on input of environmental parameters and metabolic work rate estimates. Given this input, a safe stay-time is calculated for the specific task to be performed in a work area. The most notable feature of the program is its ability to determine the insulative effect of different protective clothing. In addition to providing the evaluation of specific jobs, a table of stay-times has been generated based on work rate, air temperature, and clothing type. These guidelines are used by supervisory personnel in planning and conducting work.

While training and administrative controls provide an increased margin of safety for the worker, they do not solve a more basic problem—how to provide protection while increasing productivity. As a result, personal cooling devices in the form of vortex cooling suits and ice vests were added to the program. When mobility is required, ice vests are utilized. The ice vest consists of a tight fitting vest with 60 small pockets of ice with a total ice capacity of about eight pounds. The ice serves as a heat sink for metabolic heat produced by the body and dissipates this heat as it melts. The ice vest has been very successful in extending work times in hot environments, in some cases by a factor of two to three.

When a longer stay time is required and limited mobility can be tolerated, a vortex cooling suit is utilized. The vortex suit is a one-piece coverall made of heavy vinyl material with an internal air distribution system that directs air across the body. The vinyl material is impermeable and effectively serves as one layer of protective clothing suitable for splash and vapor protection. The key feature of the system is the vortex tube which is able to segregate compressed air into hot and cold fractions, the cold fraction being directed into the suit's air distribution system. As with the ice vest, worker acceptance of the vortex suits has been very good.

The heat stress control program developed at TMI has been well received throughout the industry and has been adopted at other power plants to mitigate the risk of heat stress among employees working in high heat environments.

Robotics Technology

Some areas of the TMI-2 plant contain sufficiently high radiation levels that worker entries have been prohibited. The Containment Building basement, for example, has radiation levels of 5 rems per hour generally, with some discrete zones of up to 1100 rems per hour. These high levels are the result of the reactor coolant water discharged to the basement during the accident. Since these conditions prevent manned entries, the use of robotics is the only feasible method of performing data gathering and decontamination.

The TMI-2 cleanup appears to be in the forefront of robotics technology to perform decontamination operations in hostile environments. The objective of robotics usage at TMI-2 has been to use existing technology to minimize research and development. For effective performance, working vehicles are needed with high strength, flexibility, and reliability. No one machine will perform all the activities required. As such, a number of units have and are being tested and used at TMI-2.

Early in the cleanup, small robots were used to enter cubicles to obtain radiation readings. The first robot to enter the Containment Building basement was the Remote Reconnaissance Vehicle (RRV) known as "ROVER" (Figure 1). "ROVER" was developed by Carnegie-Mellon University under sponsorship of the Electric Power Research Institute (EPRI). It is a 1000 lb. machine equipped with three television cameras and three radiation detectors. This unit has performed radiation monitoring of the basement and obtained sediment samples from the basement floor. ROVER is lowered into the basement from a floor hatch on the ground floor of the building. Upon being removed, ROVER passes through a shower for decontamination. The robot is tethered with a self-winding mechanism and is controlled from a remote console outside the containment.

The application of robotics technology at TMI-2 will be a valuable and essential tool for performing surveillance and decontamination of the Containment basement. A considerable amount of remote decontamination and dose reduction work must be done before manned entries can be allowed. Decontamination of the basement remains a very challenging project and one that must rely largely on robotics. The information resulting from the TMI-2 experience will prove to be highly valuable for ALARA considerations in the nuclear industry.

FIGURE 1. Remote Reconnaissance Vehicle (RRV) used in the TMI-2 Cleanup

IMPACT ON RADIATION PROTECTION PROGRAMS

To say that the TMI-2 accident has had an impact on radiation protection programs in the nuclear industry would be understating the case. For throughout the industry, the conduct of radiation protection practices has undergone significant changes resulting in an overall upward trend in radiation safety performance. Although certainly not the sole reason for the changes, TMI-2 was the impetus behind the recognition that improvements were needed in all phases of nuclear operations including radiation protection. There appear to be four major reasons that have been responsible for the dramatic changes made in radiation protection practices.

First, utility management has recognized the need for and has insisted on commitment to excellence in radiation protection. The reason for this commitment is stated quite simply: Good radiation protection practices benefit nuclear operations.

Second, with the assistance and urging of the Institute for Nuclear Power Operations (INPO), radiation protection practices are being standardized through the industry. Through INPO plant assessments, good practices and guidelines, more effective procedures have been implemented with the overall affect of improved performance by radiation workers.

Third, although there have been no significant increases in regulatory requirements in the area of radiation protection, the TMI-2 accident resulted in increased emphasis and scrutiny by the NRC in this area.

A fourth area of influence has been the impact on the health physics profession. Like GPUNC, all utilities have taken steps to achieve improvements in their radiation protection programs. This effort has required the additional staffing of professional health physics. Because of the need by the utilities to achieve technical health physics credibility, more reliance is now given to the professional health physicist than in past years. Thus more and more qualified health physicists have been lured into the nuclear industry, resulting in more credible radiation protection programs. The overall effect of these changes has been a continuing improvement in the conduct of radiation protection programs which will result in further reducing the already low risk of occupational radiation exposure.

CONCLUSION

In summary, three key points need to be emphasized. First, it is evident that the TMI-2 experience has had a considerable influence on the conduct of radiation protection practices in the nuclear industry. The changes brought about by the influence of TMI-2 have resulted in an overall improvement in radiation safety for the nuclear workforce. Furthermore, a considerable amount of information has been gained from the cleanup experience that will have practical application for the health physicist in protecting workers in the industry.

Second, although the TMI-2 accident caused considerable damage to the reactor core, only small doses of radiation were received by the off-site population. A number of independent scientific committees concluded that the health impact would be minimal. Health studies are continuing to assess the long-term impact of the accident.

Lastly, the TMI-2 cleanup has not proven to be the high dose project that was originally predicted. Through sound ALARA practices, comprehensive dose reduction measures, effective job planning and a strong management commitment, the cleanup has been performed with minimal health risk to TMI-2 cleanup workers.

REFERENCES

1. U.S. Nuclear Regulatory Commission, NUREG 0558, 1979. "Population Exposure and Health Impact of the Accident at the Three Mile Island Nuclear Station." (All NRC documents are available from Superintendent of Documents.)
2. U.S. Nuclear Regulatory Commission, NUREG 0600, 1979. "Investigation into the March 28, 1979, Three Mile Island Accident."
3. U.S. Environmental Protection Agency, EPA 600/4-80-0049, 1980. "Investigations of Reported Plant and Animal Health Effects in the Three Mile Island Area."
4. Rogovin, M. and Frampton, T., NUREG/CR-1250, 1980. "Three Mile Island—A Report to the Commissioners and to the Public."
5. The Presidents's Commission on the Accident at Three Mile Island, 1979. "Staff Reports of the Public Health and Safety Task Force to the President's Commission on the Accident at Three Mile Island."
6. Pennsylvania Department of Health, Division of Epidemiology Research, Harrisburg, PA., 1981. "Three Mile Island Population Registry, Report 1, A General Description."
7. Pennsylvania Academy of Science, 1981, "Impact of TMI Nuclear Accident Upon Pregnancy Outcome, Congenital Hypothyrodism and Mortality."
8. Commonwealth of Pennsylvania, 1985. "Cancer Mortality and Morbidity (Incidence) Around TMI."

GLOSSARY OF SELECTED TERMS

ALARA — An acronym for *As Low As Reasonably Achievable* referring to the practice of management techniques, engineering controls, and work practices designed to make individual and collective doses from radiation as low as reasonably achievable considering the economic and societal benefit derived from the doses.

EPRI — Electric Power Research Institute.

INPO — Institute of Nuclear Power Operation—an industry established institute designed to set and monitor performance standards, provide technical guidance, and act as a central clearinghouse for good practices.

NRC — United States Nuclear Regulatory Commission.

REM — A unit of radiation dose relating to the energy deposited in tissue and the consequent health risk from that deposition of energy (1 millirem = 1/1000 of a rem).

TLD — Thermoluminescent Dosimeter—Crystalline materials which store a portion of the incident radiation energy and upon controlled heating release the stored energy as light. The amount of light emitted is proportional to the radiation dose received by the dosimeter.

external dose — The dose which results from sources outside the body. Examples of sources include diagnostic x-rays, radioactive materials in pipes or sealed systems.

internal dose — The dose to individual organs or the entire body which results from radioactivity within the body. The internally deposited radioactivity may result from inhalation or ingestion of radioactive material.

person-rem — The sum of all of the individual doses for a given population—often referred to as the collective dose.

Environmental Consequences of Energy Production: Problems and Prospects. Edited by S. K. Majumdar, F. J. Brenner and E. W. Miller. © 1987, The Pennsylvania Academy of Science.

Chapter Twenty-Nine
ENVIRONMENTAL AND LEGISLATIVE CONSEQUENCES OF STRIP MINING

LEE W. SAPERSTEIN
Professor of Mining Engineering
College of Earth and Mineral Sciences
The Pennsylvania State University
University Park, PA 16802

This section discusses the present environmental controls imposed on surface mining for coal and indicates strategies for improved reclamation performance from modern surface mines. In this discussion, there is a review of significant literature references to remedies for curing the effects of past uncontrolled surface mining and then a review of the legislative controls on surface mining, along with a brief history of the development of these controls. Further, there are subsections on the specific workings of the present federal code on strip mining with respect to performance standards for new mines, restoration of abandoned mine lands, and the designation of lands unsuitable for surface coal mining. The final sub-sections deal with research in reclamation and optimization of reclamation goals.

These comments ae confined to surface mining for coal, commonly called strip mining, because most legislation for environmental control of mining is specific to coal. The technology of surface coal mining is actually common to all shallow, flat-bedded deposits: phosphate, kaolin clay, bauxite, and other similar substances.

ENVIRONMENTAL IMPACTS OF SURFACE MINING

In 1967, a special study committee of the Department of the Interior revealed the magnitude of environmentally degraded lands in the United States resulting from surface mining. Total lands disturbed were 3.2 million acres, of which approximately two-thirds required reclamation. Coal lands, included in the larger total, were 1.3 million acres. Also, 13 thousand miles of streams had been affected, as had 1.7 million acres of wildlife habitat (U.S. Department of the Interior, 1967). This report, along with an earlier, interim report (U.S. Department of the Interior, 1966), was written in response to a charge in the "Appalachian Regional Development Act of 1965" (Public Law 89-4, Section 205

(c) to study "a comprehensive long range program for the purpose of reclaiming and rehabilitating strip and surface mining areas in the United States." The report focussed the nation's attention on unreclaimed mine lands and brought together the many groups who had been working toward improved legislative control of mining and reclamation.

Periodically since that time, additional surveys have been published that refine the total numbers and add the acreage mined since 1967 (Paone, et al., 1974, U.S. Department of Agriculture, 1979, Johnson and Paone, 1982). By 1980, the total acreage "utilized" by the mining industry was 5.7 million, of which 2.7 million had been reclaimed. Coal, alone, contributed 2.7 million acres of the total area utilized.

The country responded to this focussed attention in two ways: one, extensive research on mine reclamation was performed and results published broadly; two, legislative attention concentrated on the passage of a federal surface mine law. This subsection will detail some of the significant literature of this period and the next subsection will deal with the legislation.

In many ways, the research that was performed was oriented toward finding best existing practices in reclamation. There was substantial anecdotal evidence that many mines did good reclamation work and had resolved many technical problems. Mines had been restored for farmland, recreation land, and income-producing developments; the literature compiled and organized systematically the best in reclamation technology. The environmental problems covered ranged from backfill and contour restoration to prevention of degraded drainage. The Environmental Protection Agency published a number of works during this period, several of which are referenced for their seminal nature (U.S. Environmental Protection Agency, 1973, Grim and Hill, 1974). Shortly afterwards the Bureau of Mines contracted for another major work (Skelly and Loy, 1975) that linked together mining methods, costs, and reclamation techniques. In addition, the Appalachian Regional Commission had produced a major work (Appalachian Regional Commission, 1973) that also examined the cost of restoring problems caused by mining. In this case, the analysis was of technical remedies that had been tried for the elimination of problems of past (often underground) mining.

In 1974, the National Academy of Sciences undertook, with funding from the Ford Foundation's Energy Policy Project, one of the first of a series of influential works dealing with surface mining. In this case (National Academy of Sciences, 1974), the study committee attempted to assess the capacity of the western coal lands of the United States to be rehabilitated after mining. Subsequently, the Academy, through the National Research Council, has undertaken a number of studies on the consequences of mining, the reports of which serve as excellent sources of information and further references. For example, thorough reviews of mining's effect on groundwater and on soils were published in 1981 (National Research Council, 1981a and 1981b).

In 1976, a special symposium was organized, *inter alia*, by the American Society of Agronomy to discuss reclamation of, primarily, mined lands. Speakers, who were invited, were chosen so as to complement each other and further the specific theme; the proceedings have become a byword among reclamation specialists (Schaller and Sutton, 1978). Much of the technical interest in reclaiming operating mines centers around the selection of appropriate vegetation species for the mine. To this end, the Forest Service has been extremely helpful, publishing newsletters, individual research reports, and summary guides. One of the latter is especially helpful in revegetation (Vogel, 1981). Another Forest Service report is a model for performing overburden and hydrologic analyses, particularly for western coal (Barrett, *et al.*, 1980).

As can be seen, the problem that exists today in assessing the published reports on the consequences of mining is to determine which, among the myriad of publications, will be useful references. To this end, annotated bibliographies can be very useful; the Bureau of Mines has recently published such a document (Veith *et al.*, 1985) that covers literature from 1977 to 1984. Additional bibliographies can be generated by subscribing to the reference services of NTIS (National Technical Information Service) or the Department of Energy.

INTRODUCTORY HISTORY OF MINE RECLAMATION LEGISLATION

As the technology of surface mining developed, it soon became apparent that the remedies in common (sometimes called civil or case) law were inadequate to protect individual surface owners from those mining practices that unintentionally caused harm to neighbors or, worse yet, from those operators, no matter how few, who were unconcerned with the rights of adjoining property owners. Furthermore, the severance of the mineral estate from the surface estate, which happened frequently in the assembling of minable packages, meant further loss to the surface owner as the law often favored the legally justified, but publicly misunderstood rights of the owner of the mineral extate. Severance may have occurred decades, even generations before mining began and the present owners of the surface were unaware that the right to the mineral included unrestricted rights to mine, often by surface methods. When the courts adjudicated such disputes, they read the language of the deed with care to see if there was express reference to surface mining methods, or, in the case of proposed underground mining, to the right to subside. If such language could be found, then the courts ruled in favor of the mine operator even if the surface owners' structures would be damaged or destroyed by mining. Over time, the courts required stricter and stricter language to permit surface mining and, in all but two states, declared broadly written ("broad form") deeds not to allow surface mining (Goldberg and Power, 1972). This trend, however, was too little and too late to satisfy the disaffected public; there was clamor for legislative remedies at both state and federal level.

The first legislative code for environmental protection in surface mining was passed in West Virginia in 1939, followed by Indiana in 1941, Illinois in 1943, Pennsylvania in 1945, and Ohio in 1947. The early codes often did little more than to provide of a system of licenses or permits, or both, in order that legitimate operators could be identified. Over time, the laws were amended frequently to provide for increased control over operators, stricter environmental performance standards, greater performance bond rates, mandatory exclusions from minable lands, and stricter enforcement through inspection, penalties and forfeiture of the right to mine. As will be seen, subsequent federal legislation forced all coal mining states to strengthen, yet again, existing laws or to enact new laws where none existed.

The pressure for legislative control of surface mining reached the federal level, which had toyed with but never passed surface mine legislation since 1940. There was continuing concern that the states were doing an insufficient job in enforcement and that when a state attempted to be strict it became non-competitive with those states that were lax. The solution was seen to be a national code that set threshold levels of acceptable environmental performance. As discussed in the previous subsection, the Department of the Interior had been charged by Congress in 1965 to investigate the extent of the problem (U.S. Department of the Interior, 1967).

Forty-four bills were introduced into Congress between 1949 and 1970, but none was taken up seriously. Subsequently, however, on August 3, 1977, President Carter signed into law Public Law 95-87, "The Surface Mining Control and Reclamation Act of 1977" (30 USC 1201 *et seq*.). The period between 1970 and the date of passage was, as might be imagined, filled with activity on the federal level. It included a popular but constitutionally flawed bill to ban surface mining (92nd Congress, 1st Session, H.R. 4556, introduced by Representative Ken Hechler (D-WV) on February 18, 1971.), literally hundreds of other bills, many of which were considered seriously and over which hearings were held, and two presidential vetoes of completed legislation: in the 94th Congress, 2nd Session, President Ford vetoed H.R. 25 on May 20, 1975, and in the 93rd Congress, 2nd Session, President Nixon pocket vetoed S.425. Nonetheless, the pressure for passage was inexorable and the mining industry has had to learn to live with a federal presence. The legislative history can be found in the House Report on H.R. 2 (House of Representatives, 1977, p. 140-141) and in an excellent review published on H.R. 25 (Dunlap, 1976).

The resulting federal legislation was complex; it established an Office of Surface Mining and Reclamation Enforcement (OSMRE) in Interior, provided for a program of restoring abandoned mine lands, provided for research and training, provided for individual state regulatory programs, insisted that all coal mines be permitted, and, importantly, provided performance standards for the prevention of damage to the surface and the "hydrologic regime" from both surface and underground mining. The system of state programs allows for the variations in geology, topography, and, even, demography, that exist among the states. A delegation of authority, plus certain funding independence, is provided for states that obtain "Primacy" by arranging for their State mining laws to comply with the minimum requirements of the federal law.

The federal regulations, promulgated by OSMRE under the authorities granted by the law, are even more complex and lengthy than it the law itself. The regulations are found in Title 30, *Code of Federal Regulations*, Chapter 700 (30 CFR Part 700 to end). An extensive discussion of the impact of the regulations, plus a lengthy bibliography, is contained in the "Final Environmental Statement" released for them (U.S. Department of the Interior, 1979). However, because of primacy, most coal-mining states now operate their own enforcement program. Consequently, those interested in the specific operation of Public Law 95-87 (coal mine operators, members of the public with specific interests in land, and the enforcement personnel themselves) must examine their individual State laws and the resulting State regulations. In general, all will comply with the federal law.

ENVIRONMENTAL PERFORMANCE STANDARDS FOR NEW MINES

As written, the federal surface mining (SMCRA) law is complex and, yet accommodates many compromises. An example of a basic compromise is the fact that it covers coal only and does not control non-coal minerals. Another example is that it contains separate standards, in some areas of performance, for east and west. The complexity is brought about because it contains detailed standards for methods of performance as well as for goals of environmental protection. Yet, it also contains sections that encourage basic research as well as experimental practices in mining and reclamation.

Title V, "Control of the Environmental Impacts of Surface Coal Mining," is central to the environmental protection portion of the law. Enforcement is designed around a specification-by-permit procedure. Each potential operator is required to apply for a permit, which must be granted before mining begins. Of course, the permit application goes to whichever regulatory authority has jurisdiction over the land: States with primacy, Tribes with primacy, or OSMRE itself for those limited locations without primacy. The application, which is open-ended, specifies how, for the piece of land under review, the operator intends to comply with the requirements of Section 515, "Environmental Protection Performance Standards." The law requires an assessment of the existing features of the land, including geology, coal reserves, and overburden analysis; also including water resources, both surface and subsurface; and including topography, structures, and other cultural features. It then requires a detailed mining plan, which specifies area to be mined, schedule of mining progress, equipment to be used, and probable cost of mining. Finally, it requires a reclamation plan, which assesses the mine's potential impact on the hydrologic regime and discusses the use proposed for the land after mining, including how this use is consistent with existing land use in the surrounding area. The discussion of proposed post-mining land use must also include a section on alternative land uses for the site. The reclamation plan must then address the specific environmental performance standards: restoration of approximate original contour, protection of the hydrologic regime, burial of boulders, disposal of inimical (to vegetation) material such as acid-forming strata, preservation and then restoration of topsoil, installation of erosion and sediment control facilities, revegetation, and post-mining maintenance.

Performance is assured by a combination of financial guarantees, inspections by outside agents, and civil and criminal penalties. In addition, violators of the requirements are proscribed from obtaining additional permits for new properties. An applicant, in addition to paying annual licensing fees and specific filing fees for each application, must also post a performance bond, often in the order of thousand's of dollars per acre, with each application. This bond is forfeit if the regulatory authority--the administrators of the State program under the authority of SMCRA—decides that the operator has abandoned the property in an unreclaimed state. If performance is found to be acceptable, then the bond is returned, often in stages to match the extent of reclamation: a major portion after completion of backfill and grading, another portion after initial revegetation, and the final portion after growth stabilizes and is found to be self-supporting. Periodic inspections are conducted by the regulatory authority to determine compliance with the applicant's own plans and any special conditions imposed by the agency at the time of issuing the permit. Lapses in performance are written up as notices of violation, which can result in cessation orders if the violation is not remedied. Cessation orders can be written instantly if the inspection reveals a violation that causes "imminent" danger to the environment. Financial penalties are assessed for violations; willful violations can result in criminal charges.

After the law was implemented, there was a period of adjustment by all parties, which often led to delays and some confusion. Today, the operator's concern is not with start-up delays associated with a new law but with endemic delays in granting a permit and with unequal interpretation of the rules by the various regulatory authorities who enforce the law. The federal government, on the other hand, is concerned with apparent lack of diligence on the part of some authorities, mainly states who enforce the law under primacy, and on more than one occasion has threatened to withdraw primacy.

RECLAIMING ABANDONED LANDS

The surface mining law was passed in response to the millions of acres that had been mined and not reclaimed. Title V was designed to prevent existing and new mines from joining that acreage; for the existing acreage, however, Congress wrote Title IV, "Abandoned Mine Reclamation." This section of the law is a direct result of the concern expressed in the report of Interior's study committee (U.S. Department of the Interior, 1967). Basically, Title IV states that abandoned mine lands degrade social and economic values of adjacent lands and that it is the responsibility of existing industry to contribute to the restoration of these lands. Consequently, a fund was established that is replenished by fees (severance taxes, in effect) paid by the industry. The fees of $0.35 per ton of surface-mined coal and $0.15 per ton of underground coal were collected from the date of passage of the law and for fifteen years following. By 1992, it is estimated that over three billion dollars will have been collected.

The law establishes a series of priorities for spending the money, recognizing that, although sizable, the funds represent approximately one-tenth of those needed to remedy all problems. Emergencies are to be dealt with first, where an emergency is an abandoned mine problem that will hurt people and property at any imminent time and needs to be fixed faster than normal bureaucracy permits. A subsidence pot-hole in the middle of a public road or a burning refuse pile that is shifting and threatening to inundate nearby houses are examples of emergencies. The law then establishes a series of six priorities which range from those abandoned mines that pose an extreme danger to public health, safety, general welfare, and property (priority (1)) to environmental restoration (priority (3)) to the development of public lands for recreation (priority (6)). The distinction between an emergency and a priority (1) is often the swiftness with which a remedy is seen to be needed.

Each state that sought primacy under Title V had to have a reclamation plan for Title IV. As part of these plans, data were accumulated on the extent of the abandoned mine land problems in the United States. Not surprisingly, the older coal mining states had extensive problems and the newer districts of the west had relatively few. Pennsylvania led the nation in both extent of pre-1977 mining and number of sites requiring reclamation. The attractiveness of primacy for the states was their ability to control the expenditure of half of all the monies collected in their states. In addition, they could, if needed, apply for additional funds, from the balance of the collections, for additional major or emergency projects in their state. In an extensive report by the Committee on Abandoned Mine Lands (National Research Council, 1986), it was estimated that only Pennsylvania will not have completed work on its priority (1) sites by 1992, and six states, Arkansas, Iowa, Kansas, Missouri, Ohio, and Pennsylvania, will not have completed their priority (2) sites. Priority (3) sites, those with environmental but not health, safety, and general welfare problems, will not fare so well; only those states with extensive present mining, which generates the funds, and limited past mining, which caused the problems, will reclaim all their priority (3) sites.

State reclamation programs are organized by an office within the agency that administers the federally authorized program. However, most states contract with private companies to perform the work. Even though states have control of their monies, all project proposals are reviewed by the federal Abandoned Mine Land office within OSMRE. The Committee found few problems with the process after contracts have been let; the money appears to have been spent as it was intended. However, the committee found that the assignment of priorities was inconsistent and that the diffuse bureaucratic structure sometimes stifled technical innovations. The question of what will happen to the sites, primarily priority (3) or lower, not reclaimed by 1992 is still open.

DESIGNATION OF LANDS UNSUITABLE FOR SURFACE COAL MINING

A review of the legislative history of Public Law 95-87 revealed how intimately entwined it was with legislation proposed for a national land-use planning policy (Senate, 1973). Although the latter never became a separate law, the language in it associated with protection of "Areas of Critical Environmental Concern" was incorporated into a number of bills that did pass, including P.L. 95-87. Thus, the federal surface mining law is a major step toward a mineral land-use planning policy (Saperstein, 1983).

Section 522, within Title V of P.L. 95-87, "Designating Areas Unsuitable for Surface Coal Mining," is divided into two major categories. The first, derived from the precedent of existing State laws, contains mandated exclusions from surface mining: federal parks and protected lands, publicly owned parks and sites included in the National Register of Historic Sites, and alongside roads, occupied dwellings, public buildings, churches, and cemeteries. The other category does not prohibit mining from specific lands but establishes, instead, a planning system of inventory and petition whereby lands within certain categories may be designated as unsuitable. This petition process is derived directly from the earlier land-use planning legislation. There are four classes of land that may be designated:

1. in general, where mining is incompatible with existing land-use plans,
2. fragile or historic lands,
3. renewable resource lands such as water supplies, aquifers, or aquifer recharge areas,
4. natural hazard lands.

In addition, any site may be designated as unsuitable if reclamation is not technologically or economically feasible.

Each State with primacy has established a planning process and many States have already designated sites as unsuitable. Petitions may be submitted by any person who has an interest that may be adversely affected by mining. It is incumbent upon the petitioner to establish that a case for designation exists. It is then up to the State Program to investigate the petition and to rule if designation is more beneficial than mining. At this point, the process tends to be proscriptive in that there is no process for designating lands suitable for surface mining. It is possible, however, that as experience is gained with the system, there will be an understanding, resulting from the inventory of unsuitable lands and the precedent of earlier decisions, of the lands that may be mined.

RESEARCH NEEDS

From 1963, which was the time that Pennsylvania's surface mining legislation was amended to include backfill and natural contouring of surface mines for coal, to the present, the techniques of reclamation have improved dramatically. Methods and machines for effective reclamation have been investigated by researchers and proven (or rejected) by industrial trials. Many of these techniques are included in the references listed earlier. Although, research could improve any number of the methods used in reclamation, there are several areas that are especially appealing for the amount of improvement that potentially remains to be found.

A great deal of research work has been performed on methods of isolating inimical material from groundwater that returns to a reclaimed mine after it has been completely backfilled (Phelps and Saperstein, 1982, Phelps *et al.*, 1985). The concern is to isolate undesired material and then, further, to direct groundwater flows from this isolated package. Isolation can be effected by compacting the unwanted material more than the surrounding material so that its permeability is lower and thus water will tend to flow around it. Alternatively, surrounding material can be made intentionally to be permeable by installing rock drains so that the same effect is achieved. Modeling shows these to be effective steps to reduce contamination of groundwater. However, the degree of compaction that can be expected from various types of excavating machines working on different mine spoils is not well known.

A major consideration in restoring the productivity of topsoil that has been displaced and then respread during reclamation is the extent to which the soil can reestablish its structure. Therefore, compaction of soil during reclamation becomes a significant cause of loss of productivity. Some compaction is inevitable (Holland and Phelps, 1986), but research is needed to establish operating techniques that minimize compaction as well as loosening techniques that help to reestablish structure.

Without question, there are many other important research areas, such as the continuing search for appropriate species, which will benefit the reclamation process. However, the above two represent significant gaps in knowledge, which can lead to long-term failure of mined-land reclamation.

CONCLUSIONS

There can be little doubt that the environmental consequences of unreclaimed or poorly reclaimed surface mines has led to strict legislative control of coal surface mining. The legacy of open highwalls, increased sediment load in streams, and lost topsoil was not to be pertetuated. Indeed, surface mining has become one of the most heavily regulated and inspected industries in the United States. Operating within the confines of the existing law takes knowledge, experience, and dedication to achieving a result.

Tying together the various threads in this section leads to the conclusion that efficient reclamation planning requires a good set of goals for post-mining uses of the reclaimed land. Incorporation of the techniques of land-use planning, both comprehensive, regional planning and environmental site planning, as described by Ramani in this volume, will assist in creating both an end-use for the land and a design for achieving that use that are efficient and productive. The surface mine operator and those who plan the mines need to realize that their efforts can enhance the profitability of the mine while ending the perpetuation of the legacy of lands despoiled by surface mining.

REFERENCES

Appalachian Regional Commission, 1973, *Analysis of Pollution Control Costs*, Report BAKER-ARK-73-04 prepared by Michael Baker, Jr., Inc. 436 p.

Barrett, J., et al., 1980. *Procedures Recommended for Overburden and Hydrologic Studies of Surface Mines*, U.S. Department of Agriculture, Forest Service, Intermountain Forest and Range Experiment Station, General Technical Report INT-71, 106 p.

Dunlap, L.C., 1976. "An Analysis of the Legislative History of the Surface Mining Control and Reclamation Act of 1975," *21 Rocky Mt. Min. L. Inst.* p. 11-58.

Goldberg, E.F. and G. Power, 1972, *Legal Problems of Coal Mine Reclamation*, University of Maryland School of Law under Grant No. 140010 FZU (03/72) with the U.S. Environmental Protection Agency, 237 p.

Grim, E.C. and R.D. Hill, 1974, *Environmental Protection in Surface Mining of Coal*, U.S. Environmental Protection Agency, Washington, D.C., EPA-670/2-74-093,277 p.

Holland, L.J. and L.B. Phelps, 1986, "topsoil Compaction During Reclamation: Field Studies," *1986 National Symposium on Mining, Hydrology, Sedimentology, and Reclamation*, Univ. of Kentucky, p. 55-62.

House of Representatives, 1977, House Report No. 95-218, 95th Congress, 1st Session.

Johnson, W. and J. Paone, 1982, "Land Utilization and Reclamation in the

Mining Industry, 1930-1980," *Information Circular 8862*, U.S. Department of the Interior, Bureau of Mines, 22 p.

National Academy of Sciences, 1974, *Rehabilitation Potential of Western Coal Lands*, Study Committee on the Potential for Rehabilitating Lands Surface Mined for Coal in the Western United States, Ballinger, Cambridge, MA, 198 p.

National Research Council, 1981a, *Coal Mining and Ground-Water Resources in the United States*, Committee on Ground-Water Resource in Relation to Coal Mining, Board on Mineral and Energy Resources, Commission on Natural Resources, National Academy Press, Washington, D.C., 197 p.

National Research Council, 1981b, *Surface Mining: Soil, Coal, and Society*, Committee on Soil as a Resource in Relations to Surface Mining for Coal, Board on Mineral and Energy Resources, Commission on Natural Resources, National Academy Press, Washington, D.C., 233 p.

National Research Council, 1986, *Abandoned Mine Lands, A Mid-Course Review of the National Reclamation Program for Coal*, Committee on Abandoned Mine Lands, Board on Mineral and Energy Resources, Commission on Physical Sciences, Mathematics, and Resources, National Academy Press, Washington, D.C., 221 p.

Paone, J., J.L. Morning, and L. Giorgetti, 1974, "Land Utilization and Reclamation in the Mining Industry, 1930-71," *Information Circular 8642*, U.S. Department of the Interior, Bureau of Mines, 61 p.

Phelps, L.B., J.C. Mills, and L.W. Saperstein, 1985, "Modeling Fluid Flow in Selective Placement of Toxic Mine Spoil," *Transactions of SME of AIME*, Volume 278, p. 328-333.

Phelps, L.B. and L.W. Saperstein, 1982, "Spoil Modification to Minimize the Mobilization of Formed Toxic Fluids, *1982 Symposium on Surface Mining, Hydrology, Sedimentology and Reclamation*, Univ. of Kentucky, p. 593-599.

Saperstein, L.W., 1983, "Accommodating the Land-Use Planning Provisions of The Surface Mining Control and Reclamation Act," *Transactions of SME of AIME*, Vol. 274, p. 1654-1656.

Schaller, F.W. and P. Sutton (eds.), *Reclamation of Drastically Disturbed Lands*, American Society of Agronomy, *et al.*, Madison, WI, 742 p.

Senate, 1973, Senate Report 93-197, 93rd Congress, 1st Session.

Skelly and Loy Engineers, Inc., 1975, *Economic Engineering Analysis of U.S. Surface Coal Mines and Effective Land Reclamation*, report to U.S. Department of Interior, Bureau of Mines, under contract SO 241049, 646 p.

U.S. Department of Agriculture, 1979,"The Status of Land Disturbed by Surface Mining in the United States," Soil Conservation Service, SCS-TP-158, 124 p.

U.S. Department of the Interior, 1966, *Study of Strip and Surface Mining in Appalachia*, Strip and Surface Mine Study Policy Committee, U.S. Government Printing Office, Washington, D.C., 78 p.

U.S. Department of the Interior, 1967, *Surface Mining and Our Environment*, Strip and Surface Mine Study Policy Committee, U.S. Government Printing Office, Washington, D.C., 124 p.

U.S. Department of the Interior, 1979, "Permanent Regulatory Program Implementing Section 501(b) of the Surface Mining Control and Reclamation Act of 1977," Final Environmental Statement OSM-EIS-1 from the Office of Surface Mining Reclamation and Enforcement, 288 p. plus 800 p. appendices.

U.S. Environmental Protection Agency, 1973, *Processes, Procedures, and Methods to Control Pollution from Mining Activities*, EPA-430/9-73-011, Washington, D.C., 390 p.

Veith, D.L., *et al.*, 1985, "Literature on the Revegetation of Coal-Mined Lands: An Annotated Bibliography," *Information Circular 9048*, U.S. Department of the Interior, Bureau of Mines, 296 p., 805 entries.

Vogel, W.G., 1981, *A Guide for Revegetating Coal Minesoils in the Eastern United States*, U.S. Department of Agriculture, Forest Service, Northeastern Forest Experiment Station, General Technical Report NE-68, 190 p.

Environmental Consequences of Energy Production: Problems and Prospects. Edited by S. K. Majumdar, F. J. Brenner and E. W. Miller. © 1987, The Pennsylvania Academy of Science.

Chapter Thirty
ENERGY PRODUCTION AND FOREST ECOSYSTEM HEALTH

WILLIAM H. SMITH
School of Forestry and Environmental Studies
Yale University
370 Prospect Street
New Haven, CT 06511

Historically, the combustion of fossil fuels has directly or indirectly been the source of air contaminants at three levels; local, regional, and global. Pollutants of importance at the local level have included sulfur dioxide, hydrogen fluoride and trace metals. Forest damage associated with local pollution has typically been confined to a zone of a few km immediately surrounding a facility and for a distance of several to tens of km downwind. Regional air pollutants may be deposited over expansive forest areas because they are transported tens or hundreds of km from point of release due to small size or synthesis in the atmosphere from precursors introduced into the troposphere. Regional air pollutants of potential significant influence for forests include: oxidants, trace metals, and acid deposition. Global pollutants influence the entire atmosphere of the earth, e.g. halocarbons and carbon dioxide. The latter are important because of the potential they have to influence radiation balances of the earth and global climate. Risks associated with regional and global air pollution and forest health are high. The evidence available to describe the total boundaries of the problem for all pollutants is presently incomplete.

INTRODUCTION

A challenge of developed nations throughout the temperate zone is air quality policies that protect natural resources as well as human health. Throughout Europe and North America, the balance of this Century will be recorded as a period of profound decisions regarding atmospheric contamination and natural resource quality. The central issues for those interested in forest health are as follows. Is air pollution influencing the growth of forests or individual species, changing the species composition of forest communities or destroying certain tree species, associated plants or animals over significant forest areas?

In an effort to answer this question, I shall discuss the spatial scales of air pollution stress and what we know about the response of trees to pollutants at these scales.

LOCAL, REGIONAL AND GLOBAL-SCALE AIR POLLUTANTS

Local —

During the first two thirds of the 20th century, research and regulatory efforts were focused on local air pollutants and acute vegetative effects. Pollutants of primary concern were sulfur dioxide, particulate and gaseous fluoride compounds and numerous heavy metals such as lead, copper, and zinc. Occasional interest was expressed in other inorganic gases including ammonia, hydrogen sulfide and chloride, and chlorine. The sources of these pollutants were and are typically discrete and stationary facilities for energy production, for example, fossil-fuel electric generating plants, gas purification plants; metal related industries, for example, copper, nickel, lead, zinc or iron smelters, aluminum production plants; and diverse other industries, for example, cement plants, chemical and fertilizer plants and pulp mills.

It is appropriate for us to consider the above pollutants local-scale because forest areas directly affected by these facilities are typically confined to a zone of a few km immediately surrounding the plant and for a distance of several to tens of km downwind. The dimensions of the surrounding and downwind zones of influence are variable and primarily controlled by source strength of the effluent, local meteorology, regional topography and susceptibility of vegetation. In any case, acute forest influences have typically been confined to areas generally less than a 1000 ha.

Regional —

During the past three decades we have become increasingly aware of regional-scale air pollutants. The regional designation is applied because these contaminants may affect forests 10's, 100's, or even 1000's of kms from their site of origin. The regional air pollutants of greatest documented or potential influence for forests include: oxidants, most importantly ozone; trace metals, most importantly heavy metals—e.g. cadmium, iron, copper, lead, manganese, chromium, mercury, molybdenum, nickel, thallium, vanadium, zinc; and acid deposition, most importantly the wet and dry deposition of sulfuric and nitric acids. Ozone, sulfuric and nitric acids are termed secondary air pollutants because they are synthesized in the atmosphere rather than released directly into the atmosphere. The precursor chemicals, released directly into the atmosphere and causing secondary pollutant formation, include hydrocarbons and nitrogen oxides in the case of ozone, and sulfur dioxide and nitrogen oxides in the case of sulfuric and nitric acid. The combustion of fossil fuels for energy production release hydrocarbons and sulfur dioxide. The heat of combustion causes nitrogen and oxygen to react and form nitrogen oxides. Many combustion activities generate small particles (approximately 0.1—5 μm diameter). Those activities associated with energy combustion (particularly coal burning) can preferentially contaminate these small particles with trace metals. Because the formation of secondary air pollutants may occur over 10's or 100's of km from the site of precursor release, and because small particles may remain airborne for days or weeks, these pollutants may be transported 100 to more than 1000 km from their origin. Eventual wet and dry deposition of the pollutants onto lakes, fields, or forests occurs over large rather than small areas.

The U.S. Environmental Protection Agency and the U.S. Department of Agriculture, Forest Service established a network of air monitoring stations to measure ozone concentrations in remote areas of National Forests. Analysis of selected high ozone events during 1979 suggested that long-range transport of air masses contaminated by urban centers contributed to peak concentrations at remote sites (Evans et al. 1983). In a study of rural ozone episodes in the upper-midwest, (Pratt et al. 1983) presented evidence that ozone and precursors were transported 275 km from Minneapolis-St. Paul. Studies of trace metal concentrations, in the atmosphere in remote northern and southern hemispheric sites revealed that the natural sources include the oceans and the weathering of the earth's crust, while the major anthropogenic source is particulate air pollution (U.S. Environmental Protection Agency 1983a). Murozumi et al. (1969) showed that long range transport of lead particles from automobiles significantly polluted polar glaciers. We estimated the annual lead deposition on a remote northern hardwood forest in New Hampshire to be as high as 226 g per hectare. This caused lead contamination of the forest floor 5-10 times greater than the estimated pre-industrial concentration (Smith and Siccama 1981).

Evidence is available; satellites, surface deposition of aerosol sulfate and reduced visibility (Chung 1978, Tong et al. 1976, Wolff 1981), for long-range transport of acidifying pollutants from numerous sources. During the winter, approximately 20 percent of the emissions from tall power plant stacks in northeastern United States may remain elevated and relatively coherent for more than a day and 500 km (U.S. Environmental Protection Agency 1983b).

The long distance transport of regional pollutants means they may have interstate, international and even intercontinental significance. It means further that the forests subject to their deposition exceed 10 of 1000 km^2.

Global—

In the past 25 years, we have become concerned with a third scale of air pollution—global. Global pollutants affect the entire atmosphere of the earth. Two global air pollutants of special note include carbon dioxide and halocarbons.

Careful monitoring of carbon dioxide during the past two decades in Hawaii, Alaska, New York, Sweden, Austria and the South Pole has firmly established that carbon dioxide is steadily increasing in the global atmosphere. This increase is due to anthropogenic activities including fossil fuel combustion. It may also be caused by altered land use management, such as forest destruction in the tropics. The atmospheric carbon dioxide concentration has been estimated to have been approximately 290 ppm (5.2×10^4 μg m^{-3}) in the middle of the 19th century. Today, the carbon dioxide concentration approximates 340 ppm (6.1×10^5 μg m^{-3}) and is increasing about one ppm (1.8×10^3 μg m^3) per year. In the year 2020, if the increasing rate continues, the carbon dioxide amount in the global atmosphere may be nearly two times the present value (Holdgate et al. 1982).

Naturally occurring stratospheric ozone is important because it screens the earth from biologically damaging ultraviolet radiation—light with wavelengths between 290 and 320 nm—released by the sun. Halocarbons released by humans can deplete the natural ozone layer surrounding the earth. In summary, halocarbon molecules, for example chlorofluoromethanes, released by various human activities, are slowly transported through the troposphere. They pass through the tropopause and lower stratosphere and are decomposed in the mid- to upper-atmosphere. Free chlorine, resulting from decomposition, causes a rapid, catalytic destruction of ozone. In 1979, the National Academy of Sciences estimated that release of halocarbons to the atmosphere, at rates inferred for 1977, would eventually deplete stratospheric ozone 5 to 28 percent, most probably 17 percent (National Academy of Sciences 1979). In 1982, the National Academy revised its previous estimate and suggested a depletion of from 5 to 9 percent. (National Academy of Sciences 1982a).

EFFECT OF LOCAL-SCALE, REGIONAL-SCALE AND GLOBAL-SCALE AIR POLLUTION ON FOREST ECOSYSTEMS

Local—

High deposition of local air pollutants has caused well documented forest destruction. High sulfur dioxide or fluoride doses severely injure or kill forest trees. The ecosystems, of which the trees are a part, are simplified, lose nutrients, sustain soil erosion, have microclimates and hydrologic patterns altered and ultimately they are destroyed or converted to more resistant seral stages. Miller and McBride (1975) have reviewed the studies of forests destroyed by local air pollution. Early in this century, it was clearly documented in numerous locations throughout North America that sulfur dioxide and trace metal pollution destroyed forests surrounding metal smelting facilities. Smelting centered in Ducktown, Tennessee devastated the southern hardwood forest over 27 km^2 (10.5 mi^2) surrounding the plant, converted an additional 68 km^2 (17,000A) to grassland and created a 120 km^2 (30,000A) transition zone with altered species composition. Smelters in the Sudbury, Ontario, Canada area have caused simplification of the surrounding mixed boreal forest and have caused eastern white pine mortality in a 1865 km^2 (720 mi^2) zone to the northeast.

Aluminum reduction plants have also caused local forest destruction. In Montana, fluoride pollution killed or severely injured ponderosa pine and lodgepole pine on 8 km^2 (2000A) surrounding a plant. In Washington, ponderosa pine mortality and morbidity resulted over a 130 km^2 (50 mi^2) area in the vicinity of an aluminum plant.

Local pollution has caused extensive forest mortality throughout Europe. Examples are in Austria, Germany, Hungary, Norway, Poland and Sweden. Industrial operations along the northern border of Czechoslovakia have caused extensive forest destruction.

Regional—

Deposition of regional pollutants subject forests to different perturbations than local pollutants because the doses are less. Rather than severe tree morbidity or mortality with dramatic symptoms, regional pollutants subtly change tree metabolism and ecosystem processes. Smith (1981) has provided a comprehensive review of subtle air pollution forest stress.

Regional air contaminants may influence reproductive processes, nutrient uptake or retention, metabolic rates (especially photosynthesis and respiration), and insect pest and pathogen interactions of individual trees. At the ecosystem level, regional air pollutants may influence nutrient cycling, population dynamics of arthropod or microbial species, succession, species composition, and biomass production. In the instance of high-dose local-scale pollution, the symptoms are typically acute, dramatic and obvious (severe disease, mortality, forest simplification). In the case of lower-dose regional-scale pollution, the symptoms are typically not visible (at least initially), undramatic and not easily measured. The integration of regional pollutant stresses is slower growth, altered competitive abilities and changed susceptibility to pests. Ecosystem symptoms may include altered rates of succession, changed species composition and biomass production. Symptom development is, of course, much slower at the regional scale. Evidence of the relative importance of regional pollutants is variable, caused in part by the length of time that has been devoted to the study of individual pollutants and in part by the subtleness and complexity of the pollutant interactions. The toxicity of trace metals has been studied for approximately 65 yrs, of ozone approximately 30 yrs and of acid deposition approximately 15 yrs.

Data summarized in Table 1 suggest the relative strength of evidence for forest responses to regional pollutants. A review of the column totals suggests we know most about the regional effects of oxidants, less about regional effects of trace metals and least about regional effects of acid deposition. A review of the row totals suggests tree and ecosystem processes especially vulnerable to air pollution stress. The processes with a total of five or more include: litter

TABLE 1**

Relative strength of evidence (quantity/quality)* available to support forest ecosystem interaction with regional⁺ air pollutants.

Ecosystem process/component perturbation	Air contaminants			
	Oxidants	Trace metals	Acid deposition	TOTAL
I. nutrient cycling				
1. increase nutrient availability				
a. increase input (fertilization)	0	1	2	3
b. increase soil weathering	0	0	1	1
2. decrease nutrient availability				
a. reduce litter decomposition	0	4	1	5
b. increase soil acidification	0	0	2	2
c. increase soil (cation) leaching	0	0	2	2
d. decrease microbial symbioses	0	3	1	4
II. primary producers (trees)				
1. reproductive physiology				
a. reduce flowering	1	1	0	2
b. reduce pollen				
a. reduce litter decomposition	0	4	1	5
b. increase soil acidification	0	0	2	2
c. increase soil (cation) leaching	0	0	2	2
2. foliar physiology				
a. reduce photosynthesis	4	1	0	5
b. increase (cation) leaching	0	0	2	2
c. increase necrosis	4	2	0	6
3. root physiology				
a. decrease water/nutrient uptake	0	1	1	2
b. increase necrosis	0	1	2	3
4. reduce tree growth	4	1	1	6
III. consumers				
1. arthropod pest activity				
a. increase	4	0	0	4
b. decrease	0	0	0	0
2. microbial pathogen activity				
a. increase	4	1	1	6
b. decrease	1	2	0	3
3. other pest activity (viruses, bacteria, nematodes, mistletoes, weeds)				
a. increase	0	0	0	0
b. decrease	0	0	0	0
4. wildlife (bird/mammal) activity				
a. reduce food	2	0	0	2
b. reduce habitat	2	0	0	2
c. increase morbidity/mortality	0	2	0	2
IV. ecosystem succession/species composition (cause alteration)	4	1	0	5
V. ecosystem productivity (increase/decrease biomass accumulation)	4	0	0	4
TOTAL	41	25	18	

Energy Production and Forest Ecosystem Health 437

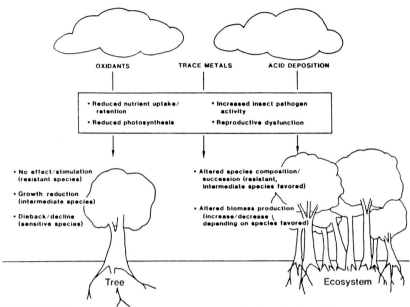

FIGURE 1.** Forest tree and ecosystem response to regional-scale air pollution stress.

decomposition, seedling survival, photosynthesis, foliar necrosis, tree growth, microbial pathogen activity, and ecosystem succession plus species composition. These are the tree and ecosystem processes at particular risk from regional air pollution. Fig. 1 provides an overview of regional air pollution influence on forest trees and ecosystems.

Global—

Increasing carbon dioxide concentration and decreasing stratospheric ozone concentration of the atmosphere may alter global radiation fluxes. Presumably a primary result of more carbon dioxide in the air will be warming. While incoming solar radiation is not absorbed by carbon dioxide, portions of infrared radiation from earth to space are. Over time, the earth would become warmer. While the forces controlling global temperature are varied and complex, the

---

*evidence scale as presented in Table 1

0—extremely limited evidence or hypotheses only
1—slight evidence
2—more evidence
3—greater evidence
4—considerable evidence including field evidence

+exclusive of local air pollution effects within several km of point sources

**Table 1 and Figure 1 are reprinted with permission from W.H. Smith In *Fossil Fuels Utilization:* Markuszewski, R. and B.D. Blaudtein, Eds., ACS Symposium Series No. 319, American Chemical Society, Washington, D.C., 1986, pp. 254-66. Copyright 1986, American Chemical Society.

increase of 0.5C since the mid-1800s is generally agreed to be at least partially caused by increased carbon dioxide. By 2000 it may increase an additional 0.5C. Numerous models advanced to estimate the average global warming per doubling of carbon dioxide project 0.7 to 9.6C. Natural impacts on climate, such as solar variability, remain important and of unclear relationship to anthropogenic causes. A mean global average surface warming, however, of 3 ± 1.5C in the next century appears reasonable (National Academy of Sciences 1982 a, b).

The consequences of a warmer global climate, with even a very modest temperature increase, on the development of forest ecosystems, could be profound. Warming, with increased carbon dioxide in the atmosphere, might enhance forest growth. Manabe and Stouffer (1980) have estimated that a doubling of atmospheric carbon dioxide would cause a 3C warming at the U.S.-Canadian border, while Kellogg (1977) has suggested that a rise of 1C in mean summer temperature extends the growing season by approximately 10 days. Other changes associated with global warming, however, may restrict forest growth. Physiological processes of plants, especially photosynthesis, transpiration, respiration and reproduction are sensitive to temperature. With warming, respiration and decomposition may increase faster than photosynthesis. Transpiration and evaporation increases may enhance stress on drier sites. Reproduction may be altered by changes in dynamics of pollinating insects, changes in flower, fruit or seed set, or changes in seedling production and survival. The geographic or host ranges of exotic microbial pathogens or insect pests may expand. Previously innocuous endemic microbes or insects may be elevated to important pest status following climatic warming. Precipitation changes are associated with global warming, and certain areas will receive more and others less. Those areas receiving less precipitation will also experience increased evaporation and transpiration. Waggoner (1984) has estimated that the projected change in weather by the year 2000, caused by increased atmospheric carbon dioxide, will cause moderate decreases of 2-12 percent in yield of wheat, corn and soybeans in the American grain belt due to increased dryness. While agriculturists may be able to adopt new crop varieties to a drier climate, forests cannot be similarly manipulated. Increased drought stress over widespread forest areas would be expected to initiate new rounds of progressive tree deterioration termed dieback/decline disease. Drought is the most common and important initiator of general forest tree decline. Forest stresses caused by other air pollutants and other agents of stress must be evaluated against this background of forest change caused by climatic warming.

A serious consequence of anthropogenic release of halocarbons to the atmosphere is the depletion of naturally occurring stratospheric ozone. Halocarbons are recalcitrant molecules and are not removed in the lower atmosphere by precipitation or other reactions. Gradually halocarbons migrate to the stratosphere and are exposed to solar UV-C (<280 nm) which causes photodissociation. Elemental chlorine released by this dissociation has an extended residence time in the stratosphere and functions as a catalyst for ozone destruction (Caldwell 1981). Some reduction in halocarbon release has been achieved in the United States and a few other countries. Immediate termination of all release worldwide, however, would still leave the world with important stratospheric ozone reductions during the next decade. Reduced upper-air ozone would increase ultraviolet radiation reaching the surface of the earth. Current understanding does not allow an inventory of the impacts of increased ultraviolet radiation on forests. Vegetation can acclimatize to changes in UV radiation. Higher plants also vary substantially in resistance to UV radiation. Because some plants are sensitive, however, reductions in biomass production and/or competitive strengths may be altered by changes in short-wave radiation flux at the surface of the earth (Caldwell 1981). Studies of more than 100 agricultural species showed that increased ultraviolet exposure reduces plant dry weight and changes the proportion of root, shoot and leaf tissue. Studies of more than 60 aquatic organisms showed that many were quite sensitive to current levels of ultraviolet radiation at the water surface (Maugh 1980). Chlorofluorocarbons can also contribute to global warming in a manner similar to carbon dioxide.

GLOBAL AND REGIONAL AIR POLLUTANTS ARE THE MOST IMPORTANT CONCERN

The effects of local air contaminants on forests have stabilized in the vicinity of existing point-sources of air pollutants. In numerous cases improvements have been achieved. In the case of sulfur dioxide, increasing stack heights and use of scrubbers have reduced ground level concentrations of sulfur dioxide. New industries and electrical plants in the U.S. can employ the best available air quality technology.

On a global-scale, the destruction of ozone by halocarbons has been addressed in the U.S. by banning certain uses of chlorofluorocarbons. This strategy has been unsuccessful in eliminating the potential for ozone destruction. The release of carbon dioxide to the atmosphere from fossil fuel combustion will continue well into or through the 21st century. Energy requirements of nations of the temperate zone will require combustion of gas, oil and coal and the atmospheric burden of carbon dioxide will continue to increase with uncertain consequences.

Regional-scale air pollutants exhibit both increasing trends and known and probable effects on forest ecosystems over large portions of the temperate region. The integration of stresses imposed by regional pollutants has the potential to cause growth reductions in some forest species and, ultimately, dieback/decline symptoms in susceptible tree species at ambient levels. At the ecosystem level this has or will cause changes in species composition and increases or decreases in biomass production depending on the specific ecosystem (Fig. 1). Documentations of decreased tree growth and increased decline symptoms solely or primarily due to air quality in the field are very limited because the changes are subtle, not continuous but patchwork in character, and extremely difficult to separate from other factors that control tree growth (e.g. age, competition, moisture, temperature, nutrients, insects, pathogens) and that induce dieback/decline symptoms (e.g. drought, other climatic stresses, insects, pathogens). In addition, species composition and patterns of forest succession are regulated by numerous determinants (e.g. vegetative site alterations, plant species interactions, insect/pathogen activities, windstorms, fires and human cultural activities) and forest ecosystem production is influenced by several variables (system age, competition, species composition, moisture, temperature, nutrients, insects, pathogens). A review of the current evidence available, to support the importance of air pollution induced forest change, has been provided by Smith (1981). The comprehensive study of oxidant pollution in portions of the San Bernardino National Forest, California demonstrated air pollution effects on forest growth and succession (Miller et al. 1982). Additional evidence of reduced forest growth imposed by oxidant pollution in the west, mid-west and east has been provided (U.S. Environmental Protection Agency 1984). For various forest ecosystems we are at, or near, the threshold of trace metal impact on nutrient cycling processes. Lead will continue to accumulate in forest floors as long as it is released into the atmosphere (U.S. Environmental Protection Agency, 1983a). Although adverse effects on forest ecosystems from acid deposition have not been conclusively proven by existing field evidence, we cannot conclude that adverse effects are not occurring. Presently, tree mortality and tree morbidity and growth rate reductions in European and North American regions do occur where regional air pollution, including acid deposition, is generally high. Temporal and spatial correlations between wet acidic deposition and forest tree growth rates have been provided. Numerous hypotheses for adverse forest effects from acid deposition, worthy of testing, have been proposed (U.S. Environmental Protection Agency 1983a, Society of American Foresters 1984). Under natural conditions, forest ecosystems are exposed to multiple air pollutants simultaneously or sequentially and interactive and accumulative influences are important. It is inappropriate to consider the effects of any regional pollutant on forests in isolation. The potential stresses imposed by regional-scale air pollutants must be viewed against the background of uncertainty associated with the changes in radiation balances associated with global scale air contaminants.

The growth reductions and decline symptoms of the forests of the Federal Republic of Germany are dramatic and should warn all nations that the resiliency of forest ecosystems has limitations. Until the cause of this decline is more clearly understood, prudent natural resource science should not reject nor indict any single stress.

CONCLUSION

Sensitive and convenient forest parameters must be found to monitor the extent and intensity of stress on expansive forest systems. Waring (1985) suggested that monitoring canopy leaf area and its duration of display is a very appropriate general index of forest ecosystem stress. Canopy quantity and quality is an indicator of productivity. Inventory techniques from the air (multispectral scanning, microwave transmission, radar, laser) and ground (correlations with stem diameter, sapwood cross-sectional area) for canopy leaf area are available. At a given site, detection of an increase in leaf area would suggest an improving environment, a decrease in leaf area would infer the system is under stress. Baes and McLaughlin (1984) have proposed that trace metal analyses of tree rings can provide information on temporal changes in air pollutant deposition and tree health.

Implementation of wide-area forest monitoring of any nature involves two challenges. First, detection of stress does not suggest cause. We are keenly aware that tree and forest health are controlled by many factors in addition to air quality—age, competition, environment (moisture, temperature, nutrition), insect and pathogen activity. We desperately need procedures to partition the relative importance of influencing variables for a given site. Fortunately we are making research progress toward this resolution (Waring 1985, McLaughlin et al. 1983, Fritts and Stokes 1975). The second challenge is to convince natural resource managers that the time and cost of systematic forest health monitoring is justified. I feel it is not only justified, but essential for intelligent decisions regarding regulation and understanding of regional and global air pollutants.

Air pollution has been killing trees locally for centuries. We have been keenly aware of this in the United States for over 100 years. We now realize that in addition to mortality, regional air pollutants may be capable of causing alterations in species composition and growth-rate reductions in certain forest ecosystems over large areas and across national boundaries. Global-scale pollutants may be shown capable of causing similar stress on woody plants.

Forests are variable in species, topography, elevation, soils and management. Air pollution deposition and influences are also variable and poorly documented in the field. Monitoring of species dynamics and productivity, necessary to detect effects of regional air pollutants, or any other environmental stress, are presently rarely available. Dendrochronological or other tree-ring analytical techniques are subject to enormous difficulty when they attempt to partition the relative importance of forces that may influence tree growth. Growth is regulated by precipitation, temperature, length of growing season, frost, drought, by developmental processes such as succession and competition, and by stochastic events such as insect outbreaks, disease epidemics, fire, windstorms and anthropogenic activities such as thinning, fertilization, harvesting and finally air quality.

For a long time, dieback and decline of specific forest species, somewhere in the temperature zone, has been common. Age, climate, or biotic stress factors have frequently been judged to be the principal causes for declines. Again, however, it is difficult to assign responsibility for specific cause and effect. Trees are large and long-lived and their health integrates all the stresses to which they are exposed over time.

The risks associated with regional and global air pollution stress and forest ecosystem health are high. The evidence available to describe the total boundaries of the problem for all pollutants is incomplete. There is enormous uncertainty about specific effects on forests of regional and global air pollutants. We do know, however, that coal combustion will provide more than 50 percent of America's electricity by 1990. We further know that without management or control, coal combustion is a source of numerous regional and global pollutants identified as important, or potentially important, to the health of forest ecosystems. Natural ecosystem health, along with human health, must be recognized in assessments, economic and otherwise, of pollution abatement strategies.

LITERATURE CITED

Baes and McLaughlin. 1984. Trace elements in tree rings: Evidence of recent and historical air pollution. Science 224: 494-497.

Caldwell, M.M. 1981. Plant response to solar ultraviolet radiation. Chap. 6. Encyclopedia of Plant Physiology, New Series 12A: 169-197.

Chung, Y.S. 1978. The distribution of atmospheric sulfates in Canada and its relationship to long-range transport of pollutants. Atmos. Environ. 12: 1471-1480.

Evans, G., P. Finkelstein, B. Martin, N. Possiel and M. Graves. 1983. Ozone measurements from a network of remote sites. Jour. Air Pollu. Cont. Assoc. 33: 291-296.

Fritts, H.C. and M.A. Stokes. 1975. A technique for examining non-climatic variation in widths of annual tree rings with special reference to air pollution. Tree-Ring Bull. 35: 15-24.

Holdgate, M.W., M. Kassas and G.F. White. 1982. World environmental trends between 1972 and 1982. Environ. Conserva. 9: 11-29.

Kellogg, W.W. 1977. Effects of human activities on global climate. Technical Note No. 156. WHO-No. 486. World Meteorol. Org., Geneva.

Linthurst, R.A. (ed.) 1984. Direct and Indirect Effects of Acidic Deposition on Vegetation. Vol. 5. Acid Precipitation Series. J.I. Teasley, ed. Butterworth Publishers, Boston, 117 pp.

Maugh, T.H. 1980. Ozone depletion would have dire effects. Science 207: 394-395.

Manabe, S. and R.J. Stouffer. 1980. Sensitivity of a global climate model to an increase of CO_2 concentration in the atmosphere. Jour. Geophys. Res. 85: 5529-5554.

McLaughlin, S.B., T.J. Blasing, L.K. Mann and D.N. Duvick. 1983. Effects of acid rain and gaseous pollutants on forest productivity: A regional scale approach. Jour. Air Pollu. Cont. Assoc. 33: 1042-1049.

Miller, P.R. and J.R. McBride. 1975. Effects of air pollutants on forests. In, J.B. Mudd and T.T. Kozlowski, eds., Responses of Plants to Air Pollution. Academic Press, N.Y., pp. 195-235.

Miller, P.R., O.C. Taylor and R.G. Wilhour. 1982. Oxidant Air Pollution Effects on a Western Coniferous Forest Ecosystem. U.S. Environmental Protection Agency, Publica. No. EPA-600/D-82-276. Corvallis, OR, 10 pp.

Murozumi, M., T. Chow and C. Patterson. 1969. Chemical concentrations of pollutant lead aerosols, terrestrial dusts and sea salts in Greenland and Antarctic snow strata. Geochim. Cosmochim. Acta 33: 1247-1294.

National Academy of Sciences. 1979. Stratospheric Ozone Depletion by Halocarbons. National Academy of Science, Washington, D.C., 249 pp.

National Academy of Sciences. 1982a. Stratospheric Ozone Depletion by Halocarbons. National Academy of Science, Washington, D.C. xx pp.

National Academy of Sciences. 1982b. Solar Variability, Weather, and Climate (Studies in Geophysics). ISBN No. 0-309-03284-9. National Academy Press, Washington, D.C., 120 pp.

National Academy of Sciences. 1982c. Carbon Dioxide and Climate: A Second Assessment. ISBN No. 0-309-03285-7. National Academy Press, Washington, D.C., 92 pp.

Pratt, G.C., R.C. Hendrickson, B.I. Chevone, D.A. Christopherson, M.V. O'Brien, and S.V. Krupa. 1983. Ozone and oxides of nitrogen in the rural upper-midwestern U.S.A. Atmos. Environ. 17: 2013-2023.

Smith, W.H. 1981. Air Pollution and Forests. Springer-Verlag, N.Y., 379 pp.

Smith, W.H. and T.G. Siccama. 1981. The Hubbard Brook Ecosystem Study: Biogeochemistry of lead in the northern hardwood forest. Jour. Environ. Qual. 10: 323-333.

Tong, E.Y., G.M. Hidy, T.F. Lavery and F. Berlandi. 1976. Regional and local aspects of atmospheric sulfates in the northeastern quadrant of the U.S. Proceedings, Third Symposium on Turbulence, Diffusion and Air Quality. Amer. Meteor. Soc., Boston, MA.

Society of American Foresters. 1984. Acid Deposition and Forest Ecosystems. Task Force on the Effects of Acid Deposition on Forest Ecosystems. Report prepared for the Society of American Foresters, Bethesda, MD, 51 pp.

Waggoner, P.E. 1984. Agriculture and carbon dioxide. Amer. Scient. 72: 179-184.

Waring, R.H. 1985. Imbalanced forest ecosystems: assessments and consequences. For. Ecol. Manage. 12: 92-112.
Waring, R.H., K. Newman and J. Bell. 1981. Efficiency of tree crowns and stemwood production at different canopy leaf densities. Forestry 54: 129-137.
Wolff, G.T., N.A. Kelly, M.A. Furman. 1981. On the sources of summertime haze in the eastern United States. Science 211: 703-705.
U.S. Environmental Protection Agency. 1983a. Air Quality Criteria for Lead. Vol. I. Publica. No. 600/8-83-028A. U.S.E.P.A., Research Triangle Park, N.C. 169 pp.
U.S. Environmental Protection Agency. 1983b. The Acidic Deposition Phenomenon and Its Effects. Vol. I. Atmospheric Sciences. Publica. No. 600/8083-016A. U.S.E.P.A., Washington, D.C. p. 3-92.
U.S. Environmental Protection Agency. 1984. Air Quality Criteria for Ozone and Other Photochemical Oxidants. U.S.E.P.A., Research Triangle Park, N.C. (in press).

Environmental Consequences of Energy Production: Problems and Prospects. Edited by S. K. Majumdar, F. J. Brenner and E. W. Miller. © 1987, The Pennsylvania Academy of Science.

Chapter Thirty-One

EFFECTS OF INDUSTRIAL CHEMICALS AND RADIOACTIVE MATERIALS IN BIOLOGICAL SYSTEMS

ANIRUDDHA GANGOPADHYAY[1] and
SAMAR CHATTERJEE[2]

[1]Author for Correspondence
Department of Pathology and Parasitology
166 Greene Hall
Auburn University, AL 36849
and
[2]School of Life Sciences,
Jawaharlal Nehru University
New Delhi—110067, India

INTRODUCTION

Biological effects of radiation and hazardous chemicals on cells, both acute effects and those having a delayed expression, continue to be important topics of scientific investigation. An impetus for this work resides in the ever-increasing potential in most countries of the world for exposure of humans, animals and other forms of life to hazardous materials. In the United States alone, approximately eighty billion pounds of hazardous wastes are produced each year,[1] a situation which creates major concerns and problems relevant to safe handling in the work-place and appropriate disposition. Also, as technological advancements continue to expand the capabilities for production of new biologically active compounds, it necessitates that an emphasis be placed on the investigation and definition of potential effects of these new compounds on living organisms. For example, the radioactive chemicals, which are unstable elements, emit charged particles that are potentially dangerous to living tissues. According to the type of radiation emitted they are classified as: gamma, neutron, beta, or alpha emitters. Different types of industrial chemicals like synthetic organic chemicals, heavy metals, mutagenic and carcinogenic agents have been shown to cause health related problems.[2-4]

Much has been written on the effects of radiation and toxic chemicals on biological systems. In this communication general considerations regarding these topics will be discussed very briefly; the major emphasis will be focused on the effects of chemicals, namely ethyl methanesulfonate (EMS) on Amoeba. Advantages to the use of amoeba for studying the effects of radiation and chemicals include the following: large mononucleate unicellular organisms having a long generation time; opportunity to study cellular organelles and biochemical and genetic alterations in a single cell system; and a long cell cycle, the stages of which can be synchronized without resorting to chemical treatment or temperature shock and thereby readily permitting study at defined stages of the cell's life cycle.

BIOLOGICAL EFFECTS OF RADIATION

The mechanism of bio-injury induced by radiation involves several complex series of physiochemical events. It is initiated with the deposition of energy by radiation in the form of ionization and subsequent excitation of atoms or molecules in living organisms. This is followed either by the inter-molecular or intra-molecular transfer of energy to produce various chemically active and short-lived species, the free radicals. These free radicals may react among themselves or with the cellular DNA and RNA of the biological system to produce alterations in biological expression.

The degree of biological damage depends on several factors, for example, dose of radiation applied; type of radiation (excitation or ionization) and type of organism involved. The energy that is absorbed from radiation in a biological system may lead to excitation or ionization. The raising of an electron in an atom or molecule to a high energy level, without actual ejection of the electron from that atom or molecule, is called *excitation*. Radiation with sufficient energy to eject one or more orbital electrons from the atom or molecule is termed *ionization*, and the radiation is called *ionizing radiation*. Although different biological effects are produced by different types of radiations[14], the cellular damage that occurs with different types of radiation has more similarities than differences. The response of organisms to radiation may vary even among closely related species due to differential tolerance. Certain chemicals are known to modify the response of radiation in biological systems. Such chemicals are, for example, radio-protectors like the protein, cysteine and radio-sensitizers like molecular oxygen.

Radiation induces deleterious effects in man which may be acute or delayed. The immediate effects appear within a very short period following irradiation and range from transient nausea and vomiting to death. Major target tissues for the immediate effect generally include bone marrow, gastrointestinal tract and nervous tissue. The delayed effects of radiation, however, may take several generations to be manifested. Prominent delayed effects include the induction of genetic damage followed by birth defects in the offspring of individuals exposed to radiation, overall shortening of the life span, development of leukemia or other types of cancer, temporary or permanent sterility (depending on the dose), and cataracts. The delayed effects from atomic radiation are well documented in the population which was exposed at Hiroshima and Nagasaki and in the descendents of the exposed persons.[5,6] Most inhabitants of Nagasaki were exposed to gamma rays whereas a substantial population in Hiroshima was exposed to neutrons. Five years after exposure to radiation the population in Hiroshima was shown to be at high risk for acute myeloid, acute lymphocytic and chronic myeloid leukemias and the irradiated population of Nagasaki to be at high risk for acute myeloid and acute lymphocytic leukemias.

Genetic changes induced by radiation are the consequence of *chromosomal aberrations, chromosomal breakage,* or *gene mutation.* The gene mutation may be dominant or recessive, leading to change in the structure of DNA involving either the base composition of the DNA or change in the sequence or both.

An interesting radiation effect has been reported in lunar astronauts. Astronauts while exploring the moon received a significant total body exposure to ionizing radiation from sudden intense sunspot activity.[12] This sudden exposure to radiation quickly caused the astronauts to develop dryness of the mouth, indicating an alteration in the function of salivary glands as an immediate and specific response to radiation exposure. Salivary gland dysfunctions also are reported to be a common sequela of radiation therapy in patients with head and neck cancer. A recent study in rhesus monkeys revealed morphological changes in acinar cells of the parotid gland to be a manifestation of acute radiation injury.[13] It is well known that most mammalian cells are particularly sensitive to radiation during mitosis, but results in the above study suggest that well-differentiated and nondividing cells like parotid acinar cells are susceptible to lethal radiation injury in interphase (nondividing phase).

The embryo and fetus are more susceptible to radiation compared with adults. Factors that impact on the effect of radiation include the dose applied, rate of administration, and stages of gestation. Exposure to radiation during developmental stages may cause embryonic, fetal or neonatal death, congenital malformations, cancer, growth retardation, or behavioral abnormalities. The most sensitive stage for the production of deleterious effects is the first trimester of pregnancy. Studies with rodents suggest that during this period of pregnancy, a dose as low as five rads is capable of producing congenital malformations and a dose lower than five rads may increase the probability of cancer.[7-11]

Effects of Radiation in Amoeba

The amoebae provide a suitable model to investigate the effects of radiation at a single cell level. Radiobiological studies on amoebae suggest that considerable similarities exist relevant to the effects of radiation, whether the irradiation is by ionization or excitation. Amoebae have been found to be more resistant to radiation exposure compared with cultured mammalian cells. The greater resistance of amoebae to x-ray may be a reflection of different effects on either the nucleus or cytoplasm or both. The LD_{50} dose of x-irradiation for *Amoeba proteus* strain PT_1 is 120 Kr[15] compared to 0.24 Kr for mouse fibroblasts. There also exists a disparity among different species and strains of amoebae regarding their resistance to radiation with x-rays and γ-rays. Delays in cell division occur in amoebae which have been exposed to radiation, such as UV, X-rays and γ-rays. Generally, the delay in division occurs between treatment and the first division following treatment. Delay in cell division increases with increasing dose.[16,17] Exposure of amoebae to x-rays may cause chromosomal damage, changes in nucleoli, protein denaturation, depolymerization of long molecules (eg. DNA), inactivation of several enzymes, gel to sol changes in the spindle and extensive vacuolization in the cytoplasm. Permanent mutation affecting the size of amoebae has been produced by x-radiation.[18]

Both cytoplasmic and nuclear damage are produced in amoebae exposed to radiation. Cytoplasmic damage by itself can produce damage to a control nucleus but in general it is rapidly reversible unless supralethal doses of radiation are used.[15] Along with the cytoplasmic damage the plasma membrane of the cell becomes fragile and its capacity for normal repair is impaired. Fragility of the cell membrane is noticeable during micrurgy when the cell frequently fails to close the created gap, thus suggesting that either the plasma membrane is itself damaged or it lacks the energy or microfibrillar activity that helps the repair process.[19] It is difficult to assess the separate extent of either nuclear or cytoplasmic damage unless nuclear transplantation experiments are performed. Transfers of irradiated nuclei to control enucleated cytoplasm and of control nuclei to irradiated enucleated cytoplasm can be done by a micromanipulator, thereby providing the opportunity to study the effects of irradiation on nucleus and cytoplasm separately. The amoeba nucleus is highly resistant to radiation damage compared to the nuclei of mammalian cells, but in amoeba the nucleus is more prone to lethal damage than the cytoplasm.

The relative resistance of amoebae to radiation compared with mammalian cells may be due to a combination of factors: (a) a highly developed repair system for radiation type damage; (b) highly polyploid nature of the amoeba nucleus; (c) a very efficient scavenging of free radicals. Radiation-induced cell death may be due to gene damage, either direct damage or damage to the repair process. Comparative DNA measurements of amoeba nuclei and tetraploid nuclei of rat liver suggested no marked difference in DNA content between the two cell types, but this measurement failed to indicate the number of gene copies. A possible explanation for the higher resistance of amoebae to radiation may be their high nuclear/cytoplasmic volume ratio. However, it is premature to draw any conclusions concerning the differential radiosensitivity of amoebae because sufficient information is lacking regarding the repair enzymes involved in radiation repair. An extensive review on radiation studies in amoebae explains in detail the various kinds of effects produced by radiation.[14]

BIOLOGICAL EFFECTS OF TOXIC CHEMICALS

The deoxyribonucleic acid in mammalian cells and in other cell types, for example, amoebae appears to be the main target for various mutagenic, carcinogenic and cytotoxic chemicals. It has been suggested that a majority of the carcinogenic chemicals or their metabolites are electrophiles that react with cellular nucleophiles such as proteins and nucleic acids (DNA and RNA).[20] Many mutagens/carcinogens are present in our environment, for example, those produced by modern industry which may be an important factor in the induction of different types of cancer.[21,22]

There exist wide differences in carcinogenic and mutagenic activity among the compounds which are closely related structurally.[23] It is known that most, if not all, of the closely related groups of substances which are active carcinogens have the ability to interact co-valently with DNA. The methylation of DNA produced by a number of alkylating agents, some of which are mutagenic but not carcinogenic, show that those which are powerful carcinogens may have small but possibly important differences in the sites of DNA bases which can be alkylated.[24]

The action of a mutagenic monofunctional alkylating agent, ethyl methanesulfonate (EMS: CH_3-Ch_2-O-SO_2-CH_3) has been studied by us at the cellular and subcellular level using *Amoeba indica* as a single cell model. The first reaction of a monofunctional alkylating agent with DNA is simply the addition of alkyl groups to the DNA,[25] although phosphates are also alkylated.[26] A monofunctional agent can occasionally cause cross-linking (either the two strands of one DNA molecule = intrastrand crosslink, or two strands of different DNA molecules = interstrand crosslink), owing to reactive DNA ends created by the alkylation—induced backbone breakage. EMS is considered to act as a mutagen mainly by the ethylation of purine base guanine at the 7—position and the subsequent induction of mutation either directly by hydrogen bonding with thymine instead of cytosine (GC → AT transition) or only after depurination.[27] Highly effective mutagenicity has been shown by EMS in a wide variety of systems, eg., *Drosophila, Habrobracon, E. coli, Bacillus subtilis, Neurospora* and *Saccharomyces cerevisiae,* to name a few. Carcinogenicity of EMS in neonatal mice has also been recorded.

Effects of EMS in Amoeba

The approach of exposing cells to radiation, different drugs, carcinogens, mutagens and other chemicals at defined phases of the cell cycle has opened up a new field of investigation. When challenged with different kinds of hazardous physical and chemical agents at specific phases of the cell cycle, significant molecular event(s) responsible for the altered phenotypic and biochemical expressions are revealed. Since its formulation by Howard and Pelc,[42] the concept of cell cycle has had a considerable importance in cell biology. The *cell cycle* time or mitotic cycle time is the time between successive cell divisions. The sequence of phases of a typical cell cycle is G_1, S, G_2 and M. A growing cell undergoes a cell cycle that is comprised essentially of two periods, the *interphase* or period of non-apparent division and the period of *division* or M phase. G_1 is the period between the end of mitosis and start of DNA synthesis, S is the phase of DNA synthesis, and G_2 is the interval between the end of DNA synthesis and the start of mitosis. The regulation of duration of the cell cycle occurs primarily by arresting it at a specific point during G_1, and the cell in the arrested condition is said to be in G_0 state. The main events of the cell cycle in *A. proteus* were summarized;[43] the main deviation from a typical cell cycle was the absence of a G_1 phase. Most of the cell cycle was occupied by G_2 period. Two hours before mitosis while in G_2, the amoebae pass T-1 or transition point one, where mitosis occurs and DNA synthesis is initiated when mitosis in completed. For our investigation we subdivided the cell cycle as follows: Mitotic phase (M, division sphere); early S phase (ES, 1 hour after division); mid S phase (MS, 3½ hours after division); late S phase (LS, 5½ hours after division); early G_2 phase (EG_2, 10 hours after division); mid G_2 I phase (MG_2I, 20 hours after division); mid G_2 II phase (MG_2II, 35 hours after division); mid G_2 III phase (MG_2III, 48 hours after division) and late G_2 phase (LG_2, 66 hours after division).

To investigate more critically the action(s) of chemicals, amoebae were exposed to EMS at discrete phases of the cell cycle to evaluate the differential response at the cellular and subcellular level.

1) Dose response and cell survival

Cells were exposed to different concentrations of EMS ranging from 0.1 to 1.5% for 15 minutes to investigate the response of cells after treatment. Sensitivity of the cells was considered in terms of cell survival after EMS treatment. A *survival curve* related the dose of EMS used to the proportion of cells which survived. Results revealed >98% cell survival when exposed to 0.1% EMS for 15 minutes versus no cell survival after treatment with 1.5% EMS for 15 minutes. The data showed a linear dose-effect relationship that was dose-rate dependent, as gradual sensitivity of the cells was achieved at progressively higher dosages. Cell cycle experiments on amoeba cells exposed to EMS showed a substantial difference in the sensitivity at different phases of the cell cycle. It was noted that cells in mitotic and S phase, especially MS phase, were more vulnerable to EMS action compared to cells in G_2 phase. Cells treated at MS phase, the period corresponding to peak DNA synthesis, showed greater sensitivity along with a prolonged delay in cell division for successive generations. A similar type of S phase sensitivity was recorded in amoebae treated

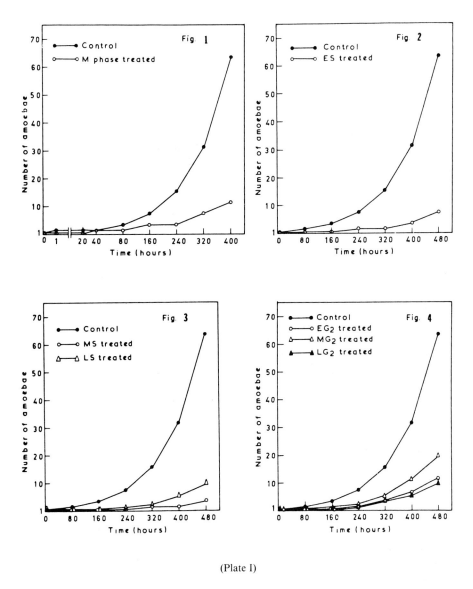

(Plate I)

FIGURE 1. *Amoeba indica* cells treated during mitotic phase (M) with ethyl methanesulfonate (EMS) illustrating prolonged inhibition in cell multiplication process.

FIGURE 2. *A. indica* cells treated during early S phase (ES) with EMS, showing mitotic arrest for successive generations.

FIGURE 3. Mitotic delay for several generations in mid S (MS) and late S phase (LS) treated cells.

FIGURE 4. Prolonged inhibition in cell replication process in early G_2 (EG$_2$), mid G_2 (MG$_2$) and late G_2 (LG$_2$) phase treated cells.

with N-methyl-N-nitrosourethane (MNU) and N-methyl-N-nitrosourea (MNUrea)[28,29] wherein nuclear transfer studies showed that the nucleus was more sensitive than the cytoplasm. Greater sensitivity of the nucleus was also documented by abnormalities in cell division. These data implicate replication as being responsible for the greater sensitivity of the S phase nucleus.

2) *Cell division time and mitotic block*

From cell survival studies it was found that treatment of cells with 0.5% EMS for 15 minutes was the most suitable dose to investigate the effects of EMS at the cellular and subcellular levels. Hence, it was the standard dose that was used in subsequent investigations.

The cell duplication time or generation time of control *A. indica* in our laboratory was 72 ± 6 hours. Prolonged inhibition in cell multiplication was evident in EMS-treated cells, continuing up to two months or more before the next mitosis occurred. There were differences observed regarding the mitotic block in cells treated at defined phases of the cell cycle. The mutagen-treated mitotic cells showed an initial delay in division of about 40 hours to produce two daughter amoebae, while in the untreated cells mitosis took approximately 35 minutes. The progeny of treated mitotic cells also showed mitotic arrest for several generations. In all cells treated at defined phases of the cell cycle, a prolonged division delay was found to occur successively for several generations. However, the mitotic block was more pronounced in cells treated in S and LG_2 phases (Figs. 1-4). DNA synthesis also was found to be depressed and it continued for a longer period. A similar situation was recorded in mouse fibroblast cells treated with nitrogen mustard (HN_2).[30] Since DNA synthesis and oxidative phosphorylation are, as a rule, prerequisites for cell division, inhibition of these processes generally results in inhibition of mitosis. It appears that a correlation exists between mitotic arrest and extension of different phases of the cell cycle.

3) *Effects on cell membrane*

The cell membrane was found to be affected severely following EMS treatment as evidenced by the alteration in cell shape, inhibition in the formation of pseudopodia, and impairment in the process of phagocytosis. Membrane functions were probably altered as a result of changes or damage to cell surface mucopolysaccharides, membrane-incorporated proteins and/or lipid bilayer of the cell membrane.

Scanning electron microscopic analysis of treated cells revealed major alterations in surface structure. It showed obliteration of cell surface protuberances which are a characteristic feature of normal *A. indica*. Appearance of several pits and mini bead-like structures also were noted (Figs. 5,6). The change in surface structure indicated damage to the mucopolysaccharide coat. Membrane

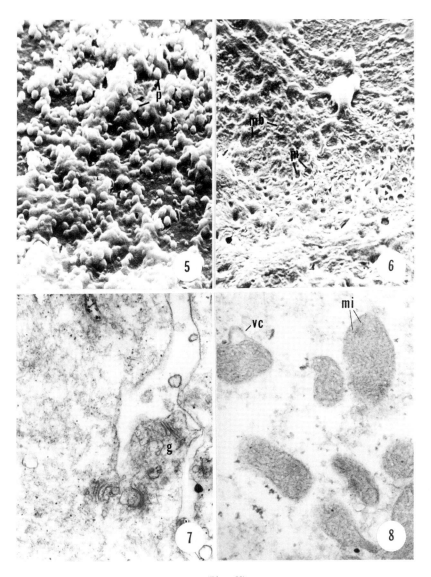

(Plate II)

FIGURE 5. Scanning electron micrograph of a randomly selected cell surface area of a nontreated *A. indica* cell showing abundant protuberances (p). X5000

FIGURE 6. Surface of an *A. indica* cell treated with EMS. Note disrupted surface structure, pits (Pi) and mini beads (mb). x 5000

FIGURE 7. Electron micrograph of an EMS-treated amoeba demonstrating disorganized Golgi body (g). x 22500

FIGURE 8. Mitochondria of an amoeba 24 hours after EMS treatment showing electron-dense mitochondrial inclusions (mi) and vesiculated crista (vc). x 30000

damage can be correlated with the damage to the Golgi apparatus which was observed following treatment. Autoradiographic studies have showed that glycoprotein in the Golgi apparatus is a precursor to cell membrane components, either present in the process of assembly or completed and awaiting transport to the cell surface.[31,32] Hence, damage to the Golgi apparatus can cause alterations in the biosynthesis of membrane components, thus leading to alterations in the surface structure.

4) *Effects on cytoplasm*

Recognizable changes in the cell body were observed in EMS-treated amoebae. A high frequency of nuclear ejection was noted within 2-3 hours following treatment. Similar effects were also recorded in *Entamoeba histolytica* subjected to radium rays and in irradiated *Acanthamoeba*. In stained cell preparations, the cytoplasm of the amoebae appeared extremely vacuolated. A similar type of vacuolation was observed in X-irradiated *Acanthamoeba* and cultured chick cells treated with nitrogen mustard. The hypervacuolation was an effect of extensive injury to the cytoplasm. It was observed that S phase cells were more vulnerable to cytoplasmic damage compared to G_2 phase cells, possibly indicating nuclear mediated cytoplasmic damage. In a study wherein nuclei were transferred between amoebae treated with MNU and control amoebae, the majority of the effects after treatment were attributed to nuclear damage.[33]

Electron microscopic observations revealed a substantial disorganization of cytoplasmic organelles in EMS-treated cells. The Golgi cisternae were disorganized, vesicular elements dispersed and certain regions of the cisternae dilated (Fig. 7). Studies of Golgi bodies in amoeba have indicated the presence of glycoprotein, acid mucopolysaccharides, acid phosphatase and thymidine pyrophosphatase in the Golgi stacks which are associated with the packaging of hydrolases to form primary lysosomes, membrane turnover, as well as other enzymatic processes such as those related to digestion.[34] It was noted that immediately following EMS treatment, along with the disorganization of the Golgi apparatus, the intensity of PAS (Periodic Acid Schiff) positive substances decreased considerably, suggesting that structural disorganization was associated with functional lesion. Increased numbers of free ribosomes in the cytoplasm were evident immediately following EMS treatment. This finding was probably the result of detachment of ribosomes from mRNA and endoplasmic reticulum due to EMS action, resulting in failure in the process of initiation of polypeptide chains. A similar type of effect has been recorded in amoebae after treatment with MNU,[35] MNUrea and UV-radiation.[36]

Distinctive changes in the form and structure of mitochondria were noted in treated amoebae. Extreme swelling of the mitochondria was observed within 12 hours after treatment. The tubular cristae often occupied a narrow peripheral region. In some cells rupture of the mitochondrial membrane was noted. Several

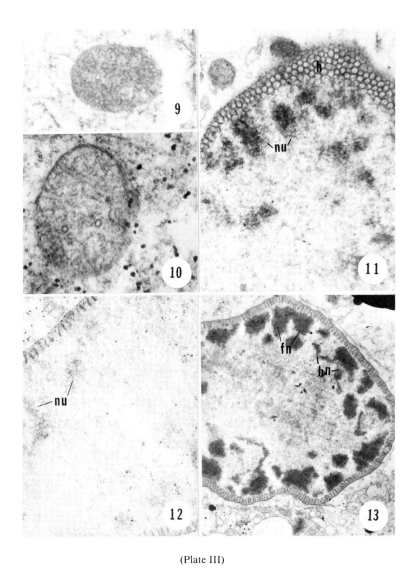

(Plate III)

FIGURE 9. Cytoplasm of a nontreated amoeba containing a normal mitochondrion. x 45000
FIGURE 10. Cytoplasm of mutagen treated amoeba at approximately 10 hours after treatment showing an enlarged mitochondrion. x 45000
FIGURE 11. Portion of nucleus in a nontreated *A. indica* showing nucleoli (nu) and nuclear honeycomb layer (h). x 17000
FIGURE 12. Portion of nucleus in an EMS-exposed amoeba immediately following treatment. Note disorganized nucleoli (nu) with loss of peripheral region. x 17000
FIGURE 13. Nucleus of EMS-treated amoeba at about 24 hours after treatment, showing nucleolar reconstitution. Note branching (bn) and fusion (fn) of nucleoli. x 8000

delayed effects also were encountered in amoebae: a) presence of vesiculated cristae in some mitochondria; b) appearance of numerous electron dense inclusions in the mitochondrial matrix; c) development of greater electron-opacity as compared with control mitochondria (Figs. 8-10). Some of these effects have also been recorded in amoebae treated with MNU,[35] N-methyl-N-nitrosoguanidine (MNNG).[36] Abnormalities in mitochondrial form and the presence of inclusions are often associated with an arrest of organelle activity[37,38] as was evident by reduced intensity of cytochemically detectable bound lipids at early periods following treatment. This suggests a change in the glycolysis and/or oxidative phosphorylation pathways of the cell. All of these findings suggest that EMS has an overall damaging action on membrane-bound organelles.

5) *Effects on nucleus*

The cell nucleus was found to be the major target of effects produced by EMS treatment. Enlarged nuclei, delays in cell division and production of giant cells were observed in EMS-treated cells which suggested that nuclear enlargement and mitotic arrest were different manifestations of common EMS toxicity. Disorganization of the nuclei and bizarre shaped nuclei also were observed. Some of these effects have also been observed in HeLa cells after hydroxyurea treatment[39] and in amoebae after treatment with nitrogen mustard, x-rays[15,16] or alkylating agents.[35] Treatment with EMS at early S phase revealed the formation of multinucleate cells which seemed to be the result of premature nuclear division.

Ultrastructural examination of the treated amoebae showed remarkable changes in the nucleoli immediately following EMS exposure. Disorganization of the nucleoli was noted along with loss of their peripheral region. Nucleoli were reconstituted at approximately 24 hours after treatment. Branching and even occasional fusion of the nucleoli were often observed as delayed effects (Figs. 11-13). Ribonucleoprotein helices, which are generally abundant in control nuclei, were found to be almost absent in nuclei of treated cells. Treatment of cells with MNU and MNUrea produces similar effects,[29,36] namely separation of nuclear helices from their attachment and subsequent elimination from the nucleus. Studies of amoebae by electron microscopic autoradiography showed that the peripheral region contained nucleolar DNA[40] and the mRNA and/or rRNA were packed in a helical form in groups near the nucleoli.[41] These findings indicate that EMS-induced nuclear damage is associated with the disappearance of RNP helices from the nucleus, suggesting lesion in the RNA synthetic activity. Shedding of the nuclear honey-comb layer was apparent in almost all cells examined after treatment.

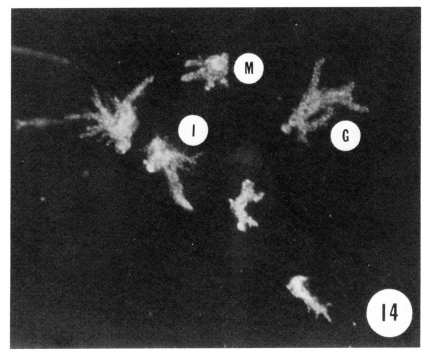

FIGURE 14. Photomicrograph of viable variant amoebae after mutagen treatment. Note giant (G), intermediate (I) and mini (M) cells. x 64

6) *Effects on macromolecular synthesis*

The profile of DNA synthesis was investigated by labelling control cells and EMS treated cells after mitosis with 3H-thymidine for 1 hour through each point. Amoebae of desired age were selected and incubated with labelled precursor. Autoradiograms were prepared to analyze DNA synthesis. The pattern of DNA synthesis is treated amoebae also was investigated by measuring ^{3}H-thymidine incorporation by liquid scintillation spectrometry. Both techniques showed that DNA synthesis was greatly inhibited in amoebae treated with EMS at early S phase. The depression in DNA synthesis continued throughout the S period. Alkylating agents such as EMS can inhibit DNA synthesis by alkylation of DNA which can be the result both of the alkylated DNA having a reduced primer activity and of a competition for sites on the DNA polymerase between normal and alkylated deoxyribonucleotide triphosphates, if the latter are unable to function as substrates in the polymerase reaction.[44] EMS also has been found to inhibit DNA synthesis and cause DNA damage in germ cells of male mice and in *Paramecium*[45,46] by ethylating several sites of DNA. It has been shown that nuclear DNA synthesis in amoebae continues at a reduced rate in the presence of, and immediately after exposure to MNU.[29] DNA synthesis has been found to be inhibited in HeLa cells immediately after treatment with methyl methanesulfonate (MMS) which blocks the initiation of replicons but the replicons which have already been initiated remain unaffected.[47]

Protein synthesis, as assayed by 3H-leucine incorporation, was found to be inhibited immediately following treatment. EMS also inhibits alkaline and acid phosphatase activities. It has been shown that alkylating mutagenic chemicals alkylate proteins resulting in the depression of protein synthesis.[48] Depression in the enzyme activity or enzyme inhibition in *E. coli* has been recorded[49] after treatment with alkylating chemical mutagens.

7) *Variant and mutant cell production*

EMS treatment caused pronounced changes in the size of amoebae. Three types of variant cells were detected, namely mini cells, intermediate cells and giant cells. The mini cells were the smallest in size; the intermediate cell types were larger than mini cells but smaller than normal amoebae; and the giant cells were considerably larger than control amoebae (Fig. 14).

The majority of the giant cells were lethal, but some returned to "apparent normal" size after several mitotic divisions and survived. Most of the giant cells were multinucleated. The formation of mini cells occurred frequently after administration of mutagen to the amoebae. In broad terms, these mini cells could be categorized into two major groups: 1) cells which after a period of time regained their normal size; and 2) cells which permanently remained smaller in size compared to controls and which were designated mini mutants for their unique unaltered mutant characteristics. The array of intermediate sized cells which were observed usually reverted back to "apparent normal" size after 8 to 9 division cycles. The size variations of different cell types and the variations in their nuclear diameter are shown in table 1.

Between 40 and 50% of the variant cells failed to survive within 15-20 days following mutagen treatment. Only a few mini cells formed stable clones. A detailed characterization of the mini mutant cells has been reported.[50] Table 2 shows the number of variants produced after exposing a known number of cells to the standard dose of EMS.

The nucleus of amoeba has long been considered to be polyploid, though the existence of polyploidy is based on circumstantial evidence. A mutation in an amoeba cell which has many representatives of each gene will require a change in all, or at least many, of them. When the amoebae are exposed to EMS for only a short period of time, it is necessary for all the copies of the genes to be vulnerable at the same time, while the majority of other genes are protected. Two activities of the chromatin are expected to leave genes vulnerable

TABLE 1

Cell types	Average cell size (μm ± SD)	Nuclear diameter (μm ± SD)
A. indica	397 ± 11	33 ± 2.5
Giant cell	594 ± 27	52 ± 3
Intermediate cell	295 ± 15	28 ± 1.5
Mini cell	220 ± 25	21 ± 2

TABLE 2

Total No. treated	No. of Variants produced	Types of Variants		
		No. of Mini cells	No. of Intermediate cells	No. of Giant cells
2018	935	238 → x 96, *142; x 96 → @ 84, **12	632 → x 565, *67; x 565 → @ 565	65 → x 7, *58; x 7 → @ 7

(X, viable cells; *, non-viable cells; @, cells reverted back to normal forms; **, formed stable clone).

to EMS, namely DNA replication and DNA transcription. Provided that during replication all the equivalent loci replicate their DNA at the same time, or that during transcription all the copies of a given gene "switch on" simultaneously, these periods are likely to give vulnerability to particular genes while other periods would be less affected. Expression of inheritable change in a polyploid system eg., *A. indica*, might indicate nonrandom change of specific cistrons, such as indicated in amoeba cells[50] and also in other systems[51,52] after treatment with EMS and MNU. The EMS induced mini amoeba cell is a size mutant which has a cell-cycle-phase-specific origin; bears a totally altered pattern of cell cycle and has specific structural, physiological and biochemical properties.

DISCUSSION

Effects produced by different types of radiation and industrial chemicals on biological systems present many health related concerns to modern society. In general, it appears that more similarities than differences exist relevant to the effects produced by radiation and chemicals at the single cell level. These effects are not only restricted to the individual or the organism exposed to the hazardous chemicals or radiation but also are carried over to the next generation. When hazardous wastes, industrial chemicals or radioactive substances are permitted to pollute the environment, a diverse population of living organisms become at risk. Pollutant air and acid rain have been major environmental concerns for several years. Pollution of underground water is another example of public concern. It has been reported by the Environmental Protection Agency that nearly one-third of the public ground water system in the United States is contaminated by chemicals. One source of this contamination is thought to be a result of waste disposal in land-fills.[53] Oceanic disposal is also becoming a concern as DDT and radioactive fallout have been detected in the deepest level of the ocean.[54]

Technological advancements may lead to safer methods of waste disposal. Possibilities include microstructural modification of radioactive waste[55,58] to immobilize them and the development of genetically engineered microbes that are capable of breaking down specific hazardous wastes. For example, genetically modified species of *Pseudomonas* bacteria are capable of detoxifying halogenated compounds.[56] The management of radioactive materials and wastes is well addressed in a report prepared by S.K. Majumdar and E.W. Miller.[57]

ACKNOWLEDGEMENTS

We gratefully acknowledge Dr. Lauren G. Wolfe, Professor and Head, Dept. of Pathology and Parasitology, for valuable discussion and editorial correction. Thanks to Betty Adcock and Lisa Johnson for typing this manuscript and personnel in the Learning Resources Center for photo- printing work.

REFERENCES

1. Epstein, S.S., L.O. Brown and C. Pope. 1982. Hazardous Waste in America. Sierra Club, San Francisco. CA., pp. 593.
2. Picciano, D. 1980. A pilot cytogenetic study of the residents living near Love Canal, a hazardous waste site. Mammal. Chromosom. Newsl. 21:86-99.
3. Wolff, S. 1984. Love Canal revisited. J. Am. Med. Assn. 251:1464.
4. Lumb, G. 1984. Health effects of hazardous wastes. In: S.K. Majumdar and E.W. Miller (eds.) Hazardous and toxic wastes: technology, management and health effects. Pennsylvania Academy of Science, Publication, Easton, Pa., pp. 388-405.
5. UNSCEAR Report. 1977. Genetic Effects of Radiation. United Nations Scientific Committee on the Effects of Atomic Radiation.
6. Hall, E.J. 1978. Radiobiology for the Radiologists. Medical Department, Harper and Row Publishers, Inc., New York. pp. 359-410.

7. Rugh, R. 1963. The impact of ionizing radiation on the embryo and fetus. Am. J. Roentgenol. 89:182-190.
8. Brent, R.L., and R.O. Ghorson, 1972. Radiation exposure in pregnancy. Curr. Probl. Radiol. 2:1.
9. Dekaban, A.S. 1968. Abnormalities in children exposed to x-radiation during various stages of gestation: tentative timetable of radiation to the human fetus, part 1. J. Nucl. Med. 9:471-477.
10. Yamazaki, J.W., S.W. Wright and P.M. Wright. 1954. Outcome of pregnancy in women exposed to the atomic bomb in Nagasaki. Am. J. Dis. Child. 87:448-463.
11. Russell, L.B. and W.L. Russell. 1954. An analysis of the changing radiation response of the developing mouse embryo. J. Cell Physiol. (suppl. 1) 43:103-149.
12. Michener, J.A. 1982. Space. Random House, New York, pp. 459-503
13. Stephens, L.C., G.K. King, L.J. Peters, K. Kian Ang, T.E. Schultheiss, and J.H. Jardine. 1986. Acute and late radiation injury in rhesus monkey parotid gland: Evidence of interphase cell death. Am. J. Pathol. 124:469-478.
14. Ord, M.J. 1973. Radiation Studies. In: The Biology of Amoeba. Academic Press, New York. pp. 401-422.
15. Ord, M.J. and J.F.Danielli. 1956. The site of damage in amoebae exposed to x-rays. Q.J. Microsc. Sci. 97:29-37.
16. Daniels, E.W. 1952. Some effects of cell division of *Pelomyxa carolinensis* following x-irradiation, treatment with bis (β-chloroethyl) methyl amine and experimental plasmogamy (fusion). J. Exp. Zool. 120:509-523.
17. Mazia, D. and H.I. Hirshfield. 1951. Nucleus- cytoplasm relationships in the action of ultraviolet radiation on *Amoeba proteus*. Expl. Cell Res. 2:58-72.
18. Schaeffer, A.A. 1946. X-Ray mutations in the giant multinuclear amoeba *Chaos chaos*. Linn. Anat. Record. 96:531.
19. Jeon, K.W. and M.S. Jeon. 1975. Cytoplasmic filaments and cellular wound healing in *Amoeba proteus*. J. Cell Biol. 67:243-251.
20. Miller, E.C. and J.A. Miller. 1976. Chemical Carcinogens. In: C.E. Searle (ed.) ACS Monograph. 173:737-762.
21. Sugimura, T. 1982. In: Molecular interactions of nutrition and cancer. M.S. Arnott, J. Van Evs, Y.M. Wang (eds.) Raven, New York, pp. 3-24.
22. Sugimura, T. 1986. Studies on environmental chemical carcinogenesis in Japan. Science. 233:312-318.
23. Brookes, P. and P.D. Lawley. 1964. Evidence for binding of polynuclear aromatic hydrocarbons to the nucleic acids of mouse skin: Relation between carcinogenic power of hydrocarbons and their binding to deoxyribonucleic acid. Nature (Lond.) 202:781.
24. Maitra, S.C. and J.V. Frei. 1975. Organ Specific effects of DNA methylation by alkylating agents in the inbred Swiss mouse. Chem. Biol. Interact.

10:285-293.
25. Strauss, B. and T. Hill. 1970. The intermediate in the degradation of DNA alkylated with a monofunctional alkylating agent. Biochim. Biophys. Acta. 213:14-25.
26. Rhaese, H. and E. Freese. 1969. Chemical analysis of DNA alterations. I Base liberation and backbone breakage of DNA and oligo-deoxyadenylic acid induced by hydrogen peroxide and hydroxyl amine. Biochim. Biophys. Acta. 155:476-490.
27. Orgel, L.E. 1965. The chemical basis of mutation. Advan. Enzymol. 27:289-346.
28. Ord, M.J. 1971. *Amoeba proteus* as a cell model in toxicology. In: W.N. Aldridge (ed.) Mechanisms of toxicology. Macmillan, London, pp. 175-186.
29. Ord, M.J. 1976. A study of the change in DNA synthesis of S phase cells treated with N-Methyl-N-nitrosourethane: A study using *Amoeba proteus* as a single cell model. Chem. Biol. Interact. 12:325-340.
30. Gelfant. S. 1963. Inhibition of cell division: A critical and experimental analysis. Int. Rev. Cytol 14:1-39.
31. Read, G.A. and C.J. Flickinger. 1980. Changes in production and turnover of surface components labelled with ^{3}H mannose in amoebae exposed to a general anesthetic. Exp. Cell Res. 127:115-126.
32. Flickinger, C.J. 1981. The presence of carbohydrate-rich material in the developing golgi apparatus of amoebae. J. Cell Sci. 47:55-63.
33. Ord, M.J. 1968. Immediate and delayed effects of N-Methyl-N-nitrosourethane on *Amoeba proteus*. Exp. Cell Res. 53-73-84.
34. Flickinger, C.J. 1975. The relation between the golgi apparatus, cell surface, and cytoplasmic vesicles in amoebae studied by electron microscopic autoradiography. Exp. Cell. Res. 96:189-201.
35. Ord, M.J. 1976. The interaction of nuclear and cytoplasmic damage after treatment with toxic chemicals. J. Theor. Biol. 62:369-387.
36. Ord, M.J. 1979. The effects of chemicals and radiations within the cell: An ultrastructural and micrurgical study using *Amoeba proteus* as a single cell model. Int. Rev. Cytol 61:229-281.
37. Oliveira, L. 1977. Changes in the ultrastructure of mitochondria of roots of *Triticale* subjected to anaerobiosis. Protoplasma. 91:267-280.
38. Smith, R.A. and M.J. Ord. 1983. Mitochondrial form and function relationship *in vivo*: Their potential in toxicology and pathology. Int. Rev. Cytol. 83:63-134.
39. Grant, D. and P. Grasso. 1978. Suppression of HeLa Cell growth and increase in nuclear size by chemical carcinogens: A possible screening method. Mutation Res. 57:369-380.
40. Minassian, I. and L.G.E. Bell. 1976. Studies on changes in the nuclear helices of *Amoeba proteus* during cell cycle. J. Cell Sci. 20:273.
41. Wise, G.E., A.R. Stevens and D.M. Prescott. 1972. Evidence of RNA in

the helices of *Amoeba proteus*. Exp. Cell Res. 75:347-352.
42. Howard, A. and S.R. Pelc. 1953. Synthesis of deoxyribonucleic acid in normal and irradiated cells and its relation to chromosome breakage. Heredity suppl. 6:261-273.
43. Prescott, D.M. 1973. The cell cycle in amoeba. In: K.W. Jeon (ed.) Biology of Amoeba. Academic Press, New York, pp. 467-477.
44. Wheeler, G.P. 1962. Studies related to the mechanism of action of cytotoxic alkylating agents: a review: Cancer Res. 22:651-688.
45. Sega, G.A. 1974. Unscheduled DNA synthesis in germ cells of male mice exposed *in vivo* to the chemical mutagen ethyl methanesulfonate. Proc. Nat. Acad. Sci. (USA). 71:4955-4959.
46. Cohen, J. 1980. Cytotoxic versus mutagenic effect of ethyl methanesulfonate on *Paramecium tetraurelia*. Mutation Res. 70:251-254.
47. Painter, R.B. 1977. Inhibition of initiation of Hela Cell replicons by ethyl methanesulfonate. Mutation Res. 42:299-304.
48. Fox, B.W. and M. Fox. 1967. Effect of methyl methanesulfonate on macromolecular biosynthesis in $P_3 88F$ cells. Cancer Res. 27:2134.
49. Lawley, P.D. and P. Brookes, 1965. Molecular mechanism of the cytotoxic action of difunctional alkylating agents and of resistance to this action. Nature (Lond.). 206:480-483.
50. Gangopadhyay, A. and S. Chatterjee. 1984. A cell-cycle-phase-specific mutant of amoeba. J. Cell Sci. 68:95-111.
51. Guerola, N., J.L. Ingraham, and E. Creda-Olmedo. 1971. Induction of closely linked multiple mutations by nitrosoguanidine. Nature New Biol. 230:122-125.
52. Guerola, N. and E. Creda-Olmedo. 1975. Distribution of mutations induced by ethyl methanesulfonate and ultraviolet radiation in the *Escherichia coli* chromosome. Mutation Res. 29:145-149.
53. US Environmental Protection Agency. 1984. Environmental progress and challenges: An EPA perspective. EPA. Washington, D.C. p. 115.
54. Duedall, I.W., B.H. Ketchum, P.K. Park and D.R. Kester. 1983. Wastes in the ocean. Vol. 1. Industrial and sewage wastes in the ocean. Wiley. New York. p. 4.
55. Ringwood, A.E., V.M. Oversby, S.E. Kesson, W. Sinclair, N. Ware, W. Hibberson and A. Major. 1981. Immobilization of high-level nuclear reactor wastes in SYNROC: a current appraisal. Nucl. Chem. Waste manage. 2:287-305.
56. Ghosal, D., I.S. You, D.K. Chatterjee and A.M. Chakrabarty. 1985. Microbial degradation of halogenated compounds. Science. 228:135-142.
57. Majumdar, S.K. and E.W. Miller, eds. 1985. Management of radioactive materials and wastes: issues and progress. Pennsylvania Academy of Science, Easton, Pa. p. 405.
58. Kindt, J.W. 1984. Radioactive wastes. Natur. Resour. J. 24:967-1014.

Environmental Consequences of Energy Production: Problems and Prospects. Edited by S. K. Majumdar, F. J. Brenner and E. W. Miller. © 1987, The Pennsylvania Academy of Science.

Chapter Thirty-Two
MAJOR ENVIRONMENTAL IMPACTS OF RESOURCE RECOVERY FACILITIES

MICHAEL D. BROWN
President, Brown, Vence & Associates
120 Montgomery
San Francisco, California 94104

INTRODUCTION

Although resource recovery facilities that process Municipal Solid Waste (MSW) greatly reduce the environmental impacts associated with landfill disposal, the processing and recovery of energy from solid waste entails a different set of environmental impacts. These effects must be carefully considered in the planning of any resource recovery plant. The recovery of energy from MSW, by combustion, creates potential health and environmental impacts that require the application of good planning and efficient pollution control technologies.

The primary paths of potential environmental impact of a resource recovery facility can be identified as follows:

- Air emissions from plant operation
- Combustion residue disposal
- Wastewater disposal

In addition, minor potential impacts can be associated with a facility in the following areas:

- Odor
- Noise
- Traffic
- Construction
- Visual aesthetics
- Plant-internal impacts (e.g., explosion hazard, dust, noise, and odor)

POSITIVE ENVIRONMENTAL EFFECTS

Incineration as a disposal method for MSW destroys most potentially hazardous agents that may otherwise persist in the environment if disposed of at landfill sites. Micro-biological contaminants (bacteria, viruses, and fungi), chemicals (household cleaners, solvents, pesticides, etc.), and disease vectors (rodents and insects) are typically present at landfills used for waste disposal. Furthermore, the anaerobic decomposition of landfilled putrescible waste produces gases that are potentially explosive, odorous, and toxic. A resource recovery facility would in fact reduce these landfill hazards. Because the facility would essentially destroy rather than merely store the wastes, and because it would be enclosed, ventilated, and equipped with emission controls, the risks from certain hazards associated with landfill would be decreased.

Siting of new landfills or expansion of existing ones has become increasingly difficult because of environmental, political, and economic factors. There is a growing scarcity of sanitary-fill sites suitable in terms of local acceptance, hydrogeology, soil parameters, acreage, etc. A resource recovery facility conserves precious landfill capacity by greatly reducing the volume and volatility of the solid waste.

Furthermore, a resource recovery facility reduces pollution derived from the plant's energy customers. Supplied with steam and/or electricity, these energy customers would consume less fuel and generate substantially less air emissions and wastewater than they would in the absence of the resource recovery facility.

AIR QUALITY

Introduction

The combustion of MSW will produce a wide range of flue gas emissions. These stack emissions constitute the major environmental impact of any combustion-based resource recovery facility and, as such, they are well regulated under federal law. Emissions of "criteria" pollutants, addressed by National Ambient Air Quality Standards (NAAQS), comprise particulate matter (TSP), nitrogen oxides (NOx), sulfur dioxide (SO_2), carbon monoxide (CO), total hydrocarbons (THC), and lead. In addition, waste-to-energy plants will emit varying amounts of "noncriteria" pollutants (i.e., with no NAAQS yet established), particularly acid gases such as hydrogen chloride (HCl) and hydrogen fluoride (HFl), as well as heavy metals (other than lead), trace toxic organic compounds, and asbestos.

Noncriteria pollutants are generated by refuse-fired plants as a result of the heterogeneous nature of the MSW fuel. Air pollution control equipment can control particulate matter, sulfur dioxide, acid gases, and a large percentage of noncriteria pollutants. In addition, proper design and control of combustion temperature, fuel residence time, and combustor turbulence substantially

TABLE 1
Resource Recovery Facility Flue Gas Emissions

Pollutant	Emissions Factor[a]		360 tpd Throughput	
	lb/ton of MSW	lb/MM Btu	lb/day	tons/yr
TSP				
Uncontrolled	29.76	3.10	10,862	1,982
Controlled[b]	0.21	0.02	76	14
SO$_2$				
Uncontrolled	2.11	0.12	760	139
Controlled[c]	1.27	0.13	456	83
NOx	4.0	0.42	1,440	262
CO	1.9	0.20	684	125
THC	0.46	0.05	166	30
HCl				
Uncontrolled	5.4	0.56	1,944	355
Controlled[d]	0.54	0.056	194	36

[a]See text for derivation.
[b]Assumes control efficiency of 99.3%.
[c]Assumes control efficiency of 40%.
[d]Assumes control efficiency of 90%.

reduce emissions of heavy metals and organic compounds.

Refuse Composition

The varying composition of refuse fuel results in emissions of great inconsistency, creating wide ranges in quantified characterizations of emissions from particular waste-burning facilities. Moreover, different technology configura- temperature, fuel residence time, and combustor turbulence substantially reduce emissions of heavy metals and organic compounds.

Several studies have correlated the chemical composition of the refuse fuel to major categories of the waste stream. Typical or average values of chemical makeup of MSW and derivative emissions from comparable facilities may not, however, be adequate to characterize emissions from a proposed plant. Actual sampling of the waste stream may be necessary to accurately predict emissions.

Characterization and Control of Criteria Emissions

Table 1 identifies emissions estimates for five criteria pollutants and hydrochloric acid gas (HCl) for a hypothetical facility combusting 360 tons of unprocessed municipal waste per day. Emission rates, derived from emission factors in pounds per ton of refuse fuel, are calculated as pounds/million Btu, pounds/day, and tons/year. Emissions of TSP, sulfur dioxide (SO$_2$), and HCl, are given in both uncontrolled and controlled quantities assuming state-of-the-art control devices are utilized; all other emissions are unabated. Discussion of each pollutant and emission factor follows. A key element in controlling exhaust emissions, assuming properly designed pollution control equipment, is combustor operation. Depending upon the system, the operator has control of underfire and overfire air, auxiliary fuel burners, and stoker speed. Proper control of these variables can eliminate much of the emissions fluctuations that result from the diversified composition of refuse fuel.

Particulate Matter. Total suspended particulate (TSP) is of particular concern because many areas are nonattainment for this pollutant. The uncontrolled emission factors of 1.4 gr/dscf (at 12% CO_2) and 3.1 lb/million Btu are derived from a California Air Resources Board (CARB) report (1984). Both factors average data from a number of refractory-wall/waterwall MSW mass-burning facilities. Fabric filters are highly efficient particulate control devices, typically capturing particles with greater than 99 percent efficiency. Based on test results reported in the CARB document, at least a 99.3 percent removal efficiency, which would reduce TSP emissions to 0.01 gr/dscf (at 12% CO_2) and to 0.022 lb/million Btu, is reasonably achievable.

Sulfur Dioxides (SO_2). This emission factor of 2.11 lb/ton of refuse and 0.22 lb/million Btu is derived from the CARB report (1984) and averages uncontrolled SO_2 emissions from comparable operating facilities. Based on representative-manufacturer information and data from the Battelle and CARB reports, state-of-the-art SO_2 control technology can conservatively be expected to remove 40 percent of the SO_2 in the flue gas stream. It should be noted that SO_2 typically emitted from the mass-burning of MSW is considerably less than that emitted from combustion of most coals (O'Connell et al. 1982).

Nitrogen Oxides (NOx). Unpublished data from a resource recovery facility in Pittsfield, Massachusetts show NOx formation at a rate of 3.2 lb/ton of MSW. The 1982 Battelle study uses a value of 1.6 lb/ton of refuse as a reasonable and conservative value for a modern waterwall incinerator (O'connell et al. 1982). The more conservative factor used here (4.0 lb/ton) takes into account higher-temperature operation to control toxic organic pollutants.

Carbon Monoxide (CO). This emission factor derives from a mass rate of 1.9 lb/ton of refuse reported by Battelle (O'Connell et al. 1982) as a conservative estimate for a modern incinerator with adequate combustion controls.

Total Hydrocarbons (THC). This emission factor of 0.46 lb/ton derives from a reported average generation rate for resource recovery facilities, including older incinerators, which frequently have higher THC emissions than newer incinerators. The Battelle study considers 0.12 lb/ton a more reasonable estimate for modern waterwall combustors (O'Connell et al. 1982).

Hydrochloric Acid (HCl). HCl, a corrosive and irritant, is not a criteria pollutant but is considered here as a primary pollutant from resource recovery facilities. HCl emissions are largely dependent on the proportion of plastic, and particularly polyvinyl chloride, in the waste stream. The emission factor used here, 5.4 lb/ton of waste, assumes an MSW chlorine content of 0.54 percent, the mean chlorine content reported by CARB (1984) for the waste streams of nine U.S. cities, including four in California and two in Pennsylvania. The

California waste streams show an average chlorine content 63% higher than that of the other cities, thus skewing the mean upward and ensuring a conservative estimate. O'Connell et al. (1982) reports that only about one-half of the chlorine in the waste is converted to HCl; thus a 50 percent conversion factor is incorporated into the emissions factor. Based on representative-manufacturer data, state-of-the-art pollution control equipment can conservatively be expected to remove 90 percent of the HCl in the flue gas stream. Also, some of the evolved HCl may be removed on ash particles as flue gases pass through the boiler.

Characterization of Noncriteria Emissions

Resource recovery facilities have the potential to emit metallic, organic, and other compounds, some of which are known to cause adverse health impacts at relatively low concentrations. These compounds can be emitted as solids, vapors, or aerosols.

Metal Compounds. Metals are present in varying quantities in MSW. Metal objects per se do not usually pose an emissions problem, but metals are released from the combustion of MSW containing paints, pigments, foils, solder, etc. Metal compounds that can be associated with emissions from MSW combustion include antimony, arsenic, beryllium, cadmium, chromium, copper, mercury, manganese, molybdenum, nickel, lead (technically a criteria pollutant but similar in behavior to other metals), selenium, tin, vanadium, and zinc. Heavy metals from refuse combustion are generally not emitted as gases but are bound with particulate matter, with a greater proportion associated with fine particles (diameter less than 2 microns). An exception is mercury, which is emitted primarily as a vapor. Heavy metal "enrichment" on fly ash is a function of particle size distribution, particle density in the flue gas, particle composition, and flue gas temperature. Heavy metals emissions would be effectively controlled along with particulates by a baghouse. Mercury can be controlled by reducing flue gas temperature to condense the metal from the flue gas stream, causing it to adsorb to the fly ash. Lead, beryllium, and mercury are currently subject to federal emission limits.

Although not a metal, asbestos may be sporadically present in MSW and can be emitted from resource recovery plants as particulate. The particulate control device will capture almost all asbestos-fiber particulates, and emissions are expected to be insignificant.

Toxic Organic Compounds. Waste-fueled combustion systems, including units such as hospital incinerators as well as resource recovery plants, have the potential to emit trace quantities of toxic organic compounds, such as polychlorinated and polynuclear aromatic hydrocarbons. The families of compounds of primary concern and increasing public visibility are the PCDDs (polychlorodibenzo-p-dioxins, often called dioxins) and PCDFs (polychlorodibenzofurans, called furans). These species comprise some of the most toxic and carcinogenic substances known. PCDDs have been found both adsorbed with fly ash and in gaseous emissions from incinerators.

The formation of PCDDs and PCDFs in combustion is complex and not yet well understood. Basically, these products are considered to result from combustion of chlorinated waste materials, such as certain kinds of plastic, or from burning simple hydrocarbons in the presence of chlorine.

Organic compounds are emitted from resource recovery plants as gas and bound with particulates. The same variables that control fly ash adsorption of heavy metals also control adsorption of exotic organic compounds. Nevertheless, the combustion environment is a primary determinant of organic emissions. Organic compounds are destroyed in a combustion system designed and operated at sufficient temperature, fuel residence time, and combustion gas turbulence (or mixing efficiency).

The U.S. EPA has not issued standards for acceptable levels of PCDDs from combustion or other sources. However, an EPA study of PCDD emissions from five operating refuse-burning facilities concluded that these emissions do not present a health hazard to those residents living in the plant vicinity.

The entire issue of dioxin effects on human health is, at present, a complicated one. Toxicity thresholds in humans have not yet been determined. Several studies have evaluated the potential health effects of dioxins from refuse-burning plants. A major 1980 study concluded that the cancer risk to the public from waste-combustion-derived dioxins is several orders of magnitude less than risk associated with home accidents, motor vehicle accidents, smoking, or influenza. The study indicates that overall health risks from other methods of solid waste disposal are at least as great as those from combustion-released dioxins. A more recent risk assessment of health hazards from waste-to-energy plants deals with the dioxins issue in depth (North County Resource Recovery Associates 1984). This California study reports that the chances of 30 years of plant operation producing cancer in residents living within 10 miles are less than one in one million.

The combustion system should be designed, operated, and maintained to minimize all hazardous trace organic substances. Organic molecules of concern have varying thermal stabilities. In general, a system operating at 1,800°F, with a fuel residence time of no less than one second and with fuel completely mixed in the combustor, should provide efficient thermal destruction of chlorinated and nonchlorinated organic compounds. Furthermore, based on epidemologial and public health studies conducted elsewhere, any very small quantities of PCDDs not destroyed by combustion and emitted with particulates or as gas do not appear to pose a measurable adverse health risk to the population in the vicinity of the plant.

Fugitive Emissions

In general, resource recovery projects will not generate significant amounts of fugitive dust. Dust generated by the disposal and handling of raw solid waste at the facility will be contained within a totally enclosed building. Dust generated by handling will be controlled by the negative pressure in the tipping area. This negative pressure is created as air for the combustion process is drawn from the refuse-storage area.

A potential source of fugitive emission is the process of transporting bottom ash and fly ash from the facility. The bottom and fly ash are usually deposited directly into a water-filled quench tank. This wet residue is then conveyed to bins and transported to an approved landfill for disposal (or sold as slag). Because the ash is kept wet, the likelihood of dust generation is remote.

Trucks carrying material to and from the site should be covered. Only during site preparation will sufficient soil be exposed to present an increased potential for air entrainment of particulates. This period will be of short duration.

RESIDUE MANAGEMENT

Solid residues from the resource recovery plant will include ash from the combustion process, wastes bypassed while the plant is out of service for repair and maintenance, and nonprocessible wastes. Each of these waste categories can be managed as follows:

Ash

Solid waste intrinsically contains ash; that is, some of the waste is noncombustible. Even the combustible portion will not burn completely. Paper, for example, is about 8 percent ash.

Two types of residual ash will be produced by a resource recovery plant. One is bottom ash, which is the noncombustible portion of the solid waste. The second is fly ash, which rises with the hot gases of combustion and is removed by the boiler and air pollution control equipment. These two types of residuals have different characteristics.

The quantity of residual ash produced is determined by the quantity of input waste, the composition of this feedstock, and the efficiency of combustion. MSW residue generated by a 360 ton/day resource recovery facility, based upon typical waste composition and mass-burning modular combustion with pollution control, is estimated as follows:

Some 93 percent of this residue will consist of bottom ash and about 7 percent of fly ash. The volume of this residue represents less than 10 percent of

Throughout, (Tons/Day)	Ash Produced*	
	Tons/day	yd^3/day
360	108	123

*Assumes that residue is equivalent to 30% by weight of input and has a bulk density of 65 lb/ft^3.

the volume of the solid waste fuel. Both fly and bottom ash fractions will be quenched with water, which also serves to stablize the residue for transport to disposal sites.

Ferrous metals may be magnetically removed from the ash for resale. Ferrous recovery will reduce total ash quantity by approximately 15-20 percent by weight. The metals-free ash has, in many instances in Europe and on a limited basis in the United States, been marketed as a fill material or as aggregate substitute. This commercial use is the preferred method for managing ash from the facility, however, if the material cannot be marketed, it must be landfilled.

Refuse combustion concentrates many of the heavy metal components of solid waste, with greater concentrations in the fly ash than in the bottom ash. Fly ash may also contain adsorbed organic compounds and other materials, such as asbestos, potentially hazardous to the public. Heavy metals contained in ash may include arsenic, barium, cadmium, chromium, cobalt, copper, lead, manganese, molybdenum, nickel, thallium, vanadium, zinc, and others. After residue is deposited in a landfill, soluble heavy metals can be leached from the ash. These leachate constituents have the potential to degrade surface water and groundwater. Heavy metals in ash, however, are not highly mobile as long as the ash is kept its typically alkaline or a neutral state. Leaching problems can arise if landfill chemistry becomes more acidic, but maintaining a high pH prevents or retards leaching of heavy metals from ash. It should be noted that potential leachate problems from heavy metals are also inherent in the solid waste and sewage sludge currently landfilled or land-applied without combustion.

By-Passed Wastes

A modular design minimizes the amount of solid waste that will require diversion to a landfill because of plant downtime. Having more than one processing line allows one to be serviced while the others are operating. Bypassed amounts can be further reduced by scheduling maintenance during periods of low waste flow. Bypassed wastes require landfill disposal.

Nonprocessibles

To the extent possible, nonprocessible wastes should be excluded from the facility and directed to a landfill. Nonprocessibles entering the facility should be separated from the waste stream by plant operators prior to combustion and hauled to the appropriate landfill.

WATER QUALITY

Resource recovery facilities use water for startup, boiler makeup, boiler blowdown, ash quenching, sanitation, and general maintenance. Plant water will be conditioned and deaerated on site before use. Assuming a "worst case" water demand in which no condensate is returned from the energy customer, operation will require approximately 400 to 500 gallons of water/ton of refuse combusted. Steam generation is the largest component of demand, followed by blowdown and ash quenching. Water-quenching of ash from boilers and reactors is necessary to cool the residue and prevent fires in ash storage receptacles. Quench water is progressively disposed of with the wet ash.

Water not recycled internally may be discharged to a municipal sewer system or to onsite treatment. Effluent from boiler blowdown and periodic ash quench tank cleaning may contain conditioning salts and some pollutants from the ash. Local regulatory agencies may require that such wastewater be neutralized or pretreated before discharge, though typically the quality of the effluent indicates that pretreatment will not be necessary.

Wastewater discharge from the plant will be offset by reductions in discharge from the plant's energy customers. Because of boiler turndown or shutdown, their process water requirements will be decreased proportionally.

MINOR IMPACTS

Odor

Odors are unavoidable in the operation of a facility receiving and processing MSW. Odor generation can be mitigated by daily sweeping, weekly steam cleaning, and periodic deodorizing. In addition, combustion air is commonly drawn from the odor-laden refuse floor enclosure, and odors are destroyed in the combustor. MSW should be transported to the site by covered trucks and stored in an enclosed receiving area before combustion. High-temperature combustion and the air pollution control equipment will also control odor-causing substances in stack emissions. Provisions for odor control should also be made for periods when the incinerators are not operating. Thus, the potential for odor impact beyond the site boundary can be considered negligible.

Noise and Traffic

External noise impacts of the facility could derive from truck traffic serving the plant as well as from occasional boiler exhaust. For those few resource recovery facilities that have exhibited significant off-site noise, measures such as redirecting exhausts and employing plenum chambers and mufflers have reduced noise to acceptable levels. The amount of additional traffic in the site vicinity will be dependent on the proportion of transfer vehicles and packer trucks hauling MSW to the facility as well as the number of daily ash-hauls. Noise and congestion impacts of generated traffic can be mitigated by careful routing and scheduling.

Construction

Construction of the facility will entail short-term impacts associated with construction traffic, noise, and airborne dust. Traffic and equipment could temporarily reduce ambient air quality and increase ambient noise levels. Earthmoving could result in localized increases in airborne dust, a constituent of particulate matter. Soil type and moisture determine the severity of this problem, which can be mitigated through wetting procedures.

The environmental impacts of construction activities would be typical of those normally associated with medium or heavy construction in an industrial area. These effects will be temporary and, for the most part, detectable only in close proximity to the project, which will be sited in an industrial zone.

Visual Aesthetics

Resource recovery plants are typically housed in common utility or industrial profile structures with a tall exhaust stack, entailing a permanent visual change of the landscape. The landscaped facility should be consistent, however, with the industrial character and zoning of the surrounding area.

Plant-Internal Impacts

Working conditions inside the plant must conform to OSHA-safety requirements. Potential for explosion of gases and vapors is generally minimal in mass burn resource recovery facilities that do not process or shred the MSW fuel before combustion. Potentially explosive materials should be removed from the waste stream by the floor operator.

In situations where an operator would be subjected to excessive noise levels, OSHA-approved ear protection devices are effective.

Internal odors will be minimized as discussed above. Internal dust generation can be mitigated by water-mist sprayers for the tipping floor, dust control devices, daily housekeeping, and personal respirators. Operating facilities have demonstrated effective dust control.

LITERATURE CITED

California Air Resources Board. May 1984. Air Pollution Control at Resource Recovery Facilities. Stationary Source Division. Sacramento, CA. Page 146.

North County Resource Recovery Associates. 1984. Risk Assessment for Trace Element and Organic Emissions: North County Recycling and Energy Recovery Center, San Marcos, California. San Marcos, California.

O'Connell, W.L., G.C. Stotler, R. Clark. 1982. Emissions and Emission Control on Modern Municipal Incinerators. Proceedings, 1982 National Waste Processing Conference. Sponsored by A.S.M.E. Solid Waste Processing Division. New York. American Society of Mechanical Engineers.

Environmental Consequences of Energy Production: Problems and Prospects. Edited by S. K. Majumdar, F. J. Brenner and E. W. Miller. © 1987, The Pennsylvania Academy of Science.

Chapter Thirty-Three

SITING LOW-LEVEL RADIOACTIVE WASTE IN PENNSYLVANIA: SOCIO-POLITICAL PROBLEMS AND THE SEARCH FOR SOLUTIONS

RICHARD J. BORD

Associate Professor
Department of Sociology
The Pennsylvania State University
University Park, PA 16802

INTRODUCTION

By January 1, 1996 the State of Pennsylvania must take title to and possession of all low-level radioactive waste generated within its borders and pay any damages resulting from failure to provide disposal.[1] While ten years appears to be a comfortable time frame within which to develop a disposal facility, the path to an operating site will undoubtedly be difficult and costly in both time and money. Public opposition to waste in general and to radioactive materials in particular makes the 1996 deadline problematic. The "not in my back yard" battle cry highlights the conflict characterizing attempts to site any kind of hazardous material. Nuclear power plants, medical facilities, manufacturing industries, and research institutes have a high stake in successful siting. Three issues must be addressed to provide a better understanding of the dilemma facing the State: first, an overview of the development of the present legal situation is necessary to understand the mandate facing Pennsylvania; second, the nature of public intransigence and their perception of the situation sheds light on approaches which may enhance cooperation; finally, probable paths to a LLRW site which are now being discussed in draft legislation provide an assessment of how the State hopes to meet the siting challenge. A discussion of general siting difficulties will conclude the paper.

THE MAKING OF A CRISIS

There are a number of publications dealing with the history of the low-level radioactive waste issue. A concise and complete treatment can be found in Jordan (1984).[2] It is impossible to adequately cover the entire relevant history of this issue in a paper of this scope but an overview is necessary to set the stage for the subsequent discussion.

Low-level radioactive waste (LLRW) is currently defined in the negative as "not spent reactor fuel, wastes from reprocessed reactor fuel, uranium mine and mill tailings, or items contaminated with specified levels of transuranic...elements".[3] It generally consists of radioactively contaminated materials that are involved in manufacturing industrial radionuclides and radiopharmaceuticals, hospital trash contaminated by the use of radiopharmaceuticals, and materials having relatively low levels of radioactivity used in the operation of nuclear power plants. In Pennsylvania approximately 70 percent of the LLRW, by both volume and curie content, is a product of nuclear power generation.[4] Radioactive wastes generated by defense industries are treated as a special case and disposed of in Departments of Defense and Energy disposal sites. Federal specifications set the maximum allowable time at which LLRW decays to the level of ordinary soil at between 300 and 500 years.[5]

Private sector LLRW became an issue in the mid 1960's. Until that time all radioactive wastes were handled by the federal government or dumped at sea. Sea dumping was discontinued in the 1960's.[6] Currently, about 3.25 million cubic feet of commercial LLRW is generated annually. This translates into 433,000, 55-gallon drums of material.[7] This amount can be expected to increase as operating reactors reach the decommissioning stage.

Although there have been six commercial disposal sites, some of which have been defined as public health hazards and subsequently shut down, three sites have accepted the brunt of the nation's LLRW since the early 1970's: Barnwell, South Carolina; Hanford, Washington; and Beatty, Nevada. In recent years the first two sites have been receiving about 45 percent of the waste each while Beatty has been getting about 10 percent.[8]

In 1979 the governments of these three states threatened to close their disposal sites unless Congress took steps to insure that other states accept responsibility for their own LLRW. In fact, Washington and Nevada did close their sites for approximately three weeks and South Carolina halved the volume of waste it was willing to accept. This created a somewhat immediate disposal problem for hospitals and research labs who have limited storage space relative to industry and power plants. Congress responded to this crisis with the 1980 Low-Level Radiactive Waste Policy Act (U.S. Public Law 96-573).

The 1980 LLRW Policy Act gave states the responsibility of disposing of LLRW generated within their borders. This act also recommended regional disposal sites as the most efficient option and encouraged states to enter into compacts to facilitate the establishment of regional sites. Compacts must receive congressional approval. That act further set January 1, 1986 as the date when disposal facilities operating with congressional approval may refuse to accept wastes from noncompact states.[9] However, as of January 1, 1986 not all states had formed compacts, none had been approved by Congress, and no state was anywhere near to establishing an operating disposal facility.

State's hesitance to move forward on this issue due to fear of public opinion and politicians' reluctance to be identified as siting proponents are the key factors in the failure of the 1980 LLRW Policy Act. For example, in response to a plea from the nuclear medicine industry that a lack of disposal capability could create a medical crisis Congressman Edward Markey of Massachusetts, an anti-nuclear activist, stated:

"The people who run this industry are going to have to get used to people participating. This bullying, this crying that there's no solution, is nothing more than a red herring to ram down the throats of some unsuspecting community a hazardous waste site".[10]

The failure of the 1980 LLRW Policy Act resulted in the Low-Level Radioactive Waste Policy Amendment Act of 1985 (Public Law 99-240). This law gives states more time to meet the demands of the 1980 law. The amended law gives utilities with nuclear-power generating stations and other waste generators access to the three operating LLRW disposal sites through December 31, 1992. This act incorporates a series of graded steps which must be met by predetermined dates. Failure to meet these deadlines results in penalty fees to be paid by the generator. States which fail to provide LLRW disposal capacity by 1996 must take title to and possession of wastes for which they are responsible.

At present, Pennsylvania is part of an approved compact, the Appalachian Compact, which includes four states: Pennsylvania, West Virginia, Maryland, and Delaware. Pennsylvania has agreed to be the host state for this compact and is in the process of drafting a radioactive waste disposal act. The following are the deadlines that Pennsylvania is attempting to meet:[11]

January 1, 1988 - the completion of a detailed siting plan.
January 1, 1990 - a license application for a new disposal facility, *or*, the governor certifies to the NRC that waste will be taken care of by 1993.
January 1, 1992 - license application filed for disposal facility.
January 1, 1993 - disposal provided.

If these deadlines are met then full-blown public opposition can be expected to surface between 1988 and 1990, precisely at that juncture when a specific area is selected as a potential site. Response to this opposition will ultimately determine the success or failure of the siting attempt. Although the state has the power of eminant domain there is little evidence that the application of that principle can succeed in the present climate of fear and litigiousness.[12]

The key issue that needs to be addressed is whether there are more or less effective methods in dealing with public opposition. The next section sketches out the reasons for public opposition and presents the results of a study done in Pennsylvania which focuses on public judgments of policy issues relating to LLRW siting. This data demonstrates what the public feels is necessary to elicit their cooperation. In addition, this research contrasts public judgments with those of major decision makers in state level environmental, civic, health, and industry related groups.

IS AN "IN MY BACK YARD" RESPONSE POSSIBLE?

It is no secret that public opposition has made the establishment of hazardous waste sites virtually impossible. McKinney states the problem succinctly: "With darn few exceptions, public participation has become a stonewall opposition to siting."[13]

In response to this "stonewall" of opposition, public participation scholars and practitioners, concerned citizen groups, and others have offered suggestions for methods of dealing with communities asked to bear the burden of waste disposal facilities. A widely accepted view is that cooperation can be enhanced by providing affected communities with the opportunity to negotiate for various kinds of compensation and incentives.[14] While it is much too early to brand this method a "failure" there are reasons to question its effectiveness.

The negotiation-for-compensation approach rests on the assumption that public opposition is a result of a violation of their sense of equity. That is, a disposal facility is a common good distributing benefits widely but concentrating costs locally. Communities near a proposed site are saddled with most of the costs but relatively few benefits. The key to public cooperations is viewed as some package of benefits which raises the reward value of the facility above that of the projected costs.

Undoubtedly equity considerations explain some portion of local opposition to disposal siting. For example, a common argument running through waste disposal siting disputes is the question of why a particular community, usually small and rural, should be held responsible for waste disproportionately generated elsewhere, usually in large urban areas. However, a substantial literature documents fear and distrust as the critical elements in public opposition. Ongoing media attention to environmental contamination from improperly dumped wastes or malfunctioning waste sites, accidents such as Three Mile Island and Chernobyl, lawsuits charging the government with carelessly exposing people to radiation from bomb tests and toxic chemicals from defoliation exercises during the Vietnam War, and entertainment movies such as "China Syndrome" and "Silkwood" maintain high levels of fear in the general population and promote distrust of those government agencies affiliated with these materials. In a survey done in a community targeted to receive a LLRW reduction and incineration facility one respondent, in response to an open-ended question on what, if anything, frightened her about LLRW answered: "Death! See Silkwood!"[15]

If fear and distrust, rather than equity, are the prime motivators of public opposition to waste disposal facilities then there is little reason to expect material incentives to be effective promoters of local cooperation. In fact, incentives can heighten distrust since they are easily interpreted as an overt manipulation technique, an attempt to buy a willing public. A more effective way to manage fear and distrust is to give those threatened some control over that which frightens them. There is an extensive literature in psychology and social psychology which documents the importance of personal control in overcoming fear.[16] Operationally, that translates into granting the affected public some means to protect their own interests. In waste siting it could mean letting the local community have a major responsibility for the decision concerning who the site operator will be and what kind of disposal technology will be used. It could mean giving adjacent communities funds to hire independent experts to check the data provided by the site operator and the government, to train and hire independent inspectors to monitor the site and report back to the community, or even grant abutting communities legal power to close the site should serious problems be encountered. In addition, real protections such as guaranteed property values, annual medical exams, or special access roads which bypass residential areas can help assure locals that their interests are being protected.

A study testing these ideas was done in mid-1985.[17] The study involved two distinct phases. Phase one was a survey of the general population of the State of Pennsylvania. One hundred and forty six telephone directories were used to randomly select a sample of the general public. Phase two involved interviewing a random sample of key decision makers in state level environmental, civic, public health, and industrial association groups. These kinds of groups will have input into the state siting process and a comparison of their views with that of the general public should help assess the climate for conflict.

The survey of the general public resulted in 846 completed questionnaires or, 59 percent of the total delivered. This rate of return indicates a substantial level of interest in the topic of LLRW. This sample has several distinguishing demographic features: there is a fairly even sex distribution, 47 percent female and 53 percent male; the age distribution is skewed toward the older end of the scale with the median age being 48; the sample is disproportionately married, 73 percent; it is disproportionately made up of home owners, 79 percent; and, respondents have higher than average levels of education, income, and occupation. In other words, this sample reflects the more stable and influential elements of local communities, precisely those who can be expected to be involved in community activities should a site be proposed in their areas. The statewide sample also includes a proportional representation from each of six regions in which the state was arbitrarily divided.

A random sample of key decision makers was obtained by using professional and state directories. These people were sent introductory letters and a copy of the interview and then contacted by telephone and interviewed either at that time or at a more convenient time later. Thirty eight interviews were completed under the following categories: nine included environmentalists and outdoor sports association leaders; six were health physicists or radiation biologists, some of whom are employed by large public utilities; seven included Pennsylvania public health officials; eight were state level civic and labor leaders; eight included regional planning and regional governmental association officers.

Both samples were asked basically the same questions. The LLRW issue, as it affects Pennsylvania, was explained then respondents were asked to evaluate various options which might be provided to communities asked to bear the burden of the site. In addition, respondents were asked who they trusted to represent their interests in the event of local siting and who they trusted to regulate the site.

Three types of concessions which could be granted affected communities were presented: options providing local control, options protecting local health and safety, and options providing material rewards. The local control options included: (1) community control over who operates the site, (2) the provision of resources to train and hire locals to inspect the site, (3) local power to shut down the site in the case of serious malfunction, and (4) the provision of funds to hire independent experts as checks on government and industry experts. Two health and safety options were provided: (1) an annual medical survey and (2) provision of a special access road which would bypass residential areas. Five types of material reward were presented: (1) guaranteed property values, (2) guaranteed agricultural prices, (3) local tax relief, (4) a surcharge on the waste to be returned to the community, and (5) a formal agreement by the site operator to buy and hire locally.

Each of these options were presented in two different formats. First, the respondent was asked how important they thought it was to provide a community with the particular concession. Then, they were asked if they thought that the provision of that concession would encourage the community to cooperate with the siting agency.

Two trust items were included. First, respondents were asked who they trusted to represent the community in negotiations with the siting agency. The following six possibilities were provided: (1) county commissioners, (2) a special task force appointed by the governor, (3) borough or township supervisors, (4) local people elected to a special committee by the community, (5) the use of referenda in all major siting decisions, and (6) general public meetings to index the will of the community. Second, they were asked who they had the most confidence in to regulate the site. The following six options were provided: (1) the Nuclear Regulatory Commission (federal), (2) the Environmental Protection Agency (federal), (3) the U.S. Department of Transportation (federal), (4) the Department of Environmental Resources (state), (5) a special state agency, and (6) specially trained local people. Note that these trust items include options ranging from those formally recognized in public policy to those focusing on local control.

The results of this research can be summarized quite simply: the general public overwhelmingly prefers those options which grant local control and protects their interests. Simple incentives such as tax relief and a surcharge to be returned to the community are perceived as relatively unimportant. All the local control options, both of the health and safety options, and the guaranteed property and agricultural price protection options were chosen as "extremely important" by over 75 percent of the respondents. The following items were selected as "extremely important" by 85 percent of the general public or more: the community should participate in selecting the site operator; money should be provided to train locals to monitor the site; the provision of an annual medical survey of the community to check for adverse effects resulting from site operation; and, guaranteed property values. In contrast, local tax relief, a surcharge to be returned to the community, and operator agreement to hire and buy locally was chosen as "extremely important" by only about 50 percent of the respondents.

The same pattern of response exists when these options are judged according to their effectiveness in eliciting local cooperation. The general public selects the "definitely will help" option relatively frequently for the local control and the health and safety options. However, while approximately 75 percent viewed these options as "very important" only about 50 percent think that they "definitely will help" promote local cooperation. In other words, these options are perceived as something that should be offered affected communities but the public is somewhat unsure that cooperation is a likely outcome. The public tends to be pessimistic about the probability of obtaining local cooperation.

On the other hand, the key decision makers are much more optimistic that the various options will produce local cooperation. They also tend to overchoose the material reward options while underchoosing the local control options. In other words, the key decision makers tend to reflect the prevailing official consensus about what is needed to promote successful siting, a view which is contradicted by the perspective of the general public.

The trust items reflect a similar distribution of responses. In deciding who they trust to represent their interests, the general public expresses high levels of trust in locals elected to a committee, referenda, and general town meetings. Similarly, the regulatory option receiving the highest vote of confidence is trained locals. In contrast, the key decision makers tend to prefer official government bodies to represent the community and to regulate the site. One option in which there is reasonably high consensus between the general public and the key decision makers is their expression of confidence in the State Department of Environmental Resources as a potential regulator: 69 percent of the general public and 87 percent of the key decision makers express high-to-moderate confidence in this agency.

There is one interesting variation in the pattern of responses discussed above. If the nine environmentalists are removed from the key decision makers and analyzed separately their responses parallel those of the general public item for item. The environmentalists also view local control as the key issue, they tend to be pessimistic about concessions promoting cooperation, and they devalue material incentives.

One further aspect of this research bears directly on the fear and trust issue: public perception of the safest type of disposal/storage technology. Respondents were asked to judge four basic types of disposal/storage arrangements for their relative safety: (1) conventional shallow land burial, (2) below ground burial with engineered barriers, (3) above ground storage facilities, and (4) abandoned mines. Both the general public and the key decision makers expressed a clear preference for below ground-engineered barrier disposal. Also, a large majority of the general public felt that they should have input into the decision concerning optimal technology.

These patterns of results clearly establish local control as the major issue for the general public. This tends to confirm that the factor driving local opposition is not a sense of inequity but is, instead, the twin elements of fear and distrust. Key decision makers in environmental organizations agree with the general public. Leaders of other organizations, however, tend to distrust local control and reflect the official view that incentives will promote local cooperation and that government agencies will satisfy the public's demand for safe site operation. It appears that the public and the decision makers are on a collision course that almost insures conflict when site selection occurs. If conflict does arise, a coalition of locals and environmentalists is a strong possibility.

It must be acknowledged that given existing institutional arrangements it would be very difficult to construct a siting process which would satisfy the public's demand for control. Furthermore, it is possible that options such as referenda and local town meetings are viewed as a means to effectively say "no" to any siting attempt. On the other hand, it is clear that cooperation will occur only if host communities are convinced that their interests are being protected.

One task remains to complete this analysis of the low level radioactive waste situation as it confronts Pennsylvania: the probable path to siting being considered by the state needs to be explicated and evaluated in light of the data presented above.

THE PENNSYLVANIA PURSUIT OF A LLRW WASTE FACILITY: A FOCUS ON FACTORS PROMOTING LOCAL COOPERATION

As of October, 1986 the development of low level radioactive waste disposal legislation for the State is well underway.[18] The purpose of this legislation is to define the siting process, not to select a site. A series of public meetings are scheduled for November and December to further discuss the draft legislation. However, since this document has already gone through several modifications the basic provisions can be expected to remain intact.

In light of the research reported above, the following aspects of the draft legislation appear to be positive elements in a process that must establish local trust: (1) the agency which is to be assigned the major responsibility for establishing the site, (2) the assignment of the actual site screening process to independent experts, (3) the proposed facility design, (4) the nature and scope of the public participation process, and (5) protections and benefits offered to the host community. Each of these will be discussed in order. Problems inherent in each will also be noted.

Public confidence in the agency responsible for overseeing the siting process is basic to establishing trust. The draft "Low Level Radioactive Waste Disposal Act" grants the authority to oversee the siting process to the State Department of Environmental Resources. In the research reported above, of six possible regulatory options the DER came in second only to "trained locals" in eliciting a "high" to "moderate" degree of confidence from the public. However, in a process spanning several years, trust is something that can vary dramatically with the manner in which the siting agency relates to the public. Ill-chosen or hostile communications from agency representatives have been responsible for scuttling otherwise well-founded siting programs.[19] Another danger, since local control is important to host communities, is the perception by locals that their interests are being ignored. A nonresponsive siting agency can be as destructive to building cooperation as a hostile or insensitive siting agency.

While the Pennsylvania program begins on a strong foundation by placing responsibility for the siting process in an agency which is generally trusted, the conduct of that agency throughout the siting process will determine its ability to maintain public trust.

The draft legislation designates responsibility for the actual site screening process to an independent contractor to be selected by the DER. The use of outside, independent experts should be a plus in the effort to build public confidence. However, the efficacy of this approach will depend on the extent to which the public is convinced of the contractor's impartiality. A history of involvement with related, unpopular issues, such as nuclear power plant siting, can be used by the opposition to charge conflict of interest or insensitivity to local concerns.

In considering acceptable disposal technology the draft legislation clearly rules out traditional shallow land burial. Any other below or above ground options are open for consideration given that they include engineered barriers. A further important stipulation is that whatever the disposal technology it must be secondary to optimal geology. The survey discussed above established a public preference for below ground disposal with engineered barriers. It appears that the proposed disposal technology will be that favored by a majority of the public.

The public participation program, which is already being implemented, can be characterized as a standard public input program which is used extensively to site everything from hydroelectric dams to nuclear power plants to hazardous waste facilities. The cornerstone of the program is the Public Advisory Committee on Low-Level Radioactive Waste Disposal which was formed on April 26, 1985. The committee consists of sixteen members representing various levels of state government, conservation and environmental groups, public health people, geologists, engineers, the League of Women Voters, and others. The Committee is to advise the DER on all aspects of the siting process. At that time that a specific site is selected the host community has the option to nominate an additional member to this advisory committee.

In addition to the Public Advisory Committee, the following opportunities are provided for public input:

(1) The private contractor doing the actual site screening must produce a report which includes at least four separate areas in the state which meet the screening criteria. This report will be made public and public meetings will be held in those areas designated by the report.

(2) Those wishing to be considered for the site operator's job must submit a proposal to the DER. These proposals will be open for public inspection and comment prior to the selection of the operator by the Secretary of the DER.

(3) Prior to the issuance of a license the applicant must prepare an environmental impact statement. This EIS will be open for written comment and a public hearing will be held in the area which is to host the site.

While this standard input participation program does provide the public the opportunity to comment at crucial junctures in the siting process, it has generic problems which have been noted frequently in the participation literature.[20] First, a public advisory committee is generally not perceived by affected communities as representing their particular interests. Such a committee tends to be viewed as representing specialized, outside interests and as allied with the siting agency. Second, the standard public input approach does not specify how, or whether, the public input is to affect decision making. In the absence of such specification the public can accuse the process as being simply a facade.[21] Finally, it leaves the affected public out of some key decisions which are central to their interests. Perhaps the most central decision is that of selecting the site operator. In the research reported above 85 percent of the general public felt that it is "extremely important" for the local public to be involved in selecting the site operator.

The draft legislation also provides a package of protections and benefits to the host community. First, money is available to allow the host community to hire independent experts to evaluate the license application. The findings of these experts will be included in the licensing proceeding. Sceond, a surcharge on the waste will be returned to any community directly affected by the facility to replace lost tax revenue, for additional local services, or for other benefits and guarantees. The "other benefits and guarantees" can include a full-time inspector who will have the right to inspect the site and site records, a water well sampling program, or other activities which the community negotiates with the DER. Finally, all inspection reports and emergency action data will be given the community promptly.

This package of protections and guarantees has some obvious strengths in terms of fostering local cooperation. The ability of the host community to hire independent experts does put actual control into local hands. Although the siting decision need not be based on the report of these independent experts, their information could provide the basis for litigation should the community judge that safety and health considerations are not being properly considered. The waste surcharge does provide the affected community with some clear protections. The provision of a local inspector is something that the previously discussed research established as "extremely important" by 86 percent of the general public. This was also one option that a strong majority of the public felt would promote local cooperation. A water well sampling program also provides the opportunity for early detection of contamination migration out of the site.

CONCLUSIONS

Even though the state is creating a siting process which has many strengths in terms of promoting local cooperation, and compares favorably with similar attempts in other states,[22] the problem of local intransigence is likely to be significant.

Establishing a hazardous waste disposal facility is a process which includes a number of factors which cannot be controlled by preplanning. Two of the most important factors involve the nature of the opposition that may arise and their mobilization skills. First, in all cases of waste disposal siting a local opposition group forms. For example, when a LLRW incineration and volume reduction facility was proposed for a community near Pittsburgh, a local group calling itself the "Kisky Valley Coalition to Save Our Children" mobilized and has effectively blocked establishment of that facility. This kind of group mobilization depends on unknowns such as local leadership, knowledge, and mobilization history. Second, anything involving radioactive hazards will certainly attract anti-nuclear groups which are attempting to use the waste issue as a lever to eliminate nuclear power generation. These groups provide leadership to local communities and help them in their attempts to block siting.

Of course, it is possible that some community may view the presence of a LLRW site as an economic benefit and volunteer to host the facility. This eventuality would solve the siting problem. However, the odds of this happening must be considered extremely small for two reasons. First, a "community" cannot volunteer, only some representative of that community. Even if respected local officials initiated such action they would encounter stiff opposition from those locals who do not share their vision. Recall that in the data reported above the general public did not express high levels of trust in local officials on the issue of LLRW disposal siting. Second, in order for volunteering communities to be selected as the host their area must pass the screening tests. The need to match volunteers with geology and other technical issues reduces the probability of such an event. Communities in both Colorado and North Dakota did volunteer to host a site and were not selected because one had inappropriate geology and the other faced effective political opposition stimulated by outsiders.

The establishment of safe disposal facilities is foundering on the shoals of a real dilemma. On the one hand our democratic values must promote practices which give affected citizens some input and control over events which can have serious impacts on their lives. That was the reason for the 1969 Environmental Protection Act and the "citizen participation revolution" which has excited so much comment.[23] On the other hand, government also has the obligation to protect its citizenry from the threats of inappropriately handled risky materials. Failing to provide optimal disposal/storage/treatment virtually insures a multiplication of risk. The unfolding of the LLRW issue in Pennsylvania is one of a number of theaters in which this drama will be played out. If siting proves impossible under existing practices there will be increased pressure to dispose of these wastes in a less than optimal manner, to modify the law with respect to citizens' rights, or to provide an "out of sight, out of mind" disposal mode such as renewed dumping at sea. In any case, our degree of responsibility, or lack of it, in dealing with this issue will have impacts on many subsequent generations. The issue cannot be ignored.

REFERENCES

1. Shillenn, J.K. 1986. Low-Level Radioactive Waste Policy Act Amended. *Postscript*, 4:1.
2. Jordan, J.M. 1984. *Low-Level Radioactive Waste Management: An Update*. National Conference of State Legislatures. Washington, D.C.
3. Jordan, op. cit., 3.
4. Witzig, W.F., W.P. Dornsife, and F.A. Clemente. 1983. *Low-Level Radioactive Waste Disposal Siting: A Social and Technical Plan for Pennsylvania*. Vol. 1, Institute for Research on Land & Water Resources, The Pennsylvania State University, University Park, PA, 16802.
5. Federal Register 47. 1984. U.S. Code of Federal Regulations (10 CFR 61). December 27, 75446-82.
6. Jordan, op. cit., 4.
7. Smith, T.P. 1982. *A Planner's Guide to Low-Level Radioactive Waste Disposal*. Report #369, Chicago, Illinois: American Planning Association.
8. Gettinger, S. 1985. Congress Again Faces Nuclear Waste Crisis." *Congressional Quarterly, Inc.* (March 16, 1985): 484-488.

9. Jordan, op. cit., 9.
10. *New England Monthly.* 1984. Very Hot Potato. November: 34-39.
11. Schillenn, op. cit., 1.
12. Morell, D. and C. Magorian. 1982. *Siting Hazardous Waste Facilities: Local Opposition and the Myth of Preemption.* Cambridge, Massachusetts: Ballinger Publishing Company.
13. McKinney, N. 1979. Siting Problems Discussed. *EPA Journal* 5: 27.
14. Morell and Magorian, op. cit.; O'Hare, M.L. Bacow, and D. Sanderson. 1983. *Facility Siting and Public Opposition.* New York: Van Nostrand Reinhold Company.
15. Bord, R.J. 1985. *Results of a Survey Done in the Appolo, PA Area: Citizen Reactions to the Proposed B & W Low-Level Radioactive Waste Treatment Facility.* Institute for Research on Land & Water Resources, The Pennsylvania State University, University Park, PA, 16802.
16. Slovic, P., B. Fischhoff and S. Lichtenstein. 1984. Perception and Acceptability of Risk from Energy Systems. in W.R. Freudenberg and E.A. Rosa (eds) *Public Reactions to Nuclear Power: Are There Critical Masses?* Boulder, Colorado: Westview Press, Inc.
17. Bord, R.J. 1985. *Opinions of Pennsylvanians on Policy Issues Related to Low-Level Radioactive Waste Disposal.* Institute for Research on Land & Water Resources, The Pennsylvania State University, University Park, PA 16802.
18. *Low-Level Radioactive Waste Disposal Act.* 1986. Draft Copy, September 18.
19. McKinney, op. cit. 27.
20. Abrams, N. and J. Primack. 1980. Helping the Public Decide: the Case of Radioactive Waste Management. *Environment* 22: 14-40.
21. Pogell, S. 1979. Government-Initiated Public Participation in Environmental Decision. *Environmental Comment* 6: 3-15.
22. Texas Low-Level Radioactive Waste Disposal Authority. 1985. *Siting a Low-Level Radioactive Waste Disposal Facility in Texas: Local Government Participation, Mitigation, Compensation, Incentives, and Operator Standards.* Texas Advisory Commission on Inter-Governmental Relations, Austin, Texas.
23. Nelkin, D. 1984. *Controversy: Politics of Technical Decisions.* 2nd edition. Beverly Hills: Sage Publications.

Environmental Consequences of Energy Production: Problems and Prospects. Edited by S. K. Majumdar, F. J. Brenner and E. W. Miller. © 1987, The Pennsylvania Academy of Science.

Chapter Thirty-Four
ECONOMICS OF ENERGY SUPPLY AND DEMAND IN DEVELOPING COUNTRIES

WEN S. CHERN
Professor
Department of Textiles and Consumer Economics
University of Maryland
College Park, Maryland 20742

This study analyzed the data related to energy demand and supply during the post oil embargo period in 15 major energy consuming less developed countries (LDC's). Despite the skyrocketing of oil and energy prices, energy demand in these countries continued to increase at very rapid rates. On the other hand, the efforts to diversify supply sources produced only a limited success. Implementation of further economic development would need additional energy; it also induces more use of energy. It would be imperative for LDC's to enhance the effectiveness of economic forces (raising energy prices) by designing and implementing more and better energy conservation strategies.

INTRODUCTION

The decade of the 1970's witnessed the importance of energy in economic and international affairs. Energy is an indispensible input in our daily life and in almost all production and consumption activities. In fact, the intensity of efficient energy use may be used as a barometer of the standard of living. There is a good correlation between energy consumption and relative well being of people.

One unique characteristic of primary energy is that it can not be produced easily. Energy is, by and large, a natural endowment and, therefore, a nonrenewable natural resource. However, energy comes from several sources. Conventional energy sources include petroleum, coal, natural gas and hydropower. The emerging unconventional energy sources are nuclear, solar, geothermal, wind, and bioresources such as ethenal. In the developing countries, a large fraction of energy requirement is met by the conventional non-commercial energy sources such as firewood, charcoal and dung cakes, etc. During the last several decades, petroleum (or oil) has been the key element in the world energy market. Its supply and demand as evolved have created a strong interdependence between oil producing and oil consuming nations. This is, of course, the source of conflicts and oil crises during the 1970's.

The purpose of this chapter is to analyze the economics of energy supply and demand. The focus of the analysis is on selected developing countries. Specifically, this study deals with the relationship between energy and economic development, and the impacts of the two oil shocks on development of domestic energy sources and diversification of energy supply in the developing countries.

ENERGY AS A SOURCE OF ECONOMIC DEVELOPMENT

The fact that energy is an important factor of production has been widely investigated and documented during the post-oil embargo period. At the aggregate level, the total production of goods and services of a nation may be measured by gross domestic product (GDP) or gross national product (GNP). The strong relationship between energy and GNP was shown in earlier studies by Hudson and Jorgensen (1974), and Institute for Energy Analysis (1979). At the sectoral level, output for a sector may be measured by value added, gross output, or a production index. The impacts of energy on sectoral output were statistically shown in numerous studies including those by Berndt and Wood (1975, 1979) and Griffin and Gregory (1976). One focus of these studies was on the substitutability between energy and capital. If energy and capital are complements as shown by Berndt and Wood, we should not subsidize capital innovation which may be counterproductive for energy conservation.

The importance of energy as a factor of production does not imply the constancy of the energy/output relationship. Although it always requires energy to produce any good; how much energy required for one unit of good may be changed. When energy was plentiful and relatively inexpensive during 1950's and 1960's, energy was used to the extent that it substituted other factors of production such as labor in most industrial and household productions. Briggs and Borg (1986) showed that the Energy/GNP ratio has been steadily declining since 1970 in the United States. This dramatic declining trend reflects the effects of the energy conservation efforts made by energy consumers in response to the two oil shocks in the 1970's.

The basic relationship between energy and economic growth applies to the less developed countries (LDC's) except that their economies are in various stages of development. The correlation between per capita real GDP and per capita energy consumption (and a corresponding set for per capita oil consumption) is illustrated in Table 1 for selected 10 oil-importing and five oil-exporting LDC's. This list includes most of the major energy consuming developing countries. It is noted that none of the selected oil-exporting countries belongs to the Organization of Petroleum Exporting Countries (OPEC). Colombia's position is not clear cut, it has been a net oil importer in recent years (except in 1984). Therefore, Colombia may be grouped as an oil-importing nation. The correlation coefficients were computed for two periods for the purpose of comparison before and after the 1973 oil embargo.

Consider first the correlations between per capita real GDP and per capita energy consumption. High correlations are observed in almost all countries. A few exceptions are generally caused by some unusual situations. For example, the low correlation for Pakistan in 1960-73 might have resulted from the secession of East Pakistan (now Bangladesh) in 1971. In the case of Argentina, a low correlation during 1974-84 may be caused by the war in 1982 and subsequent economic instability and hyper inflation in recent years. The negative correlations (in fact they are not statistically significant) observed in Chile and Egypt during 1960-73 may be caused by measurement errors because they are inconsistent with the stronger correlations observed in the later period. For some

TABLE 1

Correlation Coefficients Between Per Capita Real GDP and Per Capita Energy Consumption

Country	Per Capita Real GDP and Per Capita Energy Consumption		Per Capita Real GDP and Per Capita Oil Consumption	
	1960-73	1974-84	1960-73	1974-84
Oil-Importing LDC's				
Argentina	0.96	0.26	0.95	0.71
Brazil	0.94	0.85	0.94	0.31
Chile	−0.06	0.43	−0.15	0.44
India	0.94	0.90	0.86	0.87
Korea, South	0.97	0.95	0.99	0.90
Pakistan	0.39	0.95	0.19	−0.01
Philippines	0.97	0.53	0.97	0.17
Singapore	0.97	0.95	0.97	0.95
Taiwan	0.99	0.94	0.99	0.70
Thailand	0.97	0.96	0.97	0.67
Oil-Exporting LDC's				
Colombia	0.84	0.90	0.83	−0.35
Egypt	−0.15	0.99	−0.36	0.99
Malaysia	0.92	0.93	0.93	0.92
Mexico	0.96	0.95	0.91	0.94
Peru	0.89	0.21	0.89	0.60

Sources: (1) United Nations, *Yearbook of World Energy Statistics*. Time-series data contained in the UN computer tape were provided by the World Bank.
(2) World Bank, *World Tables*, 1986.
(3) Energy Committee, Ministry of Economic Affairs, Taiwan.

TABLE 2

Index of Average Energy/GDP Ratio by Country, Base period: 1960-64

Country	Index of Average Energy/GDP Ratio				
	1960-64	1965-69	1970-73	1974-79	1980-84
Oil-Importing LDC's					
Argentina	100	107	111	106	115
Brazil	100	110	100	93	97
Chile	100	121	142	135	121
India	100	115	110	125	139
Korea, South	100	132	159	152	170
Pakistan	100	107	77	78	81
Philippines	100	144	164	149	129
Singapore	100	106	96	72	56
Taiwan	100	100	102	114	116
Thailand	100	154	204	214	202
Oil-Exporting LDC's					
Columbia	100	109	99	101	110
Egypt	100	99	73	90	98
Malaysia	100	125	140	119	120
Mexico	100	102	97	108	121
Peru	100	112	120	130	144

Sources: See Table 1.

countries, the correlation coefficients in 1974-84 are smaller than those in 1960-73. But by and large, a strong correlation persists in most countries, implying that energy remained an important factor for GDP growth in most developing countries after the 1973 oil embargo.

The correlation coefficients between GDP and oil consumption are generally very high during 1960-73. However, in many countries, the correlations became smaller in the post embargo period. The negative correlations during 1974-1984 observed for Colombia and Pakistan reflect a very unique situation. Both countries have had a large reserve of natural gas, and consequently, their domestically produced natural gas has been substituting for imported oil during the 1974-84 period. Note that the correlation coefficient remains greater than 0.90 only in Chile and Singapore among the selected 10 oil importing LDC's while it is over 90 for three out of five oil exporting countries. Singapore is noted for its almost complete dependence on oil. The low correlations in Peru during 1974-84 are likely the result of hyper inflations during the 1970's which may have distorted the measurement of its GDP growth in real terms.

The other important observations about LDC's are their historical trends in the energy/GDP ratio. Table 2 presents the indices of the average ratios for the five partitioned periods during 1960-84. The base period is 1960-64. Note that the energy/GDP ratio reflects the intensity of energy use for the aggregate economy. While this ratio may be appropriately used to measure the trend of aggregate energy utilization efficiency in industrialized nations, it may reflect

TABLE 3

Index of Average Oil/GDP Ratio by Country Base period: 1960-64

Country	Index of Average Oil/GDP Ratio				
	1960-64	1965-69	1970-73	1974-79	1980-84
Oil-Importing LDC's					
Argentina	100	104	105	90	87
Brazil	100	106	100	88	75
Chile	100	130	145	135	113
India	100	140	164	170	183
Korea, South	100	259	501	506	514
Pakistan	100	105	70	60	49
Philippines	100	150	174	157	118
Singapore	100	106	96	72	56
Taiwan	100	175	291	378	335
Thailand	100	154	191	198	171
Oil-Exporting LDC's					
Colombia	100	112	108	100	86
Egypt	100	92	61	69	79
Malaysia	100	125	141	118	114
Mexico	100	92	90	106	123
Peru	100	107	111	116	113

Sources: See Table 1.

more of the changes in economic structure (agriculture vs industry, for example) in most LDC's. The historical trends show that in 11 out of 15 LDC's, the index in 1980-84 is higher than in 1960-64. In fact, the computed indexes show strong evidence that these LDC's have moved to a more energy intensive economies after the 1973 oil embargo. The most dramatic increases in the energy/GDP ratio are observed in India, Korea, Thailand and Peru. The picture in Pakistan (a declining trend between 1960's and 1970's) is likely disturbed by the separation in 1971. If we only examine the period after the separation (1970-84), we would find the index during 1980-84 is higher than those observed in either 1970-73 or 1974-79. In the case of Singapore, the trend is similar to those observed in the U.S. and other industrialized nations.

The historical trends of the oil/GDP ratio are summarized in Table 3. The role of oil in economic development is reflected by the dramatic increases in the oil/GDP ratio observed in India, Korea, Thailand and Taiwan. The index decreased in 1980-84 from the previous periods in many countries, apparently resulting from the efforts to reduce foreign oil dependence after the two oil shocks. The effects of the two oil shocks in 1973 and 1979 will be discussed in more detail later.

DIVERSIFICATION OF ENERGY SOURCES AS A NATIONAL STRATEGY

Ever since the oil embargo in 1973, most developing countries have followed the policies provoked in the industrialized nations to diversify the supply sources of energy. The strategies for diversification take many forms, depending upon the endowment of various energy resources, economic conditions and other

TABLE 4

Net Energy Import Quantities and Shares

Country	Year	Net Energy Imports[a] (10^6 TOE)	Net Import Shares (%) of Total Energy Consumption	Net Import Shares (%) of Total Oil Consumption
Oil-Importing LDC's				
Argentina	1971	3.5	11.2	14.1
	1984	0.7	15.9	2.9
Brazil	1971	22.9	52.2	75.6
	1984	28.5	28.8	59.6
Chile	1971	4.1	44.6	71.9
	1984	2.8	29.8	57.1
India	1971	14.3	21.8	67.1
	1984	13.1	9.4	32.4
Korea, South	1971	11.1	63.4	100.9
	1984	34.8	70.3	123.8
Pakistan	1971	3.9	47.6	95.1
	1984	5.4	30.9	91.5
Philippines	1971	8.5	87.6	92.4
	1984	10.2	70.8	97.1
Singapore	1971	6.6	97.1	97.1
	1984	11.3	103.7	103.7
Taiwan	1971	6.6	70.2	115.8
	1984	27.0	95.1	176.5
Thailand	1971	6.1	92.4	103.4
	1984	10.8	67.5	90.8
Oil-Exporting LDC's				
Colombia	1971	−5.0	−44.6	−78.1
	1984	−0.6	−2.8	−7.2
Egypt	1971	−8.6	−111.7	−141.0
	1984	−21.2	−82.2	−106.0
Malaysia	1971	−0.4	−7.4	−8.0
	1984	−15.0	−126.1	−151.5
Mexico	1971	0.5	1.2	1.9
	1984	−84.7	−84.3	−123.6
Peru	1971	1.8	27.7	36.7
	1984	−1.7	−16.2	−23.9

[a]TOE = Metric Tons of Oil Equivalent
Sources: See Table 1.

TABLE 5

Total Primary Energy Consumption and Fuel Shares

Country	Year	Total Primary Energy Consumption (10⁶ TOE)	Shares (%) of			
			Oil	Natural Gas	Coal	Hydro and Nuclear Power
Oil-Importing LDC's						
Argentina	1971	31.2	79.5	17.0	2.6	1.3
	1984	44.0	54.4	29.2	1.3	13.9
Brazil	1971	43.9	69.0	0.2	5.7	25.1
	1984	99.1	48.2	1.9	7.5	42.4
Chile	1971	9.2	62.0	13.0	13.0	12.0
	1984	9.4	52.1	9.6	13.8	24.5
India	1971	65.6	32.6	0.8	55.3	11.3
	1984	139.7	28.9	2.1	58.3	10.7
Korea, South	1971	17.5	62.9	a	34.9	2.2
	1984	49.5	56.8		36.0	7.2
Pakistan	1971	8.2	50.0	29.3	9.7	11.0
	1984	17.5	33.7	41.7	5.7	18.9
Philippines	1971	9.7	94.8			5.2
	1984	14.4	73.0		4.8	22.2
Singapore	1971	6.8	100			
	1984	10.9	100			
Taiwan	1971	9.4	60.6	7.4	25.5	6.5
	1984	28.4	53.9	3.9	21.8	20.4
Thailand	1971	6.6	90.8		1.5	7.7
	1984	16.0	74.4	13.1	5.0	7.5
Oil-Exporting LDC's						
Colombia	1971	11.2	57.1	12.5	15.2	15.2
	1984	21.5	38.1	20.5	17.7	23.7
Egypt	1971	7.7	79.2	1.3	2.6	16.9
	1984	25.8	77.2	9.7	2.7	10.4
Malaysia	1971	5.4	92.6	1.8		5.6
	1984	11.9	83.2	10.9	1.7	4.2
Mexico	1971	42.0	63.5	23.2	4.5	8.8
	1984	100.5	68.1	21.5	4.0	6.4
Peru	1971	6.5	75.4	6.2	1.5	16.9
	1984	10.5	68.3	9.6	1.0	21.1

a. Blanks indicate insignificant shares.
Sources: See Table 1.

environmental and political considerations. For example, in Korea and Taiwan, the strategies have been to develop nuclear power and to increase uses of coal and LNG. In Pakistan and Colombia, the approach to diversify energy sources has been to accelerate the domestic production of natural gas. In Brazil, the national strategies are based on the potential of hydropower and ethenal.

In general, the reduction of dependence on foreign oil has been a national goal for most developing countries. The need for reducing such foreign dependence is obvious. Since 1973, the energy import bill has skyrocketed as a result of the massive oil price increases during the two oil shocks. The balance of payments effects of raising oil prices were serious in many oil-importing developing countries. For example, in 1974, the cost of imported oil rose (in current prices) from 1973 by US$ 700 million for South Korea, US$ 400 million for Singapore, US$ 500 million for the Philippines, US$ 400 million for Thailand, US$ 93 million for Pakistan, and US$ 616 million for Taiwan. The increment in the oil import bill on an average exceeds 2 percent of GDP in all these countries except Pakistan. The impacts of the second oil shock were even more severe. In 1979 and 1980, South Korea's oil import bill rose by a cumulative figure of nearly US$ 4 billion, Singapore's by US$ 600 million, the Philippines' by more than US$ 800 million, Thailand's by US$ 1.7 billion, Pakistan's by US$ 619 million, and Taiwan's by US$ 3.1 billion.

Net energy imports and their shares of total energy and oil consumption during 1971 and 1984 in 15 LDC's are summarized in Table 4. Net energy imports increased substantially from 1971 to 1984 in Brazil, South Korea, Pakistan, the Philippines, Singapore, Taiwan, and Thailand but decreased in Argentina, India and Chile (Table 4). Mexico and Peru had positive net energy imports in 1971 but have become net energy exporters in recent years. Today, Mexico is a major oil exporter to the United States.

The net import shares of total energy consumption measuring the degree of dependence on foreign sources of energy (primarily oil but also including coal, LNG, and nuclear power, etc.). The figures in Table 4 indicate that the successes in reducing the foreign dependence rely upon the availability of domestic energy sources. In Korea and Taiwan, despite the strong pushes for energy R&D, the import shares continued to rise. The import share also increased from 1971 to 1984 in Argentina while Singapore has had a virtually total dependence on foreign energy. For all the other six oil-importing LDC's, the import shares have declined, suggesting a reduced foreign dependence in recent years. For oil-exporting LDC's, the negative import shares measures the size of their net energy exports relative to domestic consumption. For example, in Malaysia, the exports of energy (primarily oil) exceeded the domestic energy consumption in 1984.

The last column in Table 4 shows the net import shares of total oil consumption. When this share is greater than 100, the net energy imports exceeds the domestic oil consumption. This occurs in Taiwan and Korea where they imported not only oil but also substantial amounts of coal and nuclear power.

Diversification of energy supply should result in changes in fuel mix. Table 5 summarizes the total primary energy consumption and fuel shares for the 15 selected LDC's. Total primary energy consumption continued to increase very rapidly from 1971 to 1984. In terms of the fuel mix, the data clearly suggest that despite the efforts made for diversification, oil continued to be the dominate fuel in most of the countries. However, the market shares of oil declined in all countries except for Singapore where its primary energy source remained 100 percent of oil in 1984. Natural gas accounted for a very small market share in most countries except Argentina, Colombia, Pakistan, and Mexico where the increases from 1971 to 1984 have been substantial. Coal has been a dominate fuel in India where its fuel mix remained very much the same between 1971 and 1984. Coal also accounted for a significant market share in Korea and Taiwan. In the case of Taiwan, coal was mainly imported.

Consider the shares of hydro and nuclear power, there have been a substantial increase in most oil-importing countries such as Argentina, Brazil, Colombia, Korea, Pakistan, Chile, the Philippines and Taiwan. For most of these countries, the increases come from hydropower except in Taiwan and Korea where nuclear power accounted for most of the increases. These figures indicate that the development of hydropower has been more successful than the expansion of the uses of natural gas and coal for diversification in most of the oil-importing LDC's under investigation.

Energy conservation may be the cheapest alternative energy source. The effects of energy conservation have been remarkable in many industrialized nations like Japan and the U.S. In 1985, the United States consumed 1,540 million metric tons of oil equivalent (TOE) which is only slightly more than 1,475 millions of TOE consumed in 1973. In the developing countries, many have had energy conservation as a national policy. However, the results have not been very impressive. The relative ineffective energy conservation policies are reflected by the low price elasticities of energy demand estimated for the developing countries. The price elasticities measure the percentage changes in energy demand in response to a one-percent change in energy price. Based on the law of demand, the price elasticity has a negative sign. Therefore, when the price of energy increases, we can expect energy demand to decline. A high energy price elasticity (in absolute value) is generally a good indicator of a high potential for energy conservation. This is because the implementation of energy conservation programs often requires additional costs, which would, in effect, make the cost of using energy higher. If the price elasticity is high, then the reduction in energy use for the added increase of energy price (resulting from energy conservation efforts) would be consequently high.

The recent study by Chern and Soberon-Ferrer (1986) shows that the estimated energy price elasticities are only -0.078 in the short run and -0.20 in the long run for the same group of 15 developing countries. These estimates imply that a one percent increase in energy price would reduce demand by only 0.08 percent in the short run (one year) and by 0.2 percent in the long run. In this same study, Chern and Soberon-Ferrer also show that the estimated long-run GDP elasticities of energy demand exceeds unity in these LDC's. Therefore, a one

TABLE 6

Energy Production

		Production (10^6 TOE) of		
Country	Year	Oil	Natural Gas	Coal
Oil-Importing LDC's				
Argentina	1971	21.87	5.26	0.37
	1984	25.71	11.22	0.30
Brazil	1971	8.44	0.13	1.41
	1984	23.69	1.90	3.56
Chile	1971	1.81	1.19	1.04
	1984	2.33	0.86	0.82
India	1971	7.19	0.52	36.12
	1984	28.02	2.94	84.29
Korea, South	1971	a		6.66
	1984			9.33
Pakistan	1971	0.41	2.36	0.64
	1984	0.67	7.27	0.87
Philippines	1971			0.02
	1984	0.62		0.53
Taiwan	1971	0.11	0.97	2.42
	1984	0.22	1.27	1.19
Thailand	1971	0.01		0.12
	1984	1.14	2.12	0.61
Oil-Exporting LDC's				
Colombia	1971	11.39	1.44	1.69
	1984	8.80	4.39	4.06
Egypt	1971	15.00	0.08	
	1984	41.87	2.54	
Malaysia	1971	3.28	0.08	
	1984	21.67	5.65	
Mexico	1971	23.64	9.87	1.73
	1984	152.85	22.94	3.83
Peru	1971	3.14	0.36	0.05
	1984	9.15	1.02	0.04

a. Blanks indicate either none or insignificant amount of production.
Sources: See Table 1.

percent increase in GDP would result in more than one percent increase in energy demand. These estimates would strongly suggest the need for energy for further economic development in LDC's.

The reasons for the relatively ineffective energy conservation programs may lie in the lack of proper economic incentives and capital investment. Major conservation methods involve capital costs such as home insulation and installation of more efficient furnaces. In many developing countries, the major users of energy are industries not households. When there are many small scaled family factories, the energy conservation programs are not easy to implement. Information about effectiveness of energy conservation is often a serious handicap.

As earlier findings suggested, the diversification of energy supply sources has been greatly constrained by domestic endowment of energy resources in LDC's. It would seem very important for the LDC's to push for stronger energy conservation strategies to meet the growing energy demand for further economic development.

DOMESTIC ENERGY SUPPLY

Simple economic theory would suggest that energy supply is positively related to energy price. Therefore, the phenomenal increases in energy prices resulting from the two oil shocks in the 1970's should increase energy supply. The logic is, of course, very simple. High energy prices would encourage more exploration and drilling. Consequently, more oil and gas would be discovered. Furthermore, higher energy prices would make production of oil and gas with a high marginal cost of production more economical. Also, the mining of those coal with higher marginal cost may become economical under higher energy prices. Energy supply would thus increase. The extent to which a country is able to have a significant supply response depends, of course, upon the availability of their energy resources.

The production of the conventional fuels (oil, natural gas, and coal) in the selected 15 LDC's is presented in Table 6. The data for 1971 and 1984 are presented for the purpose of analyzing the impacts of the raising energy prices on domestic energy supply. Despite the limited oil resources in the oil-importing LDC's, 8 out of 10 countries increased their oil production from 1971 to 1984. The most significant increases are observed in Brazil and India. The increases in oil production in Pakistan, Chile, the Philippines, Taiwan and Thailand are only marginal while South Korea and Singapore registered no production of oil. Among the oil-exporting LDC's, all but Colombia have significant increases in oil production between 1971 to 1984. There are clearly significant positive supply responses in the oil-exporting LDC's.

With respect to natural gas, the increases in its production are observed in Argentina, Brazil, India, Pakistan, Taiwan, and Thailand among the oil-importing LDC's. These increases are only marginal except in Argentina and Pakistan. Korea and Singapore have no production of natural gas. Among the oil-exporting LDC's, all 5 countries have increased their natural gas production. However, only Mexico produced significant amounts of natural gas relative to oil production. Similar trends are also observed for coal. While most countries have increased their production of coal, only the increases in India are significant. India is perhaps the only country with abundant coal reserves among the 15 LDC's under study. Singapore, Egypt and Malaysia produce no coal at all.

Based on these production statistics, we can conclude that most oil-importing LDC's have attempted to increase production of such conventional energy sources as oil, natural gas and coal. However, only a few countries achieved measurably significant increases in domestic supply in response to the two oil shocks in the 1970's. By and large, the increases are only marginal, and therefore, they would not significantly reduce their dependence on foreign sources of energy. The situations for these oil-importing nations are not expected to change in any significant way in the future except for the countries like India, Pakistan and Brazil with significant reserves of coal, natural gas or bioresources.

There is another important observation about LDC's, and that is the growth of electricity generation or consumption. Electricity generation, in many LDC's, is the only important energy supply sector. Since electricity may be generated by all alternative primary energy sources, many countries have pushed very hard to develop the electric power sector, despite its inefficiency because of losses in generation, transmission, and distribution. Also, many LDC's have engaged rural electrification programs to raise the standard of living in rural areas. These programs continued in spite of the oil shocks in the 1970's.

The growth of electricity generation was phenomenal in Korea, India, and Taiwan during 1973-79. Although the growth rates during 1979-85 slowed down, they were still well above the growth rates for total energy supply (or demand). In fact, Korea and Taiwan have attempted to promote electricity uses in order to sustain their aggressive nuclear power programs which are not necessarily in the best interest of their economies during the current era of declining oil prices.

TABLE 7

Electricity Generation

Item	South Korea	India	Taiwan
Electricity Generation (GWH)			
1973	14,826	70,510	20,735
1979	35,600	110,130	39,547
1983	48,850	140,299	47,473
1985	58,007	a	54,803
Annual Growth Rates (%)			
1973-79	15.7	7.7	11.4
1979-83	8.2	6.2	4.7
1979-85	8.5		5.6

a. Data is not available.
Sources: (1) Energy Committee, *Taiwan Energy Statistics 1985*, Taiwan.
(2) Korea Energy Economics Institute, *Yearbook of Energy Statistics 1986*, Korea.
(3) Tata Energy Research Institute, New Delhi, India.

CONCLUSIONS

This study analyzed the impacts of the two oil shocks on the demand and supply of energy in the selected 15 developing countries. The two oil shocks resulted in dramatic increases in oil and energy prices in virtually all nations in the world. Higher energy prices should have had negative impacts on energy demand and the opposite (positive) effects on energy supply. However, these economic forces do not produce such straight forward results in the developing countries.

On the demand site, it was found that energy demand continued to increase at fairly rapid rates during the post-embargo 1973-84 period. Increasing energy use has been both the means and consequence of economic development. For the oil-importing LDC's if the high GDP elasticities of energy demand sustain in the future, the balance of payment problem would likely hamper their further economic development, even under the stabilized oil price situation. It is, therefore, imperative for the developing countries to begin to design and implement more effective energy conservation programs in order to sustain their economic growth.

On the supply side, it was found that most LDC's have attempted to diversify the supply sources of energy. The net import share of total energy consumption declined after the two oil shocks in most oil-importing LDC's. However, oil continued to be the most dominant fuel in 1984. The changes in fuel mix were, in many cases, only marginal. Even though most oil-importing LDC's have attempted to increase domestic production of such conventional sources as oil, natural gas, and coal in response to the two oil shocks, only a few achieved significant increases. By and large, the strikes for diversification have been hindered by the limited availability of domestic energy resources. The situation may get worse if oil prices continue to decline in the near future.

LITERATURE CITED

Berndt, E.R. and D.O. Wood. 1975. "Technology, Prices, and the Derived Demand for Energy." *Review of Economics and Statistics.* 57: 259-268.

Berndt, E.R. and D.O. Wood. 1979. "Engineering and Econometric Interpretation of Energy-Capital Complementarity." *American Economic Review.* 69: 342-354.

Briggs, C.K. and I.Y. Borg. 1986. *U.S. Energy-1985*, Report UCID-19227-85, Lawrence Livermore National Laboratory, July 1.

Chern, S. and Horacio Soberon-Ferrer. "Structural Changes of Energy Demand in Less Developed Countries." *OPEC Review.* Forthcoming in Winter Issue, 1986.

Griffin, J.M. and P.R. Gregory, 1976. "An Intercountry Translog Model of Energy Substitution Responses." *American Economic Review.* 66: 845-857.

Hudson, E.A. and D.W. Jorgenson. 1974. "U.S. Energy Policy and Economic Growth, 1975-2000." *Bell Journal of Economics.* 5: 4610-514.

Institute for Energy Analysis. 1979. *Economic and Environmental Impacts of a U.S. Nuclear Moratorium, 1985-2010.* Cambridge, Mass: the MIT Press.

Environmental Consequences of Energy Production: Problems and Prospects. Edited by S. K. Majumdar, F. J. Brenner and E. W. Miller. © 1987, The Pennsylvania Academy of Science.

Chapter Thirty-Five
ENVIRONMENTAL LEGISLATION AND THE COAL MINING INDUSTRY

E. WILLARD MILLER
Professor of Geography
and
Associate Dean for Resident Instruction (Emeritus)
College of Earth and Mineral Sciences
The Pennsylvania State University
University Park, PA 16802

For decades the problems of environmental degradation were ignored as the demands for fuel rose to produce the nation's industrial economy. About a half century ago the evidence of environmental degradation had become so evident that the initial voices were raised to establish quality standards. As a response the first environmental laws were enacted. For the most part these laws were ineffective but they laid the foundation for the future.

In the 1960s a national fervor demanded that environmental standards be established. Out of this growing environmental consciousness the federal and state governments passed laws to improve the quality of the nation's land, water and air. The purpose of this chapter is to review some of these laws particularly as they pertain to the coal industry, with particular emphasis on Pennsylvania.

FEDERAL AND STATE SURFACE MINING LEGISLATION

Although the ravages of the land due to strip mining have long been evident in the coal fields of eastern United States, effective control of strip mining and the reclamation of the land developed slowly. In the early days of strip mining there was the prevailing attitude that the land in Appalachia had little economic value and therefore did not justify the cost of reclamation. Because the abandonment of farmland began in the rugged regions about 1890, long before strip mining was practiced, there was little public or private interest in reclaiming spoil banks on land that had little or no economic productivity. As farmland became unproductive the practice developed for farmers to have their land strip mined in order to secure a windfall profit, frequently at the time the farmer retired.

Although Appalachia could support a forest covering there was little interest in recovering the land by reforestation. The strip-mined areas were dispersed and relatively small so that a solid forest stand was difficult to achieve and interest by major lumber companies was minimal. Finally, the rugged coal mining regions were isolated so that the tourist industry was only moderately developed. Consequently, the esthetic value of the region as a vacation land was little appreciated by outsiders.

PENNSYLVANIA BITUMINOUS COAL OPEN PIT MINING CONSERVATION LAW

During World War II strip mining was highly developed in Pennsylvania. By 1945 large areas were strip mined. Most of these areas were abandoned and it became evident controls had to be imposed if reclamation was to occur. In 1945 Pennsylvania became a forerunner in reclamation of strip-mined land with the passage of the Bituminous Coal Open Pit Mining Conservation Law. This was the most comprehensive law of the day. It required each mining company to deposit a filing fee of $100 for each stripping operation, and post a bond of $300 per acre to be stripped with a minimum of $3,000. Liability under the bond was for the duration of open pit mining at each operation, and for a period of five years thereafter. This initial act also required each strip-mine operator to cover the exposed face of the unmined coal within one year after completion of mining, and to level and round-off the spoil banks sufficiently to permit the planting of trees, shrubs or grasses. The slope of the level area was not to exceed 45 degrees. After the leveling was completed, the operator was to plant the stripped area to the specifications of the Commonwealth Department of Forests and Waters. If the operator failed to comply with these regulations, he forfeited all or part of the posted bond.

Although the initial law controlling strip mining operations appeared to provide the necessary regulations for the reclamation of strip-mined land, it was essentially ineffective. First, the forfeiture of the required bond was not a sufficient penalty to encourage land reclamation. It was estimated that land reclamation in Pennsylvania cost two to six times the posted bond. As a result, at least 80 percent of the bonds were forfeited by the mining companies. Secondly, even when the bond was forfeited there was no legal means of preventing the mining companies from securing another concession to stripmine a new area. As a consequence, violators of the law continued to stripmine new areas. Third, the reclamation of stripped land presented difficulties not previously encountered in the revegetation of the land. Because of acidic and rocky soils, many strip-mined areas where the spoil banks were leveled and planted in trees, from 60 to 100 percent of the trees died within a year after planting. Finally, state inspection of strip-mine sites was non-existent or at best superficial in many areas. As a consequence the state has a heritage of strip-mined areas from the past where nature has reclaimed the areas by a natural succession of plants. Essentially all of these areas are now covered with worthless vegetation consisting of brush and non-timber trees.

PENNSYLVANIA SURFACE MINING CONSERVATION AND RECLAMATION ACT

By the early 1960s it became evident that the initial law to control strip mining was not effective. The public recognized that strip mining was devastating the landscape in many areas and the past practices were inadequate to develop a satisfactory reclamation program. Under the leadership of H.B. Charmbury, Secretary of the Pennsylvania Mines and Mineral Industries the Pennsylvania Surface Mining Conservation and Reclamation Act of 1963 was passed by the state legislature. To make the law more effective it was revised in 1968, 1971,1972 and 1974. The present act is recognized as a model law for the control of strip mining and was used to formulate the national strip mining law of 1977.

The Pennsylvania law requires that the mining company secure an operator's license, mining permit, and establish regulations governing the methods of mining, removal of overburden, back filling requirements, blasting and other significant techniques. It also authorized on-site inspections of all surface mining operations by state mine inspectors at any time. Each operator was also required to post a bond to insure that reclamation occurs. Finally, the reclamation plan must include the following: (1) a review of the best use for which the land was put prior to surface mining, (2) the use of the land after reclamation, (3) where conditions permit, the manner in which topsoil and subsoil will be conserved and restored and, if these conditions cannot be met, what alternative procedures are proposed, (4) where the proposed land use so requires, the manner in which compaction of the soil and fill will be accomplished, (5) a complete program providing for the planting of trees, grasses, legumes, or shrubs, or a combination approved by the state Department of Environmental Resources, (6) a detailed timetable for the accomplishment of each step in the reclamation plan, and the operator's estimate of cost, (7) the written consent of the landowner, which allows access to the land for five years after mining ceases, in order to restore the land, (8) the application for a license or renewal shall be accompanied by a certificate of insurance certifying that the applicant has in force a liability insurance policy of not less than $100,000, (9) the manner in which the operator plans to direct surface water from draining into the pit, (10) no approval shall be granted unless the plan provides for a practical method of avoiding acid mine drainage and preventing avoidable siltation or other stream pollution, and (11) the application of health and safety rules necessary for the safety of the mines and the public.

FEDERAL STRIP MINING CONTROL AND RECLAMATION ACT

By the 1970s the state strip mining laws varied tremendously from those that provided effective control to those that were ineffective. The federal government recognized the need of national legislation. Finally in 1977 the federal government passed legislation providing national environmental performance standards to be applied to all coal mining operations and to be enforced by the states with backup authority in the U.S. Department of Interior (Kalt, 1984).

The need for federal legislation, as stated in the law, was, "while coal in the past has contributed significantly to the industrial and economic growth of the United States, the environmental and social costs of coal extraction have been enormous. To this day coal mining in Appalachia too often results in a legacy of polluted streams below mutilated mountain sides left treacherously unstable. In the West, permanent rehabilitation of mined areas is yet to be demonstrated. If not properly conducted, current and planned western coal development could leave behind barren wastelands susceptible to continued erosion and disrupted ground water systems, significantly diminishing the productivity of agricultural areas" (Table 1).

The federal act thus (1) covers all types of surface mining, (2) establishes administrative, environmental and enforcement standards, (3) provides authority for a federal regulatory program to augment state programs, (4) establishes a federal Office of Surface Mining in the U.S. Department of Interior, (5) establishes criteria to determine areas unsuitable for surface mining, (6) establishes a federal grant-in-aid program, (7) establishes procedures for public review and, (8) recognizes the rights of the land owners and water users.

The control of strip mine operations was contained in Title 5 of the law. It included (1) the issuing of mining permits, (2) time limitations, (3) analysis of case samples, (4) certification of insurance, (5) reclamation of the land, (6) amount of bonds, (7) hydrologic impact, and (8) environmental protection standards such as disposal of surplus spoil, blasting control, clearing and removal of siltation ponds and drainage controls. The law also recognized in Title 4 the need for abandoned mine reclamation. The bill was weak in this area for it did not provide sufficient funds nor recognize the vastness of the problem.

INTERIM IMPLEMENTATION OF THE FEDERAL STRIP MINING CONTROL AND RECLAMATION LAW

The federal strip mining law was to be implemented in two stages. During the interim period beginning December 19, 1977, proposed to last from 24 to 36 months, the Office of Surface Mining of the U.S. Department of Interior and the Pennsylvania Department of Environmental Resources jointly regulated the surface mining industry of Pennsylvania (Elliott, 1983).

TABLE 1

Status of Land Disturbed by Coal Surface Mining in the United States and Needing Reclamation as of January 1, 1974, by States

State	Reclamation Not Required by Law, Acres	Reclamation Required by Law, Acres
Alabama	57,878	118
Alaska	2,400	—
Arizona	150	—
Arkansas	9,451	494
Caribbean Area	4,687	641
Idaho	—	175
Illinois	49,748	20,891
Indiana	2,500	6,000
Iowa	25,650	—
Kansas	43,700	2,500
Kentucky	69,000	117,000
Maryland	2,250	3,851
Michigan	500	—
Missouri	75,506	1,250
Montana	300	300
New Mexico	—	25,798
North Dakota	10,000	200
Ohio	23,926	45,825
Oklahoma	13,858	6,350
Pennsylvania	159,000	33,000
South Dakota	790	—
Tennessee	20,500	5,200
Texas	5,470	—
Utah	120	—
Virginia	18,000	5,014
Washington	471	1,010
West Virginia	25,720	51,560
Wisconsin	234	76
Wyoming	3,078	2,828
Total Areas	621,887	337,081

Source: U.S. Soil Conservation Service.

The Interim Regulations established environmental standards for the use of signs and markers, backfilling and grading, disposal of excessive spoil, top soil handling, protection of the hydrologic system, construction of dams and impounding of waste material, the use of explosives and revegetation and reclamation. There was also established the warrantless inspections by the Office of Surface Mining of all surface mining operations. After inspection the Office of Surface Mining could issue a cessation of mining if conditions were found that posed imminent danger to the health or safety of the public. The work stoppage order would remain in effect until the dangerous mining practice was altered. Civil penalties could be assessed up to $5,000 per day for any violation of the federal act.

On March 13, 1979 the Office of Surface Mining established Permanent Regulations including a Basic and Purpose statement of 407 pages and 150 pages of actual regulations. During the Interim Regulatory phase all of the existing strip mining laws of Pennsylvania including the surface Mining Act, the Clean Stream Law and the regulations in Title 25 of the Pennsylvania Code remained in effect for they were more stringent than the federal Strip Mining Control and Reclamation Act.

LEGAL CHALLENGES TO THE FEDERAL STRIP MINING LAW

Many of the regulations imposed by the Office of Surface Management were challenged by the national coal industry including the Pennsylvania Coal Mining Association. While many of the challenged Permanent Regulations were withdrawn or suspended by the Office of Surface Management, many were resolved by the United States Court of Appeals and a few even reached the U.S Supreme Court.

The following illustrates the types of challenges presented in the attempts to implement the Surface Mining Control and Reclamation Act by the U.S. Office of Surface Mining. In the case of Hodel v. Indiana, which challenged the "prime farmland" provisions of the Strip Mining Control and Reclamation Act including such aspects as restoration to approximate original contour and restoration of top soil, the Supreme Court rejected the challenge of the primeland provisions as beyond the power of the Commerce Clause.

In the case Virginia Surface Mining & Reclamation Association, Inc. v. Andrus the U.S. District Court ruled that the federal act constituted an uncompensated taking of private property in violation of the Just Compensation clause of the fifth amendment. The Supreme Court initially issued an order staying the decisions of the District Court and finally repealed the decision.

These, and other cases, demonstrated the problems encountered in the pre-enforcement constitutional challenges. A number of the issues of the federal strip mining law are still to be settled. However, the U.S. Supreme Court has reversed the decisions of the District Courts as to the constitutionality of the federal law and it is now evident that the federal law is legal and enforcement will evolve over the years.

PENNSYLVANIA'S JURISDICTION OVER STRIP MINING

The Pennsylvania Department of Environmental Resources began in 1978 to prepare amendments to Pennsylvania's mining laws and regulations for submission to the U.S. Office of Surface Mining as part of the state's proposed

State Program to secure primary jurisdiction to enforce the federal Surface Mining Control and Reclamation Act. The Ad Hoc Committee of the Department of Environmental Resources consisted of representatives from the state, environmental groups, coal industry and the general public. Pennsylvania's strip mining law was considered a model law. Congressman Morris Udall, the chief sponsor of the federal law, stated, "Pennsylvania has the best law, it does the best job." It was thus the objective of the Ad Hoc Committee not to impose on the Pennsylvania strip mining industry superfluous requirements beyond the minimum standards of the federal Surface Mining Control and Reclamation Act.

On October 10, 1980 the General Assembly passed and the governor signed legislation amending the Surface Mining Conservation and Reclamation Act, the Coal Refuse Disposal Control Act of 1968, the Bituminous Mine Subsidence and Land Conservation Act of 1966, and the Clean Stream Law of Pennsylvania in order to make the state laws conform to the federal act.

The preamble to these changes stated:

> In order to maintain primary jurisdiction over coal mining in Pennsylvania, it is hereby declared that for a period of two years after the effective date of this Act which was enacted by these amendments solely to secure for Pennsylvania primary jurisdiction to enforce public law 95-87, the Federal Surface Mining Control and Reclamation Act of 1977, if the corresponding provision of that Act is declared unconstitutional or otherwise invalid due to a final judgment by a Federal Court of competent jurisdiction and not under appeal or otherwise repealed or invalidated by final actions of the Congress of the United States.
>
> It is hereby determined that it is in the public interest for Pennsylvania to secure primary jurisdiction over the enforcement and administration of public law 95-87, Federal Surface Mining Control and Reclamation Act of 1977 and that the General Assembly should amend this Act in order to obtain approval of the Pennsylvania program by the United States Department of the Interior. It is the intent of this Act to preserve existing Pennsylvania law to the maximum extent possible.

The process of drafting State Program Regulations to comply with the minimum standards of the Surface Mining Control and Reclamation Act was complicated by the many legal challenges to the federal law and the suspension of certain of the provisions by the Office of Surface Mining. There was also debate over the use of the "state window" a provision which allowed the state authority flexibility to depart from specific office of Surface Mining regulations and to adopt regulations which are "as effective" as those of the Office of Surface Mining to achieve the purposes of the Surface Mining Control and Reclamation Act. The first State Program Regulations were submitted by the Department of Environmental Resources to the Environmental Quality Board for Approval on November 18, 1980.

On November 26, 1980 the Commonwealth Court issued a preliminary injunction enjoining the Department of Environmental Resources from submitting its State Program including the State Program Regulations to the Office of Surface Mining for approval pending the resolution of judicial challenges to the Surface Mining Control and Reclamation Act and the Permanent Program Regulations. The Pennsylvania Coal Mining Association and the Bituminous Coal Association sought injunctive relief against the submission of Pennsylvania's State Program to the Office of Surface Mining pending the judicial challenges to the Surface Mining Control and Reclamation Act and the Permanent Regulations in the federal court and independently challenging several provisions of the new primary laws on constitutional grounds. The petition sought to prevent economic disadvantages to the coal industry if Pennsylvania's State Program were submitted and approved containing regulations that were later ruled invalid. The Department of Environmental Resources appealed the injunction sought by the coal industry. The injunction was in effect until November 26, 1981. During this year the Permanent Regulations were refined providing a period to establish a satisfactory working relationship between the state and the federal government. Many injunctions were issued in other eastern and midwestern states, but ironically the Pennsylvania Department of Environmental Resources was the only state regulatory authority to oppose the injunction.

PERMANENT REGULATIONS OF THE FEDERAL STRIP MINING LAW

The Interim Regulatory Program had excessive demands that were frequently impractical so that in providing the Permanent Regulations there was a serious attempt to provide a practical, operating set of regulations. In January 1983, the Office of Surface Mining issued a final Environmental Impact Statement in three volumes detailing many changes in the Permanent Program Regulations. The basic philosophy of the program, as stated in volume one, was:

> OSM's ultimate goal in proposing these revisions is to make the process of regulation work more effectively in meeting the requirements of the Act—notable, increasing the effectiveness of the regulations in protecting the environment and to revise those regulations that impose undue burden on the operators in the states with no corresponding benefit to the environment.

A number of regulations under the Federal law have required major consideration in Pennsylvania. The Department of Environmental Resources has resisted a number of changes proposed by the Office of Surface Mining insisting that the changes require the abandonment of long established practices that were successful in the state.

The federal permanent regulations proposed a limited use of the "state window." The state had to "explain how and submit data, analysis and information, including identification of sources, demonstrating—1. that the proposed alternative will be in accordance with the operable revisions and consistent with the regulations of this chapter, and 2. that the proposed alternative is necessary because of local requirements or local environment or agricultural conditions." Because of the detail of information requested, an extraordinary economic burden was placed on the coal operator. Many of the practices of the Pennsylvania law developed over the years are technically different from those in the Permanent Regulation and should be maintained in the final State Program. The strict interpretations of the "state window" in the Permanent Regulations were litigated by Pennsylvania unsuccessfully in the U.S. District Court of the District of Columbia.

The bonding requirements of the Permanent Regulations are more stringent than under the Pennsylvania law. Under the federal Strip Mining Control and Reclamation Act, a new and extended period of liability for the life of the bond was imposed which covered "a period of five full years after the last year of augmented seeding, fertilizing, irrigation or other work." Under the Pennsylvania Act the bond was released when the operator had demonstrated compliance with the basic reclamation requirements. The coal operator believed that the Office of Surface Mining regulations would have a negative impact on the industry by tying up large amounts of capital over such an extended period of time. As a response to protests the bonding regulations have been relaxed. The amount of the bond can be reduced upon providing evidence to the regulatory authority proving that the method of operation will reduce the estimated cost to reclaim the mined area.

There was also improvement in the self-bonding provisions. In the original Permanent Regulations, the self-bonding requirements were so rigorous that only the largest companies with huge capital could comply. The extended period of liability and the complex changes in legal requirements for revegetation and bond release made surety companies reluctant or unable to bond coal companies. The regulations have been relaxed so that small and medium sized companies are able to be self-bonded. The federal bonding regulations have now developed creative bonding techniques including letters of credit, escrow bonding and combined surety escrow bonding.

An additional area of improvement concerned the amount and time of the bond to be released. The Office of Surface Mining now authorizes the release of sixty percent of the bond upon completion of backfilling, regrading and drainage control.

The Pennsylvania Department of Environmental Resources has been slow in taking advantage of these relaxed bonding provisions of the Permanent

Regulations. The Pennsylvania strip mining industry has been harmed by the slow release of bonds by the state at a time when the industry is suffering economic troubles due to a declining market.

One of the most difficult technical issues of federal-state interaction created by the federal act has been the establishment of performance standards for hydrology. The federal law strongly recognizes the need to maintain clean water after surface mining ceases. A major question has been whether the Surface Mining Control and Reclamation Act authorized stricter standards than the existing regulations of the Environmental Protection Agency under the Federal Water Pollution Control Act of 1972. This issue was litigated in the U.S. District Court of the District of Columbia with the decision that the Office of Surface Management could not impose stricter standards than specifically established by the Federal Water Pollution Control Act. One of the major problems in implementing the regulations has been concern over the "catastrophic storm exemption." A basic question is, under what circumstances of extraordinary rainfall is the coal operator exempt from existing efficient limitations? It was determined that a coal operator could be exempt if a sedimentation pond was built that could contain or treat a ten year, twenty-four hour storm.

The development of Permanent Regulations to control environmental problems in strip mining has been difficult. It will require many more years to test the present regulations and modify them to make them effective in providing satisfactory economic conditions and at the same time control the environmental devastation caused by strip mining.

ENFORCEMENT OF THE STRIP MINE LAWS

The value of strip-mine legislation lies primarily in the type of enforcement that is practiced. Although Pennsylvania has attempted to control strip mining for more than 40 years, two viewpoints still persist. The one viewpoint by conservationists is that if the law is enforced, strip mining will leave no scars on the landscape and the land will be productive. But the conservationists believe that too frequently the laws are disregarded. The other viewpoint by many mining companies is that the Department of Environmental Resources and the Office of Surface Mining have been unfair and inconsistent in applying the surface mining and clean stream acts. Although much of the strip-mined land is reclaimed in Pennsylvania, there is also little doubt that reclamation is inadequate or even disregarded in certain areas.

There are a number of reasons for the lack of reclaiming the stripped land. In the early 1970s after the initial petroleum price rise there was an assumption that coal would immediately play a greater role in the nation's energy economy. As a response the number of strip mining companies increased rapidly. The market for eastern coals developed slowly in the 1970s and in the 1980s a dramatic decrease in production has occurred. A large number of companies ceased operations and many were forced into bankruptcy. As a consequence, a large amount of strip-mined land has not been reclaimed.

The implementation of the laws has also been lax. In 1981 Pennsylvania had only 51 inspectors to cover the entire coal region. In a survey in western Pennsylvania it was found that some mines had not been inspected in five years. The lack of proper inspection resulted in mining and reclamation violations such as not replacing top soil properly, silting from the mining operations and lack of control of acid mine drainage. While the state law forbids the Department of Environmental Resources to review the mining license of a company with environmental violations, this law has been violated to some extent in recent years.

There is no doubt that reclamation has improved in the past 20 years. Equally, there is little doubt that complete restoration of the land has not been achieved.

CLEAN AIR ACT

The Clean Air Act was enacted by Congress to control atmospheric pollution (Stern, 1982). As the nation's population has grown and as industry has increased not only in size of production but in its areal distribution, the quantities of solid and gaseous contaminants have increased at a pronounced rate. Consequently, there was a growing concern that the pollution of the atmosphere over considerable areas had reached such a concentration that it interfered with the comfort, health and safety of man. Of the more than 170 million tons of air pollutants emitted into the atmosphere in the United States annually, about 60 percent is attributable to motor vehicles, 30 percent to the burning of coal and petroleum, and 10 percent to natural causes. While the federal Clean Air Act encompasses all types of air pollution, the discussion in this chapter will be limited to its importance to the coal industry.

Sulfur Dioxide Emission

The Clean Air Act was most important to the coal producing regions where sulfur content was high (Hinckley, 1982). This includes most of the coal seams east of the Mississippi River. The emission of sulfur dioxide from coal-fired plants using high sulfur coals requires a complex and costly cleaning process before burning or scrubbers to remove it during the burning process.

A formula for SO_2 emissions derived by the Environmental Protection Agency was expressed in terms of weight of SO_2, relative to the Btu present in the coal. The calculations are as follows:
  Assume coal with the properties of:
    3 percent sulfur
    10,000 Btus/lb (In order to understand the relationships of SO_2 to

sulfur, it must be recognized that the weight of sulfur is twice that of oxygen. The SO_2 formed in combustion is equal to twice the weight of sulfur contained in the coal).

The sulfur content of the coal is:
.03 x 2,000 lbs = 60 pounds of sulfur.

Emission of SO_2 when coal is burned is:
2 x 60 lbs per ton = 120 lbs of SO_2 per ton.

When related to heat content, the lbs of SO_2 is:
120 lbs SO_2 per ton/20 million Btus/per ton = 6 lbs of SO_2 per million Btus.

Clean Air Emission Standards

The regulation of sulfur dioxide (SO_2) in the atmosphere from the burning of coal is the portion of the Clean Air Act that has affected the output of coal. On December 23, 1971 the Environmental Protection Agency issued pollution control regulations. These regulations classified coal into three categories of SO_2 reduction. They are:

Category 1
  All coal with an SO_2 content of equal or less than 2 lbs/million Btus must remove 70 percent of all SO_2.

Category 2
  All coals with an SO_2 content between 2 lbs/million Btus and 6 lbs/million Btus must reduce the SO_2 content so that the final emissions do not exceed 0.6 lbs/million Btus. This requires a variable percent reduction of SO_2 by 70 to 90 percent.

Category 3
  All coals with an SO_2 content between 6 lbs/million Btus and 12 lbs/million Btus must reduce SO_2 emissions by 90 percent. The maximum final emissions level, or ceiling, after 90 percent reduction is 1.2 lbs/million Btus.

Pennsylvania Compliance with the Clean Air Act

To comply with the Clean Air Act each state established specific regulations that applied to local and regional conditions (Knight, 1980). Pennsylvania established a three class level emission standard for SO_2. Because this had spatial

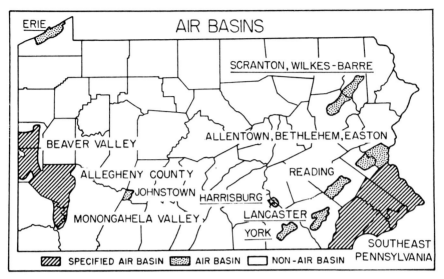

FIGURE 1. Pennsylvania Air Basins.

implications the state was divided into air basins and non-air basins (essentially a function of urbanization and population density) with discrete SO_2 emission levels described for each basin class (Figure 1). Emission levels were expressed as amounts of sulfur dioxide that may be released per quantity of energy consumed, the units being pounds of SO_2 per million (10^6) British Thermal Units (Btu) (Table 2). On August 1, 1979 the emission standards were revised. The changes permit use of higher sulfur coals over shorter time periods, but a more stringent longer time average in most areas. In general the standards were relaxed slightly. To illustrate the present standards, Allegheny County lies in a specified air basin. The pounds of $SO_2/10^6$ Btus allowable in this region range from 0.6 to 1.0. The range of restrictions corresponds to the size of the combustion units.

Since Pennsylvania coal varies considerably in both thermal (Btu) and sulfur content, it is necessary to determine emission standards as a percentage of the sulfur in relation to the Btu content of the coal. Figure 2 graphically displays the relationship between SO_2, heating value of coal, unit size of boiler, and percent of sulfur for each of the air basin types. This chart facilitates the recognition of sulfur levels in relation to Btu content necessary to meet the limits for each air basin class. For a coal with a 12,000 Btu/lb heating value, for example, being consumed in a specific air basin such as Southeast Pennsylvania or the Beaver Valley, the percentage of sulfur, by weight of the coal, may range from 0.36 to 0.60. As the Btu content of coal increases so does the percent of sulfur in the coal that can be burned legally. In Figure 2, the left-most vertical line

TABLE 2

Sulfur Dioxide Emission Standards in Pennsylvania, August 1, 1979

Air Basin Type	Sulfur Dioxide Emission Limits (#10^6 Btu) by Combustion Unit Size (10^6 Btu/hour)		
	< 250	< 250 with permit	> 250
Non-Air Basin[a]			
1 hour average	4.0	—	—
Daily average maximum	—	4.8	4.8
Daily average, 2 days in 30 day period	—	4.0	4.0
30 day running average	—	3.7	3.7
Annual average[b]	—	3.3	3.3
Air Basins			
1 hour average	3.0	—	—
Daily average maximum	—	3.6	3.6
Daily average, 2 days in 30 day period	—	3.0	3.0
30 day running average	—	2.8	2.8
Annual average[b]	—	2.5	2.5
Specified Air Basins			
Southwest Pennsylvania (Allegheny County, Beaver Valley, Monongahela Valley)	0.6-1.0[c]	—	0.6[c]
Southeast Pennsylvania Inner Zone			
1 hour average	1.00	—	—
Daily average maximum	—	1.20	0.72
Daily average, 2 days in 30 day period	—	1.00	0.60
30 day running average	—	0.75	0.45
Average annual[b]	—	0.67	0.40
Southeast Pennsylvania Outer Zone			
1 hour average	1.20	—	—
Daily average maximum	—	1.44	1.44
Daily average, 2 days in 30 day period	—	1.20	1.20
30 day running average	—	0.90	0.90
Annual average[b]	—	0.80	0.80

[a] Including Erie, Harrisburg, York, Lancaster, and Scranton-Wilkes Barre air basins.
[b] Calculations based on statistical distributions assumed for coal sulfur content by Pennsylvania Department of Environmental Resources.
[c] 2.5-50 million Btu/hour: 1.0
50-2,000 million Btu/hour: $A = 1.7E^{-0.14}$ where E = heat input
> 2,000 million Btu/hour: 0.6

Source: Applications of the Pennsylvania Coal Model.

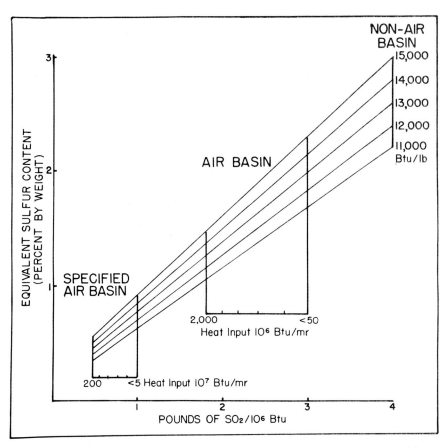

FIGURE 2. Sulfur content equivalent to Pennsylvania sulfur dioxide limitations (Source: Applications of the Pennsylvania Coal Model).

for each air basin represents the SO_2 level for large combustion units. The rightmost vertical line represents the less restrictive SO_2 levels. The smaller boilers have less restrictive SO_2 levels so that they can utilize higher sulfur coals.

Effects of the Clean Air Act on Localization of the Bituminous Coal Industry

The Clean Air Act has affected the entire bituminous coal industry of the nation. Generally, the coal fields that are free or have low sulfur content have greatly increased production and those coal fields with a high sulfur content have suffered. Sulfur occurs in several forms in coals. If it occurs in the pyritic form such as iron sulfide (FeS_2) or iron sulphate ($Fe\ SO_2$), it can be partially removed (10 to 90 percent) by precombustion cleansing processes. Organic sulfur and residual pyritic sulfurs are released upon combustion primarily in the form of SO_2 and can be removed by costly scrubbers. As a result the demand for high sulfur Pennsylvania coal was reduced but not eliminated.

In the 1970s as national demand for coal increased, users sought the low-sulfur coals of the nation to reduce costs (Navarro, 1981). In essence the Environmental Protection Agency regulations created two regional markets for bituminous coal-high sulfur versus low sulfur coals. Most western coals in such states as Wyoming, Montana, Colorado, Utah and Arizona and a few eastern coals, primarily in eastern Kentucky, are low sulfur coals and comply with the clean air standards without costly cleaning and emission controls. Most coals in Appalachia, including Pennsylvania, and in the Midwest, can only be burned when the sulfur content is reduced.

The low sulfur coals thus experienced remarkable growth. In the high sulfur coal areas of Appalachia and the Midwest the industry, at best stagnated, and in most states declined. It has survived due to (1) the utilization of the coal in electric power plants usually in non-air basins of low population, such as at Homer City in central Pennsylvania, where the high-sulfur coals can be burned after cleaning, (2) the production of coking coal, and (3) the expansion of the export market.

CLEAN STREAM LAWS

The presence of acid water draining from coal seams was recognized as early as 1803 when T.M. Morris noted that spring water issuing from hills with coal seams "is so impregnated with bituminous and sulferous particles as to be frequently nauseous to the taste and prejudicial to health." Acid mine drainage began with the first mining operations and it is estimated that 80 percent of the acid mine drainage pollution today comes from abandoned mines. Within the bituminous region acid mine water is coincident with the mining of coals from the Allegheny, Conemaugh and Monongahela groups of the Pennsylvania period and the Waynesburg Formation of the Permian Period. These formations contain the major coal seams mined in Pennsylvania.

The initial Clean Stream Laws were enacted in the 1930s, but were ineffectual in preventing an increase in the acid water pollution of the streams in the coal regions. By the 1960s it was recognized that mine drainage was a major cause of stream pollution (Hill, 1973). There was also a recognition in Pennsylvania that unpolluted streams were a necessity to attract new and expand existing industry and to develop the tourist trade.

The modern attempts to control acid mine drainage in Pennsylvania began with the passage of the Pennsylvania Clean Stream Law of 1966 and on the national level with the Federal Water Pollution Control Act of 1972 as amended in 1977. This is commonly known as the Clean Stream Act. The basic objective of both acts is to restore and maintain the chemical, physical and biological integrity of all streams that are presently polluted.

The Pennsylvania Clean Stream Act requires that mine discharge water shall have a pH of not less than 6.0 nor greater than 9.0, and that such water shall not contain more than 7.0 mg. of dissolved iron. The law also requires mine operators to submit an application for a drainage permit for their proposed operations to the Department of Environmental Resources indicating the method of controlling acid drainage during mining and how acid discharges will be prevented after completion of mining.

Acid Mine Water Control

Pennsylvania has the most widespread acid drainage problem of the Appalachian coal fields. It is estimated that 2,750 tons of acidic waters enter the state's streams daily (1,003,750 tons annually) or about 45 percent of the estimated 6,000 tons of mineral acidity entering Appalachian waters daily. In Pennsylvania about 2,300 miles of streams are polluted with acid waters. These include major portions of such rivers and their tributaries as the Monongahela, Allegheny, Conemaugh, Kiskiminetas and Clarion of the Ohio River system and portions of the Susquehanna and Schuylkill rivers.

There has been a fairly effective control of acid drainage discharges from operating mines in recent years, but there is little or no control of acid discharges from abandoned mines. Although technology is available to control acid waters by such techniques as neutralization, reverse osmosis, or ion exchange, the processes are costly. A number of pilot plants have been built in Pennsylvania to test each of the technologies. Modern estimates indicate that several billion dollars are required to clean-up the Pennsylvania streams alone (Rothfelder, 1982). The original goal of the Clean Stream Act to have unpolluted streams within 15 years of its passage has not been realized. It will be many decades before the problem of acid mine drainage is solved.

COAL REFUSE DISPOSAL CONTROL ACT

As a normal consequence of coal mining in most areas it has been necessary to deposit, on the surface, refuse material such as sandstone, shale and clay, which is removed from the underground mine along with the coal. The accumulation of these piles of refuse material has created a condition which fails to comply with the established regulations to control air and water pollution. They can also create a danger to property and public roads either by shifting or sliding of materials. To prevent environmental damage and personal injury from refuse sites, the Coal Refuse Disposal Act was passed in 1968.

To conform to the law, operators may be required to build drainage ditches, trenches and/or gullies, to build impounding dams, to remove combustible materials, to engage in spreading, compositing and/or layering, to use clay, soil and/or other inert sealing materials and to alter the slope of the material necessary to correct or prevent sliding or slipping. The Sanitary Water Board, the Air Pollution Commission and the Department of Environmental Resources are responsible for implementing the law.

The penalties for each infraction of the law include a fine of $1,000 to $5,000 plus a possible one year imprisonment. Since the passage of the Coal Refuse Disposal Act the disposal of mining waste materials has largely been controlled. Vast waste material piles exist, however, from before the law was passed. These piles of waste materials still create environmental problems and may cause personal injury.

MINE SUBSIDENCE AND LAND CONSERVATION ACT

Land subsidence can be a major conservational problem particularly in old mining regions. In underground mining the removal of vast quantities of coal and rock creates a void which, under certain conditions, may result in subsidence at the surface. The problem is exaserbated if several seams of coal are mined that lie on top of each other. As the roof of the mine falls into the void, cracking and caving of the overlying rock progresses upward. Each movement at the surface may result in many types of damage. Building foundations and walls may be cracked or displaced. Railroad tracks and roads may subside or shift out of alignment.

Subsidence damage is normally greater in urban areas than in rural areas. In many of the older mining regions towns were built directly above underground mining. In the anthracite fields of northeastern Pennsylvania the valley towns were on top of the coal seams. As millions of tons of coal and rock were removed in the mining process entire blocks of buildings have been affected by subsidence of the surface. Forest lands are little affected by subsidence and damage to croplands is minimal. The greatest effect may be the loss of ground water sources. Water from wells and springs may be diverted and sources disappear. The water table may be lowered affecting the availability and supply of water. There may be permanent damage requiring an adjustment to a different water regime.

Subsidence may begin as soon as deep mining occurs so that the problem in Pennsylvania is of long standing. For decades no governmental action was taken. If property damage occurred the owner of the property had full responsibility to repair the damage. The mining companies assumed little or no responsibility. To provide relief to today's residents in the coal regions of Pennsylvania, the Anthracite and Bituminous Coal Mine Subsidence Fund was established by the Pennsylvania legislature in 1961. Any individual could purchase insurance giving protection against mine subsidence. In 1982, 1,650 new policies were written bringing the total policy holders to 17,500 providing total insurance coverage to $645 million. Forty eight damage claims were paid in 1982 totaling $281,707.

The Mine Subsidence Insurance Section of the Department of Environmental Resources supervises the program.

In 1966 the Pennsylvania legislation passed the Mine Subsidence and Land Conservation Act. The 1966 Act was amended in 1980 pursuant to the federal Surface Mining Control and Reclamation Act. The need to control future mine subsidence was recognized because past damage from mine subsidence (1) seriously impeded land development, (2) caused danger to the health, safety and welfare of people, (3) eroded the tax base of the affected municipalities, and (4) reduced the economic welfare and growth of the region. The Act provided protection for public buildings, churches, hospitals, schools, and dwelling used for human habitation, and all cemeteries and burial grounds. The Bureau of Mining and Reclamation has established guidelines to implement the legislation and control future problems of subsidence in mines. A basic feature of the Act is that in underground mines 50 percent of the coal must be left in place in uniformly distributed pillars which cannot be smaller than 6.1 x 9.1 meters (20 x 30 feet). No mining can occur if the overburden is less than 30.5 meters (100 feet) and pillars cannot be extracted between two support areas when the distance between them is less than the support cover.

This Act has largely controlled subsidence in modern-day underground mining. However, billions of tons of coal were mined before the passage of the 1966 Act. To solve the traditional problems some attempts have been made to fill underground voids. Red ash, a non-flammable residue from culm banks, is commonly used as in fly ash, the waste from electric power generating stations. Sand is also utilized. These types of filling processes are not only costly, but subsidence is reduced only up to 50 percent. In the late 1960s eleven mine filling projects in the anthracite fields cost over $7 million. Because of prohibitive costs and limited success this procedure has been essentially abandoned in Pennsylvania (Slagel, 1985). It is less costly to pay damages through the Subsidence Fund as subsidence occurs.

CONCLUSIONS

Although federal and state legislation has attempted to control environmental degradation, the process has been more difficult than anticipated. There was much environmental degradation before attempts to provide controls and solving the problems of the past have only been partially addressed. Great strides have been made in solving the present day problems, but even here success has not been total. While the technology has become increasingly sophisticated, the costs of implementing quality environmental controls have risen astronomically. The nation has not made the difficult decisions that are necessary for maintaining a quality environment in the future. Laws without adequate economic implementation are of little or no value.

SELECTED REFERENCES

Ackerman, Bruce A. and William T. Hassler. 1981.*Clean Coal/Dirty Air: Or How the Clean Air Act Became a Multibillion Dollar Bail-out for High-Sulfur Coal Producers and What Should Be Done About It.* New Haven, CT: Yale University Press, 193 pp.

Campbell, William A. Spring 1980. The 1977 Clean Air Act Amendments: Their Potential Impact on Economic Growth.*Popular Government.* 45: 21-25.

DelDuca, Patrick. 1981. The Clean Air Act: A Realistic Assessment of Cost-Effectiveness. *Harvard Environmental Law Review.* 5 (1): 184-203.

Elliott, John M., Steven L. Friedman and Stephen C. Braverman. An Overview of Pennsylvania Coal Law in Majumdar, Shyamal K. and E. Willard Miller, eds. 1983. *Pennsylvania Coal: Resources, Technology and Utilization.* Easton, Pa.: Pennsylvania Academy of Science, pp. 523-546.

Fanelli, J. May 1985. Legislation (Annual Review 1984).*Mining Engineering.* 37: 422-3.

Gianessi, L.P. et al. 1981. Analysis of National Water Pollution Control Policies. 1. A National Network Model,*Water Resources Research.* 17 (4): 796-802.

Haigney, W.F. 1981. Federal Common Law and Water Pollution Statutory Preemption on Preservation. *Fordham Law Review.* 49, No. 4: 500-535.

Haskell, Elizabeth H. 1982. *The Politics of Clean Air: EPA Standards for Coal Burning Power Plants.* New York: Praeger, 224 pp.

Hill, R.D. 1973. Water Pollution from Coal Mines. *Proc. 45th Annual Conference, Water Pollution Control Association of Pennsylvania.* University Park, PA.

Hinckley, A. Dexter. September 30, 1982. The Scientific Foundations of the Clean Air Act, *Public Utilities Fortnightly.* 110: 23-27.

Hoover, J. May 1985. Legislation (Coal 1984). *Mining Engineering.* 37: 452-4.

Kalt, J.P. June 1984. "Capture and Ideology in the Economic Theory of Politics (Senate Voting on Coal Strip Mining Policy). *American Economic Review.* 74: 279-300.

Knight, C. Gregory and Charles B. Manula, Co-directors. 1980.*Applications of the Pennsylvania Coal Model.* University Park, Pa: The Pennsylvania State University. Prepared for the Appalachian Regional Commission, Volume one, *Survey and Discussion,* Volume two, *Exhibits.*

La Vardera, L.T. May 1985. Environmental Issues (Coal 1984).*Mining Engineering.* 37: 449.

Lave, Lester B. and Gilbert S. Omenn. 1981. *Clearing the Air: Reforming the Clean Air Act.* Washington, D.C.: Brookings Institution, 65 pp.

Majumdar, Shyamal K. and E. Willard Miller, eds. 1983. *Pennsylvania Coal: Resources, Technology and Utilization.* Easton, Pa. Pennsylvania Academy of Science, 594 pp.

Makansi, Jason. March 1982. Trends in Pollution Control: The Industry Matures. *Power.* 126: 36-38.
Marshall, B.R. and J.C. Lamb. 1982. Evolving Water Quality Regulations. *State Government.* 55 (4): 122-126.
McIlvaine, Robert W. March/April 1982. Air Pollution Control Trends in a Changing World. *Filtration and Separation.* 19: 152-155.
Melli, P. et al. 1981. Real-time Control of Sulphur Dioxide Emission from an Industrial Area. *Atmospheric Environment.* 15 (5): 653-666.
Miller, E. Willard. 1983. Federal Legislation and the Pennsylvania Bituminous Coal Industry: Production and Spatial Impact. In Majumdar, Shyamal K. and E. Willard Miller, eds. 1983, *Pennsylvania Coal: Resources, Technology and Utilization.* Easton, Pa: Pennsylvania Academy of Science, pp. 465-579.
Milliman, J.W. 1982. Can Water Pollution Policy be Efficient?*Cato Journal.* 2 (1): 165-196.
Navarro, Peter. Spring 1981. The 1977 Clean Air Act Amendments: Energy, Environmental, Economic, and Distributional Impacts. *Public Policy.* 29: 21-146.
Ramani, R.V. and R.J. Sweigard. April 1984. Impacts of Land Use Planning on Mineral Resources. *Mining Engineering.* 36: 362-9.
Rothfelder, M. 1982. Reducing the Cost of Water Pollution Control under the Clean Water Act. *Natural Resources Journal.* 22 (2): 407-421.
Schneider, M.W. 1982. Criminal Enforcement of Federal Water Pollution Laws in an Era of Deregulation. *Journal of Criminal Law.* 73 (2): 642-674.
Slagel, G.E. July 31, 1985. The Regulatory World of Subsidence Is in a State of Turmoil. *American Mining Congress Journal.* 71: 8-9.
Stern, A.C. 1982. History of Air Pollution Legislation in the United States, *Air Pollution Control Association Journal.*32 (1): 44-61.
Stevenson, Walter H. May 1980. Federal SO_2 Emission Standards: What Do They Mean? *Power.* 124: 130-131.
Wark, Kenneth and Cecil F. Warner. 1981. *Air Pollution: Its Origin and Control.* 2nd ed. New York: Harper and Row. 526 pp.
Wright, F.B. 1982. The Control of Water Pollution.*Environmental Policy and Law.* 8 (4): 116-119.
Anon. 1969. *Acid Mine Drainage in Appalachia.* Washington, D.C.: 91st Congress, 1st Session, Committee on Public Works, Home Document No. 91-180.
Clean Air Act. *United States Code Annotated.* Title 42, Sec. 1857.
The Clean Air Act as Amended August 1977. 1977. 95th Congress, 1st Session. Washington, D.C.: U.S. Government Printing Office, Serial No. 95-11, 185 pp.
Implementation of Certain Sections of the Clean Water Act: Hearings, June 23—July 1, 1980. 1980. Washington, D.C.: United States Senate, Committee on Environment and Public Works, Subcommittee on Environmental Pollution, 96th Congress, 2nd Session, 376 pp.
A Legislative History of the Clean Air Act Amendments of 1977. 1978. 95th

Congress, 2nd Session. Washington, D.C.: U.S. Government Printing Office. Prepared for the Committee on Environment and Public Works, U.S. Senate, 1607 pp.

Pennsylvania Air Pollution Control Act. *Environment Reporter.* 3 Sect 4, 491: 0101-0108.

Pennsylvania Air Pollution Control Act. *Pennsylvania Code.* Title 35, Health and Safety, Chapter 23, Air Pollution.

Pennsylvania Air Pollution Control Regulations, *Environment Reporter.* 3 Sect. 4, 491: 0501-0560.

Pennsylvania Air Pollution Control Regulations. *Pennsylvania Code,* Title 25, Part 1, Department of Environmental Resources, Chapters 121 through 141.

Pennsylvania's Coal Mining Regulatory Program. 1982. Harrisburg, Pa. Department of Environmental Resources.

Subject Index

Abandoned Mine Reclamation Fund, 47
Abandoned Mines, 481
Above Ground Storage Facilities, 481
Acid Deposition, 25, 32, 431, 432, 440
Acid Mine Drainage, 131, 132
Acid Rain, 25, 198, 232, 460
Acid Rain Problem, 246
Acid Waters, 23
Acidified Rain, 242
Acute Radiation Syndrome, 381
Adequate Screening, 327
Advisory Committee on Reactor Safeguards, 371
Aerial and Other Broadcast Sprays, 290, 293, 298
Agroecosystems, 122
Air Contaminants, 431
Air Emission, 464
Air Pollutant Deposition, 441
Air Pollutants, 1, 11
Air Pollutants, Global Scale, 432, 441
Air Pollutants, Secondary, 432
Air Pollution, 198
Air Pollution from Refuse Piles, 107
Air Pollution Stress, Global, 442
ALARA, 402, 418
Air Pollution Commission, 518
Alewife, 329
Algae, 171
Alkalinity, 312
Alluvial Channel, 301, 308
Alluvial Gravel, 354
Alternative Energy Sources, 62
Aluminum, 432
American Petroleum Institute, 155
American Shad, 319, 320, 322, 329
American Society of Testing Materials, 148
Ammonia, 432
Anadromous Fish Conservation Act of 1965, 323
Anthracite, 2
Anthracite and Bituminous Coal Mine Subsidance Fund, 518
Anthropogenic (Human) Origin, 233
Anthropogenic Sources, 233

Appalachian Compact, 476
Appalachian Mountains, 123
Application of Lime, 70
Appraisal Methodology, 202
Aquaculture, 219
Aquatic Biota, 213, 278
Aquatic Macrophytes, 392
Aquifers, 54
Area Mining Method, 81
Argentina, 489
Aromatic Hydrocarbons, 468
Ash Generation Processes, 224
Ash Production, 224
Asphalt, 404
Asphyxiation, 153
Aswan Dam, 334, 344
Atlantic Sea-Run Salmon Commission, 323
Atmospheric Deposition, 261
Australia, 3

Back Filling, 58
Baghouse, 468
Bald Cypress, 389
Bangladesh, 489
Barnwell, Barriers, 363
Barnwell, South Carolina, 404, 475
"Bathtub," 360
Bay of Fundy, 335, 337, 338, 339, 340
Beatty, Nevada, 404, 475
Bedrock, 79
Belgium, 62, 74
Below Ground Burial with Engineered Barriers, 481
Benefits of Planned Subsidence, 56
Benthic Animals, 335
Benzene, 165
Bhopal, India, 21
Biocides, 216
Biogeochemical Cycles, 335
Biological Effects of Radiation, 446
Biological Effects of Toxic Chemicals, 449
Biomass, 219
Biomass Conversion, 6, 15
Biomass Production, 439
Birds, 177

Bituminous Coal, 502
Blackberry (*Rubus allegheniensis*), 293
Blockage of Migrations, 326
Bottom Ash, 226, 470, 471
Brazil, 493, 497, 498
Bristol Channel System, 340
Bubble Policy, 253
Buffalo Creek Disaster, 112
Buffer Zone, 288
Bulgaria, 201
Burial Cells, 356
Buttermilk Creek, 354
Bypassed Wastes, 471

Cadmium, 432
Calcium Carbonate, 136
California, 468
Canada, 248
Canopy Leaf Area, 21, 441
Carbon Dioxide, 431, 433, 437, 439
Carbon Monoxide, 465
Carbon-14, 372
Carboniferous Era, 62
Centralia Mine Fire, 40
Cesium, 373
Channel Gradation, 313
Channel Morphology, 308
Channel Response, 316
Characteristics of Petroleum, 165
Chernobyl, 477
Chernobyl Nuclear Accident, 201, 381
Chernobyl Power Station, 380
Chile, 489, 490, 495, 497
Chloride, 22, 432
Chlorofluoromethanes, 434
Chlorine, 432
Chlorofluorocarbons, 439
Chromium, 432
Chromosomal Aberrations, 447
Chromosomal Breakage, 447
Classification of World Energy Reserves and Resources, 1
Clean Air Act of 1970, 250, 511
Clean Air Emission Standards, 512
Clean Coal-Dirty Air, 250
Clean Stream Law of Pennsylvania, 506, 507, 516
Clean Water Act, 248, 394
Cleanup Processes, 159
Climatology, 310
Coal, 497

Coal Combustion, 442
Coal Market, 252
Coal Mine Safety and Health Act of 1969, 45, 113
Coal Refuse Disposal Control Act of 1968, 507
Coal Refuse Pile Characteristics, 106
Coal Refuse Piles, 104
Coal Refuse Production, 102
Coal Spoil, 74
Coal Spoil Tipping, 61
Coal Technology Laboratory, 246
Coastal Bays, 334
Coastal Ecosystem, 334
Cobalt Isotopes, 373
Cobalt-58, 373
Cobalt-60, 373
Cold Shock, 219
Coleoptera, 175
Collective Dose, 408
Colliery Spoil, 62
Colluvium, 354
Colorado, 52
Combustion, 472
Commercial Power Reactors, 350
Condenser, 213
Contour Mining, 79
Control of Fires, 41
Conventional Oil, 5
Conventional Shallow Land Burial, 481
Conventional Tidal Power Development, 337
Cooling ponds, 215
Copper, 432
Cost-Benefit Analysis, 205
Crustacea, 173
Culm Bank, 519
Cumberland Basin, 338
Curative Measures for Oil Spills, 156
Cut and Stump Treatment, 290
Czechoslovakia, 199

DDT, 460
Delaware, 375
Delta T, 214
Delta, PA, 370
Demineralization, 404
Dendrochronology, 441
Department of Energy, 383
Design Information System, 87
Desulphurization, 201
Dieback, 439, 442

Dioxin, Effects, 469
Dipteria, 175
Direct Lethal Toxicity, 167
DNA, 446
DNA Content, 448
DNA Measurements, 448
DNA Replication, 459
DNA Synthesis, 458
DNA Transcription, 459
Dogwood (*Cornus florida*), 292
Domestic Energy Supply, 497
Dormant Season and Summer Basal Treatments, 290
Dosimeters, 412
Downstream Effects, 334
Drilling Mud, 146
Drilling Procedures, 165
Drinking Water, 281
Drought, 438
Dry Active Waste, 403, 404
Dry Alkaline Fly Ashes, 226
Ducktown, Tennessee, 434
Dumping Ground, 405
Dutch Delta, 335

Earthquake, 370
East Germany, 199
Eastern Canada, 345
Ecological Awareness, 204
Economic Commission for Europe, 23
Ecosystem Factors, 81
Effects of Oil on the Ecosystem, 167
Effects of Radiation in Amoeba, 447
Effluent Treatment, 403
Egypt, 4, 489, 498
Electrical Generation, 498
Electrification Programs, 498
Emission Rates, 466
Emission Reduction, 258
Emissions of Nitrogen Oxide, 238
Emissions of Sulphur Dioxide, 238
Endangered and Protected Plants, 294
Endangered Species, 158
Energy, 487
Entrainment, 216
Environmental Effects of Tidal Power, 339
Environmental Impact Assessment, 208
Environmental Protection Agency, 160, 215, 460
Environmental Studies Committee, 339
Environmentalists, 212

Ephemeroptera, 173
EPRI, 418
Erosion, 311
Estuaries, 334
Ethoxysulfate, 166
Ethyl Methanesulfonate, 449
Europe, 22
European Economic Community, 23
Eurotrac, 23
External Dosimetry, 411

Federal Water Pollution Control Act, 212
Federal Republic of Germany, 23, 62, 440
Federal Surface Mining Control and Reclamation Act, 510, 519
Ferrous Iron, 134
Ferrous Metals, 471
Fertilizers, 70
Fish, 176
Fishes, 218
Flue Gas Desulfurization, 246
Flue Gas Stream, 467, 468
Fly Ash, 226, 470, 471, 519
Fly Ash Disposal, 223
Fly Ash Slurry, 229
Forage Crops, 122
Forced Scrubbing, 253
Forest Damage, 431
Forest Ecosystem, 440
Forest Ecosystem Health, 442
Forest Ecosystem Stress, 441
Forest Lands, 122
Forest Succession, 440
Forested Ecosystems, 124
Fossil Fuels, 1, 261, 431
France, 62, 74
Fuel Waste, 147
Fuels, Conventional, 497
Fugitive Emissions, 469

Gas Injection, 166
Gaseous Pathways to Man, 373
Gaseous Waste Stream, 371
Gastropods, 218
GDP Growth Phase, 490
Gene Mutation, 447
Genetically Engineered Microbes, 460
Geology, 369, 485
Geomorphic Principles, 315
Geomorphic Systems, 307
Geomorphology, 84, 313

Global Climate, Warmer, 438
Global Observation, 234
Goldenrods (*Solidago spp.*), 293
GPU Nuclear Corporation, 407
Grassland Ecosystems, 123
Great Britain, 3
Great Lakes, 248
Gross Domestic Product, 488
Gross National Product, 156
Groundwater, 270, 361
Gulf of Maine, 340, 345
Gusher Phenomenon, 149

Habitual Losses Due to Impoundments, 328
Halocarbons, 431, 433
Handcutting, 290
Hanford, Washington, 404, 475
Hazardous Waste Disposal Facility, 484
Hazardous Wastes, 460
Health Risk Management, 410
Heat Stress Index, 413
Heavy Metals, 345, 465, 466, 471
Hemiptera, 174
Herring, 322
High Level Waste Disposal, 376
Human Health, 442
Hungary, 201
Hydraulic Changes, 308
Hydraulic Conductivity, 53
Hydraulic Geometry, 303
Hydraulic Variables, 301
Hydrocarbons, 31, 465
Hydrodevelopments, 334
Hydroelecric Development, 319
Hydrofracking, 164
Hydrogen Sulfide, 432
Hydrography, 164
Hydrologic Anomalies at West Valley, 357
Hydrology, 369
Hydropower Activity, 325
Hydropower System, 312

Illinois, 52, 53
Impact of ROW on Wildlife, 294
Impact on Soil, 296
Impact on Songbirds, 295
Impact on Visual Quality, 297
Impact on White-Tailed Deer, 295
Impacts of Hydropower on Anadromous Fish, 325
Incineration, 405, 465

India, 491, 495, 497, 498
Industrial Chemicals, 445, 460
Initial ROW Clearance, 289
INPO, 418
Insecta, 173
Institute for Energy Analysis, 488
Institute for Nuclear Power Operations, 416
Inverted Terraces, 120
Ionization, 446
Ionizing Radiation, 446
Iron, 432
Iron Pyrites, 62
Iron Smelters, 432

James River, 115
Japan, 22, 23

Kentucky, 52
Korea, 491, 493, 495, 498

L-Lake, 394
Lake Erie, 354
Land Disposal, 402
Largemouth Bass, 390
Leachate, 353
Leachate Testing, 228
Lead, 432, 440, 465
Lead Deposition, 433
Lime, 134
Limestone, 133
Liquid Pathways to Man, 374
Liquid Waste Stream, 372
Lithium, 2
LLRW, 475, 478, 484
LLRW Issue, 479
Lodgepole Pine, 435
Longitudinal Stream Profile, 304
Longwall Mining Techniques, 66
Low-Level Radioactive Waste, 402, 405
Low-Level Radioactive Waste Policy Act, 475, 482
Low-Level Radioactive Waste Policy Amendment Act of 1985, 375

Macrophyte Biomass, 393
Maine, 323
Major Pollutants, 30
Man-Made Atmospheric Pollutants, 29
Management of ROW, 289
Management of the Radiation Protection Program, 409

Manganese, 432
Maryland, 324
Mathematical Modeling, 317
Mathematical Simulation Models, 245
Meadowsweet (*Spiraea latifolia*), 293
Mercer County, Pa, 118
Meteorology, 369
Mexico, 495, 497
Microorganisms, 170
Mid-Atlantic States, 322
Minas Basin, 338, 343
Mine Drainage, 67, 503
Mine Fire Control Techniques, 41, 42
Mine Methods in the Various Regions, 92
Mine Subsidence Insurance, 519
Mineral Energy, 141
Mining Process, 139
Mississippi River, 253
Mitigation Measures, 210
Mixing Zone, 212
Mollusca, 176
Molybdenum, 432
Montour Stream Electric Station, 220
Mosquitofish, 389
mRNA, 454
Municipal Solid Waste, 464

N-methylnitrosourea, 452
N-methylnitrosourethane, 452
National Academy of Sciences, 434
National Acid Precipitation Assessment Program, 256
National Ambient Air Quality Standards, 465
National Environmental Laws, 213
National Research Council, 227
Natural Ecosystem Health, 442
Natural Gas, 2, 5, 11, 497
Natural Grasslands, 122
Natural Seeps, 164
Nekton, 216
New Coal Developments and Prospects in Britain, 64
New England, 322
New England Rivers, 323
New Hampshire, 324
New York, 324
Nickel, 432
Nitrate, 22, 237
Nitrogen, 233
Nitrogen Oxide, 262

Nitrogen Oxides, 21-23, 232, 465
Noble Gases, 373, 403
Noise & Traffic, 472
Noncriteria Emissions, 468
Noncriteria Pollutants, 466
Nonrenewable Energy Sources, 2
Nordic Council of Ministers, 23
North America, 21
NRC, 382, 418
Nuclear Fuels, 2
Nuclear Power Plants, 474
Nuclear Power Production, 369
Nuclear Radioactive Releases, 21
Nuclear Regulatory Commission, 370, 406, 418
Nuclear Waste Policy Act of 1982, 375

Odonata, 174
Odor, 472
Office of Surface Mining, 46, 116
Offshore Deposits, 5
Ohio River Valley, 261
Ohio Valley, 250
Oil Embargo of 1973, 490, 491
Oil Spill Cleanup, 158
Oil Tankers, 151
Oil-Fired Utilities, 255
Open Pit, 115
Open Pit Mining, 82
Origin of Mine Fires, 36
OSHA Safety Requirements, 473
Overburden Characterization, 135
Overview of CTL/CSTM, 251
Oxidants, 431
Ozone, 31, 432, 434, 439
Ozone Concentration, Stratospheric, 437

PA DER, 482, 484, 506
Pakistan, 489, 493, 495
Par Pond, 392
Particles From Automobiles, 433
Pawcatuck River, 324
PCDD's, 469
PCDF's, 469
Peach Bottom Atomic Power Station, 370
Pennsylvania, 52, 213, 324, 468, 474, 482, 502
Pennsylvania Air Basins, 513
Pennsylvania Department of Health, 407
Pennsylvania LLRW Waste Facility, 481

Pennsylvania Power & Light Company, 215, 219, 220
People's Republic of China, 3
Perched Water, 365
Periphyton, 169
Person-REM, 381, 418
Peru, 491, 494
Phillippines, 494-495, 497
Photochemical Oxidant Effects, 21, 23
Phytoplanktons, 376
Physiography, 310
Pipelines, 154
Pit Subsidence, 52
Planform Classifications, 308
Planned Subsidence and Modern Mining, 55
Plecoptera, 174
Poikilothermic Organisms, 216
Poland, 3
Polar Glaciers, 433
Pollution, 255, 329
Pollution Abatement Strategies, 442
Pollution Discharge Elimination System, 213
Ponderosa Pine, 435
Post-Oil Embargo, 488
Predation, 329
Pressurized Water Reactor, 371
Primary Energy Consumption, Total, 493
Productivity, Primary & Secondary, 390
Protozoa, 218
Public Advisory Committee on Low-Level Radioactive Waste Disposal, 483
Public Health Impact on the TMI-2 Accident, 406
Public Utility Regulatory Policies Act in 1978, 325
Pulp Mills, 432
Pyritic Material, 72
Pyritic Oxidation, 67
Pyritic Spoil, 74
Pyritic Sulfur, 136

Qualitative Geomorphic Techniques, 313

Radiation Exposure, 405
Radiation Exposure Management System, 412
Radiation Exposures to Cleanup Workers, 407
Radiation Protection Training, 410
Radioactive Chemicals, 445
Radioactive Fallout, 460

Radioactive Materials, 22
Radioactive Waste, 21
Radioactivity of Fly Ash, 230
Radioiodines, 373
Radionuclides, 187, 475
Radionuclides Released from TMI 2 Accident, 379
Radiopharmaceuticals, 475
Radon Barrier, 190
Radon Flux, 191
Ramping, 328
Raspberry (*Rubus idaeus*), 293
RCRA, 228
RCRA Standards, 229
Reclaiming Abandoned Lands, 424
Reclamation, 62, 70-71, 74, 419
Reclamation Goals for Uranium Wastes, 186
Reclamation of Uranium Mining, 191
Recovery Dose, 408
Recovery Dose Per Year, 408
Red Ash, 519
Regional Planning, 428
Release of Halocarbons, 438, 439
REM, 418
Remote Reconnaissance Vehicle, 414, 415
Republic of Germany, 3
Residue Management, 470
Resource Conservation and Recovery Act of 1976 (RCRA), 228
Resource Recovery, 464
Restoration Programs, 319
Revegetation, 90
Reverse Osmosis, 404
Rhine Spill at Basel, 21
Ribosomes, 454
Right-of-Way (ROW), 288, 294
River Classification, 307
RNA, 446
Robotics Technology, 414
Role of Hydropower in the North East, 324
Rumania, 201
Rotifers, 173
ROW Maintenance, 290
ROW Segment, 291

Sacramento System, 345
SAFSTOR, 382, 383
Salinity, 312
San Bernardino National Forest, California, 440
Saudi Arabia, 4

Savannah River Ecology Laboratory, 396
Savannah River Plant, 386
Savannah River Swamp, 388
Scanning Electron Microscope Analysis, 452
Scrub Scenario, 258
Scrubber Systems, 370
Scrubbing Costs, 256
Seam Mine, 139
Section 508 of PL 95-87, 87
Sediment Load, 428
Sediment Movement, 316
Sediment Transport, 305
Sediment Yield, 306
Sedimentation, 335
Seismological Stability, 369
Separation Tanks, 167
Settlement Ponds, 167
Severn Estuary, 335, 337, 341
Shallow Land Burial, 350
Shepody Bay, 338
Shifts in Regional Coal Production, 256
Siberian Coastal Basins, 6
Signatories to the Sulfur Protocol, 24
Simulation Modeling, 246
Singapore, 490, 494, 497, 498
Sink Hole, 52
Site Planning, 428
Slag, 226
Slave Routes, 153
Small Bluegill, 390
Society of American Foresters, 440
Soil Compaction, 296
Soil Erosion, 296
Soil Heating, 219
Soil Interactions, 267
Songbird Populations, 288
Sorghum, 121
Southern Toad, 391
Soviet Union, 3, 4, 402
Spill Spread Control Mechanism, 159
Spillways, 325
Spoil Heaps, 67
Sport Fishery, 128
Status of the Fisheries, 320
Steam Electric Station, 219
Steel Creek, 387, 395
Stem-Foliage Spray, 290, 293
Stepping, 328
Stratification, 339
Stratosphere, 434
Strawberry (*Fragaria spp.*), 293

Stress Factors, Biotic, 442
Strip Mining Law of 1977, 503
Strip Mining, 419
Sublethal Effects, 217
Subsidence, 49
Subsidence Control, 57
Subsidence Geometry, 50
Subsidence in General, 49
Subsidence Problems of Past Mining, 51
Subsidence Site Parameters, 50
Subsidence-Associated Damage, 53
Subsidy-Stress Model, 393
Succession, 124
Sudbury, Ontario, Canada, 434
Sulfur Dioxide, 262, 432, 439, 465, 466
Sulfur Dioxide Emission, 511
Sulfur Dioxide Emission Rates, 259
Sulfur Oxides, 21, 22
Sulphur Dioxide, 232, 245, 251
Sulphur Emission, 245, 252, 253, 255
Sulphate, 237
Sulphate in the Atmosphere, 242
Surface Mine Planning, 96
Surface Mining, 78, 504, 508
Surface Mining Control and Reclamation
  Act of 1977, 46
Surface Mining Methods, 79
Surface Subsidence, 49
Surface Water Management, 366
Surface Waters, 272
Susquehanna River, 322, 323, 326
Swamp Habitats, 386
Sweden, 22, 23

Tailings, 475
Taiwan, 491, 493-495, 497-498
Temperature, Global, 437
Terrestrial Invertebrates, 172
Texas Lignite, 257
Thailand, 491, 497
Thallium, 432
The Environmental Protection Act, 485
The Environmental Protection Agency, 479
The Middle East, 4
The National Research Council, 52
The World Energy Conference, 25, 32
Thermal Discharges, 212, 219
Thermal Ecology, 386
Thermal Effluents, 388
Thermal Gradient, 393

Thermal Plume, 216
Thermal Pollution, 386, 394
Thermoluminescent Dosimeter, 412
Thorium, 2
Three Mile Island, 402, 477
Three Mile Island Unit 2, 406, 411
Tidal Flux, 335
Tidal Influence, 335
Tidal Power, 335
Title 4 and 5, 504
TLD, 418
TMI Accident, 379
TMI-2 Plant, 411
TMI-2 Radiological Controls Program, 409
Toluene, 165
Topographic Factor, 119
Topography, 297
Total Suspended Particulates, 467
Toxic Compounds, 212
Trace Metal Analysis, 441
Trace Metals, 441
Tree Ring Analytical Techniques, 441
Tree Rings, 441
Tributaries, 312
Trichoptera, 174
Tritiated Thymidine, 457
Tritium, 373
Tropopause, 434
Types of Subsidence, 50

U.S. Congress, 245
U.S. Department of Agriculture, 433
U.S. Department of Energy, 386
U.S. Department of Interior, 504
U.S. Department of Transportation, 479
U.S. Environmental Protection Agency, 433, 469
Ukraine, 380
Ultraviolet Radiation, 434
United Kingdom, 62
United States, 3, 22, 402
Universal Scrubbing, 250
Universal Soil Loss Equation, 117
Uranium Mining, 182
Uranium Mining & Milling Wastes, 184
Uranium Mining Reclamation Problem, 183
Uranium Reclamation, 189
Uranium Revegetation, 189
Uranium Tailings, 192
Uses of Coal Refuse, 110

USSR, 199
UV-radiation, 454

Vanadium, 432
Vascular Plants, 172
Vegetation, 306
Vegetative Cover, 120
Vegetative Interactions, 265
Vegetative Reclamation, 121
Virginia, 324

Waste Generation, 402
Waste Heat, 214
Waste Policy Amendment Act of 1985, 476
Water Pollution Control Act of 1972, 510, 517
Water Pollution Problems, 107
Water Quality, 472
Water Quality Issue, 212
Water-Injection Programs, 158
Waterfowl, 218
West Germany, 62, 74
West Valley Disposal Facility, 355
West Valley, New York, 349
Western Coal, 421
Western Europe, 21, 61
Western Pacific, 21
Western Pennsylvania, 121, 173
Wet and Dry Deposition, 232, 233
Wet Deposition, 232, 234
Wet Sediments, 340
Wet Waste, 402, 404
Wetland Ecosystems, 125
Wetland Species, 126
Wetlands, 122, 394
White Pine Mortality, 435
Wildlife Habitat, 128
Witch-Hazel (*Hammamelis virginiana*), 292
Women Voters, 483
Woody Vegetation, 289
World Energy Conference, 5
World Energy Conference in Cannes—1986, 22
World Health Organization, 20
Wyoming, 52

Xylene, 165

Zinc, 432
Zooplankton, 218, 219

OFFICERS OF THE PENNSYLVANIA ACADEMY OF SCIENCE

SHYAMAL K. MAJUMDAR, President, Professor of Biology, Lafayette College, Easton, Pennsylvania 18042

KURT C. SCHREIBER, President Elect & Director, Science Talent Search, 1812 Wightman Street, Pittsburgh, Pennsylvania 15217

GEORGE C. SHOFFSTALL, Immediate Past-President/Executive Secretary, 502 Misty Drive, Suite 1, Lancaster, Pennsylvania 17603

SHERMAN S. HENDRIX, Treasurer, Department of Biology, Gettysburg College, Gettysburg, Pennsylvania 17325

RALPH A. CAVALIERE, Assistant Treasurer, Department of Biology, Gettysburg College, Gettysburg, Pennsylvania 17325

LEONARD M. ROSENFELD, Recording Secretary, Department of Physiology, Jefferson Medical College of Thomas Jefferson University, Philadelphia, PA 19127

HOWARD S. PITKOW, Corresponding Secretary, Professor of Physiology, Pennsylvania College of Podiatric Medicine, Eighth at Race Street, Philadelphia, Pennsylvania 19107

DANIEL KLEM, JR., Editor of the Proceedings, Department of Biology, Muhlenberg College, Allentown, Pennsylvania 18104

FRED J. BRENNER, Newsletter Editor, Biology Department, Grove City College, Grove City, Pennsylvania 16127

J. ROBERT HALMA, Historian, Department of Biology, Cedar Crest College, Allentown, Pennsylvania 18104

EDWARD TESTA, Director of Junior Academy, Valley View High School, Archbald, Pennsylvania 18403

JUSTICE JOHN P. FLAHERTY, Advisory Council Chairman

SISTER M. GABRIELLE MAZE, Fund Raising, Past-President, Grove and McRobert Road, Pittsburgh, Pennsylvania 15234